H. Weidner, M. Jänicke (Eds.)

Capacity Building in National Environmental Policy

Springer-Verlag Berlin Heidelberg GmbH

Helmut Weidner Martin Jänicke (Eds.)

Capacity Building in National Environmental Policy

A Comparative Study of 17 Countries

In Collaboration with Helge Jörgens

With 26 Tables

Springer

DR. HELMUT WEIDNER
Wissenschaftszentrum Berlin
für Sozialforschung
Reichpietschufer 50
10785 Berlin
Germany
weidner@medea.wz-berlin.de

PROF. MARTIN JÄNICKE
Freie Universität Berlin
Forschungsstelle für Umweltpolitik
Ihnestrasse 22
14195 Berlin
Germany
ffu@zedat.fu-berlin.de

ISBN 978-3-642-07732-6

Library of Congress Cataloging-in-Publication Data applied for
Die Deutsche Bibliothek – CIP-Einheitsaufnahme
Capacity building in national environmental policy : a comparative study of 17 countries / Helmut Weidner ; Martin Jänicke (eds.).

ISBN 978-3-642-07732-6 ISBN 978-3-662-04794-1 (eBook)
DOI 10.1007/978-3-662-04794-1

http://www.springer.de

Originally published by Springer-Verlag Berlin Heidelberg New York in 2002
Softcover reprint of the hardcover 1st edition 2002

Camera ready by authors
Cover design: E. Kirchner, Heidelberg
Printed on acid-free paper SPIN 10863298 32/3130/as 5 4 3 2 1 0

PREFACE

This book is the second collection of systematic case studies describing national environmental policies in 17 countries in terms of capacity building (see Appendix). The OECD defines environmental capacity building as "a society's ability to identify and solve environmental problems." While various institutions, including UNEP, FAO, World Bank and OECD, have hitherto used the terms environmental capacity and capacity building almost exclusively with reference to developing countries, we have extended the concepts to industrialized countries, as well. The first collection, edited by Martin Jänicke, Helge Joergens (both Free University Berlin) and Helmut Weidner (Social Science Research Center Berlin), was published in 1997 under the title "National Environmental Policies – A Comparative Study of Capacity-Building" (Berlin, etc.: Springer Verlag). It included 13 studies of countries.

As in the first volume, chapter 1 presents the conceptual framework underlying the national case studies. It is a slightly shorter version of the corresponding chapter in volume 1. The design of all case studies in the two volumes is largely congruent with this conceptual framework. Although the various sections of the studies do not always have identical titles and subtitles, the central elements of the capacity-building approach have been applied in all cases.

There are two main reason why we decided to use the same conceptual framework for both project phases: (1) it proved very helpful for analyzing complex national environmental policies; (2) it very much facilitated the cross-national comparison of environmental policies in a total of 30 countries. Furthermore, the largely identical design of the country studies allows our readers to use the two volumes as manuals.

This volume covers a broad spectrum of different types of countries, ranging from advanced industrial countries, newly industrializing and so-called transition countries to developing countries. While highly developed countries and the most advanced frontrunners in environmental policy ("the pioneers") were, not unintentionally, over-represented in volume 1, the present volume puts more emphasis on countries belonging to the group of laggards or those which, owing to systemic restraints and contexts, find it particularly difficult to develop and pursue an effective and stable environmental policy. This not only offers a broader empirical basis for generalizing statements, but also the opportunity to make problems and capacity deficits visible that often play only a secondary role in advanced industrialized countries. In particular, it permits identification of the kinds of capacity that need particular strengthening if an effective global environmental policy is to be attainable. This information is particularly relevant for international organizations, whose strategies and support often do not sufficiently reflect such factors.

In developing and transition countries, for example, insufficient monitoring and reporting capacities as well as underdeveloped democratic structures and processes often play a much more important role than in advanced countries. The same is true of capacity deficits at the implementation level. Such deficits may lead to insubstantial real outcomes with respect to emission reduction or environmental quality in spite of sufficient institutional capacities at the national level. This means that, in principle, it is necessary to include the implementation level or even full policy cycle in the analysis of national environmental policies if the quality of environmental capacities is to be correctly assessed. The case studies presented here and in volume 1 may also be considered a first necessary step towards the design of such a complex and elaborate integrative analytical approach.

The new case studies largely confirm the main findings of our first project phase: there is an acceleration in the global process of environmental innovation diffusion and in global environmental policy learning; the "second wave" of environmental institutionalization – stimulated by the Rio Conference – continues, particularly in developing countries, but also in developed countries; networking, negotiation and co-operation among stakeholders is increasing; "soft" or so-called New Environmental Policy Instruments and policies are becoming more important, but conventional regulatory policies retain their dominant position; the access of environmental NGOs to policymaking bodies and their political influence is on the increase. Furthermore, the great importance of democratic system structure for capacity building and effective environmental policy is demonstrated. *And, finally, it became more evident that in times when the influence of structural restrictions is augmenting, environmental policy success depends very much on how skillful proponents are in using and systematically creating situative opportunities to win broad support for unpopular measures or against powerful opponent groups.*

Two general trends just emerging at the time of our first study are now clearly visible: (1) the worldwide convergence in the basic pattern of environmental policy, largely irrespective of the (sometimes strong) differences between countries in political capacities and problem situations; (2) the globalization of environmental policy, i.e., transnational and global networking among and across the various environmental proponents. All this suggests there is no justification for general pessimism about globalization in environmental policy; on the contrary: globalization, in the sense of facilitated and growing worldwide communication and networking between groups of actors and organizations generally seems to support environmental interests. We have found no evidence for a systematic "race to the bottom" or any creation of "pollution havens."

Even the shift in the dominant environmental policy paradigm towards sustainable development – with notable exceptions like the United States – is taking place on a global scale and is increasingly backed by institutionalization at both the local and national levels, and within international and regional organizations. This is worth mentioning, although it is often no more than the institutionalization of an environmental policy learning process. The growing commitment to sustainable development and the broadening institutionalization of environmental capacities does not automatically engender sustainable environmental policy outcomes and impacts. What we pointed out in our first study still holds true: existing environmental capacity, even in the most advanced countries, has so far been sufficient only for – predominantly technological – standard solutions, such as end-of-pipe treatment of air and water pollution, waste disposal, nature conservation, the control (substitution) of toxic substances, or more efficient use of energy and water. A fundamental change is still needed in the nature and scale of production and consumption across economic sectors. Additional capacities are needed to develop effective strategies for land use or material flow management, for soil or climate protection; and to this end structural restrictions still need to be overcome. Especially in policy areas where no (marketable) pollution control technologies are available (for example, land use management), there is an urgent need for additional capacities which, however, have yet to be established anywhere.

In contrast to our first study, no outstanding environmental pioneer was ascertained among the 17 countries included in this volume. Some years ago it seemed as if Canada and New Zealand, and perhaps Austria, would play this role, but this has not materialized. Among transition countries, Poland also seemed to be on its way to becoming a more advanced case (at least within this group of countries). However, the pace of development there has slackened substantially over the last few years. The pioneer countries identified in the previous book thus retain their status. There is, however, one new and promising development: the frequently positive and proactive role the European Union plays with respect to both member states and other countries, especially candidates for accession to the Union, and also in global environmental politics. The EU has become the most important global player in environmental policy, while the United States is increasingly applying the brakes in global environmental matters, underusing its existing capacities for effective environmental policy.

The 17 case studies in this book are presented in alphabetical order. This seemed a better systematic solution than a more or less arbitrary grouping of countries, e.g., in terms of progressiveness or inaction, or by geographical region. Alphabetical order also facilitates the intended use of the reader as a manual. The first chapter presents the analytical framework underlying the national case studies. The final chapter summarizes the main findings of this and the previous book, giving an empirical basis of 30 national case studies for this analysis. An appendix provides an overview of the 13 countries studies (and their authors) in the 1997 volume.

This project would not have been possible without generous financial support from the United Nations University/World Institute for Development Economics Research (UNU/WIDER) for the first phase (13 countries), and the Social Science Research Center (WZB) for the second phase (17 countries). We would like to offer our sincere gratitude to both institutions. In addition we wish to extend our thanks to Professor Wolfgang van den Daele, Director of the WZB-Department "Standard-setting and the Environment" for his wide-ranging support of the project, Rhodes Barrett for language editing and translations, and Dagmar Kollande (WZB) for her organizational support. We are also deeply indebted to Friederike Theilen-Kosch for editorial work and for preparing the camera-ready copy of the manuscript. Last but not least, we sincerely regret the somewhat lengthy delay in completing this reader and would like to thank our contributors for their patience and understanding.

Berlin, August 2001 HELMUT WEIDNER, MARTIN JÄNICKE, HELGE JÖRGENS ronmental capacity, even in the most advanced countries, has so far been [illegible]

[illegible]

CONTENTS

LIST OF CONTRIBUTORS

Arun Agrawal	Yale University, Princeton, U. S. A.
Christof Amann	Institute for Interdisciplinary Studies of Austrian Universities (iff), Vienna, Austria
Magnus Andersson	Uppsala University, Baltic University Programme, Sweden
Susan Baker	University of Cardiff, U. K.
Ton Buehrs	Lincoln University, Canterbury, New Zealand
Joanne Caddy	OECD Paris, France
Le Thac Can	Environment & Sustainable Development Center (VNESDC), Hanoi, Vietnam
Bruno Dente	Istituto per la ricerca sociale (irs), Milan, Italy
Adam Fagin	University of Portsmouth, U. K.
Maria Fauconnet	Bureau Urbaplan, Lausanne, Switzerland
Marina Fischer-Kowalski	University of Vienna, Institute for Interdisciplinary Studies of Austrian Universities, Vienna, Austria
Kathryn Hochstetler	Colorado State University, Fort Collins, U. S. A.
Martin Jaenicke	Department for Environmental Policy, Free University Berlin, Germany
Helge Joergens	Department for Environmental Policy, Free University Berlin, Germany
Peter Knoepfel	Institut des Hautes Etudes en Administration Publique, Lausanne, Switzerland
Corinne Larrue	Université de Tours, France
Rudolf Lewanski	Università degli studi di Bologna, Italy
Donna Lybecker	Colorado State University, Fort Collins, U. S. A.
Stephen P. Mumme	Colorado State University, Fort Collins, U. S. A.
Robert Paehlke	Trent University, Ontario, Canada
Elim Papadakis	The Australian National University, Canberra, Australia
Eduardo Silva	University of Missouri, St. Louis, U. S. A.
Ching-Ping Tang	National Chung Cheng University, Taiwan
Shui-Yan Tang	Department of Political Science, University of Southern California, Los Angeles, U. S. A.
Anna Vari	Hungarian Academy of Sciences; Institute of Sociology, Budapest, Hungary
Helmut Weidner	Research Unit "Standard-setting and the Environment", Social Science Research Center Berlin (WZB), Germany
Noriko Yokozuka	Yale University, Princeton, U. S. A.

THE POLITICAL SYSTEM'S CAPACITY FOR ENVIRONMENTAL POLICY: THE FRAMEWORK FOR COMPARISON*

Martin Jänicke

> "The ability of a country to follow sustainable development paths is determined to a large extent by the capacity of its people and its institutions as well as by its ecological and geographical conditions. Specifically, capacity-building encompasses the country's human, scientific, technological, organisational, institutional and resource capabilities".
> (Agenda 21, Chapter 37)

1 Capacity for Environmental Policy and Management

What is capacity for environmental protection? The OECD defines it broadly as "a society's ability to identify and solve environmental problems" (OECD, 1994: 8). More specifically, the concept points to the objective limits to (and necessary preconditions of) successful solutions of a given type of problem, limitations beyond which failure sets in, even in cases of good luck, skill and highly motivated actors. Lack of ecological, technological or administrative knowledge, lack of material or legal resources, the weakness of environmental organisations or institutions in relation to vested interests are well-known examples of such limitations. The main implication for the debate on sustainable development is that long-term strategies must include concepts for improving the conditions of environmental action (capacity-building).

Generally speaking, the capacity problem is not specific to the field of environmental policy. It can be found also in other policy fields. Environmental policy, however, seems to be the most prominent field for the discussion of limitations on government intervention (Luhmann, 1990). Some authors see such limitations – or "government failure" – as regards more preventative forms of environmental policy (Jänicke, 1990). In some areas even a "capture of the system of control (...) by those which it is meant to control" (OECD, 1992: 46) has been discussed. There is at least a capacity problem for new environmental problems: "(...) in some cases existing capacity is rendered irrelevant or made inadequate by the emergence of new challenges, and awareness of these challenges has grown faster than the ability of many institutions to solve them" (OECD, 1994: 11).

---

* The following chapter was conceived and used as a framework for the 13 case studies of "National Environmental Policies I" (1997). To make the additional 17 cases of the present book comparable to the first volume we asked the authors again to use this chapter as a starting point. Therefore, aside from minor reductions, the original chapter has not been revised.

A country's capacity for environmental protection is not and cannot be restricted to government policies. Increasingly it depends on societal forces of all kinds. The OECD therefore prefers the formulation of a societal "capacity for the environment" (OECD, 1994).

2 Capacity Limits in the Advanced Industrial Countries

The debate on capacity-building (or development) in environmental protection is at present mainly restricted to developing countries. They are the candidates for transfer of expertise, etc. from the more successful "donor countries" (OECD, 1995). One might believe that the advanced countries provide an indubitable model to be imitated. But is this really so?

Without any doubt there has been a development of capacity in the industrialised countries that can be empirically demonstrated by cases of best practice and successful environmental policy measures in fields such as air and water pollution control or the establishment of protected areas. In some advanced democracies there has been a remarkable increase in environmental policy capacity. However, the restrictions to ecologically sustainable strategies should not be ignored. Therefore, capacity-building for the environment is no less relevant for the industrialised world.

In the advanced democracies, too, success in environmental policy and management depends on the type of problem to be solved. Success is more or less restricted to problems that can be handled (mainly by "additive" technical standard solutions) without restricting markets or relevant societal routines. As regards the big environmental problems – especially the problems of long-term resource management – government failure and capacity overload can be frequently observed.

A brief look at the environmental record of the industrial countries shows that between 1970 and 1990 there was a general improvement in terms of such indicators as SO_2 or suspended particulate emissions, the proportion of protected areas, or the share of the population served by sewerage systems. However, most industrial countries showed a deterioration with respect to domestic waste, the effects of motor traffic, or the proportion of built-up areas (Jänicke, Mönch and Binder, 1996, 1996a). For soil and ground water pollution or toxic waste, the same picture would be likely to emerge, but internationally comparable data are not available here. And this is a typical situation in areas where environmental capacity is low. For important environmental problems there is no "getting rich – getting clean". This points to the idea that the structure of problems is important for the evaluation of environmental policy and its capacity.

In the politically difficult field of sustainable development in particular, the conditions in advanced industrial countries primarily give rise to the question of how capacities for environmental policy and management might be developed and increased.

3 The Usefulness of the Concept of Environmental Capacity

The usefulness of the concept of environmental capacity may be summarised as follows:

The first relevant function of the concept is to stress the importance of objective limitations on successful intervention. Failure in environmental policy (or other policies) cannot be explained just by the wrong choice of instruments or other kinds of "intervention failure" (OECD, 1992) as many analysts have done. It is not only the kind of action, but also its structural conditions which matter.

A second advantage of the concept of capacity might be its close relationship to the concept of development. Political development or modernisation can be described as a process of capacity-building. This was already the case in publications on developing nations in the 1960s and 1970s (Pye, 1966; Binder et al., 1971), a tradition that has been revived by the debate on sustainable development. Political development or modernisation might be defined in general as institutionalisation and internalisation of new stages of problem-solving capacities in reaction to (or anticipation of) societal challenges or crises (Jänicke, 1973). This definition could also be applied to the industrialised world. And there is no reason to restrict the concept of political development to developing countries. The concept of capacity-building also makes sense if we analyse the development of specific policies. The relatively new field of environmental policy is a good example.

This leads to the third advantage: The concept of capacity for environment and management provides a tool for comparison between the different abilities of different countries as well as the degrees of development of these abilities within a country.

Since capacity analysis deals with the structural preconditions for solutions to specific problems, the final advantage of the concept may be its practical orientation as well as its greater realism regarding the difficulty of problems that have not yet been solved.

The main function of the concept is to change the focus of environmental policy analysis. Therefore it is firstly a heuristic analytic tool. We should not overestimate the possibilities of its operationalisation. On the one hand capacity relates to concrete problems, where operationalisation may be possible. On the other hand, if we speak for example of institutions, we use the concept for a broader field or the sum of its concrete tasks. Here operationalisation creates difficulties. We should also remember, that all kinds of evaluation – the diagnosis of success or failure – depends upon explicit criteria (Bartlett, 1994).

Another reason for caution is the difference between capacity and its utilisation by concrete actors in concrete situations. But this is an important enough point to be explained in more detail later on.

4 An Analytical Framework

4.1 Categories for the Explanation of Environmental Policy Outcomes

In a number of altogether 36 case studies on the preconditions for successful environmental policy and management carried out by the Environmental Policy Research Unit at the Free University of Berlin (Jänicke and Weidner, 1995; Conrad 1995; Jänicke, 1996) a set of categories to explain differences in policy outcomes has been developed. When reformulated, it may also be useful for the analysis of environmental capacities:

Usually, perceived environmental problems lead actors to develop and implement strategies (typically against opposing target groups) under systemic conditions and within situative contexts. The outcome, then, is influenced mainly by the following factors (see Figure 1): (a) actors, (b) strategies (c) systemic framework conditions, (d) situative contexts and (e) problems. The structure of problems as well as the capacity to respond to them is strongly influenced by economic performance. All factors, of course, have to be seen in an international context.

Figure 1. Model of Policy Explanation.

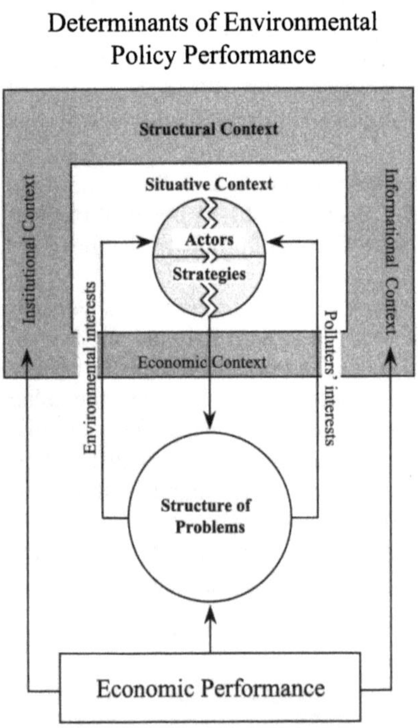

Source: Martin Jänicke, FFU (1998)

The main purpose of this systematisation is to make clear that successful environmental protection is brought about by a complex interaction of influences and not by a single, isolated factor, nor a favourite instrument, nor a single type of actor, nor a particular framework condition or institution. The literature on environmental policy and management is full of such proposals of a particular measure, regardless of its context.

The shortcomings of such a "mono-factorial" view may be illustrated by the example of the time-consuming debate on environmental policy instruments (sometimes taking on the character of a debate on the "instrument of the year"). Of course, we need assessments of the general advantages and weaknesses of our policy tools. But recent research makes clear that this has been a simplistic debate using a mechanistic top-down model of policy and ignoring the complex interaction dynamics between quite different factors (Sabatier and Jenkins-Smith, 1993).

The combined application of at least the following (however broad) categories seems to be indispensable to the explanation of environmental policy outcomes:

a) *Actors:* the proponents and opponents of special issues, their support groups, but also third parties. Usually the actors are representatives of organisations and coalitions. But they are often coalitions of relevant micro-actors across organisations and institutions. They have stable general interests and "core beliefs" (Sabatier and Jenkins-Smith, 1993) as well as variable definitions of these, or situative opinions and attitudes that may change more readily. Their capacity for action depends on their strength, competence and constellation, but also on the structural framework conditions. Personal will and skill also make a difference (see also Shonfield, 1969; Schmidt, 1993; OECD, 1995: 5). Schumpeter emphasised this with respect to technical innovations. But it is also important for policy innovation.

b) *Strategy:* the general approach to the problem. Strategy is the purposeful use of instruments, capacities, and situative opportunities to achieve long-term goals (see also Bartelmus, 1994; Carew-Reid et al., 1994; Hey and Brendle, 1994; Fiorino, 1995). It includes necessary sub-goals, the flexible application of instruments over time, and, last but not least, the purposive improvement of capacities. Strategies depend on capacities such as available knowledge or the possibility of strategic and coordinated action. Environmental protection needs strategy also to compensate for the generally weak position of its proponents. For this the intelligent use of time is important. An environmental measure which is impossible to implement at present may be feasible within five years. Failure in the short term has often turned out to be a success in the long run.

c) *Structural Framework Conditions:* the systemic conditions of environmental action or the opportunity structure of the relevant actors (Kitschelt, 1986). Policy analysis generally differentiates between the influence of the political, economic and cultural framework conditions of policy action (Feick, 1992; Kern and Bratzel, 1996). The author prefers a slightly different typology:

   Cognitive-informational framework conditions: the conditions under which environmental knowledge is produced, distributed, interpreted and applied. Without knowledge there is no (perceived) problem, no public awareness and consequently no policy process, and therefore, it is the first factor here. The cognitive opportunity structure (Jahn, 1993: 178), the societal interpretation of the environmental situation is the second. While culture is more of a background condition, the available knowledge about threats and options, combined with environmental consciousness, is an immediate resource for environmental actors.

Political-institutional framework conditions: the constitutional, institutional and legal structure, the institutionalised rules and internalised norms, constituting the framework for interaction. The conditions for participation, integration and long-term action should be given special emphasis in this context.

Economic-technological framework conditions: the performance, technological standard, sectoral composition, or general availability of raw materials. GNP per capita is an important determinant of existing capacities. But – as for culture – its influence is mainly indirect (contrary to its influence on the structure of problems). And – as in culture – it is no easy target for environmental capacity-building. Factors such as price structures, however, have a more immediate impact and may also be more easily influenced by (e.g. fiscal) policy.

d) *Situative contexts:* the short-term variable conditions of action (Feick, 1992; Sabatier and Jenkins-Smith, 1993). Their importance was already there to be studied with respect to the situative change that the French Revolution caused for European neighbour countries (changing the opportunities of relevant actors). In environmental policy and management, headlines sparking off public debate on a concrete pollution problem are a strong motor of environmental change. Events such as the Bhopal, Seveso or Chernobyl accidents offered sudden situative opportunities for proponents of environmental protection. In about half of the above mentioned 36 success cases analysed in 12 OECD countries, direct public pressure played an important situative role. Situative changes sometimes have an immediate influence. German industry, for example, started reducing the use of cadmium mainly in reaction to a Swedish law on the substance, provoking a public debate on the issue in Germany (Jänicke and Weidner, 1995). In times of recession, on the other hand, environmental policy and management act under restrictive conditions.

e) *The character of the problem* (Solesbury, 1976; Sabatier and Jenkins-Smith, 1993): is it easy or hard to solve? It makes a difference whether a pollution problem is urgent and experienced by the public or latent and only threatening future generations. Similarly it makes a difference if the polluter is economically relevant, has strong influence on society and no easy way out, or if the target group is weak and/or a technical or other solution is available. "The systemic or structural nature of many problems" (Carew-Reid et al., 1994) causes restrictions that may limit a given capacity.

As mentioned above, successful environmental policy measures so far have been restricted to a certain type of problem. As a first step, the advanced democracies were generally able to manage the "dilution of pollution", solid waste collection, or the creation of protected areas. Later on, end-of-pipe measures of all kinds, risk management and the substitution of dangerous materials took place. Here success in terms of reduced pollution was possible. However, there is a broad range of environmental policy failures in fields such as: waste reduction, nuclear waste disposal, soil and ground water protection, road traffic volume stabilisation, sustainable land-use and absolute reduction of resource inputs into the production process (see above). These problems have been identified, but not solved.

4.2 Capacity for the Environment and its Utilisation

The above defined five categories of policy explanation could be used also for capacity analysis, if one differentiates between capacity, as a relatively stable condition of action, and its utilisation, which leads to the subjective and situative aspect of environmental policy and management. The above model, then, can be reformulated as follows:

The capacities for the environment are constituted by:

1) the strength, competence and configuration of organised governmental and non-governmental proponents of environmental protection and
2) the (a) cognitive-informational, (b) political-institutional and (c) economic-technological framework conditions.

The utilisation of the existing capacity depends on:

3) the strategy, will and skill of proponents and
4) their situative opportunities.

This has to be related to:

5) the structure of the environmental problem: its urgency as well as the power, resources and options of the target group.

Capacity, therefore, defines the necessary structural conditions for successful environmental policy as well as the upper limit beyond which policy failure sets in even in case of skilful, highly motivated and situatively well-placed proponents.

The category of "actors" has to be differentiated: If we try to explain cases of success or failure (policy outcomes) in environmental protection, we must analyse the dynamic interaction between proponents and opponents and their conditions of action. However, in this context we are concerned with capacities for environmental protection. And here we are interested first and foremost in the possibilities available to the active part: the organised proponents of environmental policy and management (thus treating the opposing target group of polluters as a restrictive factor and part of the problem). It is true that the target groups of polluters themselves can also learn and sometimes they even become ecological modernisers. This important type of ecological innovation can be influenced not only by particular events but also by framework conditions such as a high general environmental awareness or an institutional framework of interaction that integrates ecological interests. Such framework conditions which create opportunities for proponents of environmental interests (and often originate from former environmental activities) might also explain the polluters' innovative reactions. But the structural chances of the target group of successfully opposing environmental protection measures must be analysed separately, because they have their own dynamic. This separate analysis of the target group's structural chances and the restrictions they might place on proponents is indeed necessary for evaluating whether or not the capacity of proponents is adequate to solve a certain problem.

5 Capacities in Environmental Policy and Management

5.1 The Capacities of Proponents

Proponents of environmental policy can be specialised government environmental protection institutions as well as non-governmental green interest organisations. Comparative research in environmental policy shows that within the organisational context individual "people make a difference" also (Jacobson and Brown Weiss, 1994: 7; Jänicke and Weidner, 1995; OECD, 1995).

Following Agenda 21, studies on environmental capacity tend to stress not only the organisational and institutional strength but also the abilities of agents or proponents: the role of human capital, technical expertise, functional skills etc. (OECD, 1995).

Government institutions for environmental protection. The establishment of such institutions may plausibly be regarded as the necessary condition for successful measures. "Before pollution policy can be developed, it first has to be invented, and this in turn involves constructing the relevant institutions" (Weale, 1992: 14). Indeed, many improvements in the 1970s and 1980s can be at any rate loosely traced back to the new environmental protection institutions being created from 1970 onwards. Our 36 case studies mentioned above showed that in cases of environmental improvements at national or regional level, government regulation was by far the most important immediate factor in change (Jänicke and Weidner, 1995; Low, 1992; see also OECD, 1996). This dominant role of government is also confirmed by surveys on the factors perceived as influential by enterprises (Gothenburg Research Institute, 1995; Brinkmann and Kirchgeorg, 1995). It can reasonably be assumed that this influence could only be exerted by a competent administration (Kitschelt, 1983; Knoepfel, 1993; Jacobson and Brown Weiss, 1994; OECD, 1994). In more recent studies the importance of institutions for environmental planning is also stressed (Andrews, 1993; OECD, 1994, see also Weale, 1992; OECD, 1995a).

Unfortunately it proves difficult to test the hypothesis that early institutionalisation leads to a better environmental policy record (Jänicke, 1996): Most industrial countries created an environmental agency or ministry at nearly the same time, around the Stockholm Conference of 1972. Small countries such as Luxembourg or Norway started later, but then used the knowledge of neighbour countries. East European countries such as the GDR, Hungary and Poland also started in the early 1970s, but have been very unsuccessful in many areas. This suggests a differentiation between formal and substantive institutionalisation. In democratic systems it may be easier to create institutions in the strict sense, resulting in stable, calculable, internalised and valued patterns of behaviour (Huntington, 1965). In the former communist countries environmental institutions neither had stringently protected jurisdiction nor support from a strong ecology movement or an independent media.

Non-governmental proponents. Today there are three important groups of non-governmental proponents of environmental protection, all of them able to influence polluters directly:

- environmental organisations,
- the media and
- ecologically innovative firms, which have become very important factors in environmental improvements within industry, albeit only in the last few years.

Environmental policy studies emphasise not only the growing direct influence of environmental organisations on target groups (e.g. Weale, 1992; Lester, 1995; Jänicke and Weidner, 1995) but also point to the increasing number of enterprises responding pro-actively to the challenge of environmentalism (e.g. Steger, 1992; Shrivastava, 1992; Fischer and Schot, 1993; Conrad, 1995; Porter and van der Linde, 1995).

5.2 The Structural Framework of Action (Systemic Opportunity Structure)

5.2.1 Cognitive-Informational Framework Conditions

Without a doubt, environmental knowledge and public awareness are important factors influencing environment policy and management. Only in 5 out of our 36 environmental success stories was public awareness not at least a relevant background factor (Jänicke and Weidner, 1995; Conrad, 1995).

Culture and the value system in a country are also highlighted in studies on environmental policy (especially OECD, 1994). A high degree of post-materialism (Inglehart, 1989) for example could plausibly constitute a good *cultural precondition* for environmental policy success. Differences in environmental awareness (e.g. if Germany and France are compared) can often plausibly be explained by cultural differences. But culture is not the only factor influencing public awareness. As mentioned above, cultural traditions are a background condition. Post-materialism, on the other hand, is statistically connected with GDP per capita, which is even more closely related to environmental improvements (Binder, 1996a). Regarding capacity-building in environmental protection the most important task would not be to change cultural traditions but rather to improve informational and communicative capacities. There are therefore reasons not to overestimate cultural factors in this context.

Environmental knowledge and the conditions under which it is generated, distributed, interpreted and applied, however, are of the utmost importance (OECD, 1994). The openness of the scientific community to new problems and paradigms as well as the openness of the media to new issues are important to environmental policy. The role of environmental reporting in environmental capacity has been emphasised in a comparative study by Weidner et al. (1992). Knowledge about environmental dangers is the necessary condition for public awareness. Its non-existence, on the other hand, can be seen as the functional equivalent of the suppression of respective interest articulation.

But it is not only the scope of knowledge that matters. The leading paradigm of policy actors or the "structure of available knowledge and thinking" (Vowe, 1994: 424; see also Nullmeier, 1993; Keck, 1993) is seen as being increasingly important in policy research. The leading paradigm of environmental protection has crucial influence on the strategy. Even the availability of environmental statistics follows the leading pattern of policy orientation.

5.2.2 Political-Institutional Framework Conditions

The importance of institutional framework conditions for environmental policy outcomes has often been emphasised in the literature, though cross-national research has proven difficult (Binder, 1996, 1996a; Strom and Swindle, 1993). Three aspects seem to be most important, and may be structured as participative capacity, integrative capacity, and capacity for strategic action (Jänicke, 1992):

Participative capacity. The openness of the input structures of the policy process is an important aspect of the opportunity structure of environmental interests. Thus it is a common theme in related studies (Kitschelt, 1983; Jänicke, 1990; Weidner, 1996; Weale, 1992; Hey and Brendle, 1994; OECD, 1994; Carew-Reid et al., 1994). Decentralisation and strong local communities are seen as a favourable condition for participation. The role of cities in climate protection policy may be cited as an example (OECD, 1994a). Sometimes the pressure from environmental movements is mentioned as a relevant factor for the general opening-up and modernisation of political systems (Paehlke, 1990; Jänicke, 1993; Weidner, 1996). The openness of the legal system to "protective" interests is also mentioned as a kind of catalyst to participation (Weidner, 1996). In terms of resources, environmental laws as well as constitutional mechanisms such as plebiscites (which changed nuclear energy policy in at least three countries) can be regarded as important tools for proponents.

Integrative capacity. Participation, decentralisation and especially cross-sectoral strategies require a high level of integration. In the literature volunteer cooperation – "consensual capacity" (Jänicke, 1992) – is generally stressed as an institutional condition for success in environmental policy (Lundqvist, 1980; Kitschelt, 1983; Brickman et al., 1985; Jänicke, 1992; OECD, 1994; Carew-Reid et al., 1994). According to Badaracco, "co-operation is a superior approach" and corporatism is good for environmental policy (Badaracco, 1985; see also Ricken, 1995). In OECD countries, low rates of unemployment correlate with improvements in the environment (Jänicke, 1992; Binder, 1996a). This may indicate that in countries with better pollution control outcomes there is no issue competition between environmental and employment policy. But it may also mean that effective steering mechanisms in employment policy – e.g. neo-corporatism, concerted action or other cooperative mechanisms – are not only helpful in industrial and employment policy (Schmidt, 1986; Traxler and Unger, 1991), but also in pollution control.

Integrative capacities are important at different levels. Firstly, intrapolicy cooperation (Knoepfel, 1993), i.e. the internal integration of the policy field, is important. This includes the coordination of environmental policy at different levels (and in respect to different jurisdictions) of the political system. The second field of integration is interpolicy coordination, the cross-sectoral integration of conflicting policies (Carew-Reid et al., 1994). Thirdly, the external integration of environmental policy institutions and non-governmental actors, including target groups is needed. "Network management" (Carley and Christie, 1994; Dutch Committee for Long-Term Environmental Policy, 1994; Glasbergen, 1995; Ricken, 1995) has today become a highly sophisticated, professional strategy.

A cooperative policy style is a prerequisite for effective policy integration. It is also good for policy innovation, because innovators are integrated earlier into the decision-making process than is the case in countries with a more confrontational tradition (Jänicke, 1992). Obviously, if the environmental policy goal is anchored in a broad consensus, implementation will be easier and more successful.

Integrative capacities together with participative ones offer possibilities for describing types of political systems (Kitschelt, 1983). Examples might be the open, but fragmented US-American system, or the closed and highly integrated French system, or the open and integrated systems of smaller democracies such as the Netherlands, or Norway.

Capacity for strategic action (Scharpf, 1987; Jänicke, 1992) can be defined by the ability to implement long-term common policy objectives against short-term special interests. This capacity may at present be seen as the highest stage of institutional capacity-building in environmental policy and management (Carew-Reid et al., 1994; OECD, 1994, 1995a). If we compare countries such as Italy or Greece on the one hand and Japan or the Netherlands on the other, we find great national differences in this capacity.

5.2.3 Economic-Technological Framework Conditions

Per capita GDP is most closely connected with environmental policy outcomes. But the statistical result is somewhat paradoxical or ambivalent, because GDP per capita correlates with environmental improvements in some fields, but with deterioration in others. If a standard technological solution – clean-up technology, substitutes, higher efficiency, or recycling – is available, the probability of improvement is highest in rich countries (Jänicke, 1992). Wealthier countries both demand and can afford stronger environmental measures (Ringquist, 1993: 107). On the other hand, if a standard technological solution has not been developed (examples are land-use, soil contamination, or road traffic) GDP per capita is usually a good predictor of ecological deterioration (Jänicke, Mönch and Binder, 1996).

As mentioned above, economic performance is a difficult aspect of environmental capacity. This is because it has not only a strong, but a contradictory impact on the environment, influencing both the structure of problems and the capacity to solve them. Its influence on capacity for environmental protection is complex and more indirect, implying a connection to research, educational, communicative or administrative capacities. And the general state of the economy is not a reasonable target for environmental policy.

However, the technological standard of a country's economy may be influenced more easily and directly, e.g. by transfer of technology and expertise (OECD, 1995). The prices of environmentally sensitive goods are another economic factor that can be influenced by policy (OECD, 1993, 1996).

6 Capacity-Building

If capacity in environment relates to the ability of societies to identify and solve ecological problems, "capacity development in environment relates to the 'process' by which those abilities are developed" (OECD, 1994: 9). This includes "the capacity for capacity development" (OECD, 1995: 46). And, of course, capacity development can only be conceived of as a comprehensive strategy.

Political science has its own tradition of the concept "capacity". The idea of development through overcoming crises (formulated by Erikson in 1968) played a certain role in the debate. Along these lines political modernisation can be defined as the institutionalisation of a higher capacity to identify and solve problems. This is usually promoted by crises (i.e. a threat of severe loss or even collapse) and made possible by already existing capacities and former policy experiences. Institutionalisation includes the internalisation of new routines of cognition and policy action. Pye (1966), Binder, Almond and Coleman (Binder et al., 1971), or Rokkan used the concept of development crisis as potentially leading to higher capacities of the political system: crises of penetration, identity, legitimacy, distribution, participation and integration. In our context the capacities of participation and integration are especially important. The capacity of a system for strategic action or the capacity to provide relevant knowledge could be added to the list.

If we examine the process of capacity-building in environmental policy since 1970, we see typical developments. As a rule, however, systemic capacity development is no steady, linear process. In the UK (under Thatcher), the US (under Reagan, and after the 1994 Congressional elections) and also in Germany (after unification) or Japan (since 1985) we observe phases of downturn in environmental policy connected partly with the reduction of institutional capacities (e.g. curtailment of financial resources or participative opportunities). However, this has sometimes strengthened the position of environmental organisations.

6.1 Proponents

The long-term process of capacity development usually begins with the institutionalisation of the new policy area. In the early 1970s most industrial countries founded an environmental agency and/or a ministry for the environment (Jörgens, 1996). This institution was again differentiated, as a rule, according to the growth of staff, budget and scientific competence. The process of capacity-building also implied policy integration throughout dispersed jurisdictions (intrapolicy coordination) and a closer cooperation with "polluting" policy areas such as transport or energy (interpolicy coordination). The existence of a pluralistic and cooperative national institution for environmental planning represents a high level of institutional development.

A process of capacity development can also be observed with respect to non-governmental actors:

The role of environmental organisations developed from more traditional nature conservation to an active representation of environmental interests, usually increasing their membership, financial resources and competence. They develop from scattered, lay grass-roots organisations to a kind of national institution (Roth,

STAGES IN CAPACITY DEVELOPMENT: PROPONENTS OF ENVIRONMENTAL PROTECTION

1. Environmental Policy Institutions:
a) No central ministry/agency for environment,
b) weak and isolated national ministry/agency for environment,
c) environmental institutions at all levels of the political system,
d) mechanisms for environmental policy integration into different policy sectors; long-term environmental planning.

2. Green NGOs:
a) Local organisations,
b) weak or non-professional national green NGOs,
c) strong and highly competent green NGOs, playing a consultative role in environmental policy decisions,
d) green NGOs playing a consultative role in industry also.

3. The Media:
a) Few, largely official reports on the environmental situation,
b) environmental reports in some critical sections of the media,
c) environmental problems widely reported in all the media,
d) frequent direct media attacks against polluters.

4. The "Green" Business Sector:
a) Environmental interests are scarcely articulated within the economic system,
b) environmental interests are articulated mainly by the producers of clean-up technology,
c) environmental interests are articulated by a particular green business organisation,
d) the "green" business sector has a strong impact on the whole economy (e.g. by demand).

1994) which often plays a consultative role in government decisions and finally increasingly in enterprise decisions also. Today just the five largest environmental organisations in the EU (of the 12) have about 10 million members, with nearly 20% of the adult population in the Netherlands belonging to some environmental organisation (Hey and Brendle, 1994: 642).

In the media a few newspapers usually start by publicising some critical information, then the issue becomes established in a larger segment of the media, with an opposing segment of media remaining friendly to polluters. In most advanced democracies today the environmental problem is by and large accepted as an issue by most publishers. The media, however, also represent an area from where it can be seen that capacity development is not always a linear process. Highly concentrated commercialised mass media with close ties to economic power may in recent years have worsened conditions for environmental awareness in some countries.

Capacity development can also be observed in the universities, which were for a long time important generators of knowledge for the polluting sections of the economy, and in some countries made a relatively late changeover to new ecological paradigms, often following government initiatives (and the associated research budgets).

The development of a green business sector – at first formed only by the eco-industry, but then also by consultancy firms, insurance companies and ecological pioneer enterprises with their own green organisations – can also be described as part of the process of capacity-building.

In recent times capacity-building in environmental policy and management also has been described in terms of actor constellation (Knoepfel, 1993). This can be shown for the development in the United States (Hoffman, 1994):

In a first stage (1960-70) industry acted relatively autonomously within its institutional sphere. In a second stage (1970-82) the state intervened in this sphere, under pressure from environmental organisations. In a third stage (1982-88) – under the presidency of Reagan – state and industry cooperated under the common pressure of green organisations. In the fourth stage (since 1988) the cooperation between state and industry partly includes environmental organisations. In addition, an intra-market proponent of environmental protection has become influential: insurance companies, "green" capitalists or pioneer enterprises.

This kind of capacity-building by the improvement of actor constellation could also be shown for the case of Germany, Switzerland or other advanced countries.

6.2 The Systemic Framework of Action

The opportunity structure, or the sum of systemic resources available to the proponents of environmental protection also shows typical lines of development.

The participative capacity of the political system as well as its integrative capacity have usually increased during the process of environmental policy development. The crucial stage is (or would be) the opening of the closed policy networks which make de facto decisions in ecologically problematic policy fields, before proposals are decided upon by the representative institutions. The pluralistic structures of Public Utilities Commissions in the US are an example of such

DEVELOPMENT OF PARTICIPATIVE CAPACITIES IN ENVIRONMENTAL POLICY:

a) Participation is restricted to general elections
b) Participation rights for green organisations regarding licensing of local sites as well as national environmental policy formulation
c) Participation rights for green organisations extended to other policies
d) General participation of green NGOs in early stages of the policy formulation process of all environmentally relevant laws/decrees and in long-term environmental planning.

high-level participation. The process of developing integrative capacities within government can be described in similar steps.

The scope as well as the structure (paradigm) of available knowledge is another level of capacity development:

SELECTED INDICATORS OF CAPACITY BUILDING: ENVIRONMENTAL KNOWLEDGE AND AWARENESS:

1. Statistics (as indicators of the leading paradigm):

a) Scattered or irrelevant data
b) Mainly air and water quality data
c) National emissions and solid waste data also
d) Data on relevant material flows also, comprehensive indicator systems

2. Science and Research:

a) No systematic research on pollution problems
b) Few independent environmental research institutes (typically with disputed professional standards)
c) Erosion or the traditional "productionist" research paradigms and priorities
d) Environment as a relevant theme in universities and research institutes

This introduction has tried to make clear that the concept of capacity-building could be a useful heuristic tool in the analysis of environmental policy. The selected policy indicators are seen as examples to underline the assumption that environmental policy is developing in typical stages that can be described as capacity development.

References

Andrews, R. N. L., 1993: Long-Range Planning in Environmental and Health Regulatory Agencies, Ecology Law Quarterly 20, 515-582.

Badaracco, J. L., 1985: Loading the Dice. A Five-Country Study of Vinyl Chloride Regulation. Boston and London: MIT Press.

Bartelmus, P., 1994: Environment, Growth and Development – The Concepts and Strategies of Sustainability. London, New York: Routledge.

Bartlett, R. V., 1994: Evaluating Environmental Policy – Success and Failure, in: N. J. Vig and M. E. Kraft (eds.), Environmental Policy in the 1990s. Toward a New Agenda (second edition). Washington, D.C.: CQ Press.

Binder, L., J. S. Coleman, J. LaPalombara, L. W. Pye, S. Verba and M. Weiner, 1971: Crises and Sequences in Political Development. Princeton, N. J.: Princeton University Press.

Binder, M., 1996: Die Operationalisierung umweltpolitischen Erfolges: Probleme und Lösungsansätze, in: M. Jänicke (ed.), Umweltpolitik der Industrieländer. Entwicklung – Bilanz – Erfolgsbedingungen. Berlin: edition sigma, 133-152.

Binder, M., 1996a: Mögliche Erfolgsbedingungen der Schwefeldioxidminderung im internationalen Vergleich, in: M. Jänicke (ed.), Umweltpolitik der Industrieländer. Entwicklung – Bilanz – Erfolgsbedingungen. Berlin: edition sigma, 153-201.

Brickman, R., S. Jasanoff and T. Ilgen, 1985: Controlling Chemicals: The Politics of Regulation in Europe and the United States. Ithaca, New York: Cornell University Press.

Brinkmann, J. and M. Kirchgeorg, 1995: Umweltorientiertes Unternehmensverhalten, *Zeitschrift für Umweltpolitik und Umweltrecht* 18, 377-390

Carew-Reid J., R. Prescott-Allan, S. Bass, and B. Dalal-Clayton, 1994: Strategies for National Sustainable Development. A Handbook for Their Planning and Implementation. IUCN/IIED, London: Earthscan.

Carley, M., I. Christie, 1994: Managing Sustainable Development. London: Earthscan.

Conrad, J. (ed.), 1995: Successful Environmental Management in European Companies, Forschungsstelle für Umweltpolitik/Freie Universität Berlin (final report of a project for the European Commission, to be published in 1996).

Dutch Committee for Long-Term Environmental Policy, 1994: The Environment: Towards a Sustainable Future. Dordrecht, Boston, London: Kluwer Academic Publishers.

Feick, J., 1992: Comparing Comparative Policy Studies – A Path Towards Integration, *Journal of Public Policy* 12, 257-285.

Fiorino, D. J., 1995: Making Environmental Policy. Berkeley, Los Angeles, London: University of California Press.

Fischer, K. and J. Schot (eds.), 1993: Environmental Strategies for Industry – International Perspectives on Research Needs and Policy Implications. Washington, D. C.: Island Press.

Glasbergen, P. (ed.), 1995: Managing Environmental Disputes – Network Management as an Alternative. Dordrecht, Boston, London.

Gothenburg Research Institute, 1995: The Nordic Business Environmental Barometer. Oslo: Bedriftsoekonomens Forlag.

Hey, C. and U. Brendle, 1994: Umweltverbände und EG. Strategien, Politische Kulturen und Organisationsformen. Opladen: Westdeutscher Verlag.

Hoffman, A. J., 1994: The Environmental Transformation of American Industry: An Institutional Account of Environmental Strategies in the Chemical and Petroleum Industries. Unpublished paper, MIT, Dep. of Civil and Environmental Engineering.

Huntington, S. P., 1965: Political Development and Political Decay, *World Politics* 17, 3, 386-430.

Inglehart, R., 1989: Kultureller Umbruch – Wertwandel in der westlichen Welt. Frankfurt/M., New York: Campus.

Jacobson, H. K. and E. Brown Weiss, 1994: Strengthening Compliance with International Environmental Accords: Some Preliminary Observations from a Collaborative Project. Paper presented at the XVI World Congress of the International Political Science Association Convention, 21-25 August, Berlin.

Jahn, D., 1993: The Rise and Decline of New Politics and the Greens in Sweden and Germany, *European Journal of Political Research* 24, 177-194.

Jänicke, M., 1973: Die Analyse des politischen Systems aus der Krisenperspektive, in: M. Jänicke (ed.), Politische Systemkrisen. Köln: Kiepenheuer & Witsch.

Jänicke, M., 1990: State Failure. Cambridge: Polity Press.

Jänicke, M., 1992: Conditions for Environmental Policy Success: An International Comparison, *The Environmentalist* 12, 1, 47-58.

Jänicke, M., 1993: Über ökologische und politische Modernisierungen, Zeitschrift für Umweltpolitik und Umweltrecht 16, 159-175.

Jänicke, M. (ed.), 1996: Umweltpolitik der Industrieländer. Entwicklung – Bilanz – Erfolgsbedingungen. Berlin: edition sigma.

Jänicke, M., H. Mönch and M. Binder, 1996: Umweltindikatorenprofile im Industrieländervergleich: Wohlstandsniveau und Problemstruktur, in: M. Jänicke (ed.), Umweltpolitik der Industrieländer. Entwicklung – Bilanz – Erfolgsbedingungen. Berlin: edition sigma, 113-131.

Jänicke, M., H. Mönch and M. Binder, 1996a: Getting Rich – Getting Clean? Umweltindikatorenprofile im Industrieländervergleich, Zeitschrift für angewandte Umweltforschung (ZAU) 9, 1, 41-55.
Jänicke, M. and H. Weidner (eds.), 1995: Successful Environmental Policy. A Critical Evaluation of 24 Cases. Berlin: edition sigma.
Jörgens, H., 1996: Die Institutionalisierung von Umweltpolitik im internationalen Vergleich, in: M. Jänicke (ed.), Umweltpolitik der Industrieländer. Entwicklung – Bilanz – Erfolgsbedingungen. Berlin: edition sigma, 59-111.
Keck, O., 1993: Information, Macht und gesellschaftliche Rationalität. Das Dilemma rationalen kommunikativen Handelns, dargestellt am Beispiel eines internationalen Vergleichs der Kernenergiepolitik. Baden-Baden: Nomos Verlagsgesellschaft.
Kern, K. and S. Bratzel, 1996: Umweltpolitischer Erfolg im internationalen Vergleich: Zum Stand der Forschung, in: M. Jänicke (ed.), Umweltpolitik der Industrieländer. Entwicklung – Bilanz – Erfolgsbedingungen. Berlin: edition sigma, 29-58.
Kitschelt, H., 1983: Politik und Energie. Frankfurt/M., New York: Campus.
Kitschelt, H., 1986: Political Opportunity Structures and Political Protest: Anti-Nuclear Movements in four Democracies, *British Journal of Political Science* 16, 1, 57-85.
Knoepfel, P., 1993: New Institutional Arrangements for the Next Generation of Environmental Policy Instruments: Intra- and Interpolicy Cooperation. Cahiers de l'IDHEAP no 112. Lausanne: Institut des Hautes Etudes en Administration Publique.
Lester, J. P., (ed.) 1995: Environmental Politics and Policy – Theories and Evidence. 2nd edition. Durham, London: Duke University Press.
Low, P., 1992: International Trade and the Environment: An Overview, in: P. Low (ed.): International Trade and the Environment, World Bank Discussion Papers 159, Washington, D. C.
Luhmann, N., 1990: Ökologische Kommunikation (third edition). Opladen: Westdeutscher Verlag.
Lundqvist, L. J., 1980: The Hare and the Tortoise: Clean Air Policies in the United States and Sweden. Ann Arbor: University of Michigan Press.
Nullmeier, F., 1993: Wissen und Policy-Forschung. Wissenspolitologie und rhetorisch-dialektisches Handlungsmodell, in: A. Héritier (ed.), Policy-Analyse. Kritik und Neuorientierung, *Politische Vierteljahresschrift*, Special Issue 24. Opladen: Westdeutscher Verlag, 175-196.
OECD, 1992: Market and Government Failures in Environmental Protection: The Case of Transport. Paris: OECD.
OECD, 1993: Taxation and Environment: Complementary Policies. Paris: OECD.
OECD, 1994: Capacity Development in Environment. Paris: OECD.
OECD, 1994a: Urban Energy Management. A Handbook of Good Local Practices, Heidelberg Conference, Sept. 7-9, 1994. Paris: OECD.
OECD, 1995: Developing Environmental Capacity. A Framework for Donor Envolvement. Paris: OECD.
OECD, 1995a: Planning for Sustainable Development. Paris: OECD.
OECD, 1996: Environmental Performance in OECD Countries. Paris: OECD.
Paehlke, R., 1990: Democracy and Environmentalism: Opening a Door to the Administrative State, in: R. Paehlke and D. Torgerson (eds.), Managing Leviathan. Environmental Politics and the Administrative State. Ontario, Lewiston, NY: broadview press, 35-51.
Porter, M. and C. van der Linde, 1995: Green and Competitive: Ending the Stalemate, *Harvard Business Review*, September-October 1995, 120-134.
Potier, M., 1994: Agreement on the Environment, *OECD-Observer* 189, 8-11.
Pye, L. W., 1966: Aspects of Political Development. Boston: Little, Brown and Co.
Ricken, C., 1995: Nationaler Politikstil, Netzwerkstrukturen sowie ökonomischer Entwicklungsstand als Determinanten einer effektiven Umweltpolitik – Ein empirischer Industrieländervergleich, Zeitschrift für Umweltpolitik und Umweltrecht, 18, 4, 481-502.
Ringquist, E. J., 1993: Environmental Protection at the State Level – Policies and Progress in Controlling Pollution. London: Armonk.

Roth, R., 1994: Demokratie von unten. Neue soziale Bewegungen auf dem Wege zur politischen Institution. Köln: Bund-Verlag.

Sabatier, P. A. and H. Jenkins-Smith (eds.), 1993: Policy Change and Learning: An Advocacy Coalition Approach. Boulder/CO: Westview.

Scharpf, F. W., 1987: Sozialdemokratische Krisenpolitik in Europa. Frankfurt/M., New York: Campus.

Schmidt, M. G., 1986: Politische Bedingungen erfolgreicher Wirtschaftspolitik. Eine vergleichende Analyse westlicher Industrieländer (1960 – 1985), *Journal für Sozialforschung* 26, 3, 251-273.

Schmidt, M. G., 1993: Theorien in der international vergleichenden Staatstätigkeitsforschung, in: A. Héritier (ed.), Policy-Analyse – Kritik and Neuorientierung, *Politische Vierteljahresschrift*, Special Issue 24. Opladen: Westdeutscher Verlag, 371-393.

Shonfield, A., 1969: Modern Capitalism. The Changing Balance of Public and Private Power. London et al.: Oxford University Press.

Shrivastava, P., 1992: Corporate Self-Greenewal: Strategic Responses to Environmentalism, *Business Strategy and the Environment* 1, 3, 9-21.

Solesbury, W., 1976: Issues and Innovation in Environmental Policy in Britain, West Germany, and California, *Policy Analysis* 2, 1-38.

Steger, U. (ed.), 1992: Handbuch des Umweltmanagements – Anforderungs- und Leistungsprofile von Unternehmen und Gesellschaft. München: Beck.

Strom, K., S. Swindle, 1993: Political Parties, Institutions and Environmental Reform. University of California, Center for German and European Studies: Working Paper 2.17.

Traxler, F. and B. Unger, 1991: Institutionelle Erfolgsbedingungen wirtschaftlichen Strukturwandels, *Wirtschaft und Gesellschaft* 16, 189-223.

Vowe, G., 1994: Politische Kognition. Umrisse einer kognitionsorientierten Ansatzes für die Analyse politischen Handelns, *Politische Vierteljahresschrift* 35, 423-447.

Weale, A., 1992: The New Politics of Pollution. Manchester, New York: Manchester University Press.

Weidner, H., 1996: Basiselemente einer erfolgreichen Umweltpolitik – Beschreibung, Analyse und Evaluation der Instrumente der japanischen Umweltpolitik. Berlin: edition sigma.

Weidner, H., R. Zieschank and P. Knoepfel (eds.), 1992: Umwelt-Information. Berichterstattung und Informationssysteme in zwölf Ländern. Berlin: edition sigma.

ENVIRONMENTAL CAPACITY BUILDING IN AUSTRALIA

Elim Papadakis

To understand more fully the character of environmental problems in Australia it is important to consider the continent's geography and its distinctive historical development as a nation state. Australia covers a vast territory (7,682,000 square kilometres) but has a low population density. Most people are concentrated in several cities, largely on the eastern coast. Compared to other OECD countries the current population of 18.4 million is likely to grow at a fairly rapid rate. The unique aspects of Australia's political economy (Bell and Head, 1994) are also important factors, and have been pivotal in debates about greenhouse gas emissions. Government sources point to heavy reliance on exporting agricultural products and resource processing industries. Another important consideration is the shape of the political system with its three tiers (federal, state, and local). The states have extensive powers to legislate, manage and regulate the environment. Local government is responsible for land use planning and aspects of development. However, over the past two decades, the main shift in power in this area has been towards the federal government.

Taking a broad historical perspective, the settlement of this continent by Aboriginals about 50,000 years ago has also shaped the environment. Though the relationship between the Aboriginals and nature was generally benign, the practice of setting fire to the bush (to hunt animals and to renew vegetation and plant life to attract more animals) reshaped the environment in several ways. Unchecked fires probably destroyed several species of wildlife and had a major impact on the landscape, destroying certain types of forest and replacing them by others.

The encounter between Aboriginals and nature was far less devastating than between white settlers and land they sought to conquer. According to Philip Toyne, former Director of the Australian Conservation Foundation, as a nation and a continent, Australia leads the world in rates of mammal extinction (Toyne, 1994: 3–4). A 1996 official report noted that: "For the land animals and plants about which we know enough to assess their current state, the trends are disturbing. Some 5 per cent of higher plants, 23 per cent of mammals, 9 per cent of birds, 7 per cent of reptiles, 16 per cent of amphibians and 9 per cent of fresh-water fish are extinct, endangered or vulnerable. Australia has the world's worst record of mammal extinctions. In the past 200 years, we have lost 10 of 144 species of marsupials and 8 of 53 species of native rodents" (State of the Environment Advisory Council, 1996a: 24).

The predominant vision of the conquest of nature in the nineteenth century was influenced by several factors. Australia was seen as an adjunct of the British Empire and, apart from serving as a penal settlement, was exploited for raw materials. Another important factor was the introduction of new species of plants and animals by the settlers, often with devastating consequences. The problem of soil erosion following the destruction of trees and the introduction of sheep and cattle remains one of the greatest concerns facing the country. At any rate the soil is generally of poor quality. Still, the environment is immensely diverse and includes tropical rainforests, wetlands, alpine regions and rangelands.

While many of these factors contribute to the unique character of environmental policy in Australia, many aspects of policy development and capacity building are comparable to experiences of other countries.

1 Main Environmental Problems and Improvements Since 1970

There are several ways of identifying major environmental problems over the past three decades. One is to examine the adoption of environmental issues as policy by major political parties. Another is to examine the pattern of media coverage of issues. Finally, one can enumerate the topics covered in opinion polls and any evidence of concern about these questions.

If we take opinion polls as a guide there is widespread consensus about the urgency of tackling environmental problems. The 1996 Australian Election Study found the vast majority of people regarded as very or fairly urgent problems the greenhouse effect (86 per cent), soil degradation (90 per cent), destruction of wildlife (87 per cent), logging of forests (74 per cent), waste disposal (93 per cent) and pollution (93 per cent). The figures from the 1990 and 1993 Studies were either very similar or almost identical. Likewise, in the 1998 Study 91 per cent felt that "stronger measures should be taken to protect the environment'. Opinion polls mainly capture what those who frame the questions have included as legitimate concerns. Another form of agenda-setting occurs through initiatives of established political organizations. If we consider electoral platforms, all these issues have become an important part of political debate. For instance, land use has long preoccupied political associations (Papadakis, 1993; 1996; Toyne, 1994; Doyle and Kellow, 1995). As early as 1933 the Country Party (today known as the National Party) adopted soil conservation in its electoral platform. The Liberal Party took up this issue in its 1948 platform and the Australian Labor Party (ALP) in 1953. This remains a key issue for Australian governments.

The thrust behind attempts to address this problem have been economic (land productivity) and environmental (preservation of eco-systems). Attempts by Labor governments in the late 1980s and early 1990s to address the issue through a National Soil Conservation Strategy were significant for several reasons. They highlighted the neglect of the land by governments, even though information had been available for decades (see New South Wales Soil Conservation Act, 1938). Second, they recognised soil degradation as one of the most pressing problems facing Australia, especially economically. Third, the government participated in a major initiative to tackle the problem in cooperation with the National Farmers Federation and the Australian Conservation Foundation. This represented a significant attempt to integrate development and environmental protection and influence patterns of coalition formation.

The government attempted to change perceptions by focusing on economic aspects of the problem, arguing that around $20 billion, or 40 per cent of gross export earnings came from "activities based on the soil" (Commonwealth of Australia, 1989: 40). The government recognised that detailed knowledge about these problems was already available in the mid 1970s, when studies showed that "about 45 per cent of non-arid lands and some 55 per cent of arid lands in use required treatment for land degradation' (Commonwealth of Australia, 1989: 40). Some writers argued that land degradation has been costing Australia $2.5 billion a year in lost rural production (Beale and Fray, 1990). A government study painted an alarming picture of damage in different regions, destruction of productive land through over-clearing, increases in salinity and disappearance of soil nutrients (Eckersley, 1989).

Another measure of the importance of land and water resources to the Australian government has been the establishment of a Natural Heritage Trust (see below) with a principal focus on the National Landcare Programme, revegetation, rehabilitating the Murray-Darling River Basin, creating a national reserve system and tackling pollution of the coast and seas. The Natural Heritage Trust is regarded as the foundation for conserving biodiversity and ecologically sustainable management of land and water resources.

While protecting land and water resources has become part of government agendas, environmentalists have campaigned vigorously for protecting forests. In announcing the National Soil Conservation Strategy, the Labor government presented data used by environmental groups to describe changes that had occurred over two centuries:

When the Europeans first came to Australia, forests covered about 10 per cent of the continent. Woodlands, where trees are often smaller or more widely spaced, covered 23 per cent. Since then, some 50 per cent of the original tall and medium forest and 35 per cent of the woodlands have been cleared or severely modified, mainly for agricultural purposes. Three quarters of Australia's rainforests have been destroyed (Commonwealth of Australia, 1989: 46).

The government added that these trends had not yet been reversed, mainly because of ongoing land clearing for agriculture and urban development, and poor management (including damage by livestock of seedlings and the use of fertilisers, herbicides and pesticides). Still, by 1992, following a decade of pressure by environmental groups, the Labour government was able to claim that it had prevented the construction of the Franklin Dam and created a huge World Heritage area in Tasmania, had increased threefold the size of Kakadu national park and secured its inscription on the World Heritage List and protected Uluru national park, the Queensland Wet Tropics, Fraser Island and Shark Bay.

Analysis of electoral platforms shows that concern about preserving forests can be traced to the 1940s and 1950s, though intense preoccupation with their preservation and enlargement began in the 1970s (Papadakis, 1996). Since the late 1970s there have been bitter controversies over issues like woodchipping, including the export of woodchips to countries like Japan. The issue of wilderness protection reflects the particular geography of Australia and its unique historical development.

Over the past few years there has been a steady increase in the number of protected areas and recent data indicate that protected areas cover about 7.8% of Australia's land area. The number of terrestrial and marine protected areas is 4,187 and 306, respectively. There are 11 World Heritage Areas, 12 Biosphere Reserves and 42 Wetlands of International Importance under the Ramsar Convention. There are 264 recovery plans in preparation or being implemented to protect threatened species. The number of threatened species is 1,031, of which 312 are in the "endangered" category and 719 in the "vulnerable" category.

Other key issues include pollution and the greenhouse effect. In order to address the problem of climate change, the federal government, in October 1990, agreed to a proposal by the Minister for the Environment to set the target of a 20 per cent reduction in the emission of greenhouse gases (from 1988 levels) by the year 2005. However, the ALP appeared to back off from this commitment, and in 1996 the new National-Liberal government followed suit by launching a major diplomatic effort to persuade other OECD countries to adopt the principle of differentiated rather than uniform greenhouse gas targets. According to recent governments the costs to Australia of adopting a uniform approach would be much greater than those borne by other countries. The disadvantages for Australia are seen in terms, relative to the European Union, of a more rapid population increase and greater reliance on export industries requiring high levels of energy. (Commonwealth of Australia, 1997a). The Australian economy relies more than most OECD countries on fossil fuels – this affects both the energy supply industry and industries exporting coal, natural gas and oil. Australian agriculture also accounts for a significant proportion (17 per cent) of CO_2 emissions. The government pointed out that the European Union itself adopted a differentiated approach enabling countries like Portugal to increase emissions. Some critics have nonetheless argued that the government could introduce mandatory standards for fuel efficient vehicles and restructure taxation systems (Hamilton, Hundloe and Quiggin, 1997) and reduce land clearing (Hamilton, 1994), which would enable Australia to meet international targets.

The outcome of the 1997 Kyoto Conference, when Australia achieved an agreement allowing it to raise levels of greenhouse emissions 8 per cent (in 2010 from 1990 levels) was regarded as a success by the government. Iceland was the only other wealthy nation granted this concession. Following the Kyoto meeting, Australia continued to work closely with countries that shared its views on issues like international emissions trading. This "Umbrella Group" includes the United States, Japan, Canada, New Zealand, Russia, Norway, Iceland and the Ukraine. The Australian government's approach to the 1998 Buenos Aires Conference was partly in alignment with that of other members of this group. The government saw as important decisions at Buenos Aires "the process for elaborating rules for the Kyoto (flexibility) mechanisms – international emissions trading, the Clean Development Mechanism (cooperative projects with developing countries with provision for the transfer of emission credits), and joint implementation (cooperative projects with other developed countries)" (Commonwealth of Australia, 1998). The government undertook to analyse the impact of the Kyoto protocol on Australia before deciding on whether or not to ratify it.

The government responded with various initiatives to the problem of climate change. A National Greenhouse Gas Inventory for 1988 and 1990 was published in

Table 1. Evolution of Air Pollution Emissions (per capita emissions)

Pollutant	1975	1980	1985	1990	1995
CO_2	13.75	14.63	14.19	15.51	15.85
NO_X	–	–	139.22	131.87	117.96
CO			1167.41	1059.70	833.56
VOC			122.31	114.84	98.45

Source: OECD, 1997. The calculations of per capita emissions were kindly provided to the author by Professor Detlef Jahn.

1994 and inventories for the period 1988–1994 were compiled and released in 1996. The inventory showed that carbon dioxide (CO_2) emissions increased by 3.9 per cent, from 394 million tons in 1990 to 410 tons in 1994. Methane (CH_4) emissions remained virtually unchanged (at around 136 million tons) and nitrous oxide (N_2O) emissions rose by 3.6 per cent during that period.

Overall, the outcomes of efforts to tackle air pollution have been mixed. In some areas (nitrogen oxide, carbon monoxide and volatile organic compounds) there have been improvements, in others (notably CO_2) there has been a deterioration (Table 1). Trends in sulphur oxide (SO_X) emissions were about 1.8 million tonnes in 1988, rising to 2.15 million in 1994 and "projected to increase to 2.4 million tonnes by 2000" (OECD, 1998: 107).

The figures for CO_2 emissions reflect one of the largest per capita in the OECD: "Between 1985 and 1995, the OECD average for emissions per unit of GDP fell by over 10 per cent but Australia's were stable; during the same period, Australia's energy-related CO_2 emissions per capita grew four times faster than the OECD average" (OECD, 1998: 175). In 1997 the government issued a discussion paper on the national greenhouse strategy with proposals on how to reduce emissions (Commonwealth of Australia, 1997b; see also 1997c). This effort follows earlier attempts to develop and monitor a national greenhouse strategy (Commonwealth of Australia, 1992a; 1995). A wide range of measures were anticipated to reduce emissions, including energy performance codes and standards for domestic appliances and industrial equipment as well as housing and commercial buildings, stricter emission standards for vehicles, energy market reforms, a Renewable Energy Innovation Investment Fund, revegetation and plantations (Commonwealth of Australia, 1997c: Appendix). One important innovation has been cooperation with industry through the Greenhouse Challenge initiative, a program of cooperative agreements with companies across various industry sectors and the promotion of sustainable agricultural practices. By November 1997 a hundred companies and industry associations had signed agreements under this plan to reduce emissions by about 22 million tonnes of CO_2 by 2000, and these companies accounted for over 45 per cent of industrial emissions in Australia (Commonwealth of Australia, 1997c).

Other government initiatives include the creation of Cooperative Research Centres (for Southern Hemisphere Meteorology, and Renewable Energy) to investigate the greenhouse question. In 1998 the government established an Australian

Greenhouse Office to coordinate and implement the national greenhouse strategy and manage partnership programs like Greenhouse Challenge.

As regards ozone-depleting substances, consumption of CFCs, methyl chloroform and carbon tetrachloride were phased out by 31 December 1995, though, in line with the Montreal Protocol, utilisation was permitted for essential purposes like metered dose inhalers and laboratory uses. Halon consumption was phased out by 31 December 1994 and, in accord with the Protocol, consumption of HCFCs and methyl bromide will be eliminated over a 10–20 year period. The government has contributed A$10 million from 1994 to 1996 to the Montreal Protocol for the transfer of ozone-friendly technology to developing countries.

The question of pollution associated with emissions of carbon monoxide, nitrogen dioxide, photochemical oxidants, sulphur dioxide, lead and particles, will partly be addressed by the National Environment Protection Council which began work in 1996 on developing national ambient air quality standards.

2 Development of Environmental Policy and Principal Regulations

Another important factor is the three-tiered shape of the federal political system. The states and territories have extensive powers to legislate, manage and regulate the environment. Local government is responsible for land use planning and aspects of development. However, over the past two decades, the main shift in power in this area was towards the federal government.

The main stimuli to developing modern environmental policy included initiatives by political and administrative elites, their response to concerns expressed by activists in the 1960s, a growing awareness of international concerns about issues like the "limits to growth', and the willingness, particularly of the ALP in the late 1960s and early 1970s, to promote environmental concerns both in its policy platforms and speeches and in the formation of new institutions. The focus by political and other elites on integrating consideration for the environment with traditional concerns about economic growth can be traced to the 1960s, with the introduction of the Environmental Impact Statement (EIS) as a way of identifying environmental impacts of new projects. The goal of the EIS was to achieve a balance between economic and ecological imperatives. In practice it has often been perceived as falling short of achieving this.

Another stimulus was the rise in opposition to development projects within local communities, notably over the flooding of Lake Pedder in Tasmania (see below). A further impetus for modern environmental policy came from the 1965 ALP platform which included some of the most significant statements by a major party. Notwithstanding the lasting attention by the ALP on economic issues, it exercised leadership on environmental matters. This may partly derive from its opposition to nuclear weapons testing since the 1950s and its proximity to protest movements on this issue. Furthermore, as an opposition party, the ALP could detach itself more easily from institutional practices and inertia that conservative (Liberal and National) parties had created through electoral victories in the 1950s and 1960s and their almost exclusive attentiveness to economic development rather than environmental protection (Papadakis 1996).

The conservative regime created a Federal Office for the Environment in 1971, with a small budget, very few staff and as part of the portfolio for Aborigines and the Arts. In 1975 the ALP government established a separate Department of the Environment and Conservation. This drew the federal government into the centre of environmental policy, which had previously been exclusively a matter for the states. In the early 1970s, particularly following the advent of the ALP, political elites recognised the importance of both the polluter pays principle and international cooperation. In the 1970s Australia became a signatory to international conventions like the United Nations World Heritage Convention (1974). This initiative led to decisions that had a powerful impact on relations between state and federal governments, and a de facto redefinition of the powers of the states to legislate, manage and regulate the environment.

Australia participated in the United Nations Environment Programme, established in 1972. Pressure by environmentalists led to numerous reforms in the 1970s, including the creation of the Australian National Parks and Wildlife Service (National Parks and Wildlife Conservation Act 1975) and the Australian Heritage Commission (Australian Heritage Commission Act 1975).

During this period all state governments established departments of the environment or conservation. In New South Wales a State Pollution Control Commission was created in 1971. In Victoria a Ministry for Conservation was formed in 1973 to coordinate activities of several agencies already working on environmental protection, fisheries and wildlife, land conservation and national parks. Similar agencies were established by other states.

Developments in South Australia highlight the significance of state initiatives. Between 1962 and 1972 the number of significant parks and reserves increased from 19 to 99, comprising about 233,620 hectares and 3,546,465 hectares, respectively. By 1983 coverage increased to 4,521,584 hectares. The National Parks and Wildlife Act of 1972 has been described as the "Magna Carta of wildlife conservation" in South Australia (Whitelock, 1985: 143). The act established institutional structures for preservation and management of wildlife and historic sites and areas of geographical, natural or scenic interest. Similar acts were passed in other states. In New South Wales a National Parks and Wildlife Service was created following The National Parks and Wildlife Bill (1967).

Though much legislation fell short of the demands of conservation groups, it represented significant gains. Take the creation of national agencies, like the Australian National Parks and Wildlife Service (ANPWS) in 1975 whose functions are similar to those of organizations created in the states: to manage parks, protect, conserve and control wildlife and conduct research on these goals. However, the ANPWS was different in two respects. It was authorised to cooperate with other countries to protect and conserve animals and plants in those nations. Second, it recommend to government the establishment and naming of parks and reserves and protection and conservation of wildlife throughout Australia. One of its earliest recommendations was for the creation of national parks at Kakadu and Katherine Gorge.

The formation of the Australian Heritage Commission (AHC) in 1976 was another milestone in the transformation of the institutional order. The AHC was to advise governments and make recommendations for preservation of the national estate, including the natural environment. The rationale for creating the AHC anticipated concerns in the 1990s about the value of environmental protection to the rest of the world. The AHC had to consider protecting places because of their "outstanding world significance'.

In 1974 the Commonwealth passed the Environment Protection (Impact of Proposals) Act. This was to designed to guarantee that the environment became a consideration in any actions taken by or on behalf of the federal government and its agencies. According to Moss Cass, Federal Minister for Environment and Conservation, an important consequence was to increase the capacity to gather better information on potential environmental damage. The act would also "enable the public to argue a case publicly, to have the case published, and to force government to justify their decisions" (Debate in the House of Representatives, cited by Bates, 1983: 40).

Although most environmental legislation is still enacted by the states, the influence of the Commonwealth has grown considerably over the past three decades. Important legislation enacted by the Commonwealth includes the Great Barrier Reef Marine Park Act 1975; Environment Protection (Alligator Rivers Region) Act 1978; Environment Protection (Nuclear Codes) Act 1978; Antarctic Treaty (Environment Protection) Act 1980; Whale Protection Act 1980; Antarctic Marine Living Resources Conservation Act 1981; Environment Protection (Sea Dumping) Act 1981; Wildlife Protection (Regulation of Exports and Imports) Act 1982; World Heritage Properties Conservation Act 1983; Sea Installations Act 1987; Ozone Protection Act 1989; Hazardous Wastes (Regulation of Exports and Imports) Act 1989; Resource Assessment Commission Act 1989; Motor Vehicle Standards Act; 1989; Antarctic (Environment Protection) Legislation Amendment Act 1992; Endangered Species Protection Act 1992; Murray-Darling Basin Act 1993; Native Title Act 1993; Wet Tropics of Queensland World Heritage Area Conservation Act 1994 (Australian and New Zealand Environment and Conservation Council, 1994).

The 1980s were a turbulent period in relations between federal and state governments, particularly over World Heritage and protection of native forests. The parties had recourse to the High Court, which tended to adjudicate in favour of the federal government, and there appeared the prospect that the judicial system and courts would assume a far greater role than ever before in conflict resolution. Environmentalists were also keen for federal governments to frame legislation. In the 1990s governments have opted for a more cooperative approach with the states, and in 1992 both parties signed the Intergovernmental Agreement on the Environment (IGAE). The government presented a "new approach to intergovernmental dealings on the environment" (Commonwealth of Australia, 1996) which specified the roles of various layers of government and the "ground rules" (including principles and schedules for cooperation). Among the aims of the agreement were to clarify their respective roles as regards the environment, reduce disputes and achieve greater certainty in decision making.

Specific proposals included commitment by each side to a procedure of "accreditation" of processes or systems introduced by either tier of government in order to "to avoid the arbitrary revisiting of environmental issues by the parties". On the question of international conventions the agreement attempted to improve consultation between the parties. Other proposals included collection of data on the environment, like "the development of consistent standards for the description and exchange of land-related information and improved mechanisms for making data more accessible across spheres of government"; collaboration in efforts to consult those affected by land use decisions; eliminating duplication in applying procedures for environmental impact assessment and accreditation of processes adopted by different governments; developing national environmental standards, guidelines and goals; cooperation on the development of a National Greenhouse Response Strategy and implementation of the Convention on Biological Diversity; and arriving at agreements on nominations to the World Heritage List (Commonwealth of Australia, 1996).

In outlining guiding principles and core objectives of the National Strategy for Ecologically Sustainable Development (NSESD), the government focused on integration of concerns about the economy, environment and social equity; recognition of the "global dimension" of impacts on the environment; need for "a strong, growing and diversified economy which can enhance the capacity for environmental protection" and promoting "international competitiveness in an environmentally sound manner"; improvements in policy instruments for valuation, pricing and incentives; a precautionary approach; and community participation (Commonwealth of Australia, 1992b). Government departments were expected to incorporate these principles into their mission statements. In sum, the federal government, in dealing with the states and territories, adopted a "whole of government" approach to implementing sustainable development.

3 Evaluation

The first independent and extensive review of the state of the environment in Australia was released in 1996 and prepared by seven reference groups which reported to an independent advisory council (State of the Environment Advisory Council, 1996b). Another major report has been published by the OECD (1998), and a draft report issued by the Productivity Commission (1999). Other assessments include those by Christoff (1995) and Papadakis (1996), who shows how issues such as sustainable development have been used by political organizations to forge close links between economic and environmental concerns. Though some of these accounts have shared the view that there has been "considerable progress in developing a framework for the integration of environmental and economic policies" (OECD, 1998: 28), they also point to obstacles to reform, difficulties in implementing policies, institutional barriers to policy integration policy across tiers of government and inadequate knowledge of the state of the environment. Commitment by government agencies to environmental protection is very uneven.

Several reports point to the inadequacy of data and research. The rationale for this criticism is that, without better knowledge, Australia cannot demonstrate whether its pattern of development is sustainable. Lack of knowledge makes it much more difficult for policy makers to identify the causes of particular problems. It also means that there is sometimes "a long time lag between changes in human activities and any observable differences in natural systems" (State of the Environment Advisory Council, 1996a). Another factor is the structure of governmental institutions which have blocked coordinated approaches.

The 1996 report recommended a "systems approach" and considers the complexity of the natural world and the cultural values associated with it: "There is little likelihood of a coherent policy emerging from the traditional compartmentalised approach in which different departments or different levels of government each handle different, small parts of the problem" (State of the Environment Advisory Council, 1996a: 11). Although the report acknowledged huge increases in awareness of environmental problems at all levels of society, the authors perceived a lack of integration of environmental factors into decision-making, despite political rhetoric: "It is assumed that the first priority should be a healthy economy, and that problems can always be solved using the wealth created. The economy is a subset of human society which, in turn, is part of the environment. Progress towards sustainability requires recognition of this fundamental truth, and a willingness to build environmental thinking into our economic planning" (State of the Environment Advisory Council, 1996a: 15).

The OECD review found that there was "significant potential" for improvement in setting environmental standards, use of economic instruments (product charges, deposit refund systems and emission trading), and application of the user pays principle in waste management and waste water treatment (1998: 2). The report recognised achievements like environmental auditing of development assistance proposals, promotion of sustainable forest management in the South Pacific and Asia-Pacific regions, and the "special circumstances" of Australia as a "large exporter of fossil fuels and energy-intensive products" (1998: 10–11). It also pointed to the inadequacy of existing measures and funding to prevent damage to land and water resources (1998: 4).

However, the 1996 State of the Environment Advisory Council found that compared to many other regions, Australia does not have serious difficulties associated with sulphur dioxide and acid rain. Government legislation and industry initiatives have led to a decrease in recent years in levels of pollution from lead, carbon monoxide, nitrogen oxides and large particulates. Significant measures include mandatory use of catalytic converters and unleaded petrol in motor vehicles manufactured or imported since 1986, price differentials between leaded and unleaded petrol and reducing the amount of lead in leaded petrol. Water quality in urban areas is generally of a high standard. Contamination of food with chemicals and residues is relatively low.

The report also praised governments for listing natural and cultural areas under the World Heritage Convention, coordinated institutional approaches to protecting the Great Barrier Reef (Great Barrier Reef Marine Park Authority) and the Murray-Darling rivers (Murray-Darling Basin Commission), soil degradation (the Landcare program which encouraged participation by landowners and the wider community); initiatives to phase out substances that diminish the ozone layer; recycling schemes; and efforts to tap into renewable sources of energy like solar power; and to achieve cleaner production methods.

The report identifies the most significant problems requiring attention from policy makers as the loss of biological diversity following pollution or destruction of wetlands, saltmarshes, mangroves, bushland, inland creeks and estuaries; waste disposal; salinity and algal blooms in water systems; ozone depletion; soil erosion and salinity. Another concern is continued logging in old growth forests "despite several findings that this practice violates sustainability principles" (State of the Environment Advisory Council, 1996a: 33).

The report also criticised governments for being tardy in coordinating activities between federal and state levels; failing to establish policies for dealing with threats to biodiversity; continued focus by some agencies on economic development without due attention to consequences for the environment; not dealing adequately with declining fish stocks; allowing land clearing in some states; ineffective efforts to deal with urban sprawl; failing to tackle transportation and energy consumption issues; and falling well short of prior commitments to reduce greenhouse gases. The research team questioned reforms of the electricity industry which meant lower prices but increases in CO_2 emissions.

The question of biodiversity is regarded as pivotal since a vast number of species are found only in Australia including 85 per cent of flowering plants, 84 per cent of mammals, more than 45 per cent of birds, 89 per cent of reptiles, 93 per cent of frogs and 85 per cent of inshore, temperate-zone fish. A crucial area for preserving biodiversity is "the vast and less visible world of invertebrate animals and micro-organisms'. According to the report there are more than one million species in Australia, but descriptions exist of only 15 per cent of them (State of the Environment Advisory Council, 1996a: 22). Although Australia does have a National Strategy for the Conservation of Biological Diversity, the report argues that governments have failed to provide adequate resources to render policies effective.

The study also identified as potential problems the predicament of some marine species, threats to some types of forest; the system of reserves designed to protect biodiversity; and air and water quality in urban areas. As regards air pollution, the report pointed to sporadic formations of ozone and photochemical smog in Sydney and Melbourne. Their main concern was over the "lack of a consistent national set of ambient air quality standards', "often inadequate monitoring in major cities'; and "lack of data on the effects of air quality on flora and fauna" (State of the Environment Advisory Council, 1996a: 29).

In a climate of insecurity about employment prospects and Australia's capacity to compete in a global economy, the focus on the environment has declined. The explanation for all this may lie in the fact that "most decision makers" believe "that the wealth created by economic activities will overcome environmental effects" (OECD, 1998: 8). Not surprisingly, implementation of many policies is subject to adequate funding and circumstances prevailing in particular states. Policies in one area can easily contradict those in another. More could be done within the existing framework to refine economic and regulatory instruments for implementing sustainable development (see OECD, 1998).

Overall, there is considerable scope for improvement, though there has been progress in developing and implementing environmental policy. Some initiatives in policy development are noted below, including the increase in capacity of public institutions. Among the noteworthy accomplishments over the past two decades have been the greater involvement by federal governments in determining the listing of World Heritage areas, and mobilizing public support for initiatives like Landcare which have drawn an unprecedented degree of attention to the problem of soil erosion.

4 Capacity of Proponents of Environmental Policy

4.1 Public Institutions

As noted earlier, though the states have extensive powers to enact legislation and manage and regulate use of the environment, involvement by federal governments has increased markedly. At the federal level the primary responsibility for environmental policy is vested in the Department of the Environment, which has a wide range of programmes including:

- the Environmental Resources Information Network which gathers information on the state of the environment through a computer network of many agencies
- the Environment Protection Agency which, among other things, gathers data for a national pollution inventory and national solid waste database
- a programme for reporting on the state of the environment and creating environmental indicators
- statutory authorities like the ANPWS, the AHC and the Great Barrier Reef Marine Park Authority
- the Bureau of Meteorology, responsible for a national climate data bank
- the Antarctic Division which conducts research on oceans and marine life, the atmosphere, climate change and the Antarctic environment
- significant initiatives in ecologically sustainable development, biodiversity and climate change.

The range of activities extends beyond one particular department, as in setting guidelines for purchasing "environmentally-responsible products and standards", heritage programmes, the creation of a "green corps" programme for young people to work in environmental conservation projects, providing educational material to schools and professional development for teachers. The Department of Foreign Affairs and Trade oversees the Australian Agency for International Development (AusAID) manages an overseas aid programme. The aim is to consider environmental impacts in the design and implementation of activities undertaken by AusAID, and subject them to independent environmental audits. AusAID has completed Country Environment Profiles on Papua New Guinea, Indonesia, Laos and China. It also backs international environmental organisations and programmes that address environmental questions transcending national borders. The Department of the Prime Minister and Cabinet was engaged in coordinating forest policy activities, and provided the secretariat for the Intergovernmental Committee on Ecologically Sustainable Development, though this was wound down in 1997. The Department of Primary Industry and Energy plays a pivotal role in gathering data on natural resources (Bureau of Resource Sciences) and developing a National Forests Inventory. The Australian Bureau of Agricultural and Resource Economics carries out research on minerals, energy, agriculture, forestry and fishing. The Department of Transport plays a central role in evaluating the environmental impact of transport policy, and, with the Australian Maritime Safety Authority, monitors marine and coastal environments. Finally, the Treasury oversees agencies like the Australian Bureau of Statistics (which conducts environmental accounting), the Productivity Commission (which considers environmental issues in public inquiries) and the Australian Taxation Office (which manages taxes, levies, deductions and rebates linked to activities affecting the environment) (see Commonwealth of Australia, 1997b: 9, Table 1.5).

Apart from attempts by government departments and agencies to increase the capacity for environmental policy, there have been several other significant initiatives. In order to overcome resistance by the states to federal intervention and address conflicts between developers and environmentalists in the 1980s, ALP governments established new institutional mechanisms like the Resource Assessment Commission, a National Soil Conservation Strategy and the formulation of a strategy for Ecologically Sustainable Development (ESD groups).

The Resource Assessment Commission was established in 1989 to identify environmental, cultural, social, industry, economic and other values involved in using resources. Apart from investigative and advisory roles, it emphasised public participation and education. Prior to its dissolution in 1995, the Commission produced significant reports on mining in conservation areas and protection of forests and coastal areas.

The creation of ESD working groups was an attempt to defuse confrontation between environmentalists and developers. The groups represented one of the most comprehensive attempts by government to control the agenda and warring factions, to shift from competing paradigms of developers and environmentalists. The involvement of many government agencies and interest groups in defining ESD principles was itself an achievement. Participation in the ESD process moved some environmentalists to give greater recognition to market-based strategies. There was a marked shift towards more dialogue and cooperation. This applies, for instance, to issues like soil degradation and involvement by government in bringing together the National Farmers Federation and the Australian Conservation Foundation. This was a significant attempt to integrate development and environmental protection and influence patterns of coalition formation. It should, however, be emphasised that the government was partly responding to factors like:

- willingness by interest groups to cooperate on environmental matters
- lowering of resistance by business and unions to adapting green policies
- the emergence of electoral competition (and the ability of the ALP in the 1983, 1987 and 1990 elections to attract "green preferences" which are credited with having a crucial impact on the final outcome)
- the sharp rise in membership of powerful environmental groups in the 1980s
- favourable coverage of environmental concerns by the media
- decisive shifts in public opinion, especially around 1989, following concern about climate change and ozone depletion and the resilience of public opinion on environmental issues even during an economic recession (Papadakis, 1994).

Moreover, most proposals by the ESD groups have not been implemented (see Productivity Commission, 1999), and governments have been slow to tackle the emission of greenhouse gases.

The Liberal-National governments of the 1990s endorsed many objectives set by previous governments, though relationships with green groups has changed in subtle ways. An important innovation by the Liberal-National government was the proposal, now enacted by parliament, to create a $1.25 billion Natural Heritage Trust, funded through the partial privatisation of the national telephone company,

Telstra. The sale of Telstra allows the government to realise a national vegetation plan, rehabilitate the Murray-Darling River Basin, conduct a national land and water resources audit, establish a national reserve system and tackle pollution of the coast and seas. The trust fund should generate enough revenue to support further initiatives.

An important attempt to coordinate efforts by public institutions has been the creation of the National Strategy for Ecologically Sustainable Development (NSESD). The process of formulating a national strategy was facilitated by a November 1991 agreement by the Heads of Government to create an intergovernmental Ecologically Sustainable Development Steering Committee. On 7 December 1992 the Council of Australian Governments (COAG) endorsed the NSESD with two qualifications: implementation would depend on financial backing and circumstances prevailing in particular states. The recommendations of the ESD working groups and the IGAE (1 May 1992) were regarded as the basis for developing policy. Membership of COAG comprises the leaders of federal, state and territory governments, and President of the Local Government Association. COAG is therefore the most authoritative association for intergovernmental cooperation, though there is a long way to go in achieving effective coordination between agencies (Productivity Commission, 1999).

The government has identified the connections between the NSESD and other environmental policies and initiatives. These initiatives include the IGAE as well as the National Greenhouse Response Strategy (and the recent formation of the Commonwealth Greenhouse Office) and the National Strategy for the Conservation of Australia's Biological Diversity. Other significant processes linked to the NSESD are the National Waste Minimisation and Recycling Strategy, the Commonwealth Major Projects Facilitation initiative, and the National Forest Policy Statement. Further work is being undertaken on frameworks for tackling questions arising from international agreements like Agenda 21, the Framework Convention on Climate Change and the Convention on Biological Diversity. Policies are also being developed to reflect the precautionary principle and realise the principle of intergenerational equity.

As with all these policies, the stipulation remains that they are subject to availability of funds, and the present administration has repeatedly emphasised the difficulties facing all levels of government in this regard.

At any rate, new initiatives like the Natural Heritage Trust which is responsible for allocating vast sums to projects for environmental protection and sustainable development will be largely controlled by central government agencies like the Department of Primary Industry and Energy and the Department of the Environment, which plays a primary role in trying to coordinate the various strategies outlined in this chapter.

Though the impact of legislative enactments on practices is often limited (Productivity Commission, 1999), there has been a reduction in major conflicts between developers and environmentalists. This partly reflects shifts in perception and a new ethos for cooperation. It may also mirror waning interest in reform at the highest political levels, and the absence of a powerful political champion. Moreover, at the local and regional level there remain important conflicts.

Among the most interesting efforts to improve coordination between federal and state governments in dealing with environmental problems has been the

Murray-Darling Basin Commission (established in 1985). The Commission is responsible for administering the Murray-Darling Basin Initiative, seen as "the largest integrated catchment management program in the world, encompassing 1.6 million square kilometres" (State of the Environment Advisory Council, 1996a: 36). A key feature of this initiative is involvement by federal and state governments, and by community groups, local government and other stakeholders in addressing land degradation, water quality, salinity and conservation.

4.2 Environmental Organizations

Support for conservation groups that have influenced changes in legislation, for instance, the creation of national parks and reserves, originates from the nineteenth century. However, the 1960s were a watershed for political activism that challenged prevalent concepts about economic growth and exploitation of natural resources. Challenges to established approaches to land use have only come to fruition in the past two decades, following a period, especially in the 1950s and 1960s, of sustained economic growth and development.

The 1960s saw a burgeoning of voluntary environmental groups, and new arguments about the importance of institutional change entered the political agenda. Among the new groups, the Australian Conservation Foundation, formed in 1965, was the first truly national organization. Prominent politicians, administrators and dignitaries (like the Duke of Edinburgh and the Chief Justice of the High Court, Sir Garfield Barwick) played a key role in its formation, and tried to change perceptions and the organization of environmental protection.

In the 1970s the prospects for placing the environment on the public policy agenda appeared bleak. Though a Department of the Environment was created in 1971, legislative and other powers remained with the states. However, social movements and political organizations built on increasing opposition to development projects within local communities like a major conflict over the flooding of Lake Pedder in Tasmania by the Hydro-Electric Commission. This prompted the formation of the Save Lake Pedder Committee in 1967, a much more radical group than existing organizations, and attracting support from mainland Australia and other countries. The flooding of Lake Pedder in 1972 led to the formation of two groups which exerted a decisive influence on environmentalism: the United Tasmania Group in 1972 and the Tasmanian Wilderness Society in 1976. The latter was subsequently renamed The Wilderness Society. In 1972 state elections the United Tasmania Group nearly won a seat in the Tasmanian parliament.

Another campaign in Tasmania, to prevent the construction of the Franklin Dam between 1979 and 1983, gave further impetus to The Wilderness Society. Branches were formed in all states and campaigns for preserving forests conducted across Australia. The Wilderness Society was an expression of the entry of a new generation of environmentalists into politics. It also reflected the distinctive focus by environmentalists in Australia with "wilderness" issues, which was seen as a strength by some (Hay and Haward, 1988) but not others (Kellow, 1990).

Apart from mobilizing people, new groups raised substantial financial support. The radicalism of The Wilderness Society contributed to shifts in direction by established organizations like the Australian Conservation Foundation. In the

1980s the latter attracted many more members and supporters as did the World Wide Fund for Nature and Greenpeace.

The rise in support for these groups contributed to stronger articulation of environmental issues by the Australian Democrats, a party formed in 1977 by Don Chipp, a former minister in a federal Liberal government. In 1990, with 11.2 per cent of first preferences for the House of Representatives, the Democrats, secured the best vote for a party other than the ALP or the Liberals and Nationals since federation. The Western Australia Green Party secured 7.5 per cent of the vote in that state. The success of Green Independents in Tasmania, who held the balance of power in the state parliament between 1989 and 1991, represented an extension of protest movements that emerged in the late 1960s.

The influence on policy of environmental movements lies partly in their mobilization of social actors. Campaigns over flooding the Franklin and lower Gordon Rivers provided opportunities for coalitions between environmental groups, including the Australian Conservation Foundation, state conservation councils, national parks associations and seventy branches of the Tasmanian Wilderness Society. The coalition rallied support from around 800 conservation groups with up to half a million members.

Apart from direct electoral pressure exercised by certain groups, the most striking manifestation of environmentalists' influence lies in support for environmental groups. Data from the Australian Election Studies (AES) shows that in 1990 3 per cent of a national sample indicated that they were members of an environmental group. The figure rose to 5 per cent in 1993 and dropped to around 3 per cent in 1996. Data from the 1995 National Social Science Survey shows that almost 10 per cent of a national sample indicated they were members of a group "whose main aim is to preserve or protect the environment" (Kelley et al. 1995). This suggests that anywhere between 300,000 and 1,000,000 voters were members of environmental or conservation groups or saw themselves as members, because they may have made donations to groups (like Greenpeace).[1] In the 1996 AES 70 per cent of the sample approved (strongly or very strongly) of environmental groups. The figure for 1998 was 76 per cent.

Another indicator is the growing concern about environmental issues among the mass public. Despite radical shifts in perception among elites and significant changes in institutional responses to environmental problems, the great movement in public opinion only occurred in the late 1980s. Data from the Roy Morgan Research Centre between 1975 and 1995 reveal that until 1988 only a small minority (less than 5 per cent) of people identified the environment (or conservation or pollution) as an issue about which they were most concerned or as an important issue that the federal government should be doing something about. Between 1988 and 1989 there was a rise from 5 to 26 per cent in the number of people seriously concerned about this issue. Though there has been a decline in this figure (for

---

1 This figure appears somewhat higher than that recorded in other surveys (like the Australian Election Survey) though the wording of the question is slightly different and may have affected response rates. In the Australian Election Survey people were asked whether they were already members or how likely they were to join environmental groups campaigning to protect the environment or concerned with stopping the mining and export of uranium or anti-war and anti-nuclear weapons movements. In this survey the focus was purely on membership of "any group whose main aim is to preserve or protect the environment".

instance, to 15 per cent in 1995), the rise in the relative importance of environmental issues represents an important shift and an attempt to work through the implications of this new awareness.

4.3 Economic Proponents of Environmental Protection

Despite initial resistance by industrial and business groups to arguments advanced by the green movement, there has been a steady shift over the past decade to adopting the concept of sustainable development and identifying opportunities for economic growth coupled to environmental protection. The Business Council of Australia (1990) adopted the Brundtland notion of sustainable development (World Commission on Environment and Development, 1990) and tried to promote dialogue rather than confrontation with environmentalists.

This is partly in response to the vast potential for developing technology in waste management, sustainable agriculture and energy production. Australian corporations have monitored developments which demonstrate how it is far cheaper to recycle than dump materials like steel. The rehabilitation of landscapes is another area in which business has a claim to innovation. For instance, the Broken Hill Proprietary company, which had one of the largest mineral sand mining operations in Australia at Bridge Hill Ridge in New South Wales during the 1970s and early 1980s, claimed a major success in rehabilitating the site. The mined area became part of the Myall Lakes National Park, thereby demonstrating "that mining and environmental commitment need not be incompatible" (Business Council of Australia, 1990: 31). In 1990 Alcoa was praised by the United Nations Environment Programme for rehabilitating bauxite mines. In Western Australia Alcoa has developed a programme for planting trees to serve its own needs and those of other mining companies, local authorities and farmers.

Although there is little reliable data on aggregate levels of investment in environmental technologies, there are indications that some sectors of business are aware of opportunities and there has been a significant increase in expenditure on such technologies in the 1990s. Participation by business and industry groups in ESD working groups initiated by the ALP government is an important indicator of growing awareness and participation by business in capacity building (see Toyne, 1995). The main source of data is from a study by the Australian Bureau of Statistics and shows that, in 1993–94, total environmental protection expenditure was A$6.5 billion (1.5 per cent of GDP) (see OECD, 1998: 135–6). The largest share came from government (A$2.9 billion or 45.2 per cent of the total). Industry accounted for A$1.6 billion (24.4 per cent), and the remainder is expenditure by households (mainly for sewerage and waste collection).

Other important initiatives include Greenhouse Challenge and the EnviroNET scheme, sponsored by government to provide, via the Internet, information on environmental technologies and services. A similar function is performed by the Environmental Management Industry Association of Australia which highlights the work of firms involved in environmental protection and the development of environmental technologies (see Environmental Management Industry Association, 1997.

5 The Changing Constellation of Actors

From being a peripheral consideration on the political agenda, the articulation of environmental problems by established political organizations began in the late 1960s (though political organizations like the Country Party of Australia expressed concern about soil conservation in the 1920s). In the 1960s the impact of new ideas about the limits to growth were strongly articulated by the ALP. In the 1970s the Liberal Party took the lead on issues like the impact of the exploration of mineral resources on the environment. Depending on the issue one finds various parties took the lead at different times (Papadakis, 1996).

The most rapid changes occurred in the 1970s with the rise of social movements opposed to development projects in Tasmania and metropolitan areas like Sydney (Roddewig, 1978). After a flurry of activity in the early 1970s established political organizations, like the Liberal-National Coalition government, focused on the impact of rising oil prices and tackling an economic recession. Apart from the ALP, new political parties like the Australian Democrats (founded in 1977) and the Nuclear Disarmament Party (founded in 1984) played a pivotal role in setting the agenda. Prior to the 1983 election the ALP declared against construction of the hydroelectric power scheme on the Gordon or Franklin Rivers and tried, after its victory, to accommodate concerns of environmental groups on issues like protecting rainforests. There were also serious conflicts between federal and state governments over World Heritage listing and logging of forests. The government came under intense pressure from green organizations, since sympathisers of the green movement seemed to have a significant impact on electoral outcomes. The ALP struggled to balance economic development and environmental protection in the 1980s, even though it was widely perceived as the most likely, among the established parties, to promote environmental policies.

This tension affected the Cabinet and deep divisions emerged between the Ministers for the Environment and for Resources. The government was forced to frame environmental policy in innovative ways. Though it sided with environmentalists over the Franklin Dam, it was unable to respond to rising expectations for a more interventionist approach. The Federal government averted several challenges by the mining industry which took issue with initiatives to protect some areas through World Heritage legislation. However, the Cabinet remained divided over questions like the logging of forests and mining.

The government therefore tried to incorporate the environmentalist challenge, by getting the ESD working groups to adapt the ideas of the Brundtland Commission, creating agencies to tackle environmental problems and enacting legislation. Though economic imperatives remained pivotal, in its attempt to incorporate environmentalism the government transformed its own agencies, which attempted to synthesise apparently conflicting approaches. Until the mid-1980s industry groups, for example, generally paid little attention to environmental factors. Two other considerations were important in preventing effective environmental policies: the division of powers between federal and state governments and "lack of knowledge about the economic costs and benefits of environmental protection" (Christoff, 1995: 165). As Christoff notes, the problem of integration between economic and environmental policy was clearly identified by a House of Representatives Standing Committee Report (1987).

To address these difficulties and ensure integration of environmental and economic imperatives governments initiated institutional processes centred on ESD working groups, the Resource Assessment Commission and the IGAE. Business was more willing to engage in dialogue with environmental groups and vice-versa. A new era of cooperation began between the federal government and the states. There was also growing awareness about possibilities for sustainable development.

For a brief period under the ALP regime of the late 1980s, the Australian Council of Trade Unions, the peak union organization, issued several key statements which embraced the concept of sustainable development and recommended strategies to tackle problems in a wide range of areas (including forests, energy and transport), as in the policies and resolutions passed under the heading "Environment and Sustainable Development Policy" at the 1989 Congress). Clause 1.5 of the policy called for the promotion of job creation, training and retraining for workers in industries changed because of environmental issues (http://www.actu.asn.au/national/about/policy/89envir.htm). These policies were elaborated at the 1991 Congress which supported the ESD working groups (http://www.actu.asn.au/national/about/policy/91envir.htm). The Congress also endorsed the original targets by government for a 20 per cent cut in CO_2 emissions. However, there is little evidence of trade unions launching major campaigns since then to promote the objectives set out at those two congresses.

Overall, the process of effecting change is fraught with difficulties and setbacks. There is uncertainty about whether the earlier momentum of input by environmental groups can be maintained and whether the Liberal-National government will include environmental considerations in all aspects of development. A recent draft report commissioned by the Treasury has issued a challenge to government departments and agencies to revive the process of implementing, coordinating and monitoring the process of ESD (Productivity Commission, 1999).

Although institutional inertia and concern about economic development may present a formidable barrier to change, over the past twenty years established political organisations have responded to pressure from environmental organizations, the media and other social actors and contributed to a significant shift in the national political agenda. This has entailed shifts in the leading policy paradigms from end-of-pipe solutions to some elements of ecological modernisation. Take the efforts by the Environmental Management Industry Association of Australia, which has moved from a predominant focus on end-of-pipe technologies to greater interest in advocating "cleaner production and pollution prevention approaches" (OECD, 1998: 138).

6 Capacity Building

Bearing in mind the interpretation of capacity-building presented by Jänicke and Weidner (1997: 148), there has been an expansion and deepening of commitment to environmental protection over the past 15 years.

Federal governments have enacted comprehensive legislation for environmental protection, and recently presented new laws like the Environment Protection and Biodiversity Conservation Bill 1998, which combines several parts of existing laws and several new stipulations which enshrine in legislation for the first time the promotion of ESD. There has been a growing recognition of the need to overcome impediments posed by prevailing institutional structures like potentially conflicting roles of levels of government. A "whole of government" approach has been adopted in principle. Governments have contributed to improving the knowledge base about the state of the environment, and the economic costs and benefits of pursuing certain actions. Several institutional innovations have broadened and intensified the capacity to integrate various considerations in the policy making process. These considerations pertain to economic, cultural, social, scientific, ecological and recreational values. In this respect mechanisms like the Resource Assessment Commission, the ESD working groups and the Murray-Darling Basin Initiative have all been important. There have been efforts to consider the environment in policy areas which traditionally did not place much emphasis on this factor. There has been a significant response to international initiatives like the World Heritage List and to the United Nations Conference on the Environment and Development (see Papadakis, 1999).

Still, in all cases the pace of change and integration of policy has been retarded by institutional and other obstacles, particularly across tiers of government. There remain difficulties associated with fundamental considerations like adequate knowledge of the state of the environment. The commitment by many government agencies to environmental protection remains superficial, and their competence falls well short of expectations (Productivity Commission, 1999). At both national, state and local levels there is a long way to go if one considers as important the introduction of economic instruments to tackle environmental problems (see OECD, 1998). On the positive side, some tax incentives (subsidies, deductions and credits) have been introduced at the national level for development of clean technologies and capital expenditure on soil and water conservation; state and local governments have introduced user charges for industrial waste and sewerage treatment; local governments levy charges for municipal waste; and state governments have instituted tradeable resource entitlements and emission rights in industries like fishing, timber harvesting and the use of water by farmers (Christoff, 1995: 175). However, Australia trails countries like the Netherlands, Sweden and the United States on the introduction of charges for carbon dioxide emissions and aircraft noise; product charges (on items like petrol, electricity, coal, fertiliser and vehicles); and schemes for deposit-refunds on bottles and cans (see Christoff, 1995: 182–4 based on OECD, 1994), and a large proportion of OECD countries on environmental performance generally (see Jahn, 1998).

As regards development of research infrastructure there has been an increase in capacity to monitor and evaluate progress, even if there is still considerable room for improvement. Major initiatives include the 1996 State of the Environment Report, creation of Cooperative Research Centres like those focused on Waste Management and Freshwater Ecology, and establishment of new research centres and institutes in universities.

Many agencies gather and manage data on the state of the environment including the Commonwealth Scientific and Industrial Research Organisation, Australian Bureau of Statistics, Bureau of Meteorology, Australian Institute of Marine Science, Australian Geological Survey Organisation, Bureau of Resource Sciences, Australian Nature Conservation Agency, Australian Heritage Commission, Environmental Resources Information Network, Environment Protection Agency, Great Barrier Reef Marine Park Authority, Australian Bureau of Agricultural and Resource Economics, and Australian and New Zealand Environment and Conservation Council. Many of these agencies operate as specialised organizations within federal government departments. The Australian Science and Technology Council and the Prime Minister's Science and Engineering Council are also pivotal in providing policy advice. A similar role has been performed by expert groups, like the National Greenhouse Advisory Committee.

The emergence of new agencies or reshaping of established organizations reflects pressure applied by green groups. Green organizations have influenced policy in several ways. The growing support for non-government organizations among the mass public prompted governments in the 1980s to become more accessible to them. Leading figures in the green movement were employed as government policy advisers. Green groups were influential in issuing recommendations to their supporters as to which of the major parties was most likely implement policies. In Tasmania Green Independents held the balance of power in the state parliament. At the federal level, green candidates, in conjunction with the Australian Democrats, have also exerted significant influence in the Senate (or Upper House). Still, the electoral system for the House of Representatives (the Lower House) is restrictive: the preferential voting system is based on a majoritarian model which favours large well-established parties.

While some institutional structures favour established political organizations, their character, in terms of their agenda and the significance of the environment portfolio under all governments, has changed in important ways. Apart from the influence of rising awareness about environmental issues, documented through opinion surveys, the power of the media is another significant factor. Coverage of environmental issues intensified over the past two decades and much of the reporting is pro-environment (Papadakis, 1996). Overall cognitive-informational capacities, including the ability of the political and administrative system, as well as of NGOs, to make strategic use of information has increased significantly, but there is still considerable room for improvement. Communication channels like the national press play an important role in articulating the agenda, evoking further comment from key actors like government ministers or leaders of NGOs. As indicated above (see also section 4), numerous government agencies are involved in identifying or analysing problems. The capacity of agencies like universities and special research centres has also increased over time. The challenge remains how to deliver effective solutions (see Productivity Commission, 1999).

7 Overcoming Barriers to Effective Policy Making

Australia has developed distinctive responses to problems, some of which are unique to this vast continent. Over the past two decades some political leaders have seized the initiative, notably with respect to ESD policies, and at various times Australia has set ambitious goals, for instance, on reducing greenhouse gas emissions. However, there has been a withdrawal from or reconsideration of these initiatives. It is therefore difficult to categorise governments' approaches other than point out that at times they have been highly interventionist and even corporatist in their approach, especially around 1990–92, that the recent focus has been on voluntary agreements between government and industry and a "whole of government approach" that includes the states in decision-making. Many possibilities have emerged for implementing sustainable development, and, to some degree, commitment has been institutionalized (via the IGAE and NSESD). There have been significant financial pledges to initiatives like the Natural Heritage Trust, and some cooperation between agencies, and progress on developing measures for and monitoring sustainable development.

The main barriers to effective policy change include waning interest among political elites; lack of resources for analysing long-term results of traditional approaches to industrial development and economic growth, gathering information about the state of the environment and developing clear indicators to measure performance; failure to develop more fully economic instruments and apply the user pays principle; the strong focus on defining the national interest largely in economic terms, thereby threatening a revival of adversarial debates over environment and development. In a climate of insecurity about employment prospects and Australia's capacity to compete in a global economy, the focus on the environment has declined among political elites. Hence globalization is seen, as presently restrictive on environmental policy, though it may also offer a way out. For instance, as noted earlier, decision makers tend to believe that economic development will generate wealth which can then be used to tackle environmental problems (OECD, 1998: 8). Paradoxically, there may be some truth in this since researchers have found a strong correlation between wealth measured by GNP per capita and environmental performance (Jahn, 1998: 118; see also Jänicke, 1992).

Another obstacle is the tendency to delegate responsibilities for sustainable development from central to state or local governments. Apart from factors like a philosophical position that questions the role of "big government', this may reflect the emphasis on balancing the budget. Policies for implementing sustainable development are subject to adequate funds being available. The abiding institutional arrangements continue to play a potentially obstructive role. Policies in one area contradict those elsewhere. An official assessment of policies has concluded that some agencies, notably those responsible for natural resource management and environmental protection, were treating ESD as "a core policy concern', while the majority were not performing well because they were unclear about what constitutes ESD, lacked relevant information, fell short of "good practice" guidelines for policy making, did not have committed decision makers, failed to coordinate activities across areas of government, and neglected long-term horizons on sustainable development (Productivity Commission, 1999). A key aspect of the report was the reproach, issuing from environmental organizations and from industry groups like the National Association of Forest Industries of Australia, of failure to monitor and report on environmental performance (1999: 111–113). However, the report also highlighted initiatives to develop environmental indicators by several government agencies, including the National Land and Water Resource Audit as part of the Natural Heritage Trust.

Other obstacles to successful policy implementation include the threat to cooperative relationships between environmental organizations and the federal government following disputes over greenhouse emissions, even though in many areas there is still extensive consultation. While Australian governments have used notions like sustainable development to reduce conflicts among interest groups (and their own supporters and representatives) and reconsider policy directions, serious impediments have arisen because of perceived short-term costs associated with improving environmental performance.

Other long-term problems are how to fulfil the promise of preserving biological diversity in a vast but sparsely populated continent, and an economic structure so dependent on exports of minerals and fossil fuels. Apart from geography, demography and economic structure, there is the enduring issue of defining the national interest, which is partly conditioned by electoral considerations (including a short electoral cycle), and an ideology privileging economic measures of success and interests. One of the challenges in future will be how to conceive "economic success'. At present, economic interests, notably resource industries, play a pivotal role in shaping government understandings and discouraging implementation of aspects of sustainable development.

Overall, national environmental plans have been helpful not so much in guiding environmental policy but in highlighting its unrealised potential. Most government agencies or political leaders have not fully explored possibilities for using technological and other innovations in environmental policy to modernize both the economic and political system. Yet, there is an awareness in some quarters about the need for doing so (see Productivity Commission, 1999).

Australian approaches to environmental policy reflect complex responses to national electoral and economic imperatives as well as to international pressures. The possibility remains that the interests of the nation state will be construed in a manner that intersects with notions of international responsibilities and best environmental practice. Often, awareness of this potential, or the willingness to articulate it, has been stronger outside rather than within government.

References

Australian and New Zealand Environment and Conservation Council, 1994. *Guide to Environmental Legislation in Australia and New Zealand* (fourth edition), Report No. 29. Canberra: Department of the Environment, Sports and Territories, Commonwealth of Australia.

Bates, G. M., 1983. *Environmental Law in Australia*, Sydney: Butterworths.

Beale, B., and P. Fray, 1990. *The Vanishing Continent.* Sydney: Hodder and Stoughton.

Bell, S., and B. Head. 1994. Australia's Political Economy: Critical Themes and Issues, in: S. Bell, and B. Head (eds.), *State, Economy and Public Policy.* Oxford: Oxford University Press.

Business Council of Australia, 1990. *Achieving Sustainable Development. A Discussion Paper by the Business Council of Australia.* Canberra: Business Council of Australia

Christoff, P., 1995. Market-based Instruments: The Australian Experience, in: R. Eckersley (ed.), *Markets, The State and The Environment: Towards Integration.* Melbourne: Macmillan.

Commonwealth of Australia, 1989. *Our Country, Our Future, Statement on the Environment*, Canberra: AGPS.

Commonwealth of Australia, 1992a. *National Greenhouse Response Strategy.* Canberra: AGPS.

Commonwealth of Australia, 1992b. *National Strategy for Ecologically Sustainable Development*, (http://www.environment.gov.au/portfolio/esd/ nsesd/intro.html).

Commonwealth of Australia, 1995. *Progress in Implementing the National Greenhouse Response Strategy*, Canberra: Department of the Environment, Sport and Territories, Report by the Intergovernmental Committee for Ecologically Sustainable Development.

Commonwealth of Australia, 1996. *Summary of the InterGovernmental Agreement on the Environment*, (http://www.environment.gov.au/ portfolio/esd/nsesd/appndxa.html)

Commonwealth of Australia, 1997a. *Future Directions for Australia's National Greenhouse Strategy*. Canberra: Department of the Environment, Sport and Territories, for the Intergovernmental Committee for Ecologically Sustainable Development.

Commonwealth of Australia, 1997b. *Investing in Our Natural Heritage. The Commonwealth's Environment Expenditure, 1997–98*. Canberra: AGPS.

Commonwealth of Australia, 1997c. *Safeguarding the Future: Australia's Response to Climate Change. Statement by the Prime Minister of Australia, the Hon John Howard*, (http://www.greenhouse.gov.au/ago/ safeguarding.html) (last modified 29 September 1998).

Commonwealth of Australia, 1998. *Outcomes from Buenos Aires and the Implications for Australia*, (http://www.greenhouse.gov.au/pubs/factsheets /fs_buenos.html)

Doyle, T., and A. Kellow, 1995. *Environmental Politics and Policy Making in Australia*. Melbourne: Macmillan.

Eckersley, R., 1989. *Regreening Australia: the environmental, economic and social benefits of reforestation*, Melbourne: CSIRO.

Environmental Management Industry Association of Australia, 1997. *The Green Book. The guide to Australia's environmental technologies and resources*. Moorabbin, Vic: Hallmark Editions.

Hamilton, C., 1994. *A Comparison of Emission Sources and Emission Trends Among OECD Countries*, Background Paper No. 1. Canberra: The Australia Institute.

Hamilton, C., T. Hundloe, and J. Quiggin, 1997. *Ecological Tax Reform in Australia: Using taxes, charges and public spending to protect the environment without hurting the economy*, Discussion Paper No. 10. Canberra: The Australia Institute.

Hay, P., and M. G. Haward, 1988. Comparative Green Politics: Beyond the European Context. *Political Studies*, 36, 433–448.

House of Representatives Standing Committee on Environment and Conservation, 1987. *Fiscal Measures and the Achievement of Environment Objectives*, Report of the House of Representatives Standing Committee on Environment and Conservation. Canberra: AGPS.

Jahn, D., 1998. Environmental performance and policy regimes: Explaining variations in 18 OECD countries. *Policy Sciences* 31, 107–131.

Jänicke, M., 1992. Conditions for Environmental Policy Success: An International Comparison *The Environmentalist* 12, 47–58.

Jänicke, M., and H. Weidner (eds), 1997. *National Environmental Policies. A Comparative Study of Capacity-Building*. Berlin: Springer-Verlag.

Kellow, A., 1990. Spoiling for a Fight or Fighting for the Spoils? Resources and Environment Politics and Policies Towards 2000, in: B. Hogwood (ed), *Australia Towards 2000*. London: Macmillan.

OECD, 1994. *Managing the Environment: The Role of Economic Instruments*. Paris: OECD.

OECD, 1997. *OECD Environmental Data Compendium 1997*. Paris: OECD.

OECD, 1998. *Environmental Performance Reviews: Australia. Conclusions and Recommendations*. Paris: OECD.

Papadakis, E., 1993. *Politics and the Environment. The Australian Experience*. Sydney: Allen and Unwin.

Papadakis, E., 1994. Development and the Environment, in: C. Bean (ed), *The_1993 Federal Election. Australian Journal of Political Science*, special issue 29, 66–80.

Papadakis, E., 1996. *Environmental Politics and Institutional Change*. Cambridge: Cambridge University Press.

Papadakis, E., 1999. "In the National Interest': Australian engagement with the UNCED process, in: W. Lafferty, and J. Meadowcroft (eds), *Bringing Rio Home*. Oxford: Oxford University Press, (forthcoming).

Productivity Commission, 1999. *Implementation of Ecologically Sustainable Development by Commonwealth Departments and Agencies*, Draft Report. Canberra.

Roddewig, R., 1978. *Green Bans. The Birth of Australian Environmental Politics*. Sydney: Hale and Ironmonger

State of the Environment Advisory Council, 1996a. *Australia: State of the Environment 1996. Executive Summary*. Canberra: AGPS.

State of the Environment Advisory Council, 1996b. *Australia: State of the Environment 1996.* Canberra: AGPS.

Toyne, P., 1994. *The Reluctant Nation*. Sydney: ABC Books.

Toyne, P., 1995. Ecologically Sustainable Development in Australia, Presentation to The President's Council on Sustainable Development. Canberra: Department of the Environment.

Whitelock, D., 1985. *Conquest to Conservation: History of Human Impact on the South Australian Environment.* Cowandilla: Wakefield Press.

AUSTRIA: AN "ECO-LAND"?

Christof Amann and Marina Fischer-Kowalski

A recent advertisement of the Ministry of Agriculture and Forestry labelled Austria an "eco-land". On the international stage, Austria likes to see itself as a pioneer in environmental protection (Lauber, 1997a: 86). This article discusses whether this self-assessment is overstated or whether it is well-founded. We briefly describe Austrian environmental policy in the period 1960 to 1990, the problems it tried to solve, its successes and failures. We then analyse the preconditions and the perspectives for environmental policy in the 1990s.

1 Environmental Problems and Policy Responses 1960–1990

1.1 Introduction

Passing the last three decades in review reveals certain several common features to environmental policy. First, the predominant environmental problem was considered to be "pollution". The stage was populated by "polluters" (mainly industry, sometimes households) acting irresponsibly, by the public in general whose health was put in jeopardy, and by environmental authorities seeking to control polluters and pollution. The environment was compartmentalised into the media suffering pollution: water, air, and soil. A less brilliantly lit side stage was reserved for nature conservation. While the play proceeded, the audience usually followed attentively. Sometimes, however, it would suddenly storm the stage and join the play, adding some drama and sometimes giving it quite a new turn. Second, the role of the protagonist was ill defined at the beginning of the play: there is no ministry for the environment. In the case of Austria, the first act was quite bereft of a leading character. It is not until the second act that an environmental minister was summoned to play a role, with little more to start with than the title. For the rest of the play he was sometimes distracted from the struggle against his opponents by the need to gain enough power to act at all. Most power, territory and competencies were held by other players, either further down the feudal order (provinces, municipalities), or laterally by other federal ministers. And, of course, gradually some staff had to be assembled. By the end of the play the scenery looked different: the political battles over the environment have helped in the structural reorganisation of the whole political set-up.

Our presentation describes the setting and tries to reconstruct what the actors were like, where they succeeded and failed, and what the cast and audience were able to learn during the performance.

Act number one started in 1959 with water legislation and continued with the connection of most of Austria to a sewerage system. In act number two a minister for the environment entered (in 1972) and soon started a battle against air pollution. During this act the public was particularly agitated. In 1978 it forced a referendum to ban nuclear power as a source of energy generation, and by a small margin the one and only Austrian nuclear power plant in Zwentendorf was prevented from going into operation. The issue was beyond the pollution theme: it involved a much wider horizon of acceptable risk taking in industrial society. Six years later another energy issue again provoked mass mobilisation. This time the controversy focused on a hydroelectric dam project at Hainburg which would have destroyed some of the Danube's last wetlands. And again the opponents of the project triumphed over the energy lobby and the united social partners (organized business and labour). The end of this act was marked by an apocalyptic climax: the Ukrainian nuclear power facility in Chernobyl exploded, contaminating large parts of Austria. The third act was more or less an epilogue: it centred on soil and waste policy. Gradually it becomes obvious that the setting of the play was no longer adequate: pollution was not all that matters and the domestic stage was becoming too narrow.

1.2 A Play in Three Acts[1]

Act I: Water Management and Protection

Long before Austria established its own ministry for the environment, environmental problems had entered on the political scene. One of the first activities to deal with environmental problems was the clean-up of Austrian lakes. The effects of water pollution were obvious and negative consequences on tourism could be foreseen. Steps taken included canalisation, and the construction of water treatment plants and circular sewerage systems around lakes to preserve them from toxic and eutrophicating substances. By the end of the 1980s, the condition of some lakes was again comparable to that in the 1930s. These activities were mostly financed by the water management fund, introduced in 1959 by amendment to the Water Act. Subsidies were given to build water supply facilities, sewerage systems, and municipal and industrial waste water treatment plants. The proportion of households connected to sewerage networks increased from 39% in 1968 to 77% in 1996. The proportion of households connected to biological waste water treatment plants rose from 3% to 74%. This programme was very successful but also quite expensive: from 1959 to 1986 more than 90 billion ATS (about 6.5 billion €, 61% of which was public subsidies) were spent on municipal waste

---

1 Sources of the data on the situation of the environment: Bundesministerium für Land und Forstwirtschaft, 1995; Bundesministerium für Umwelt, 1996; Bundesministerium für Umwelt, Jugend und Familie, 1998; Gerhold, 1996; Kind/Welan, 1996; OECD, 1995; Umweltbundesamt/Arbeiterkammer, 1996

water treatment projects and almost 12 billion ATS (0.9 billion €, 51% of it public subsidies) on industrial projects (Fischer-Kowalski, 1988: 125).

In the 1970s, attention was also paid to the quality of river and stream water polluted by industrial waste waters (mainly from the pulp and paper, textile, food, and chemical industries), but also by households not connected to a biological waste water treatment plant. River water quality could be improved significantly. In a 4-part classification of biological water quality, by 1995 72% of total length were in the best two classes, and by 1995 no river was in the worst class (1971: 5%). However, groundwater contamination by pesticides like atrazine, especially in the eastern part of the country, continued to be a problem, as did large toxic loads (heavy metals etc.) in river sediments.

About 99% of drinking water in Austria is taken from groundwater, which for the most part is in quite good condition. Half originates from gravel aquifers, and in some regions it is contaminated by nitrate and chlorinated hydrocarbons. The source of these substances is leakage from contaminated and unsecured waste sites, and the inappropriate use of fertilizers and pesticides, especially atrazine, which is now prohibited. As a response to sporadic discoveries of contamination, an extensive monitoring system for groundwater (with about 2.000 measuring points in 1996) was installed.

At the same time, all major rivers were regulated, be it for flood protection or hydroelectric power generation. Only about one-third of the 1.900 km of major rivers in Austria are still unregulated.

The Ministry of Public Health and Environmental Protection, established in 1972, played no important role in this process due to its lack of power, competencies and resources. The main actors in water management were the Ministry of Agriculture and Forestry, which still is responsible for water legislation and regulations, for water management, and hydraulic engineering (including hydropower plants and torrent and avalanche control), and administrative authorities on the provincial and the municipal levels. According to Austrian legislation, provincial governments carry out day-to-day water management functions on behalf of the federal government, including permitting and enforcement. Municipalities provide water supply, sewerage systems and waste water treatment.

Act II: Air Pollution

In the 1970s, the problem of air pollution was discussed internationally. In Austria clean air policy was first initiated by a widely publicised report of the Austrian Academy of Sciences *(Österreichische Akademie der Wissenschaften)* in 1973 suggesting regulatory policies against air pollution. But for almost a decade this report brought no results. In the late 1970s and the 1980s, public concern about air pollution increased. Forest dieback *(Waldsterben)* was supposed to be a result of acidifying air pollutants. This led to strong public pressure on the government. First measures in air management were not taken before 1980. Then the Steam Boiler Emission Act was passed that is broadly seen as the first environmental law in Austria using a modern set of instruments. The law passed the federal parliament with the votes of the Social Democrats only: the People's Party opposed the measure on the grounds that it was anti-business. Emission standards for steam boilers were bound to the state of engineering, even if no direct danger to public health was to be expected (Lauber, 1997b: 611). In 1988 it was substituted by the Clean Air Act for Steam Boilers. This act in combination with the Industrial Code provided the legal framework for effective improvements in reducing air pollution.

To protect forests, an ordinance pursuant to the 1975 Forest Act was passed in 1982 (strengthened in 1984), which sets ambient air quality standards for SO_2, hydrogen fluoride, hydrogen chloride and ammonia.

In 1983, the environmental fund was created to finance "clean air investments". Investments that go beyond the state of engineering in air pollution abatement were to be supported in every case (Glatz, 1995: 13).

In 1989 the Smog Alarm Act passed after a long constitutional fight. It obligated the *"Landeshauptmann"* to identify loaded areas, to measure ambient air pollution (SO_2, particulate matter, CO, and NO_2) and to make plans for the event of limit values being exceeded (smog alarm).

The 1992 Ozone Act regulated reduction of NO_X and VOCs emissions. They were to be reduced by 40 per cent by the end of 1996, 60 per cent by 2001, and 70 per cent by the end of 2006 (on the basis of 1985 for NO_X and 1988 for VOC). These targets have so far not been met (OECD, 1995: 44; BMUJF, 1998), and how this will be achieved in the future remains to be seen. In the summer months the burden of ground-level ozone is quite high compared to western and northern European countries. Another important aspect of this law was the obligation to inform the public of measures to be taken at the provincial level when a limit is exceeded. Such measures could be a driving ban, a prohibition on burning biogenous matter, or using solvents (Kind/Welan, 1996: 29f.).

Since 1980, some important improvements have been made in reducing air pollution. The emission of SO_2, for long the predominant air pollution problem, could be reduced by more than 80% by 1996. Today Austrian emissions of SO_2 per unit GNP are among the lowest in OECD countries. Emitters of SO_2 are industry (40%), power and heating installations (24%), households and commerce (18%), and traffic (11%). The remarkable reduction was due to the use of desulphurisation equipment, the lowering of permissible sulphur content in fuel oil, the substitution of coal and oil by gas, and the use of district heating.

NO_X could only be reduced by 30% between 1980 and 1994. Road traffic is responsible for 55% of total emissions. The prescribed use of catalytic converters in cars since the mid-1980s brought a substantial reduction of NO_2 per car, but an 80% increase in mileage between 1980 and 1991 offset most improvement in performance. There is still no sign that car traffic will stabilize at its current high level, or even decrease.

Particulate matter is still a problem, especially in cities and in winter, although it could be reduced about 60% by rigorous standards for emitters, like industry and heating installations. Main sources are households (40%), industry (34%), and traffic (22%).

No significant reduction was achieved for volatile organic compounds (VOCs) in the past two decades. In the 1990s, a weak downward trend was recorded. Emission by cars was reduced by 39% between 1980 and 1994 through the use of catalytic converters and activated charcoal filters in tanks.

As regards forest dieback, in 1979 about 200 sq. km were classified as "smoke-damaged areas". The classified area increased to 11.530 sq. km by 1986, mainly owing to the new awareness of the problem. In the period from 1989 to 1994, the number of trees without any visible damage to their crowns (defoliation) increased from 49.4% to 59.9%, the number of trees with medium to high damage decreased from 10.8% to 7.8%. This overall picture of improvement, though, does not hold for the most sensitive tree species: Just 30.8% of firs show no visible damage, and the amount of highly damaged or dead oaks increased from 1.1% to 4.4%.

The institutional structure in air management policy in this period is a good example of high complexity and low integration. After a constitutional amendment in 1988, the federal government gained overall responsibility for air quality management. Implementation and related legislation was distributed among the federal, provincial, and municipal levels. The Ministry for the Environment, Youth and Family has to develop overall air pollution strategies and policies. This has to be done in conjunction with other ministries, notably the Ministry of Economic Affairs, the Ministry of Womens' Affairs and Consumer Protection, the Ministry of Social Affairs and Public Health, and the Ministry of Agriculture and Forestry.

Authority to set standards is dispersed among administrations. The Ministry for the Environment, Youth and Family has to formulate ambient air quality standards and fuel specifications for mobile combustion engines. Standards for steam boilers are regulated by the Ministry of Economic Affairs in agreement with the environment ministry.

Provinces are in charge of regulations on heating installations, and of licensing facilities under the Industrial Code. They are also responsible for the inspection, monitoring and enforcement of license conditions (OECD, 1995: 36).

Act III: Soil Protection and Waste Management

For a long time, no attention was paid to the pollution of soil. The rapid industrialization of agriculture led to an increased use of mineral fertilizers and pesticides. The result was eutrophication of rivers and contamination of soils and groundwater. In 1986 a tax was imposed on fertilizers. Federal revenue from this tax was to be used to subsidize grain exports, and the environmental effect was a reduction in fertilizer consumption. The tax was repealed in 1994, but consumption has not since increased (OECD, 1995: 92). Another problem is the contamination of soil by heavy metals, either air-borne, or from waste, fertilizers, and sewage sludge.

Landfills and other contaminated sites are a major source of soil pollution. First estimates counted 80.000 old contaminated sites in Austria. 29,500 such sites had been registered by the Federal Environmental Agency by 1998. Under the Contaminated Site Clean-up Act (1989), 142 have been classified as dangerous. 200 to 300 million ATS are invested every year on investigations and cleaning-up. This money comes from the contaminated site clean-up charge, to be paid for by current users of landfills.

Before 1990, when the Waste Management Act came into effect, there was no clear distribution of competences between the federal government and the provinces in dealing with household and hazardous waste. A 1988 amendment to the Federal Constitution Act assigned responsibility for hazardous waste management to the federal government, and for non-hazardous waste to the provinces and municipalities. The federal government may issue regulations on non-hazardous waste if uniform legislation is necessary. This was done with the 1990 Waste

Management Act, which provided a comprehensive legal framework for waste management and recycling, hazardous waste management, and waste exports and imports. The aims of this act were said to be primarily directed at preventing waste generation and at recycling; environmentally compatible treatment of the remaining waste implied that only inert waste was to be deposited in landfills. The minister of the environment has to publish a Waste Management Plan every three years, which has been done for 1992, 1995, and 1998. A couple of decrees regulate specific areas such as the separate collection of hazardous waste and the waste from construction.

From 1975 to 1996, domestic processed output from the Austrian economy, encompassing emissions to air, waste deposition, dissipative uses (fertilizer, pesticides etc.) and loss of materials (abrasion from tyres and roads), and material loads in waste water, increased by 17% from 92 to 107 million tonnes (12.2 to 13.3 tonnes per capita) while GDP rose by 68% in the same period. Therefore a relative decoupling of emissions from the environment and economic growth took place. Domestic processed output is dominated by emissions to air (mainly CO_2), 90 million tonnes, followed by waste deposition, and dissipative uses and losses (18 million tonnes). Material loads in waste water are negligible. The proportion of emissions to air has increased steadily while material outputs to land and material loads in waste water have decreased in the last 20 years. This shift in affected media corresponds to political response. The problem of increasing CO_2, the quantitatively most important emission, has been one of the challenges for environmental policy in the 1990s (Matthews et al., 2000).

1.3 A New Setting of the Stage

These three decades can be said to represent the successful development of environmental policy: the emergence of a certain view of the problem, the establishment of institutional structures, public pressure and successful problem solving. Each decade corresponds roughly to a phase with a particular policy focus and a different constellation of actors.

The first, concentrating on water management and the cleaning-up of lakes, was characterized by a clear constellation of proponents and opponents, of instruments, and policy styles. Tourism and the construction industry both had a strong economic interest in solving the problem of polluted lakes. Tourism was afraid of losing visitors, and the construction industry saw a new area of activity. Public pressure was very weak, and there was no need for the government to change its policy style or strategy, consisting mainly in subsidizing investment in end-of-pipe technologies. This policy was well integrated across hierarchical levels (i.e., federal government, provinces, districts, and municipalities), but hardly integrated horizontally. At the federal level, it was the exclusive concern of the very traditionally organized Ministry of Agriculture and Forestry. Institutional innovations were confined to the creation of a financing instrument, the water management fund. More than a decade later, consumer mobilisation led to the Detergents Act in 1984, which limited the use of phosphates in detergents causing eutrophication of surface water. This was the first law in the competence of the Ministry of Public Health and Environmental Protection that had been established 12 years before.

Air quality management was the major focus of the second phase. This phase was accompanied by an economic interest in environmental protection, e.g., on the part of forest owners (often public) worried about their forests and (sometimes) on the part of municipalities concerned about historic buildings endangered by acid rain, and about wildwater disasters, expected to follow forest dieback. Contrary to the first phase, strong public interest and media coverage put pressure on the actors in the political arena. Main policy instruments were the setting of emission standards and (again) subsidizing end-of-pipe technology equipment. For this reason, too, industry developed an economic interest (i.e., in the production of filters and other air treatment facilities). Air-quality management powers were widely distributed both hierarchically (federal government, provinces, municipalities) and horizontally (Ministry for the Environment, Ministry of Agriculture and Forestry, and the Ministry of Economic Affairs), which led to long delays in legislation.

The third phase concentrated on issues of waste management and soil protection. The industrialization of agriculture, increasing problems with landfill capacity, and the contamination of soil and groundwater by the leakage of toxic and eutrophicating substances made it clear that things had to change. Increasing requirements in waste treatment, in landfill equipment, and in end-of-pipe technology (filters) stimulated economic interests in these technologies. Powers in waste management are vertically distributed among the federal, provincial and municipal levels. Beside the Ministry for the Environment, Youth and Family, also the Ministry of Agriculture and Forestry has competence in this area as part of its water management function. Regulatory instruments played the major role in soil protection and waste management, but economic instruments like a tax on fertilizers, the contaminated site clean-up charge or deposit-refund systems as well as voluntary agreements were also used.

From the 1960s to the early 1990s, environmental policies brought significant improvements to the environment, but some major conflicts of interest over environmental concerns could not be handled within the existing political framework. This led to a power clash and a number of major steps towards restructuring the Austrian political system: One of the key characteristics of Austrian policy style is so-called "concordance", co-operation between the two big parties, institutionally supported by the strong position of organized business and labour, the so-called social partners *(Sozialpartner),* in pre-setting issues. The frequent "grand coalitions" between the Social Democrats (SPÖ) and the People's Party (ÖVP) fostered a tendency to avoid open conflict or confrontation (Gerlich, 1997: 507). Environmental issues gave more or less no ground for concern. This policy style was severely shaken by conflict over environmental issues. After the oil crisis in 1973/74, the Austrian government planned to build a number of nuclear power plants. This led to the greatest wave of mobilisation since World War II and finally to a referendum in 1978 in which nuclear power was rejected, nuclear power generation subsequently being banned by law (Lauber, 1997a: 82). The immediate reason for this referendum was the Zwentendorf nuclear power plant, already constructed but not yet commissioned. This was a victory of the "anti-nuclear-power movement" over the government (Social Democrats at the time) and the social partners who had strongly supported Zwentendorf. Six years later, in 1984, a similar constellation occurred: the conflict about an already approved hydroelectric power plant in Hainburg[2] brought another victory for the environmental movement via mass mobilisation, this time with the support of major newspapers. "For the ecology issue, this was the major breakthrough, and now the government clearly felt the need to respond to a concern that enjoyed such strong support" (Lauber, 1997a: 83). The federal government was not able to handle this public mass protest against core investments in energy production. They had greatly underestimated the extent of resistance, and there were no effective participatory problem-solving instruments in place. One may say these two conflicts were the main reason for basic re-arrangements in the overall political arena.

The political scene had hitherto been "duo-polized" by the People's Party plus organized business and agriculture on the right and the Social Democrats plus organized labour on the left, but now both parties had to establish a critical distance to their social partners (who were both viewed by the public as solidly anti-environmental). These events opened up opportunities for a "Green Party" to enter the federal parliament in 1986 as a "religious force", which was the beginning of the end of the "duo-polistic" political landscape in Austria. Citizen initiatives and environmental NGOs had firmly established their presence on the political stage and were to be reckoned with in the future. After this period it was clear that a more participatory policy style and more transparency in public information on environmental issues was inevitable.

2 Environmental Politics in the 1990s

2.1 The Change in Policy Paradigm and the Introduction of New Instruments

From "Environmental Protection" to "Sustainable Development"

The focus of the first 30 years of environmental policy had been more or less on single substances that polluted water, air, and soil. Improvements had been achieved without changing the dominant policy style by subsidizing investments in environmental protection equipment and by setting standards. This policy is actually the opposite of the polluter-pays principle, especially when private enterprises are subsidized by public authorities. This situation has not yet changed significantly.

The concept of sustainable development, the central paradigm of the United Nations Conference on Environment and Development (UNCED) in 1992 (Agenda 21), was brought up for the first time in 1987 by the so-called Brundtland-report. It resulted in an important change in the perception of environmental problems. The Austrian National Environmental Plan (1995) takes account of the recommendations of Agenda 21 and tries to achieve the integration of environmental

2 The economic usefulness of this plant was doubtful and some of the Danube's last wetlands would have been destroyed. During the conflict, the area was occupied by some thousand people. When the police tried to clear the occupied area, environmental organisations and the mass media organized a demonstration against the project in which about 40.000 persons participated. The plant was not built and the area was converted into a national park instead.

issues in sectoral policies. Reduction of material flows and energy consumption, the closing of material cycles, and the abandonment of fossil fuels are main targets (Österreichische Bundesregierung, 1995). Nevertheless, traditional environmental policy, where every sector bothers about "its" media, continues.

The concept of sustainable development implies an extensive, more systemic view of environmental problems:

- Environmental problems cannot be handled in isolation from the question of social welfare and economic stability.
- Society is seen as an input-output system, where not only output (i.e., waste or emissions of pollutants) counts but also resource consumption. Everything used by the socio-economic system as input is ultimately output (*social metabolism*).
- At the same time, the view of responsibilities and the constellation of the actors changed. It is no longer "industry", as the irresponsible polluter, but a systemic logic that causes certain outcomes, and this logic has to be handled differently by the responsible activity of many actors. The systemic logic has several implications for policy instruments. It is clear that single-substance regulation alone is not adequate to achieve sustainable development.
- Natural environment is no longer seen as "compartmentalized", e.g., the reduction of waste by incineration is not viewed as a "success", but as a mere shifting of loads.
- Environmental problems are increasingly seen not as local, regional or "domestic", but as international and global problems, and also problems of equity between the underdeveloped South and the North have to be resolved in this context.
- The insight that long-term processes matter is gaining importance (the equity between present and future generations is one of the major elements of sustainable development).
- Corresponding to this, a change in the view of time can be observed: the feeling of urgency (everything should have happened yesterday) is now replaced by a feeling of the need for planning structural change over the medium and long term.

The most important question for environmental policy in the 1990s can be formulated as follows: Is the capacity developed in the first phase of Austrian environmental policy, the structure of actors, institutions, instruments, environmental legislation, and policy style, sufficient to handle the challenges linked with this new framing of the problem?

Instruments

The 1990s brought not only a new perspective on environmental problems, they also saw application of new instruments: less regulation, but more economic instruments, improved public information, more civil participation and voluntary agreements were signs of a changed approach to environmental problems and the necessary responses.

During the previous decades of environmental policy in Austria, public spending and regulatory instruments had played the major role. Economic and societal instruments had hardly been used. These policy investments had been quite successful, but in the second half of the 1980s regulatory instruments increasingly became the focus of public criticism (even if they were better than their reputation, see Glatz, 1995: 6). They were considered effective but not efficient, and they also imply the need for a large bureaucracy and detailed and complex legislation.

A major economic instrument discussed internationally is an ecological tax reform. In Austria several recommendations about taxes on energy and resources were made. A suggestion put forward in 1993 and 1994 by the Austrian CO_2-Commission was evaluated by the Austrian Department for Economic Research *(Österreichisches Wirtschaftsforschungsinstitut, WIFO)*. Core parts of the concept were a tax on energy, a reduction in the extra cost on labour, and a program for the reduction of energy use by enterprises and households. According to the model the WIFO applied, the realisation of these suggestions would have led to higher employment, higher growth in GDP, and a decrease in net lending. CO_2 emissions would have decreased by 5.4% after a modelling time of 5 years (Köppl, 1995). Ecological tax reform was discussed intensively by all parties in 1995, without a satisfying result, the budgetary crisis overshadowing discussion (Lauber, 1997a: 85); but an explicit energy tax, the tax on gas and electricity, was introduced in 1996 without any compensation in other areas and without the potency to have a marked effect on energy consumption.

The National Environmental Plan (NUP) was prepared from 1992 to 1995. The main objective was to operationalize and implement the concept of sustainable development for relevant policies in Austria in a middle and long run perspective (Österreichische Bundesregierung, 1995: 12). 250 civil servants, scientists, and other experts were involved, nominated by the federal ministries, the social partners, and the provinces (Payer, 1997: 128). Seven working groups were to develop targets and measures for the following areas: industry; energy and the oil industry; traffic and transportation; agriculture, forestry and water resources; natural resource management; and consumers. The plan is not binding on anyone and did not even became part of the subsequent government agreement between the Social Democrats and the People's Party (Fischer-Kowalski, 1995: 63). The importance of this plan was to be sought elsewhere: a new perspective, a new mind-map of environmental problems was introduced. This new quality can also be seen at the level of commitment to targets, which is reflexive, systemic and operational. The dimension of analysis is resource-economical, actor-related, and methodically integrated. The measures are primarily economic in nature. Not all parts of the plan have the same quality, but one may take it as a sign of increasing competence in dealing with environmental problems (Fischer-Kowalski, 1995: 69).

Several new instruments were discussed and also introduced as a result of the Austrian EU membership: voluntary agreements, auditing, participation and public information. An important step towards participation and precautionary environmental protection is the Environmental Impact Assessment and Citizen Participation Act of 1993. The procedure, obligatory for specific projects, requires the applicant to describe the project and the expected impacts on the environment. On the basis of this environmental impact declaration an expert opinion is procured. The state government has to decide. Concerned municipalities, the environmental ombudsmen, and citizen initiatives (larger than 200 persons) have appealing rights, and there is also a comprehensive participation process (Kind/Welan, 1996: 34f.).

Other types of so called "soft instruments" also began to play a bigger role. Over 17 voluntary agreements had been signed by 1995. Most of them were negotiated between the Ministry of Economic Affairs and business or industry. Such voluntary agreements exist for recycling activities, on the import of wood from tropical forests, and on the use of vapour recovery systems in service stations (OECD, 1995: 95). This instrument is seen positive in principle, but there are some critical points in practical application: the main problem is the lack of information of and participation by the public and affected groups (consumers, employees, environmental organisations). The lack of a minimal standard of legal protection is another important point. In most cases compliance is not monitored and there are no sanctions (Glatz, 1995: 17ff.). Another instrument that is voluntary in nature is the eco-audit, which is regulated in a European Union regulation (Community eco-management and audit scheme: EMAS). Firms give a declaration of their environmental activities and environmental data. A group of authorized experts examines this declaration and if everything is correct the firm is allowed to use an eco-audit label. The label must not be used on products as it is not a label for environmentally friendly products. In 1995, the first group of experts was authorized, and the first firms in Austria were registered in February 1996 (Umweltbundesamt/Arbeiterkammer, 1996: 116). By September 1997, 80 sites had been registered in Austria, the second largest number per head of population in the European Union (BMUJF, 1998: 654).

The instruments of information, labelling and publicity also gradually became more important. Consumers were to decide if they preferred environmentally friendly products or not. But the increase in "green" labels made it quite difficult for consumers to decide whether terms like "eco" or "environmentally friendly" were just marketing ploys, or if they indeed implied any environmental benefit. This led to a discussion on labelling. Since 1990 the use of the Austrian Environmental Label *(Österreichisches Umweltzeichen)* is granted by the Ministry of Environment after a procedure involving business, consumer unions, environmental organisations, and the administration. The label is subject to very rigorous criteria and was accordingly not a big success: in 1996 only about 80 products had been allowed to use it. Another example is the label for ecological farm products, which is quite successful. It is well known and the number of labelled products is increasing steadily. Increasingly, of course, the issue of product labelling policy is being transferred from the national to the EU level. This is particularly obvious with the highly controversial issue of genetically altered food, where labelling regulations are still in the make.

2.2 Main Actors

Governmental Institutions at the Federal, Provincial, and Municipal Levels

When the Ministry of Public Health and the Environment (BMGU) was founded in 1972, most environmental tasks were in the hands of several other ministries and the provinces, and it suffered from a lack of power and resources for over 15 years. The ministries that had the powers (and the resources) were not willing to renounce these issues, and hardly allowed the environmental ministry to extend its influence (Lauber, 1997b: 609). In an amendment to the constitution in 1988, the Ministry of the Environment, Youth and Public Health was granted considerable powers for the first time. The staff of three dozen in the mid-1980s then increased by a factor of ten in the following decade.

The Austrian Federal Institute of Public Health (ÖBIG) was founded in 1973 and provided the scientific backbone for federal environmental policy on the base of a highly committed and motivated staff. Later some of its tasks were transferred to the Federal Environment Agency (UBA), established in 1985 as a consequence of the Hainburg conflict, to keep government science under somewhat tighter control. The Federal Environment Agency has the function of collecting and measuring environmental data, and advising the Environmental Ministry. As one of several federal institutions, it was privatised in early 1999. A change in the quality of its tasks or a loss of independence is not apparent to date.

In the early 1990s, the Interministerial Commission for Co-ordination of Measures Concerning Global Climatic Change, and the Austrian CO_2-Commission were established by the Ministry of the Environment, Youth and Family. In the CO_2-Commission, economists, ecologists, technicians, and others worked out suggestions for Austria's global climate policy, encompassing a new tax on energy, a reduction in contributions on income, and an environmental investment program (Köppl, 1995: 31 and 155). This organisation was transformed into the Austrian Climate Advisory Board in 1996 *(Klimabeirat)* (Huber/Schramm, 1996: 72). While Austrian representatives on the international stage were usually in support of stricter CO_2-emission reduction targets, the advice of these commissions was not taken very seriously for internal Austrian environmental politics. Austrian CO_2-emissions in the second half of the 1990s were thus as high and higher as they ever had been since 1980 (BMUJF, 1998).

There is also a legal basis for parliamentary commissions of inquiry to assist in the preparation of decisions on comprehensive and important matters. So far, use has been made of this institution only once. In 1991, a commission of inquiry was asked to advise on the possible effects of genetic engineering and to discuss the chances, risks, and effects, and to give recommendations on responsible handling (Gottweis, 1997a: 668).

In general, Austrian politics does not tend to make use of strong and independent advisory instruments. None of the commissions mentioned was equipped with appreciable resources or an independent voice. Contrary to other political cultures, bodies of experts in Austria still tend to be viewed as a possible irritant to the political process rather than a help.

During the first phase of environmental policy, the complexity of the vertical distribution of powers across the federal, provincial, and municipal levels, but also the horizontal division of powers between the various federal ministries was frequently criticized, but survived basically unaltered. In Austria, environmental powers are still distributed among many public actors who often do not co-operate well. Whenever a question of redistribution of competence is raised, it turns into a problem of federalism and the constitution. So every actor tries to preserve his own legislation, which results in highly complex regulations. Thus legal structures for the environment can be characterized as extensive, but far too complex, and the structure of regulations is often criticized as muddled (Kind/Welan, 1995: 14).

Presently the Ministry of the Environment, Youth, and Family (BMUJF) is responsible for general environmental policy, air quality, waste, chemicals and the allocation of federal funding to the provinces, and it holds partial responsibility for environmental inspection (OECD, 1995: 24f.). The minister of agriculture and forestry presides over a kingdom of his own covering many important environmental areas such as agriculture and forestry policy, water legislation, water management, and hydraulic engineering. But the following ministries also have considerable responsibilities in environmental affairs: the Ministry of Economic Affairs (energy, mining, tourism, and federal roads), the Ministry of Science and Traffic (environmental research, public traffic), the Ministry of Women Affairs (genetic engineering, consumer protection), and the Ministry of Foreign Affairs, which co-ordinates Austrian international environmental policy. EU membership produced a partly new constellation in environmental policy, which from now on was also an EU-wide issue. The influence of the federal parliament and environmental organisations decreased. The winner in this process was the minister of the environment, who has to co-ordinate Austria's position in EU environmental policy, where tasks are also discussed that are not within the ambit of the Ministry for the Environment, Youth, and Family in Austria, e.g., water (Lauber, 1997b: 614).

All nine Austrian provinces have their own administrative units for environmental protection. They are in charge of nature and landscape protection, waste management (except for hazardous waste), emissions to air relating to heating equipment, physical planning, and construction. They also administer much federal legislation, and most civil servants dealing with environmental issues are therefore at the provincial and municipal levels (OECD, 1995: 86).

Most of the provinces established environmental ombudsmen, who may take up environmental concerns in administrative proceedings governed by provincial law. They have access to all relevant files, may formulate objections, and in some provinces they may appeal to the higher courts. In practice, they do not always play an important role in environmental policy because of their weak political position (Lauber, 1997a: 90).

Municipalities, while administering federal and provincial laws on behalf of higher authorities, also perform functions directly assigned to them by the constitution. They handle waste collection and disposal, water supply, sewerage networks, waste water treatment, land use planning, and – in some cases – public transport. These services are run by the municipalities themselves or by companies owned by municipalities (OECD, 1995: 28; Lauber, 1997a: 88).

Most of the activities necessary for implementating the National Environmental Plan (NUP) are in the competence of provinces and municipalities. Eight of the nine provinces and 215 cities and municipalities with more than 3.6 inhabitants are members of the Climate Alliance *(Klimabündnis),* an international organization created in 1993. Provinces, cities, and municipalities undertake to reduce CO_2-emissions by 50% by 2010, and to help alliance partners in the Amazon region to preserve their rain forests (Klimabündnis Österreich, 1997). The organization prompted several measures in the areas of energy use, reduction of motorized private transport, environmental awareness, as well as in directing the consumption of public households to environmentally friendly products. Local Agenda 21-

processes are another example, where municipalities are important actors. Such activities are expected to become more important in the coming years (Kanatschnig et al., 1998).

Political Parties, and Organized Business and Labor ("Social Partners")

After World War II, the Social Democratic Party (SPÖ) and the Austrian People's Party (ÖVP) long dominated the political stage. Together they gained 90% and more of the vote, and this duopoly of power became a characteristic of post-war politics in Austria. In the first half of the 1980s, they began to lose members and votes. This situation was aggravated by the 1984 Hainburg environmental conflict, which helped the Greens enter the federal parliament in 1986. Both parties, the SPÖ and the ÖVP, had to develop political concepts of their own for the environment. The SPÖ started to address the "ecological reorganisation of industrial society", the ÖVP went for a concept of "eco-social market economy", an idea developed some years before (Lauber, 1997b: 611ff.). Today the topic "environmental protection" can be found in the programmes of all five parties in the federal parliament, even in the right-wing Freedom Party of Jörg Haider, which won 27% of the vote in the 1999 election.

The Liberal Forum, a party with a stringent liberal programme (free market, low bureaucracy), is the youngest party in the federal parliament. It takes a quite pro-environment stance on several issues, especially in pushing through an eco-tax reform. It was founded in 1993 by 5 former Freedom Party MPs and received 5.5% of the vote in 1995. By the election of 1999 it had dropped out of parliament again, having just failed to take the 4% hurdle.

Green Parties had emerged in Austria around 1980. They obtained their first seats in a provincial parliament in 1984 and entered the federal parliament in 1986. But the Greens were divided into two different political organisations, one leftist, the other conservative. This division and the resulting infighting considerably weakened their impact (Lauber, 1997a: 91). In the 1986 elections, the two organisations formed an alliance, and won 4.8% of the vote and 8 seats (out of 183). From 1990 on, the two organisations presented separate candidates, and the conservative group became insignificant. In 1994, the Greens gained 7.3% of the vote and 13 seats, but then fell back to 4.8% in 1995, to regain almost 9% of the vote in 1999. Since 1986, a clear trend towards professionalization and consolidation is apparent. At least from 1992 on, the Greens tried to escape the image of an "inherently obstructionist party" and present themselves in constructive and practicable proposals as a party with competence. Despite their small size, they played a considerable role in identifying and articulating environmental problems. For several years, they formulated the issues the other parties then had to address. In the second half of the 90s, political agenda-setting has been taken over by the xenophobic rightist party FPÖ, led by Jörg Haider, a party that never had a strong say in environmental issues.

A very important (and mostly retarding) institution for the development of environmental policy in Austria is the so-called "social partnership" (cf. Tálos, 1997: 432), an informal association of the Chamber of Commerce, the Chamber of Labour, the Chamber of Agriculture, and the Austrian Trade Union Federation. These four groups are closely associated with the two big parties in the federal parliament, the government, and the administration. The Chamber of Commerce and the Chamber of Agriculture have close links with the People's Party, the Chamber of Labour and the Trade Union Federation with the Social Democrats. The structure of the social partners and their relations with government, administration, political parties, and federal parliament is referred to as an "all-channel-network" (Lehmbruch, 1985, cit. in Tálos, 1997: 440). The presidents of the four interest groups are also (mostly) members of the federal parliament (Tálos, 1997: 443).

The privilege of obligatory membership in the Chamber of Commerce, the Chamber of Labour, and the Chamber of Agriculture, and their representation in the social partnership give them great influence on political decision-making. The social partners focused on certain special, but very important areas: income, economy, social policy, and employment. But their role in environmental policy should not be underestimated.

In environmental policy, especially in energy policy, political consensus could not always be obtained between the political system, the public, and the social partners. The Zwentendorf and Hainburg conflicts were examples of the role played by the social partners as opponents of environmental and nature protection. This was not always consistently so. For several years the environmental subdivision of the Chamber of Labour became a prominent promoter of environmental concerns, even if its proposals did not always conform to the organization's official line. In late 1994, the Chamber of Labour even supported an ecological tax reform, provided that compensation could be found for the poor and energy-intensive industries (Lauber, 1997a: 92). The Trade Union Federation, though, has mostly been uninterested in environmental issues, unless they seemed to compete with investments and jobs: then it adopted a thoroughly anti-environmental position. The Chamber of Commerce supports economic instruments for environmental investments, voluntary agreements, and eco-audits. Members are provided with information and training for greater environmental awareness and better technology (OECD, 1995: 95). Typically the Chamber of Commerce opposes binding environmental regulations and public participation. The Association of Austrian Industrialists *(Industriellenvereinigung)* has very strong political and also personal links with the Chamber of Commerce, and, similarly, usually resists further environmental reform, which is viewed as a threat to Austrian competitiveness. Further progress, they argue, should be achieved only in step with the EU and in taking account of competition from eastern Europe (Lauber, 1997a: 92f.).

Environmental Movement and Environmental Organisations

Environmental organisations first arose in the second half of the 19th century. These "traditional" nature protection organisations are the Alpine Club, founded in 1862, the Friends of Nature (1895) and the Nature Protection Society (1913), and they all still exist. They have hundreds of thousands of members, with a membership peak in the early 70s now declining. Since 1973, they have been associated in the Austrian Society for the Protection of Nature and the Environment (ÖGNU).

A second wave started in 1963, when the Austrian office of the World Wildlife Fund for Nature (WWF, now World Wide Fund for Nature) was set up. Other international environmental organisations like Greenpeace, Friends of the Earth, and Global 2000 founded in the late 1970s were able to gain quite strong footholds (Fischer-Kowalski/Payer, 1995: 565). They play an important role as actors in environmental policy. They have a very good image, and one precondition of their power is that they finance their activities only by private donations. About 15% of all Austrian households now regularly donate to environmental NGOs, a very high figure by international standards (Greenpeace Austria, 1999: personal communication). They are seen as an important factor in the formation of environmental awareness and in mobilizing the public (OGM, 1993).

Like other social movements, the environmental and anti-nuclear movements took shape in the early 1970s, when under the impression of the energy crisis in 1973, the Austrian government decided to build a number of nuclear power plants. In a political climate of very low accessibility to the system and a strong tendency to avoid conflicts (Gottweis, 1997b: 344) social movements arose to exert influence on official politics. An important element of the movements concerning environmental issues were the citizen initiatives, which usually operated at the local level, especially in cities, where their activities were quite successful, but they were "single-issue movements" with a very low degree of organisation at a higher level.

This situation changed rapidly in 1978, when the referendum against the completed Zwentendorf nuclear power plant was held. A nation-wide combination, the Initiative of Austrian Nuclear Power Opponents (IÖAG) was founded after the government prepared an information campaign to put this nuclear power plant into operation. At its peak the anti-nuclear power movement mobilized about 500,000 people, one in ten voters (Gottweis, 1997b: 347). In the November 1978 referendum, 50.5% voted against commissioning Zwentendorf and a law on banning nuclear power in Austria was enacted in December of the same year. Plans for building nuclear power plants were finally abandoned following the Chernobyl disaster in 1986 (Lauber, 1997a: 82). Another important victory of the environmental movement was the prevention of the Hainburg hydro power plant in 1984, where prominent ecologists, elements of the anti-nuclear power movement and the WWF were able to show their mobilisation potential. In 1997, the environmental NGOs supported a public referendum against genetically modified food, and this referendum gained 1.2 million supporting votes (one fourth of the Austrian electorate) and turned out to be one of the strongest referendums in Austrian history.

In the 1990s, the environmental movement encompassed the environment protection movement, the nature conservation movement, and the animal protection movement. Increasing group overlap and specialisation was apparent. "Classical" issues like chemicals, energy, waste, traffic, water and nature protection were discussed beside new tasks such as climatic change, genetic engineering, and cruelty to animals. For a long time, environmental organisations were seen as a troublesome factor for the interests of employers' and employees' associations. Environmental tasks were not taken seriously by the latter. The founding of the Austrian Society for the Environment and Technology (ÖGUT) in 1985 had tried to change this situation. Environmental organisations were accepted as equal and environmental protection came to be an issue that could not be dismissed.

Both international environmental organisations (such as Greenpeace) and single-issue local or regional initiatives have provided quite a substantial challenge to the political system. As a consequence of their activities and their power to mobilize the public, prime ministers resigned (such as Kreisky in 1979 and Sinowatz in 1985), and public referendums forced changes of policy against the will of broad party coalitions and against the united efforts of the social partners (such as with atomic energy generation in 1978, or permission to plant genetically engineered crops and liberalize the market for genetically altered food in 1997).

In 1993 even a chamber of the environment was under discussion to obtain more influence on environmental policy, but it came to nothing. "The governing coalition parties were afraid of the potential power of such an institution; in turn, the more activist environmental organisations were concerned about the compromises that such a semi-official status might require." (Lauber, 1997a: 94). Instead, a number of environmental organizations (Greenpeace, WWF, Global 2000, Eco-Institute, Forum of Austrian Scientists for Environmental Protection, Austrian Traffic Club, and others) formed an alliance and are now organised in the "Öko-Büro", while more traditional organisations stayed in the Austrian Society for the Protection of Nature and Environmental Protection (ÖGNU). Both umbrella organisations are members of the European Environmental Bureau (EEB) in Brussels. While environmental issues have faded away in public debate, environmental organisations do not seem to have lost support.

The Media

Since the 1970s, environmental issues have gained a firm place in the media. The issues have changed (nuclear energy, forest dieback, climatic change, hazardous waste, genetic engineering), but the presence of the environmental issues increased steadily (Glatz, 1991: 52).

In some cases the "Neue Kronen Zeitung", a tabloid read by almost 45% of Austrians, played a crucial role. It acted as spokesman for the campaign against the Hainburg hydro power plant in 1984, and in 1997 it supported the initiative against the use of genetic engineering in agriculture, selling foodstuffs produced by genetic engineering methods, and patents on life.

Austrian radio and TV (ORF) report regularly on environmental issues. While nature conservation related films like those of the "universum" series are among the most popular programmes on TV, environmental matters are only occasionally dealt with in the context of general politics. Radio programmes regularly focus on environmental issues and often support environmental group activities. Radio broadcasting also provides as an important link between the public and the scientific community. The reports are of high quality, and very specialised issues are discussed, too.

In general, environmental issues have a considerable position in the Austrian media landscape. However, the quality of presentations varies considerably. In a poll conducted in 1993, 63% of the respondents felt radio and TV (ORF) reporting was adequate, while only 22% felt the same about "Neue Kronen Zeitung" reports, 41% of the respondents stating that the paper tended to exaggerate (SWS-Rundschau, 1994: 63).

Economic Actors with a Vested Interest in Environmental Policy

In most cases organized economic actors like the Chamber of Commerce or the Association of Austrian Industrialists cannot be seen as proponents of environmental protection but usually as opponents. But the rigorous pollution limiting values, and the use of the best available technology demanded by most Austrian environmental laws made it necessary for the industry to build up a high level of know-how in environmental technology, especially in "end-of-pipe" technologies, which developed remarkably. "Increasing efficiency using new technology was more and more identified as a singularly appropriate way to satisfy both economic and environmental criteria." (Lauber, 1997a: 97) With 250 firms and 11,000 employees, the environmental technology industry had a business volume of € 1.5 billion in 1995. Another € 1 billion are earned in waste management and environmental consulting. "In relative terms, this branch is more important in Austria than it is in Germany or Switzerland. This industry supports strict regulation." (Lauber, 1997a: 93, referring to Köppl/Pichl, 1995).

A second quite important group of interested economic actors with a vested interest in environmental policies, even if ambivalent, is Austrian agriculture. For geographical reasons, a high proportion of agricultural land in Austria can support little intensive cultivation. At the same time there is a shift towards medium and small holdings. Austrian agriculture is thus interested not so much in further industrial intensification than in a development towards quality production and – in one way or another – organic farming. In terms of the number of production units, Austria has the highest proportion of organic farming in the world. Many Austrian farmers therefore support reorienting agricultural policy and subsidization towards non-polluting production techniques. This interest strongly influences the overall policy of the People's Party. Franz Fischler, European Commissioner for Agriculture, is just one symbol of this constellation.

Another relevant economic actor, in terms of pressure on the environment, but also in terms of environmental protection, is tourism, a large and prosperous economic sector in Austria (earning almost one fifth of Austria's foreign exchange, as compared to the average 5% for the OECD). In the 1960s and 1970s, the interests of tourism have been among the main driving forces behind the rapid cleaning-up of Austrian lakes and the expansion of public sewage systems. The idea of "safeguarding the environmental capital as the central condition for the qualitative growth of tourism" as stated in the federal "Tourism Policy 2000" represents the spirit of many objectives stated by provincial and municipal tourism boards. The "sustainable tourism" propagated by regional tourism boards may include moratoriums on the further expansion of tourism infrastructure in high-intensity areas, an increase in public transport facilities at the expense of private road traffic, the promotion of landscape protection and national parks (OECD, 1995: 121f), and strong opposition to truck road traffic (particularly international transit traffic).

Paradoxically, the construction industry also has an interest in environmental protection, very directly in providing municipal sewage and waste water treatment facilities (which are also a major source of income for small and medium size local enterprises), as well as to torrent and avalanche control systems[3]. More recently, construction companies have been facing strong competition from urban sprawl as prohibitive to sand and gravel extraction and a pressure on prices for rubble deposition. This makes them more apt to support reductions in construction material targets and "quality not quantity" approaches in structural engineering.

3 The Capacity for Environmental Policy and Management

We now consider how Austria's overall capacity for environmental policy and management can be assessed in terms of Jänicke's (1997) guidelines. According to Jänicke, capacity for the environment is constituted by 1) the strength, competence and configuration of organized governmental and non-governmental *proponents* of environmental protection, 2) the *framework* for their activities in terms of cognitive-informational, political-institutional and economic-technological *conditions*. This analysis of capacities has to be related to the *kind of problem*. Within the time span we have covered with our description of environmental institutions and policies a major change in the "kind of problem" occurred. While in the first phase reported on, problems were perceived as sectoral environmental challenges to which policies could respond, mainly by promoting end-of-pipe measures, the second phase was characterized by a shift from "environmental protection" to "sustainable development". For sectoral environmental policies, capacities had been built up quite systematically during the 1970s and the 1980s.

If we classify these capacities according to Jänicke's criteria, we find Austria has developed quite remarkable preconditions for environmental policy.

As far as sectoral environmental politics *(first-generation environmental problems)* are concerned, quite a potent set of actors seems to be firmly in place. On the part of the government structures on federal, state, and municipal level, appropriate structures have been developed. Environmental monitoring and regulation are routinely performed. On the part of industry (and perhaps less so on the part of industrial agriculture) environmental concerns have been introduced into managerial concerns. Compliance with high environmental standards is the standard repertoire of industrial self description. A large and active non-governmental scene of environmental organisations guards against environmentally destructive activities. The media and public are concerned about environmental quality, even if their attention has been somewhat distracted in recent years by social and economic issues. Moreover, a new type of actor has entered the scene, the consumer. Consumers were very effectively mobilized against the use of environmentally problematic products such as tropical timber, PVC, or genetically modified food.

Table 1–3: Assessment of Capacity for Environmental Protection in Austria (classified according to Jänicke, 1997: 16ff.)

3 Contrary to international standards, the Austrians include torrent and avalanche control among environmental protection measures. This amounts to about 10% of public environmental expenditure (OECD, 1995: 87f).

Table 1. Proponents of Environmental Protection

Austria's Capacity	Stages in Capacity Development: Proponents of Environmental Protection
	1. Environmental policy institutions
	• no central ministry/agency for environment
	• weak and isolated national ministry/agency
X	• environmental institutions at all levels
X	• environmental divisions in most other ministries also
	• institutions for integrated environmental planning
	2. Green interest organizations
	• local organizations, no clear national interest organization
	• weak or non-professional national interest organizations
X	• strong and highly competent green organizations, playing a consultative role in political decisions
X	• green organizations playing a consultative role in industry also
	3. The media:
	• few, largely official reports on the environmental situation
	• environmental reports in some critical sections of the media, but dominant media resistance
X	• environmental problems widely reported in all the media
X	• frequent direct media attacks against polluters
	4. The "green" business sector:
	environmental interests are scarcely articulated within the economic system
X	• environmental interests are articulated mainly by the eco-industry (producers of clean-up technology)
X	• environmental interests are articulated by particular green business organizations
	• ecological pioneer enterprises (e.g. trade or insurance companies) have a strong impact on the whole economy

The cast is well selected and trained for sectoral environmental politics, but not for more integrative approaches to sustainable development. While the classic actors in environmental protection at the governmental and non-governmental levels have shown interest in more integrative approaches, other actors required to implement such policies are shirking their roles. While Austria has agreed on firm standards for CO_2-reduction, both industry and labour unions oppose the implementation of the appropriate policies; the last large tax reform, despite previous declarations to the opposite, completely omitted a socio-ecological tax reform. None of the major parties was willing to put its weight behind this agenda, and even the Greens seemed to concentrate on other issues. Below the federal level, on a more regional and local basis, a lot of motivated actors seem to be committed to making Agenda 21 and related activities a success (Kanatschnig et al., 1998).

Table 2. Cognitive-informational Framework Conditions

Austria's Capacity	Selected Indicators of Capacity-Building: Environmental Knowledge and Awareness
	1. Statistics (as indicators of the leading paradigm)
	• Scattered or irrelevant data
	• Mainly air and water quality data
X	• National emissions and solid waste data also
X	• Data on relevant material flows (such as heavy metals) also
	• National eco-balances, green GDP
	2. Science and Research
	• No systematic research on pollution problems
	• Few independent environmental research institutes typically with disputed professional standards
	• Erosion of "productionist" research paradigms and priorities
X	• Environment becomes a relevant theme in universities and research institutes
	3. Public attitudes towards the environment
X	• environment among top priorities in public opinion[4]

Table 3. Political-institutional Framework Conditions

Austria's Capacity	Development of Participative Capacities
	• Participation is restricted to general elections
X	• Participation rights are instituted regarding licensing of local sites and urban planning
X	• Participation exists in the formulation of national environmental policy
X	• Participation is possible in policy formulation in fields, e.g. energy of transport
X	• Participation in ecologically relevant long-term planning (Agenda 21)
Austria's Capacity	Development of the Integrative Capacities of Government
	• Only additive environmental institutions, fragmentation of jurisdictions
	• Intrapolicy co-operation
X	• Selected ad-hoc interpolicy co-operation
X	• Environmental subdivisions in all relevant ministries and administrations
X	• Strategic interpolicy co-operation
	• Long-term intersectoral planning on a broad societal basis

4 This criterion has been added to Jänicke's table: In several polls around 1990, the environment was a more important concern to the public than, for example, unemployment (Plasser/Ullram, 1990: 31ff., SWS-Rundschau 1994: 64, Pleschberger, 1995: 92f.). Asked about their willingness to pay, a majority of Austrian consumers were ready to pay a higher price for environmentally friendly products (Simhandl, 1996: 636f.). This public climate is also reflected in the environmental awareness of local decision makers across Austria: according to a survey in the early 1990s (Hogl/Pleschberger/Reiger, 1996: 321), practically all respondents display a rather high or very high awareness of environmental problems.

Another element of political-institutional framework conditions not explicitly introduced by Jänicke has characterized much of environmental policy in Austria over the past three decades: the readiness of the Austrian population to join in spontaneous public protest. While this is very common on a small scale for various infrastructural (traffic, energy, ...) and natural protection issues, it may become a majority issue that also expresses itself formally in referendums.

Major weaknesses are discernable in the integrative capacity of government. While political and administrative institutions for sectoral environment policy at the federal, provincial and municipal levels are firmly institutionalized, most linkages across different administrative units are rather weak horizontally and vertically. This is particularly apparent in the attempts to formulate a coherent energy policy and climate protection strategy.

Ambitious reduction targets for CO_2 emissions, such as the so called Toronto target agreed in 1990 by the Federal Government (20% reduction in CO_2 emissions within 15 years) required an extent of policy integration and strategic coherence obviously beyond the capacity of the Austrian political system (Steurer 1999: 206ff). Similarly, the effort invested in a "National Environmental Plan" (NUP) in the early 1990s did not bear fruit in actual implementation[5]. This lack of integration in strategic policy brought climate policy and sustainable development implementation to a virtual standstill in the second half of the 1990s.

From a current perspective it is difficult to decide whether this standstill was due more to a lack of strategic environmental policy capacity or to unwillingness to make use of existing capacities. At the beginning of the 1990s, there were indications that Austria had joined the leading European countries in the seriousness with which it took environmental concerns. In the last few years much of this vigour has abated.

References

Bundesministerium für Land- und Forstwirtschaft (ed.), 1995: *Österreichischer Waldbericht* 1994. Vienna.

Bundesministerium für Umwelt (ed.), 1996: *Umweltsituation in Österreich. Vierter Umweltkontrollbericht des Bundesministers für Umwelt and den Nationalrat.* Teil A. Vienna.

Bundesministerium für Umwelt, Jugend und Familie (ed.), 1998: *Umweltsituation in Österreich. Umweltkontrollbericht des Bundesministers für Umwelt, Jugend und Familie an den Nationalrat.* Vienna.

Bundesministerium für Wissenschaft und Verkehr, 1997: *Forschungsbericht 1997.* Vienna.

Fischer-Kowalski, M. (ed.), 1988: Öko-Bilanz Österreich. Zustand, Entwicklungen, Strategien. Vienna/Köln: Falter Verlag/Verlag Kiepenheuer & Witsch.

Fischer-Kowalski, M., 1995: Österreich – auf dem Weg zu einer nachhaltigen Entwicklung, in: Altner, G. et al. (eds.), 1995: *Jahrbuch Ökologie 1996.* München: Beck.

Fischer-Kowalski, M.; Payer, H., 1995: Fünfzig Jahre Umgang mit der Natur, in: Sieder, R.; Steinert, H.; Tálos, E. (eds.), 1995: *Österreich 1945 – 1995.* Vienna: Verlag für Gesellschaftskritik, 552–566.

Gerhold, S., 1996: Problemorientierte Umweltindikatoren – Weiterentwicklung und Aktualisierung, *Statistische Nachrichten* 12/1996, 950–966.

5 This was confirmed in an interview conducted with one of the government officials responsible for the NUP in September 1999.

Gerlich, P., 1997: Politische Kultur der Subsysteme, in: Dachs, H. et al. (eds.), 1997: *Handbuch des politischen Systems Österreichs. Die Zweite Republik.* (3rd Edition) Vienna: Manz, 506–513.

Glatz, H., 1995: Österreichische Umweltpolitik. Eine kritische Einschätzung der Instrumente. *Informationen zur Umweltpolitik* 111. Vienna: Kammer für Arbeiter und Angestellte für Wien.

Glatz, H., 1991: Defizite und Perspektiven des Umweltjournalismus, in: Glatz, H.; Kronberger, H. (eds.), 1991: *Umweltjournalismus. Zwischen Resignation und Aufbruch.* Vienna: Uranus, 51–62.

Gottweis, H., 1997a: Biotechnologiepolitik, in: Dachs, H. et al. (eds.), 1997: *Handbuch des politischen Systems Österreichs. Die Zweite Republik.* (3rd Edition) Vienna: Manz, 664–669.

Gottweis, H., 1997b: Neue soziale Bewegungen in Österreich, in: Dachs, H. et al. (eds.), 1997: *Handbuch des politischen Systems Österreichs. Die Zweite Republik.* (3rd Edition) Vienna: Manz, 342–358.

Hogl, K.; Pleschberger, W.; Reiger, H., 1996: Umweltprobleme und Einstellungen lokaler Eliten in Mitteleuropa: Ein Vier-Länder-Vergleich, *Österreichische Zeitschrift für Politikwissenschaft* 3/1996, 319–337.

Huber, M.; Schramm, B., 1996: *Durchführungskriterien der Treibhauspolitik in der Europäischen Union und Österreich.* Studie im Auftrag des Bundesministeriums für Wissenschaft, Verkehr und Kunst. Endbericht. Hamburg/Vienna.

Kanatschnig, D. et al., 1998: *Regionalisierung des Umsetzungsprozesses einer nachhaltigen Entwicklung unter besonderer Berücksichtigung des Nationalen Umweltplanes für Österreich.* Vienna.

Kind, M.; Welan, M., 1996: *Umwelt und Recht in Österreich. Stand 1996.* Diskussionspapier des Instituts für Wirtschaft, Politik und Recht Nr. 55-R-96. Vienna.

Klimabündnis Österreich, 1997: *Klimabündnis-Rundbrief* 1c/97. Vienna.

Köppl, A. et al., 1995: *Makroökonomische und sektorale Auswirkungen einer umweltorientierten Energiebesteuerung in Österreich.* Gutachten des Österreichischen Instituts für Wirtschaftsforschung. Schriftenreihe der Sektion I des Bundesministeriums für Umwelt, Band 24/1995. Vienna

Lauber, V., 1997a: Austria: A Latecomer Who Became Pioneer, in: Andersen, M. S.; Liefferink, D. (eds.), 1997: *European Environmental Policy: the Pioneers.* Manchester: Manchester University Press, 81–117.

Lauber, V., 1997b: Umweltpolitik, in: Dachs, H. et al. (eds.), 1997: *Handbuch des politischen Systems Österreichs. Die Zweite Republik.* (3rd Edition) Vienna: Manz, 609–618.

Matthews, E.; Amann C.; Bringezu, S.; Fischer-Kowalski, M.; Hüttler, W.; Kleijn, R.; Moriguchi, Y.; Ottke, C.; Rodenburg, E.; Rogich, D.; Schandl, H.; Schütz, H.; van der Voet, E.; Weisz, H., 2000: *The Weight of Nations: Material Outflows from Industrial Economies.* Washington D.C.: World Resources Institute.

OECD, 1995: *OECD-Environmental Performance Reviews: Austria.* Paris: OECD

OGM, 1993: *Grundlagenstudie Greenpeace.* Zusammenfassung. Unpublished. Vienna.

Österreichische Bundesregierung (ed.), 1995: *NUP Nationaler Umwelt Plan.* Vienna.

Payer, H., 1997: Der Nationale Umweltplan (NUP) für Österreich, in: Jänicke, M. et al., 1997: *Nationale Umweltpläne in ausgewählten Industrieländern.* Berlin/Heidelberg/New York: Springer, 121–139.

Plasser, F.; Ulram, P. A., 1990: Bruder Baum und Vater Staat: Meinungsforschungsergebnisse über die Einstellungen zur Umweltpolitik in Österreich, in: Riegler, J. (ed.), *Antworten für die Zukunft. Ökosoziale Marktwirtschaft.* Vienna: Verlag für Geschichte und Politik, 31–42.

Pleschberger, W., 1995: Environmental Concern of the Local Elite, *The Annals of The American Academy of Political and Social Science,* 540, July 1995, 90–104.

Simhandl, G., 1996: Umweltbedingungen, Umweltverhalten. Ergebnisse des Mikrozensus Dezember 1994, *Statistische Nachrichten* 8/1996, 624–639.

Steurer, 1999: Klimaschutzpolitik in Österreich. Bilanz der 1990er Jahre und Ausblick, *SWS-Rundschau* 3/1999, 197–219.

SWS-Rundschau, 1994: SWS-Bildstatistiken: Umweltbewußtsein in Österreich, *SWS-Rundschau* 1/1994, 57–64.

Tálos, E., 1997: Sozialpartnerschaft. Kooperation – Konzertierung – politische Regulierung, in: Dachs, H. et al. (eds.), 1997: *Handbuch des politischen Systems Österreichs. Die Zweite Republik.* (3rd Edition) Vienna: Manz, 432–451.

Umweltbundesamt; Bundesarbeiterkammer (eds.), 1996: *Umwelt aktiv. Zahlen • Daten • Fakten zur österreichischen Umwelt.* Vienna.

BRAZIL*

Kathryn Hochstetler

In terms of the categories used to classify the world's political systems, Brazil usually occupies a middle position. Economically, for example, Brazil is often classified as a newly industrializing country, between the wealthier, increasingly postindustrial countries and the poorer agrarian economies. Politically, Brazil is a new liberal democracy, having returned to civilian rule in 1985 after a period of military rule. As such it shares some characteristics with the established liberal democracies, but also with less liberal regions. The key to understanding Brazil, however, is to realize that it reaches these middle positions by embodying opposite extremes simultaneously, and not by actually occupying a single, average position. Brazilians themselves sometimes call their country "Belgia," half Belgium and half India, in one country. Thus, Brazil's per capita income, which places it between the richest and poorest countries of the world, hides the fact that many Brazilians are among the world's poorest while a few are among the richest. The newly industrializing economy is actually a dual economy, which has extensive industrial enterprises – and the largest remaining tropical forests in the world. The new liberal democracy carries substantial authoritarian legacies from the previous regime, which itself had liberal elements.

All of these dualities carry consequences for Brazil's environmental capacity, which, not surprisingly, also displays contradictory characteristics. In this domain, the dualities often appear as formal environmental capacity which is underutilized in actual decision making affecting the environment.

1 Brazil's Main Environmental Problems

Full documentation of Brazil's environmental problems could fill volumes. Brazil's large geographical size and diversity are overlaid with equally diverse political and economic practices. The end result is a matrix of environmental degradation that extends from the degradation of poverty to the degradation of wealth, with multiple actors, causes, and dynamics.

* Early stages of the research for this chapter in 1989-1991 were funded by the Institute for the Study of World Politics, the Midwest Universities' Consortium on International Affairs, and a MacArthur Pre-Dissertation Grant. Research in 1996 was supported by the Professional Development Fund of the College of Liberal Arts, Colorado State University. I would like to thank Helge Jörgens for helpful comments on an earlier draft of the manuscript. Any remaining errors of fact and interpretation are, of course, my own.

1.1 Deforestation, Biodiversity, and Land-Use Change

Deforestation is a very old activity in Brazil, dating back 3900 years (Dean 1995: 25). Deforestation accelerated after the arrival of the Portuguese and other Europeans, who largely cleared Brazil's Atlantic coastal forest, leaving scattered remnants in nine states that sum to only about 5% of the presumed original forest (Republic of Brazil, 1997: Ch. 11).[1] Despite the international fame of the Amazon rainforest, many Brazilians are more concerned with the fate of the Atlantic forest. Rates of deforestation are much higher than they are in the Amazon, with 11% of the tiny remaining Atlantic forest destroyed in 1985–1995. Recent years have also seen an increase in efforts to protect the remaining Atlantic forest, with the area under some form of preservation nearly quintupling from 1981 to1990 (Dean, 1995: 338) and a national decree law (#750, of 1993) limiting nearly all cutting of specific kinds of native forests within the domain of the Atlantic Forest (Hodge, Queiroz, and Reis, 1997: 341). The continuing deforestation shows the limited power of the new legislation. Hodge, Queiroz, and Reis even argue that the strict new laws have led to quicker deforestation, since they do not consider the needs of small farmers.

Deforestation is also a significant problem in the Amazon rainforest, as is well known (Foresta, 1991; Kolk, 1996). There has been considerable controversy over the exact rate of deforestation in the Amazon (Kolk, 1996: 61–68), but none of the proffered figures deny several related facts. First, the quantity of land being deforested annually over the last 35 years is huge, and shows no overall decline, although individual years are higher and lower than others. A recent study reports that an average of 19,610 km^2 of the Amazon forest were cleared annually from 1993–1995, and 18,140 km^2 were cleared in 1996 (Nepstad et al., 1999: 506). The same study cautions that deforestation figures underestimate the level of forest degradation by nearly half, given the forest impoverishment that comes from logging and fire processes that do not fully clear ground (Nepstad et al., 1999: 505). The second consensual fact about the Amazon, however, is that proportionally huge quantities of forest remain. However measured, Brazil is a global forest giant: in 1990 Brazil had more than three times the total rainforest area of Indonesia, which had the next largest forest area (Kolk, 1996: 67), and it still had 59% of its total massive land area in forests and woodlands (UNEP, 1993: 157). Finally, all observers agree that the fate of Brazil's forests is key to maintaining biodiversity, both in Brazil and globally. Brazil is one of a handful of countries classified as "megadiverse", and much of that diversity is found in its extensive forests.

As an environmental problem, deforestation has features that make it both easy and hard to solve. Deforestation is a prototype of graphic, urgent, media-friendly environmental problems. The sweeping flames of forest fires, exotic faces of human and animal inhabitants, and stark devastation of a clear-cut forest all easily

1 This report provides a summary of Brazil's progress on the various commitments of Agenda 21, the final action document of the United Nations Conference on Environment and Development, held in Rio de Janeiro in June, 1992. It is organized by the chapters of Agenda 21. The Brazilian report is incomplete in many respects, but is the best summary of the government's current assessment of environmental issues there. Non-governmental organizations (NGOs) provided critical commentary in their own report (Brazilian NGO Forum, 1997), although it is less detailed.

make the point that deforestation is a problem requiring immediate action (Hecht and Cockburn, 1990; Lewis, 1990). They also make the less tangible impacts of deforestation on long term climate and diversity plausible. In a 1997 survey, 45% of the Brazilian population recognized deforestation as an environmental problem in Brazil, more than any other environmental problem.[2]

Yet deforestation continues because other features of the problem make it difficult to solve. Most importantly, this is not a problem with a simple technical solution. Much analysis has rightly focused on the ways that the Brazilian government encouraged and even required deforestation through the 1980s (Mahar, 1989), but a recent study argues that subsequent deforestation has been driven "mostly by local economic forces" (Anderson, 1996: 325).

Ultimately, the deforestation issue is connected to larger debates about appropriate land use. Real estate speculation, agricultural expansion, and agrarian reform have fueled the recent high rates of deforestation in the Atlantic states (Capobianco, 1998), and similar causes are at work in the Amazon deforestation. Brazil's problems with defining appropriate uses of its many kinds of natural ecosystems do not stem from any quantitative measure of overpopulation for its landmass. Instead, historically highly unequal patterns of land distribution contribute to conflicts over land and the increasing economic use of vulnerable ecosystems. Since 1990, the federal government has been promoting a nation-wide economic/ecological zoning project (ZEE) that seeks to provide technical environmental criteria to guide a process that is still largely political and economic.

1.2 Air Pollution

The problem of air pollution showcases the other, urban, side of Brazil's dual economy. Brazil's large and medium sized cities suffer from poor air quality. Table 1 demonstrates that most kinds of atmospheric emissions continued to grow through the 1980s and 1990s, from various sources.[3]

For half a century, the Brazilian state led an ambitious program of industrialization, which transformed the Brazilian economic structure from a rural, primary-production economy to the eleventh largest economy in the world, with a strong base in industrial manufacturing. One measure of the failure of this model, however, is its lasting legacy of industrial pollution. Guimarães notes that "the massive late industrialization of the Brazilian economy brought in technological patterns which were advanced for the national [economic] base but were outmoded in terms of the environment" (Guimarães, 1992: 50). The environmental impact of this imported technology was greater because of the spatial concentration of industrial production in Brazil. The Southeast Region, which includes São Paulo and

2 Results of this survey "O Que o Brasil Pensa Sobre o Meio Ambiente, Desenvolvimento e Sustentabilidade" are available at http://www.mma.gov.br/port/SE/pesquisa. The Museu de Astronomia (MAST) and the NGO ISER (Institute for the Study of Religion) conducted the survey in 1992 and 1997, with support from Brazilian national environmental agencies. All survey results cited in this chapter come from this source, and are not individually attributed to it.

3 As discussed in the section on deforestation, there is considerable controversy over actual deforestation rates. Most sources do not measure the steady decline that the Brazilian government reports here. See Kolk (1996: 61-68).

Table 1. Emissions and their main sources

Emissions and their main sources	1980	1990	1995
Carbon dioxide (million tons) – deforestation	1400	969	850
Sulfur dioxide (million tons) – industry	1.5	2.5	3.0
Nitrogen dioxide (million tons) – vehicles	5.0	6.0	6.5
Consumption of ozone depleting substances (tons)	.018	.016	.020

Source: Republic of Brazil, 1997: Ch. 9.

Rio de Janeiro, dominates Brazil's industrial sector, although there has been some deconcentration of industry since 1970 (Martine and Diniz, 1997). As a consequence, much of the industrial pollution is also concentrated in the Southeast, with 41% of the potentially polluting industries located there (Guimarães, 1992: 51).

The city of Cubatão, near the São Paulo coast, was the most extreme version of the historic disregard for the impact of industrial production on the environment. A 1983 analysis counted 320 sources of pollution in the region, and traced many of them to the petrochemical fertilizer, iron, and steel industries, which all had production facilities there (Government of São Paulo, 1987: 88–89). A year later, Cubatão was also the site of Brazil's worst industrial accident, when as many as 500 people died after a gasoline spill by Petrobras, the state-owned petroleum company (Government of São Paulo, 1987: 92). Following extensive mobilizations by residents and civil society, the state environmental agency, CETESB, charged over 500 fines for pollution and eventually managed to have 288 of 320 of the biggest pollution sources controlled (Lemos, 1998; Romanini, 1995). Low-cost loans, partially funded by the World Bank, paid for facilities for treating effluents, controlling emissions, and disposing of solid wastes (Redwood III, 1993: 46–47).

Nevertheless, the newsletter *Environment Watch Latin America* reported in November, 1995, that high air pollution levels had returned in Cubatão, after a decade of improvement. Acid rain and toxic waste dumps are other legacies of pollution there (Romanini, 1995).

Table 2. Reductions Achieved for Major Pollutants Emitted in Cubatão (in metric tons per day)

Type of Pollution	Emissions 1984	Emissions 1992	% Reduction
Particulate material	236.6	49.7	79.0
Sulfur dioxide	78.4	49.5	37.0
Nitrogen dioxide	61.1	52.7	14.0
Fluorides	2.6	1.1	58.0
Ammonia	8.7	2.6	70.0
Hydrocarbons	90.0	27.5	69.0

Source: Lemos (1998: Table 1), from data compiled by CETESB, the São Paulo state environmental protection agency.

A survey of industrialists in São Paulo and Minas Gerais states concluded that many industries continue to be largely reactive on pollution issues. If there is strong pressure from economic agencies, environmentalists, or consumers, they will try to solve industrial pollution at the end of the production process (Gutberlet and Segura, 1996/97: 26–27). Thus, although this problem has some technical solutions, they are inconsistently pursued. Only a very small subsection of the Brazilian population is engaged in more profound rethinking of the industrial model itself. In fact, only 12% of the Brazilian population recognized area pollution as an environmental problem in 1997, and only 5% traced pollution to an industrial source.

In the 1990s, the continuing poor quality of the air in Brazil's large cities is increasingly attributed to the transportation sector rather than to industry (Earthwatch, 1992: 186–194). The number of motor vehicles in use rose sharply between 1980, with 48.1 per 1000 inhabitants, to 90.4 in 1990/91 (Republic of Brazil, 1997: Ch. 4). A 1993 law sets limits on auto emissions, and eight metropolitan regions regularly monitor their air quality. Since 1994, São Paulo's State Secretary of the Environment has instituted an annual *rodizio,* or rotation, of cars to try to control traffic during the winter months when thermal inversions are common. Despite such measures, vehicle-based emissions continue to rise (see Table 1).

1.3 Waste and Sewage Disposal

For poor Brazilians asked to prioritize their public service needs, lack of connection to the sewage system is one of the two most important issues (Jacobi, 1993: 49). According to the Brazilian government's own assessment of the provision of basic sanitation services, "Of the 113 million people who live in urban Brazil today, 75 million have no sewage treatment, 20 million do not have running water and 60 million do not have their trash collected ... (Republic of Brazil, 1995: 4.05)."[4] Not surprisingly, access to basic sanitation services is highly segmented by class, according to the same report. As Gilbert states bluntly, "Governments have given priority to the servicing of areas that could afford to pay for them (Gilbert, 1996: 94)." Over the last five decades, all sources show steadily but slowly increasing sanitary coverage of the Brazilian population, however.

Ironically, greater success at collecting sewage and trash has simply moved the location of the urban sanitation problems to less visible locations. Generally speaking, Brazil's city administrations have done better at collecting household wastes than at handling them once they are collected. The Brazilian Association of Sanitary Engineering (ABES) reports that "90% of the sewage collected receives no treatment, 60% of the trash collected is dumped in open pits, [and] 99% of municipalities have no sewage treatment system whatsoever ... (cited in Republic of Brazil, 1995: 3.20)." A few cities, such as Porto Alegre, Curitiba, and Belo Horizonte have experimented with strategies like recycling and bioremediation, but São Paulo has focused on incineration as its strategy (Caccia Bava and Mullahy, 1995; Teixeira, 1996/97), with predictably negative effects on air quality.

4 The sanitary infrastructure for rural Brazilians is even less adequate, but less pressing since the population is more dispersed and more likely to have individualized solutions like septic systems and wells.

Governments at the municipal level continue to hold most of the responsibility for decision-making on and implementation of solid waste policies. In 1996, all three levels of government began to coordinate several proposed pilot programs on solid waste management and minimization. As of 1997, hazardous wastes were still handled by a patchwork of regulations and practices at all three levels, under a series of directives which the National Council on the Environment (Conama) has issued since 1988 (Republic of Brazil, 1997: Chs. 20, 21). The regulations followed Brazil's worst nuclear accident, in 1987, when five people were killed and dozens injured after they handled radioactive hospital waste (cesium 137).

Brazil's urban environmental infrastructural problems are exactly opposite in nature to its deforestation problems. While some of the urban problems are visible to anyone who walks down a street in one of Brazil's large cities, they are not urgent and eye catching to the national and international media. Only 4% of Brazilians recognized these as environmental problems in 1997. On the other hand, many of these problems do have readily available technical solutions, already long in place in developed countries (Jänicke, 1997: 8). The limiting factor for Brazil is the financial cost of these solutions. The Brazilian government estimates that it would need to spend US$100 billion to make up the basic infrastructural deficit, and then 4.5% of the GNP annually to maintain it (Republic of Brazil, 1995: 4.96).

1.4 Water and Coastline

Finally, all of the environmental problems outlined above inevitably affect the quality and availability of water resources. In 1990, potable water was piped to 79.5% of the population, while 18.5% had waste water piping inside their houses (Republic of Brazil, 1997: Ch. 17). That water has become increasingly unreliable, however. While the country as a whole has huge amounts of freshwater, 70% of it is in the less-populated Amazon region, while the South and Northeast suffer from water scarcity. As much as 40% of the piped water across the country disappears in "losses and unaccountable uses," creating a need for upgrading and streamlining water provision services (Republic of Brazil, 1997: Ch. 18). In addition, the quality of the water provided is in question. Rio's State Secretary of Sanitation recently examined hospital records, and claimed that 60% of hospital stays might be the result of diseases caused by poor water quality (*Jornal do Brasil* 4 June 1999). This may help explain why Brazilians were more likely to recognize water pollution as an environmental problem, with 26% doing so in 1997.

Land-based sources of pollution also damage the environmental quality of Brazil's long Atlantic Ocean coastline. Poor sanitation services mean that raw sewage eventually finds its way into the ocean, especially with the population concentrated along the coast. At least eight major industrial districts are also located along the coast, which further contribute to environmental degradation. By the Brazilian government's own admission, nearly all the activities that would help to sustainably manage its coastlines are "poorly covered" by current policies, except for a few integrative planning activities (Republic of Brazil, 1997: Ch. 17).

2 Environmental Administration and Policy Development

Environmental policy has developed over almost 70 years in Brazil, and is the product of many different kinds of policy-making processes. A clear conclusion must be that Brazil does not lack environmental policy tools, as it has consistently developed new institutions and regulations in pace with many countries which are recognized as environmental leaders. Because of space limitations, this chapter discusses only national policies and institutions. Adding sub-national layers would further multiply the plenitude of environmental policies and institutions, although they are quite unevenly distributed across Brazil's states and municipalities (Ames and Keck, 1997–98; Zulauf, 1994).

2.1 Institutional Development: 1973–1998

Table 3 shows the evolution of Brazil's federal environmental institutions. The most noteworthy characteristics are the many reformulations of the institutions, as well as their increasing formal status and autonomy.

Table 3. Federal Environmental Institutions in Brazil since 1970

Year	Content
1973	Creates the *Special Secretariat of the Environment* (SEMA), for pollution abatement and natural resource management, within the Ministry of the Interior.
1981	Creates the *National System of the Environment*, integrating and reorganizing existing environmental institutions. The *National Council on the Environment* (Conama) is the primary deliberative agency, whose resolutions have the force of law. SEMA is now its technical staff.
1985	Creates a *Ministry for Urban Development and Environment*, and transfers Conama and SEMA to it, continuing their previous functions.
1989	Eliminates SEMA and other smaller environmental agencies; replaces them with a new environmental agency, the *Brazilian Institute for the Environment and Renewable Resources* (Ibama).
1989	Reorganizes existing environmental institutions, and adds a new *Superior Council of the Environment* as the top environmental decision-maker, above the Conama, with Ibama just below it, as the enforcement agency, as it continues to be in later reorganizations.
1990	Reorganizes existing environment institutions, creating a *Secretary of the Environment* (SEMAM) with four major departments. SEMAM is an autonomous agency below ministerial level, whose Secretary reports directly to the President.
1992	Transforms the Secretary of the Environment into a *Ministry of the Environment*, and reorganizes its competencies.
1993	Reorganizes the Ministry of the Environment as the *Ministry of the Environment and Legal Amazônia.*
1995	Reorganizes the Ministry of the Environment and Legal Amazônia as the *Ministry of the Environment, Water Resources, and Legal Amazônia.*
1997	Creates an additional Committee, the *Committee on Sustainable Development Policies and the Brazilian Agenda 21* (CPDS) to oversee the Brazilian Agenda 21.
1999	Renames the ministry the *Ministry of the Environment.* There is now a Secretariat for Amazon Coordination within the Ministry.

Sources: Guimarães, (1995); Fernandes, (1995); http://www.mma.gov.br/.

The military government which ruled Brazil from 1964 to 1985 created Brazil's first environmental peak organization a year after the 1972 Stockholm Conference on the Human Environment.[5] This was the Special Secretariat of the Environment (SEMA), located as a division within the Ministry of the Interior. SEMA had quite a bit of autonomy during its first decade, but it was an autonomy born of indifference. The Ministry of the Interior was focused on large public works projects during this time, and SEMA was far outside its circles of power. SEMA had to compete with other divisions within the ministry for funding and personnel, and fared poorly: one estimate sets Brazilian spending on the environment during this period as a mere 0.065% of GNP (Guimarães, 1995: 167,185,191). The three employees and two rooms SEMA had at its creation (Woreman, 1990: 46) were a cheap institutional "solution" to Brazil's environmental problems. SEMA was originally created to deal with pollution issues (Guimarães, 1995: 160–161), but also took up natural resource conservation (Foresta, 1991: 115). In both issue areas, SEMA was ostensibly a policy-making agency but retreated to a technical and scientific role to protect itself from the vagaries of authoritarian rule (Foresta, 1991: 122).

In 1981, Law 6.938/81 introduced an ambitious effort to coordinate and reorganize the existing environmental institutions and legislation into a National System of the Environment. In this reorganization, SEMA officially became a technical rather than policy-making agency, under a new National Council on the Environment (Conama) which included representatives of different social sectors in addition to government ministries. Conama was a consultative and deliberative body, whose resolutions had the force of law (Zulauf, 1994: 7). It reported directly to the national president. Although SEMA remained within the Ministry of the Interior, its new relationship with Conama helped to link it to other government agencies in new and more authoritative ways (Foresta, 1991: 126–127). Together, SEMA and Conama established and monitored new norms and licensing procedures for activities with potential environmental impact.

To this point, the Brazilian environmental policy making process still showed a technical orientation to the environmental issue, with very little development of the participative capacity. Technocrats in government agencies, state-run corporations, and private corporations interacted with each other, but rarely with outsiders who did not share their technical and scientific expertise. In this regard, the environmental bureaucracy was simply an extension of the generally technocratic orientation of the Brazilian military (Guimarães, 1995: Ch. 5).

A third stage of environmental decision making began under the new civilian regime in 1985, with a series of policy innovations that broadened the range of participants in environmental policy making. In 1985, Conama and SEMA were moved to a new Ministry for Urban Development and Environment. While the military government had chosen the representatives of environmental organizations in Conama's first years, environmentalists were now allowed to choose their

---

5 Many sources credit the timing of this first agency to the international conference, but Guimarães (1995:160-161) traces its exact timing to a foul odor from a wood-pulp plant in the home state of the president's chief of staff.

own representatives, and their numbers increased from three to five.[6] In its new institutional home, Conama blossomed, and issued some of its most important directives (see discussion below). SEMA, on the other hand, felt increasingly marginalized, and long-time director Paulo Nogueira Neto resigned shortly after the reorganization (Guimarães, 1995: 200).

Brazilian environmental arrangements might usefully have slowed their evolution at this point, but exactly the opposite happened. In the late 1980s, the environment suddenly moved out of the restricted realm of scientists, technocrats, and environmentalists. By linking deforestation of the Amazon to Brazil's need for international financing, international environmentalists and their allies in the Brazilian Amazon made environmental degradation a major public policy issue (Keck, 1995; Kolk, 1996). While the first Brazilian governmental reaction was hostile rejection of the so-called "internationalization" of the Amazon (Hurrell, 1992), successive Brazilian governments have generally cooperated with the new environmental demands since 1989 (Kolk, 1998). One side of this is a number of specific laws and decrees designed to protect the Amazon and biodiversity. The Brazilian government counted seven major pieces of legislation on such issues, which it passed between 1993 and 1996 (Republic of Brazil, 1997: 42).

The institutional side of the new attention to the environment showed a similar frenzy of activity, although its impact is less salutary. Between 1989 and 1995, the main environmental agencies at the federal level were shuffled six times, or, on average, annually (see Table 3). Virtually all of these new institutions included important integrative and participative capacities on paper, but the rapid turnover prevented actual use of them. For the last three years, Brazil has had a Ministry of the Environment, Water Resources, and the Amazon, but competencies and lines of accountability are still being sorted out. For example, the Ibama (Brazilian Institute for the Environment and Natural Resources) was officially the primary federal agency charged with environmental enforcement, but until recently found itself developing legislation as well, with the Ministry of the Environment virtually one of its appendices instead of the intended reverse.[7] One important shared characteristic of all of the post-1989 environmental agencies is that they continued to report directly to the national president, even before they were given ministerial-level status. This enhanced their autonomy and status.

2.2 Environmental Policy Instruments

Environmental ideas have a long history in Brazil (Pádua, 1991), and environmental legislation also came early there. The first environmental laws were two codes decreed in 1934: the Code on Forests and the Code on Water. In the 1960s, additional laws introduced some conservation concerns to the public administration of the private use of natural resources, including wildlife, forests, water, minerals (Fernandes, 1995).

---

6 Author interview with Celso Petrillo of the São Paulo environmental federation APADEMA, then the elected representative of environmental organizations in Southeast Brazil to the CONAMA, São Paulo, 22 October 1990.

7 Interview with then new Minister of the Environment, José Sarney Filho, published in *Parabólicas*, No. 47, January/February 1999, http://www.socioambiental.org/parabolicas.

Table 4. Selected List of Major Brazilian Environmental Legislation after 1970

Year	Common Name/Area of Coverage
1975	industrial pollution
1977	nuclear installations
1977	water quality norms
1979	regulates National Parks
1979	zoning of urban land for reasons including environmental
1980	industrial zoning in high-pollution areas
1980	agricultural chemicals
1981	Ecological Research Stations and Areas of Environmental Protection
1981	National Environmental Policy and National Environmental System
1985	Law of Diffuse Interests. Allows public civil action suits for environmental degradation.
1986	studies (EIAs) and reports (RIMAs) of environmental impact
1988	Federal Constitution: Article 225 is a chapter on the environment.
1988	National Plan of Coastal Administration
1989	National Environmental Fund
1989	criminal penalties for environmental infraction
1993	air pollution from automobiles
1994	National Biodiversity Program (Pronabio)
1994	National Integrated Policy for the Amazon Region
1994	ecotourism
1995	biosafety law
1995	Green Protocol
1995	ozone regulation
1996	desertification
1997	National Policy of Water Resources; National System of Water Resource Management
1998	Environmental Crimes Law
1999	National Policy of Environmental Education

Sources: Wainer (1991); Republic of Brazil (1991, 1997); Conama (1991); http://www.mma.gov.br/

After 1970, the Brazilian government began to write quite a bit of legislation specifically on environmental issues. Some of the more important examples are summarized in Table 4, which is only a partial list, and again includes only national-level legislation.

Interestingly, this legislation began during the military regime, which is usually considered to be quite hostile to environmental concerns. The regime is well-known for its strong emphasis on industrial development (Guimarães, 1995) and determined occupation of the Amazon rainforest (Mahar, 1989; Zirker and Henberg, 1994). Yet the military regime also issued 19 federal laws or decree laws and 20 decrees on the environment during its years in power (Republic of Brazil, 1991). Unlike the Chilean military, which had a strong ideological bias against a large state role, which restricted its environmental capacity building (Silva, 1997), the Brazilian military government generally increased the presence of the state in society.

After the transition to civilian rule in 1985, the new civilian governments continued to write additional environmental legislation, including legislation which brought additional participants into the environmental decision-making process. The Law of Diffuse Interests gave legal standing to collective actors, including environmental organizations, on issues where interests are broadly distributed across society (Fernandes, 1995; Fuks, 1994). In the next year, Conama issued what is usually considered its most important contribution to environmental legislation in Brazil, Resolution 001/86, which established the basic criteria and processes for evaluations and reports of environmental impact. In keeping with the increasing openness of the policy making process, Article 11 required that the reports be made public, and subject to public hearings. Finally, in 1988, the new Brazilian federal constitution collected many of these policy advances in a chapter on the environment (Article 225). It guaranteed Brazilians a healthy and sound environment and committed the collectivity as well as the government to this end. The constitution did not add many innovations, nor did it need to; Brazilian environmental legislation was essentially complete even before the constitutional process, comprising "a wide range of flexible and interesting legal instruments, which were sufficient to give legal support to public policies and private actions on the matter of environmental protection (Fernandes, 1995: 92–93)."

Through 1994, most of the legislation written by both military and civilian governments shared certain characteristics. First, it was primarily command-and-control regulation, which introduced increasingly more stringent government guidelines for economic activity. Second, many regulations tended to set quite general standards of environmental quality, although Conama's resolutions also specified precise emissions limits or conditions that potential polluters must meet (compare Conama, 1991; Republic of Brazil, 1991). Finally, most of the legislation is sectoral, addressing one specific environmental problem. The exceptions to this last generalization are the 1981 National Environmental Policy and the 1988 constitution, but they simply list a number of guiding environmental principles.

With the Cardoso presidency which began in 1994, Brazil may be entering a new stage where policy-makers are once again trying to sort, reconsider, and consolidate the floods of new legislation produced over the last two decades. Several broad pieces of recent legislation indicate that its overall impact on environmental capacity will be mixed. Former environmental congressman Fábio Feldmann has suggested that the environmental gains of the 1980s, and the 1988 constitution in particular, reflected the typical congressional representative's ignorance of the importance of environmental issues.[8] Environmental legislation now faces much more contentious political debates, and the results are not always propitious for the environment. For example, in 1997, the Conama revisited its 1986 resolution on environmental impact assessment, and reduced the requirements, timetables, and costs for receiving an environmental permit, under pressure from industrialists (Campanilli and Leitão, 1998). An extensive new Environmental Crimes Law passed Congress in 1998. It establishes criminal penalties (fines and imprisonment) for environmental degradation and had strong support from environmental organizations. While this appears to be a clear gain for environmental capacity, President Cardoso weakened the law immediately by signing it with ten vetoes, under pressure from industrialists, the conservative rural delegation, and evangelical Christians *(Folha de São Paulo* 13 February, 1998).[9] In August, 1998, Cardoso

---

8 Fábio Feldmann at a campaign forum in Campinas, São Paulo, 30 August 1990.

9 While the interests of industrialists and agriculturalists should be obvious, those of evangelicals may be less clear. They asked Cardoso to veto the bill's provisions on noise pollution, which he did.

used a provisional measure to give businesses up to ten years to come into compliance with the new legislation. Environmentalists then erupted, and Cardoso hastily edited the measure to cut the maximal extension to seven years.

Perhaps in response to the increasing complexity of the politics of the environment, the Cardoso administration has also experimented with two new kinds of environmental directives. First, the government introduced economic instruments as incentives for environmental protection, in tandem with its generally greater focus on market mechanisms relative to previous administrations. State governments led the way first, but the national government followed with the Green Protocol in 1995. The Green Protocol aims to prevent environmental degradation by linking official government financing "to those projects with major characteristics of self-sustainability and that cause smallest damages to the environment (Ministério de Meio Ambiente, 1995: 13)." At its creation, the protocol applied to the major national banks and to the regional banks for the Amazon and Northeast regions, but the government aims to include state-level public banks as well as private lenders eventually. In addition to the Green Protocol, clauses in other legislation introduce most kinds of economic incentives ("subsidized credits and incentives, taxes on solid wastes, pollution, natural resource use, eco-labelling [sic], and container deposit-return schemes") except for tradable permits (Republic of Brazil, 1997: Ch. 8).

Second, an effort to achieve holistic environmental planning of an integrated kind is underway. It is worth noting that this is the second effort of its kind. The 1992 United Nations Conference on Environment and Development in Rio de Janeiro called for such a national strategy for sustainable development in its action document, the Agenda 21. The Franco administration duly created an Interministerial Council on Sustainable Development in 1994, but it was never installed. Instead, shortly before the 1997 Rio+5 Conference, a presidential decree created the Commission of Sustainable Development Policies and the Brazilian Agenda 21 (CPDS).[10] The Ministry of the Environment oversees this committee, which includes representatives of four other ministries and a secretariat, as well as five civil society representatives and other participants. NGOs and experts have written draft texts on six themes: sustainable cities, sustainable agriculture, infrastructure and integration, natural resource management, reduction of social inequity, and scientific and technical development. As of this writing (June, 1999), the CPDS is soliciting responses from the public before completing the final version of the Brazilian Agenda 21 – probably in the first year of that century. This would be the first document to set out a detailed and integrated plan for sustainable development in Brazil. The slow pace of this effort may help explain why only 5% of Brazilians had heard of Agenda 21 in 1997, while 41% remembered the Rio conference.

10 Information about the CPDS can be found on the webpage of the Ministry of the Environment, http://www.mma.gov.br/.

3 Evaluation of Environmental Policy and Institutions

This section outlines some of the major factors which prevent full utilization of the environmental capacity which exists on paper in Brazil. Because of the rapid multiplication of policies and institutions, this section does not focus on specific pieces of legislation, but on the overall pattern of underutilization of Brazil's emerging environmental capacity. It concludes by proposing that the rapid and ambitious pace of policy and institutional change may itself prevent better utilization of the policies.

3.1 Enforcement

Enforcement of environmental legislation in Brazil is a weak link that explains much of the gap between Brazil's environmental capacity and its actual environmental performance. The size of the gap varies for different environmental issues, however. Enforcement gaps are, overall, largest in rural and forested areas, while enforcement can be more effective for urban industrial projects (Ministério de Meio Ambiente, 1995: 14).

In rural areas, more than six decades of forest policy and conservation efforts have largely failed to stem continuing deforestation and degradation of Brazil's remaining tropical forests (see discussion above). This failure stems in part from inadequate management of the 4.5% of Brazilian land in national parks and reserves. A recent study by the World Wildlife Fund charges that only 8% of those areas meet reasonable maintenance standards, while 87% lack management plans and 73% need more personnel (Kepp, 1999).[11] While the Environment Ministry and its associated agencies responded that the report does not consider recent initiatives (Kepp, 1999), past structural problems with enforcing environmental protection in national parks and reserves are still unresolved. Many of these were created in the 1970s, when the Brazilian Institute for Forest Development (IBDF) had set aside whole ecosystems for conservation, blithely ignoring the presence of indigenous peoples, farmers, and other inholders (Foresta, 1991: 115–120). These often contributed to forest degradation from within the reserved areas, while even otherwise successful areas of conservation are now "protected areas enclaves on hostile landscapes" (Foresta, 1991: 189).

Outside officially protected rural areas, forest and land degradation continue apace. Until the late 1980s, government policies officially encouraged deforestation (Mahar, 1989). Since then, the relatively distant rural lands continue to be largely outside Brasilia's reach. For example, in a report on two recent Amazonian dam projects, the authors generalize that national requirements for environmental assessment, public hearings, ecological zoning, and constitutional protections all fall in the face of projects that gain a high local political priority in the region (Fearnside and Barbosa, 1996a; Fearnside and Barbosa, 1996b). Another study of the soybean and pulp and paper industries concurs, noting that initial reviews of new projects are followed by virtually no monitoring activity, despite an abundance of applicable laws and procedures (May and Vinha, 1998: 73). Recent im-

11 See also http://www.wwf.org.br.

provements of environmental performance in these industries seem to depend on market rather than government incentives (Gentry, 1998: 64).

In urban industrial centers, the case of Cubatão shows that government-led environmental improvement is possible. The improvement there depended on highly focused attention from the government, the public, and the media (Lemos, 1998), as well as financial support for basic environmental improvements. When these disappeared in the 1990s, environmental emissions began to creep upwards again. The broader pattern of environmental quality shows growing emissions of various kinds, despite the ever-larger body of environmental legislation. At least until recently, businesses could assume that enforcement and fines for environmental infractions would usually be minimal. Before the 1998 Environmental Crimes Law, fines topped out at about US$5000. Even so, fines were often not paid: 11,000 firms had unpaid environmental fines in late 1996 (May and Vinha, 1998: 71). Now potential fines reach US$50 million, and many loopholes to their enforcement have been closed. Whether new legislation like the Environmental Crimes Law and the 1995 Green Protocol work to provide negative and positive incentives for environmental protection depends on whether these laws are better-enforced than their predecessors.

As should be clear, broader patterns of legal looseness underlie specific problems with the lack of environmental enforcement. Political influence and corruption lie behind many failures of government agencies to enforce their own laws, and the Environmental Crimes Law now includes sanctions for such failures as a result. The many environmental laws also sometimes left important loopholes in their coverage. For example, the Ibama tried repeatedly to punish steel producers for burning Amazon trees as charcoal in their factories, but lost every time due to a legislative gap that allowed this kind of deforestation (Hinchberger, 1998: 3). Laws that are otherwise complete often languish for years without full regulation. Thus, the 1988 constitution provided for the designation of environmental crimes, but the Environmental Crimes Law came a decade later; the 1981 National System of the Environment created new kinds of conservation units – ecological stations and areas of environmental protection – but they were not regulated until 1990; and the 1997 National Policy of Water Resources cannot be regulated and implemented because the Ministry of the Environment and the National Agency of Electric Energy are debating who controls water resources *(Jornal do Brasil* 2 May 1999). Finally, inadequate budgets and personnel also limit the ability of government agencies to enforce their own legislation. Total spending for Ibama, the primary federal environmental enforcement agency, was only US$77.4 million in 1998 (Kepp, 1999: 4).

As in many countries, the background reason for the failure of Brazil's environmental legislation is that its wealth of environmental legislation and institutions exists side by side with policies and institutions designed for economic growth – and the former are virtually always trumped by the latter. A major justification for the recent Environmental Crimes Law was that economic actors enjoyed virtual impunity for environmental degradation, despite several decades of environmental legislating. Fernandes (1995: 103) even suggests that Brazil's extensive environmental legislation exists to be "ideological manipulation" for a state which passes the laws, but then subverts them for economic interests. Guimarães (1995: 184–5) supports this position, pointing out that neither a lack of legislation nor the absence of planning can be blamed for Brazil's environmental problems. Environmental units are attached to virtually every public enterprise or project, and they are present for decisions on issues like nuclear energy or mining, but they cannot claim to have much impact on the decisions: "Never have major revisions been promoted or projected because of their work" (Guimarães, 1995: 185, 124–5).

3.2 Financing

Another facet of the priority placed on macroeconomic policy making over environmental issues can be seen in the budgets of these same environmental agencies. For the last 15 years, economic stabilization policies have dominated Brazil's macroeconomic policy making, with policy makers attempting nine different plans from 1985 to 1992 (Lopes, 1996: 198) before the tenth, the Plano Real, finally brought inflation under control (Flynn, 1996). During these years, environmental budgets were an easy target for budget cutters, as they wrote paper budgets but then embargoed funds from actually being used.

In 1996, the federal government's economic policy makers only released 13% of the funds originally budgeted for the Ministry of the Environment *(Parabólicas* January/February 1997). In 1998, the Ministry of the Environment was allowed to spend half of its original budget.[12] Similarly, when the government announced a new round of budget cuts to meet International Monetary Fund conditions in the fall of 1998, the budget line which took the largest percentage cut was the department which handles housing and sanitation, which lost 54% of its annual budget *(Jornal do Brasil* 10 November 1998). The Environment Ministry and its agencies were expected to lose 40–50% of their non-payroll and pension spending (Kepp, 1999: 4), but the Congress restored US$ 250–$ 500 million (R$ 500 million) of the budget.[13] Cuts in Brazilian environmental spending multiply negatively, since it receives financing from several large foreign environmental projects, which depends on Brazilian matching funds. Thus when Brazilian spending is cut, it can receive less of the foreign funding.

Brazilian spending has been comparatively low even for its peer countries. In a survey of 35 high-biodiversity developing countries, Brazil ranked third in per capita income, but fifteenth in its annual expenditure per km^2 of protected area, at US$97.08 per km^2 in 1995 (Financial Resources ..., 1997: 12).

Given the dearth of funding, it is perhaps ironic that legislative gaps appear here as well. The 1985 Law of Diffuse Interests created a fund where fines resulting from convictions under the law could be placed. These funds were to be used for environmental projects. Thirteen years later, however, the accompanying legislation to disburse the funds had not been passed, and the money languished in the account (Santilli, 1998b).

3.3 Intergovernmental Relations and Decentralization

Administrative decentralization is often proposed as a mechanism for improving decision-making and implementation. Arguably, decentralization is especially

12 World Wildlife Fund press release, 18 February 1999, www.panda.org/forests4life/news
13 Ibid.

important for environmental issues in a country like Brazil, since environmental problems are quite different in different parts of this huge country. Drought and desertification in the Northeast contrast with the rainforests of the Northern Amazon region, while both are sharply different from the agricultural and industrial issues of the rest of the country. Nonetheless, decentralization has had a mixed effect on environmental performance in Brazil.

Brazil, the size of the continental United States, has been administered as a federal system since it became a republic. Souza (1997: 25) notes a historical oscillation between centralization and federalism there, with successive efforts in one direction or the other. The military regime of 1964–1985 concentrated a great deal of power in the national government, including control over environmental issues, although its environmental provisions were more decentralized than other issue areas (Zulauf, 1994: 6). Still, the 1975 legislation on industrial pollution gave only the central government the right to carry out environmentally-based suspensions of economic activities (Guimarães, 1995: 163), and a constitutional amendment in 1964 gave the federal government control over public lands (Dean, 1995: 289).

The 1988 constitution reversed this trend: Article 24 gives all three levels concurrent legislative responsibilities over natural resources and pollution (VI) and general responsibility for damage to the environment (VII). This has opened up the possibility of effective environmental legislation at all levels, but it also allows a situation where none of the three levels accepts full responsibility and accountability. One of the oddities of the Brazilian decentralization process is that it has much more successfully distributed revenues than responsibilities, leaving it with a central government that no longer has adequate resources for its responsibilities (Martinez-Lara, 1996: 120). Moreover, without clearer assignment of responsibilities, much of the new revenue at state and local levels is being swallowed up in politicized spending on personnel, both active and retired: 74% of total state spending in December, 1996, and still 67% by late 1997 (Selcher, 1998: 40).

The population increasingly expects decentralization of environmental administration. The most notable change between two surveys in 1992 and 1997 was that in the later survey, the percentage of respondents who considered the federal government most responsible for environmental solutions dropped from 51 to 43%. Local governments and communities and ecological organizations, in contrast, all increased 4–5 percentage points.

By its nature, decentralization does not produce equal results. The extreme geographical disparities in Brazil, where the GNP per capita in industrialized São Paulo is 7.4 times that of Piaui, the poorest state (Selcher, 1998: 33), exacerbates this inequality. Environmental capacity also varies widely, with São Paulo's environmental budgets and capacity routinely outstripping national capacities, while environmental agencies in the North and Northeast limp along with few resources. As recently as 1991, the whole set of seven state environmental agencies in the Amazon region had only 23 computers, for example, while the nine Northeastern states shared 25 (Zulauf, 1994: 17, 25). Across the two regions, only three had even minimal state forest police forces, and none pre-dated 1989 (Zulauf, 1994: 15, 23). In contrast, all seven states in the South and Southeast regions had extensive forest police forces by 1991 – despite having much less forest than the Northern Amazon region – and a total of 174 computers (Zulauf, 1994: 37, 39, 45, 46). Even public opinion varies across regions. Respondents in the North were most likely (21%) to agree with the statement that "Nature is not sacred and humans may use it according to their needs," versus a national average of 13% in agreement. These regional differences lie behind some of the disparities in environmental enforcement, since industrial pollution issues, which have been most successfully addressed, are most important in the states with higher environmental capacity, while the rural regions have the lowest environmental capacity.

Even within particular states, however, environmental performance depends quite a bit on the commitment of changing administrations. The relative success and then decline in Cubatão's pollution control efforts matches the turnover from the Montoro to the Quércia state administrations in São Paulo (Lemos, 1998). Similarly, local political dynamics were critical for explaining the differences in environmental outcomes in a recent study of four Brazilian states (Ames and Keck, 1997–98).

3.4 Judicial and Congressional Oversight

The legislation of the 1980s and 1990s has created formal opportunities for significant judicial oversight of environmental policy making and compliance in Brazil. Environmental issues are treated in both administrative and criminal law, and both administrative and criminal penalties exist for environmental infractions. In addition, the Law of Diffuse Interests provides legal mechanisms that go beyond even class action provisions in the United States. Environmental and other related associations as well as government agencies, and the District Attorney/Attorney General *(Ministério Público)* have standing to challenge government agencies, businesses and individuals for environmental infractions in court. They may do so even without showing a direct personal affect (hence, "diffuse interests"). The 1998 Environmental Crimes Law also targets government agencies and individual bureaucrats as well as private actors.

Despite advanced legislation, the underdeveloped Brazilian judiciary has largely not met its potential for oversight. The Brazilian judiciary is quite conservative, and has historically defended individual private property rights over collective or social rights, such as environmental rights (Fernandes, 1995: 100–101). Still, this situation may be changing. An NGO newsletter counted 11 court cases in 1996 and 1997 where judges ruled for greater environmental protections, levying preliminary injunctions, fines, and other punishments against the projects of a wide array of actors. These included the National Congress, the national indigenous agency (Funai), Ibama, a municipal government, and various private businesses and property owners (Santilli1998a).

Legislative assemblies at all three levels also have mechanisms for oversight in their parliamentary inquiry commissions (CPI). These are often convened, but are rarely effective (Ames and Keck, 1997–98: 9).

3.5 Political Transition

Finally, the constantly changing policies and institutions themselves may encourage disregard of them. Brazilian social scientist Bolívar Lamounier has coined the phrase "hyperactive paralysis" to describe politics in Brazil since the transition to civilian government in 1985. He concludes that Brazilian politicians were

"... biting off more than they could chew: engaging in too much proposing and too much reforming, all of it through lengthy and cumbersome procedures ... (Lamounier, 1996: 171)." Although Lamounier was writing about quite different aspects of recent Brazilian politics, his words could equally well apply to Brazil's recent environmental policy making. It seems unlikely that the annual institutional changes in federal environmental agencies from 1989 to 1994 improved environmental policy making more than they disrupted it. Similarly, the several modifications of the Environmental Crimes Law might well convince a rational economic actor to lobby for changes or exceptions to the law, rather than to actually take steps to comply with it.

4 Key Actors in Brazilian Environmental Politics and Policy

Another important component of environmental capacity is that attributable to specific individual and collective actors. They themselves have varying levels of capacities, resources, and skills. Their alliances, or lack thereof, may alter their individual effective capacity. Whether their relations are cooperative or conflictual, whether they are flexible learners or rigid dogmatists, their characteristics are the final piece necessary to evaluate Brazilian environmental politics. This section begins with the constellation of proponents for stronger environmental policies, and then considers their array of opponents.

4.1 Proponents of Environmental Policies and Their Capacities

Despite the many weaknesses of Brazilian national environmental agencies outlined above, they remain here, as elsewhere, centerpieces of the coalition that works to increase environmental capacity. Numerous dedicated individuals have worked in the agencies at all levels. Paulo Nogueira Neto, for example, held the Special Secretary of the Environment together on a shoestring for over a decade. During Nogueira Neto's tenure, Foresta (1991: 224–225) writes, "When a problem arose, a frequent response in the division's ranks was 'Dr. Paulo is working on it,' a phrase that conveyed at least optimism and normally absolute faith." Other energetic leaders included Maria Tereza Jorge Pádua of the Brazilian Institute for Forest Development and later the Ibama and environmental pioneer José Lutzenberger, who spent a tumultuous period as Secretary of the Environment at the beginning of the 1990s. Although often criticized by both proponents and opponents of environmental policies, these individuals brought purpose and energy to their jobs; they have made careers of environmental activism in many forms, both before and after their governmental stints. Proponents of building environmental capacity within the environmental agencies have built especially strong links with two sets of actors: scientists and international environmental proponents.

The relationship with scientists is of longer duration, dating back to the very beginning of national environmental agencies. Even during the military years, environmentalists regularly traveled to Brasilia to provide SEMA with their technical expertise, if they had any. More recently, the Ministry of the Environment's report to the Rio+5 conference seems to conceive of capacity building largely in terms of scientific data development and university research and teaching initiatives. For example, its report on the capacity building that has taken place in the biodiversity issue area counts, "There were about 50,000 scientists engaged in biodiversity research as of the early 90s, as well as 306 courses at master degree level, and 158 at doctoral level, which have enhanced capacity building in assessment, systematic observation, evaluation and biodiversity protection enforcement" (Republic of Brazil, 1997: Ch. 15). Brazil has multiple scientific organizations, which have contributed to cataloguing and analyzing environmental data. Some of the more active groups include the Brazilian Society for the Advancement of Science (SBPC), the Brazilian Institute of Geography and Statistics (IBGE), the Brazilian Council for Technological and Scientific Development (CNPq), and the National Institute of Amazon Research (INPA). None the less, the Brazilian government reported it had good information for decision making in only 5 of the 40 environmental issue areas it assesses, and poor information in 15, with some gaps in the remaining areas (Republic of Brazil, 1997: Ch. 40).

Both Brazilian environmental agencies and scientists have extensive ties with international proponents of building environmental capacity in Brazil. While international observers have long had an interest in Brazil's environmental politics, due to the scale of Brazil's environmental problems and promise, their active engagement increased substantially during the 1980s. Activities around Brazil's role as host of the United Nations Conference on Environment and Development in 1992 also focused the attention of the world – Brazilians and non-Brazilians – on the Brazilian environment. Both individuals and institutions were a part of this process. Multi-state organizations such as the World Bank, the Inter-American Development Bank, the Global Environmental Fund, the United Nations Development Program, and the European Union all have sent funds and advisors for various projects, as have individual countries like Germany, the United States, Japan, France, the United Kingdom (Kolk, 1998; Republic of Brazil, 1997). Foreign funds have been especially important for Brazil's efforts to combat deforestation, with a major initiative called the Pilot Plan for the Amazon which is supported by the G-7 and others. Non-governmental organizations have sent some funds as well, but have been most important as allies for Brazilian citizens' groups who seek to use international pressure to influence their own government's policies (Bramble and Porter, 1992). International allies were recruited recently, for example, to support Brazilian environmentalists' efforts to shore up the Environmental Crimes Bill.[14] Yet only 39% of Brazilians agree that the Brazilian environment should be protected with the aid of international actors.

Brazilian national environmental agencies have had few direct ties with citizens groups. During the military government, even environmental education was off limits until after 1979, when the regime had already opened up considerably. After that, a former SEMA coordinator of social communication and environmental education explains that the various agencies have never had the money and organization to maintain an ongoing relationship with citizens' groups, although the

---

14 Steve Schwartzman of the U.S.-based Environmental Defense Fund and João Paulo Capobianco of the Brazilian Instituto Socioambiental sent a joint Urgent Action request for letters to Brazilian president Fernando Henrique Cardoso to, among others, recipients of the BrasaNet mailing list, on 25 February 1998.

post-military agencies saw a role for citizens in public hearings and occasional campaigns, and as participants in environmental education efforts.[15] More recently, the Ministry of the Environment reported that it had "no information" on the participation of citizens groups in more than half of the substantive areas dealt with in its report to the Rio+5 conference (Republic of Brazil, 1997).

Nonetheless, citizens' groups are an important part of Brazil's national environmental capacity, and have continuously grown in number and capacity over the last 30 years. Brazil's non-governmental sector is very diverse, and thus contributes to Brazil's environmental capacity in different ways. The participation of all of them inherently contributes to Brazil's participative environmental capacity. The work of these organizations was supported by 80% of Brazilians, although only 1% claimed membership and few could name any existing organizations.

At one end of the organizational spectrum are the environmental foundations (often called non-governmental organizations or NGOs in Brazil), which began to appear in the mid to late 1980s. Unlike the grassroots environmental groups, the environmental NGOs have permanent staffs and relatively large budgets. The scientists on their payrolls have been important contributors to the cognitive capacity of Brazil's environmental policy making. For example, the SOS Atlantic Forest Foundation has collaborated in a series of important data gathering exercises that provide much of the information that is known about developments in the Atlantic Forest (SOS Mata Atlântica, 1993). Preparations for the 1992 United Nations Conference on Environment and Development, held in Rio de Janeiro, brought in another set of NGOs, originally interested primarily in development issues (Hochstetler forthcoming). These NGOs, such as the Brazilian Institute for Social and Economic Analysis (IBASE) and the Social-Environmental Institute, also helped build integrative capacity for environmental actors, with their links to urban popular groups and unions. NGOs of all kinds have become an important part of the governing structure in Brazil: 70% reported some kind of partnership with government agencies in 1996, and NGOs influenced or controlled almost US$1.4 billion that same year, in funds administered by the Ministry of the Environment, the Ministry of Social Security and Welfare, and international banks *(Folha de São Paulo* 9 June 1996). Many international actors have made citizens' participation a condition of their funding for Brazilian environmental programs.

At the other end of the spectrum are numerous grassroots environmental groups, with small memberships and smaller budgets. Because they are spread over a very large area geographically and often rise and fall with local environmental problems, virtually every count of the number of environmental groups in Brazil misses many of these.[16] They have also contributed to Brazil's cognitive environmental capacity, but more in diffusing environmental information than in generating new information. They take environmental education seriously, for themselves and for their communities. As local activists, they also contribute to local integrative capacity, sitting on numerous local councils and raising environ-

15 Interview with Regina Gualda, Brasília, 16 September 1991.

16 During 1990, I was able to attend state-level conferences of environmental groups in four Brazilian states, as well as sub-national gatherings in two, and can vouch for the numbers, enthusiasm, and diversity of these actors even though they are unknown outside of their localities.

mental issues when other actors do not. It is difficult to argue that they contribute in large ways to Brazil's environmental capacity, but they do broaden and deepen its effects.

Both grassroots environmental groups and environmental NGOs have worked with selected legislators to try to improve Brazil's environmental legislation. These legislators, scattered across numerous parties, have been an important force behind Brazil's environmental legislation since 1982, when the first three environmental legislators were elected at state and local levels. Environmentalists used a "Green List" of environmental candidates to help support environmentalist Fábio Feldmann's successful run for the National Congress in 1986, where he was able to put together an entire coalition of legislators behind the environmental chapter in the 1988 constitution and other laws (Hochstetler, 1997). The tiny Green Party is the most consistent supporter of environmental legislation, although individual representatives from the Workers Party (PT) and Feldmann's Brazilian Social Democratic Party (PSDB) frequently support environmental legislation as well. In the Chamber of Deputies, many of them sit on the Commission in Defense of the Consumer, the Environment, and Minorities (Viana, 1996/1997).

A relatively new group of contributors to Brazil's environmental capacity are its "green" industries and corporations. As of 1997, 48 Brazilian firms and professional associations were members of the International Chamber of Commerce's Business Charter for Sustainable Development, and three firms had joined the World Business Council for Sustainable Development (Republic of Brazil, 1997: Ch. 30). By August, 1998, 102 Brazilian chemical firms were participating in the Responsible Care initiative of the International Council of Chemical Associations (Roberts, 1998: 2), while three pulp and paper firms have launched environmental initiatives of various kinds (May and Vinha, 1998: 73). By June 1999, 100 firms had pledged to follow the voluntary ISO 14001 environmental guidelines.[17] One recent study concluded that many of Brazil's environmentally-concerned industries are responding to foreign consumers of their products, since domestic consumers do not yet often demand environmentally sound practices and products (Gutberlet and Segura, 1996/97).

In addition, both of Brazil's largest peak labor organizations, the Unified Workers' Central (CUT) and Union Force *(Força Sindical),* have environmental committees. These have pushed hard on workplace environment issues, and the CUT has played a leadership role in the Brazilian NGO Forum for Environment and Development.

Finally, environmental reporting has become wide-spread over the last twenty years in Brazil. In this period, most large papers have had regular pages or entire sections devoted to the environment and/or science, including daily newspapers such as the *Folha de São Paulo, Jornal do Brasil,* and the business newspaper *Gazeta Mercantil.* TV Globo, the largest television network in Brazil and one of the largest in the world, has a weekly ecology program (Ames and Keck, 1997–98: 10). Non-traditional media are also expanding, especially since preparations for the UNCED. Just one source catalogues 433 Brazilian organizations and 84

17 http://www.mma.org.br/novidades

government agencies related to the environment, all with electronic mail addresses and/or webpages.[18]

4.2 Opponents of Stronger Environmental Capacity and Their Capacities

Notwithstanding the environmental initiatives of some industries and corporations, the strongest opponents of Brazil's environmental policies continue to be economic actors and their political allies. Business leaders received the lowest evaluation as defenders of the environment, with 41% of the Brazilian population considering them a negative or very negative influence on the environment. This stance unites both urban based industries and rural extractive and agricultural powers, both large and small economic actors, and economic producers oriented to both the domestic and international economies. Thus, anti-environmentalism serves to unite a powerful set of actors who are divided amongst themselves on many other issues.

In the 1990s, urban industrialists face increasing environmental legislation as well as new pressures to be competitive in a turbulent global marketplace, after decades of state protection. In this context, environmental regulations are interpreted as impossible new costs. As a federation of the chemical industries explained,

> The firms in the chemical sector are passing through a process, without precedents, of cutting costs to maintain their activities and to be competitive internationally. Implementing a program of environmental administration would require qualified personnel, investments and installations. These resources are, at the moment, very scarce (cited in Gutberlet and Segura 1996/97: 27).

Other sources concur, noting that "in most industrial sectors in Brazil, environmental investment is perceived as a financial onus rather than as a potential source of competitiveness" (May and Vinha, 1998:73). Consequently, many industries lobby hard against environmental legislation, as in the examples cited above. Once it exists, they fight to weaken it and to prevent its implementation. Some have also begun to consider environmental regulations in their location decisions, moving from regions like São Paulo which have relatively high environmental capacity, although evidence of actual decision-making on these grounds is still anecdotal (Martine and Diniz, 1997: 219).

Rural economic powers are squeezed between new environmental considerations and a growing and combative movement to redistribute land, the Landless Movement (MST). One response has been for rural landowners to join with industrial actors to resist new environmental and distributional initiatives in the legal, political sphere. However, the new pressures often seem like life or death threats to these rural actors, and they have also responded with increasing levels of violence against local environmental or distributional proponents like indigenous peoples, rubber tappers, and the landless (Dimenstein, 1996). The death of rubber tapper Chico Mendes in 1988 at the hands of local cattle ranchers is the best-

18 http://www.infolink.com.br/~peco

known example, but is only one of many. A total of 3374 land conflicts between 1985 and 1990 left 563 dead; from 1964 to 1991 only 29 cases of rural killings went to trial with 13 convictions – for 1630 dead (Santos, 1992: 6–7). The Pastoral Land Commission of the Brazilian Catholic Church, which compiled these figures, concluded in 1993 that the state has in essence granted impunity for rural violence (Comissao Pastoral da Terra, 1993). Figures such as these contribute to the widespread cynicism about the judiciary, and hamper efforts to address such rural environmental issues as deforestation. From an ecological and social perspective, options like extractive reserves make sense (Hall, 1997; Hodge, Queiroz, and Reis, 1997), but they are unacceptable to many rural economic elites.

Business-related ministries and agencies support at least the legal efforts of urban and rural economic actors. More tellingly, the economic actors find allies within the environmental agencies as well, who serve as a counterpoint to the individuals who have built environmental capacity there. After forest fires destroyed large parts of the Xingu National Indigenous Park and threatened the lives of the 5000 indigenous people who live there, two high-ranking local Ibama employees were accused of issuing irregular permits for burns in the area (*Jornal do Brasil* 5 September 1998). A few months before, the national Ibama office was forced to replace a corrupt leadership in its regional office in Rio de Janeiro, and to suspend all permits they had given for cutting lumber in the Atlantic Forest in the state (*Jornal do Brasil* 26 and 30 May 1998). Some environmentalists believe that lower-level employees in the state environmental agencies routinely give businesses copies of complaints filed against them by environmentalists (Fischer, 1999). These recent examples serve to illustrate an ongoing pattern of corruption that makes some employees of the environmental agencies prominent opponents of environmental capacity.

Finally, the Brazilian military continues to wield influence over environmental policy, as it does over many aspects of policy-making in Brazil. Environmental regulations challenge two of the military's cherished priorities, industrial development and geopolitical strength. The challenges to industrial development have already been discussed, so I will concentrate here on the geopolitical issues. Put simply, the same Amazon forest which spills across national boundaries in a profusion of biodiversity is a national security nightmare from the perspective of the Brazilian military. The national boundaries are unmarked and indefensible, crossed every day by drug traffickers, contraband, illegal migrants, indigenous peoples, and international ecotourists (Escola Superior de Guerra, 1990). Zirker and Henberg (1994: 270–273) argue that successive new civilian governments, anxious to keep the military happy and out of Brasilia, have offered it continuing prerogatives in the Amazon. There it supports large mining and agricultural initiatives and resists land distribution – with itself as Brazil's largest landowner – and thus becomes a continuing limitation on environmental capacity, intricately linked to the other opponents of environmental policy.

5 Conclusions

The only certain conclusion about Brazilian environmental capacity is that it will continue to be one of the more contentious policy domains of Brazilian politics.

The opponents of developing Brazil's environmental capacity are both powerful and feeling threatened. This combination makes them even more formidable than if they were only some of the most powerful actors in Brazilian politics. They will let few opportunities go by to make their positions more secure, and they stand largely united in doing so. Seen from the other side, those recent threats to environmental opponents are the recent gains of the proponents of building environmental capacity in Brazil. These actors, both Brazilian and international, also have important connections and capacities. Their characteristics often make them best suited to a guerrilla-type struggle for increased capacity, and, here, their diversity may be a strength. Initiatives to improve environmental capacity may come in legislation, but they may also come as an urban neighborhood's demands for trash collection, an international consumer boycott, or a scientific report. The final result also depends quite a bit on developments that are only indirectly related to the environment: in Brazilian politics, its economy, and in global developments.

References

Ames, B., and M. Keck, 1997–98. The Politics of Sustainable Development: Environmental Policy Making in Four Brazilian States. *Journal of Interamerican Studies and World Affairs* 39 (4),1–40.

Anderson, L. E., 1996. The Causes of Deforestation in the Brazilian Amazon. *Journal of Environment and Development* 5 (3), 309–328.

Bramble, B. J., and G. Porter. 1992. Non-Governmental Organizations and the Making of US International Environmental Policy, in: A. Hurrell and B. Kingsbury (eds.), *The International Politics of the Environment.* Oxford and New York: Oxford University Press/Clarendon Press.

Brazilian NGO Forum, 1997. *National Consultation Process of Brazilian NGOs and Social Movements for the Rio+5 Forum.*

Caccia Bava, S., and L. Mullahy, 1995. Making Cities Livable: Local Initiatives in Solid Waste and Public Transportation Management in Brazil, in: C. A. Reilly (ed.), *New Paths to Democratic Development in Latin America: The Rise of NGO-Municipal Collaboration.* Boulder: Lynne Rienner.

Campanilli, M., and S. Leitão, 1998. Conama Facilita Licenciamento Ambiental. *Parabólicas* 37.

Capobianco, J. P., 1998. Rio é o Campeão Nacional do Desmatamento. *Journal do Meio Ambiente, <http://www.jornal-do-meio-ambiente.com.br/index.htm>* (July).

Comissao Pastoral da Terra, 1993. O Campo Brasileiro em 1992. *Cadernos do CEAS* 148, 76–84.

Conama, 1991. *Resoluções Conama 1984 à 1990.* Brasilia: Secretaria do Meio Ambiente da Presidência da República.

Dean, W., 1995. *With Broadax and Firebrand: The Destruction of the Brazilian Atlantic Forest.* Berkeley and Los Angeles: University of California Press.

Dimenstein, G., 1996. *Democracia em Pedacos – Direitos Humanos no Brasil.* Sao Paulo: Companhia das Letras.

Earthwatch, 1992. *Urban Air Pollution in Magacities of the World.* New York: WHO/UNEP/ Blackwell Reference.

Escola Superior de Guerra, 1990. *Estructura do Poder Nacional para o Ano 2001.* Rio de Janeiro: Escola Superior de Guerra.

Fearnside, P. M., and R. I. Barbosa, 1996a. The Cotingo Dam as a Test of Brazil's System for Evaluating Proposed Developments in Amazonia. *Environmental Management* 20 (5), 631–648.

Fearnside, P. M., and R. I. Barbosa, 1996b. Political Benefits as Barriers to Assessment of Environmental Costs in Brazil's Amazonian Development Planning: The Example of the Jatapu Dam in Roraima. *Environmental Management* 20 (5), 615–630.

Fernandes, E., 1995. *Law and Urban Change in Brazil.* Brookfield, VT: Ashgate.

Financial Resources for Biodiversity Conservation, 1997. *Environmental Policy and Law* 27 (1), 9–13.

Fischer, G. R., 1999. Carta – Alerta para os Ambientalistas. *Jornal do Ambiente* 41.

Flynn, P., 1996. Brazil: The Politics of the 'Plano Real'. *Third World Quarterly* 17 (3), 401–426.

Foresta, R. A., 1991. *Amazon Conservation in the Age of Development: The Limits of Providence.* Gainesville, FL: University Press of Florida.

Fuks, M., 1994. Theoretical and Practical Considerations on Environment-Related Litigation. *International Sociology* 9 (4), 447–462.

Gentry, B. S., 1998. Agriculture – Cases from Brazil and Costa Rica, in: B. S. Gentry (ed.), *Private Capital Flows and the Environment: Lessons from Latin America.*. Cheltenham, UK: Edward Elgar.

Gilbert, A., 1996. Land, Housing, and Infrastructure in Latin America's Major Cities, in: A. Gilbert (ed.) *The Mega-City in Latin America.* New York: United Nations University Press.

Government of São Paulo, 1987. *A Batalha do Meio Ambiente no Governo Montoro.* São Paulo: State of São Paulo.

Guimarães, R. P., 1992. Development Pattern and Environment in Brazil. *CEPAL Review* 47, 47–62.

Guimarães, R. P., 1995. *The Ecopolitics of Development in the Third World: Politics and Environment in Brazil.* Boulder: Lynne Rienner.

Gutberlet, J., and D. S. Baena Segura., 1996/97. Tendências de Gerenciamento Ambiental nas Indústrias Brasileiras. *Debates Sócio Ambientais* 2 (5), 26–28.

Hall, A., 1997. *Sustaining Amazonia: Grassroots Action for Productive Conservation.* Manchester and New York: Manchester University Press.

Hecht, S., and A. Cockburn, 1990. *The Fate of the Forest: Developers, Destroyers and Defenders of the Amazon.* New York: HarperCollins.

Hinchberger, B., 1998. Brazil Overhauls its Environmental Laws. *EcoAméricas* 1 (1), 3.

Hochstetler, K., 1997. The Evolution of the Brazilian Environmental Movement and Its Political Roles, in: D. A. Chalmers, C. M. Vilas, K. Hite, S. B. Martin, K. Piester and M. Segarra (eds.), *The New Politics of Inequality in Latin America: Rethinking Participation and Representation.* Oxford: Oxford University Press.

Hochstetler, K., forthcoming. Democratizing Pressures from Below? Social Movements in New Brazilian Democracy, in: P. R. Kingstone and T. J. Power (eds.), *Democratic Brazil: Actors, Institutions, and Processes*, edited by. Pittsburgh: University of Pittsburgh Press.

Hodge, S. S., M. Hering de Queiroz, and A Reis, 1997. Brazil's National Atlantic Forest Policy: A Challenge for State-Level Environmental Planning. The Case of Santa Catarina, Brazil. *Journal of Environmental Planning and Management* 40 (3), 335–348.

Holston, J., and T. P.R. Caldeira, 1998. Democracy, Law, and Violence: Disjunctions of Brazilian Citizenship, in: by F. Agüero and J. Stark (eds.), *Fault Lines of Democracy in Post-Transition Latin America.* Miami: North-South Center Press.

Hurrell, A., 1992. Brazil and the International Politics of Amazonian Deforestation, in: A. Hurrell and B. Kingsbury (eds.), *The International Politics of the Environment.* Oxford: Clarendon Press.

Jacobi, P., 1993. A Percepção de Problemas Ambientais Urbanos em São Paulo. *Lua Nova* 31, 47–56.

Jänicke, M., 1997. The Political System's Capacity for Environmental Policy, in: M. Jänicke and H. Weidner (eds.), *National Environmental Policies: A Comparative Study of Capacity Building.* Berlin and New York: Springer.

Keck, M., 1995. Social Equity and Environmental Politics in Brazil: Lessons from the Rubber Tappers of Acre. *Comparative Politics* 27 (4), 409–424.

Kepp, M., 1999. Brazil's Environment Budget to Take a Hit. *EcoAméricas* 1 (3), 4.

Kolk, A., 1996. *Forests in International Environmental Politics: International Organisations, NGOs and the Brazilian Amazon*. Netherlands: International Books.

Kolk, A., 1998. From Conflict to Cooperation: International Policies to Protect the Brazilian Amazon. *World Development* 26 (8), 1481–1493.

Lamounier, B., 1996. Brazil: The Hyperactive Paralysis Syndrome, in: J. I. Domínguez and A. F. Lowenthal (eds.) *Constructing Democratic Governance: South America in the 1990s*. Baltimore and London: The Johns Hopkins University Press.

Lemos, M. C. de M., 1998. The Cubatão Pollution Control Project: Popular Participation and Public Accountability. *Journal of Environment and Development* 7 (1), 60–76.

Lewis, S., 1990. *The Rainforest Book: How You Can Save the World's Rainforests*. Venice, CA: Living Planet Press.

Lopes, J. B., 1996. Obstacles to Economic Reform in Brazil, in: A. Lijphart and C. H. Waisman (eds.), *Institutional Design in New Democracies: Eastern Europe and Latin America*. Boulder: Westview.

Mahar, D. J., 1989. *Government Policies and Deforestation in Brazil's Amazon Region*. Washington, DC: The World Bank.

Martine, G., and C. Campolina Diniz, 1997. Economic and Demographic Concentration in Brazil: Recent Inversion of Historical Patterns, in: G. W. Jones and P. Visaria (eds.), *Urbanization in Large Developing Countries: China, Indonesia, Brazil, and India,*. Oxford: Clarendon Press.

Martinez-Lara, J.., 1996. *Building Democracy in Brazil: The Politics of Constitutional Change, 1985–93*. New York: St. Martin's Press.

May, P., and V. Gonçalves da Vinha, 1998. Financing Expansion of Brazil's Pulp and Paper Industry, in: B. S. Gentry (ed.), *Private Capital Flows and the Environment: Lessons from Latin America*. Cheltenham, UK: Edward Elgar.

Ministério de Meio Ambiente, 1995. *Protocolo Verde 1995*. Brasilia: Ministério de Meio Ambiente.

Nepstad, D. C., A. Veríssimo, A. Alencar, C. Nobre, E. Lima, P. Lefebvre, P. Schlesinger, C. Potter, P. Motinho, E. Mendoza, M. Cochrane, and V. Brooks, 1999. Large-scale Impoverishment of Amazonian Forests by Logging and Fire. *Nature* 398 (April 8), 505–508.

Pádua, J. A., 1991. O Nascimento da Política Verde no Brasil: Fatores Exógenos e Endógenos, in: H. Leis (ed.), *Ecologia e Política Mundial*. Rio de Janeiro: FASE, Editora Vozes, and AIRI/PUC-Rio.

Redwood III, J., 1993. *World Bank Approaches to the Environment in Brazil, A World Bank Operations Evaluation Study*. Washington, DC: The World Bank.

Republic of Brazil, 1991. *Meio Ambiente (Legislação)*. Brasília: Senado Federal.

Republic of Brazil, 1995. *Relatorio Brasil: Sobre os Assentamentos Humanos*. Brasília: Itamaraty.

Republic of Brazil, 1997. *Brazil: Country Profile: Implementation of Agenda 21: Review of Progress Made Since the United Nations Conference on Environment and Development, 1992*. Brasília: Republic of Brazil.

Roberts, J. T., 1998. The End of 'Pollution Haven' as 'Comparative Advantage'? Emerging International Environmental Standards and the Brazilian Chemical Industry. Paper read at Space, Place, and Nation: Reconstructing Neo-Liberalism in the Americas, at University of Massachusetts at Amherst.

Romanini, V., 1995. O Fantasma de Cubatão. *Problemas Brasileiras* 307, 34–40.

Santilli, J., 1998a. Meio Ambiente: Vitórias no Judiciário. *Parabolicas* 36.

Santilli, J., 1998b. Multas por danos ambientais se perdem na burocracia. *Parabolicas* 38.

Santos, J. V. T. dos, 1992. Violencia no Campo: O Dilaceramento da Cidadania. *Reforma Agraria* 22 (1), 4–11.

Selcher, W. A., 1998. The Politics of Decentralized Federalism, National Diversification, and Regionalism in Brazil. *Journal of Interamerican Studies and World Affairs* 40 (4), 25–50.

Silva, E., 1997. Chile, in: M. Jänicke and H. Weidner (eds.), *National Environmental Policies: A Comparative Study of Capacity-Building*. Berlin and New York: Springer.

SOS Mata Atlântica, 1993. *Atlas da Evolução dos Remanescentes da Mata Atlântica*. São Paulo: SOS Mata Atlântica.

Souza, C., 1997. *Constitutional Engineering in Brazil: The Politics of Federalism and Decentralization*. New York: St. Martin's Press.

Teixeira, M. A. C., 1996/97. São Paulo e as Alternativas de Tratamento e Distinação Final do Lixo. *Debates Sócio Ambientais* 2 (5), 17–18.

UNEP, 1993. *Environmental Data Report 1993–94*. London: Blackwell Reference.

Viana, G., 1996/1997. A Atuação da Comissão do Consumidor, Meio Ambiente e Minorias. *Debates Sócio Ambientais* 2 (5), 21–22.

Wainer, A. H.. 1991. *Legislação Ambiental Brasileira: Subsídios para a História do Direito Ambiental*. Rio de Janeiro: Forense.

Woreman, N. B., 1990. Brazil's Thriving Environmental Movement. *Technology Review* 93 (7), 42–47.

Zirker, D., and M. Henberg, 1994. Amazônia: Democracy, Ecology, and Brazilian Military Prerogatives in the 1990s. *Armed Forces and Society* 20 (2), 259–281.

Zulauf, W., 1994. *Brasil Ambiental: Síndromes e Potencialidades*. São Paulo: Fundação Konrad-Adenauer-Stiftung.

ENVIRONMENTAL PROTECTION CAPACITY IN POST-COMMUNIST BULGARIA

Susan Baker

The capacity of Bulgaria to effectively manage its environment continues to be shaped by the legacies of over four decades of communist rule. It is also shaped by the difficulties Bulgaria has in finding a direction to the transformation process set in motion after the collapse of communist rule in 1989.

1 Structural Constraints and Conditions Framing Environmental Capacity

The Republic of Bulgaria lies in southeastern Europe, on the east of the Balkan Peninsula. Following occupation by Soviet troops, the Bulgarian Communist Party (BCP) came to power in 1944. In 1954 Todor Zhivkov became First Secretary of the BCP and remained in power until forced to resign from office in 1989. During the period of communist rule Bulgaria was regarded as the Soviet Union's "most faithful servant". Its economy was closely modelled on the Soviet style of industrialisation, collectivisation and urbanisation. The rule of Zhivkov was remarkable in the degree to which it followed the policies of the Soviet Government, especially in foreign affairs (Crampton, 1997: 225). Efforts were also made to shape public culture in strict ideological terms (Pickles et al, 1998: 7).

The collapse of Communism and the break-up of the Soviet system have left Bulgaria politically and economically vulnerable. Since 1989 the country has been unable to make an effective response to the externally imposed demands of transition. In Bulgaria, and indeed throughout the Balkans, many elites are either unable or unwilling to adopt Western-style models of democracy and market-driven economic reform. At the same time, Bulgaria has failed to find alternative models to guide its transformation process. Consequently there is a great deal of political instability, a high turn-over rate for government (with seven governments between 1990 and 1999) and a political discourse that is sharply contested between support for communist and for anti-communist ideologies. The polarised political climate has hindered the emergence of mature political debate in Bulgaria.

Bulgaria's party system can be characterised as one of "polarised pluralism" (Gallagher, 1998: 47). The two main political parties are the communist successor party, the Bulgarian Socialist Party (BSP) and the Union of Democratic Forces (UDF). The BSP has exhibited much resilience in holding on to political power and continues to enjoy popular support, particular among rural communities. It is the party with the largest membership, best-developed structure and most substantial financial support. The BSP held power in 1990-91 and again between 1994 and 1997. It is only weakly committed to Western-inspired reform projects. In 1994-95 the party, for example, reversed some of the policies of openness towards the West that had crept in since 1989 and made efforts to slow down the penetration and influence of Western institutions and capital in transforming business and government (Pickles and Mikhova, 1998: 109). There have also been allegations that the BSP made attempts to restore authoritarian rule during its periods in office, particularly in 1994-95 (Crampton, 1997: 230).

In contrast, the UDF originated as an anti-communist coalition of some 15 parties, including the Conservative Ecology Party and the Ekoglasnost environmental group. Its membership is, to a large measure, made up of nomenklatura elites and elites whose origins can be traced back to the pre-1994 period (Andreev, 1996: 27). Attempts to turn this opposition movement into a conventional political party have proved difficult. As a heterogeneous coalition, the UDF does not posses a distinctive ideological profile. However, its politics can be defined within certain basic parameters: unyielding anti-communism, commitment to speedy market reform, pro-Western orientation and a focus on the democratic rights and liberties of the citizen (Andreev, 1996: 32).

Before the first general election under the new constitution in 1991, a series of splits occurred in the UDF. Part of Ekoglasnost, for example, with its then chairman Peter Slabakov, left the UDF. It was the most right-wing of the three factions that carried the UDF flag into the elections. Despite this, however, they emerged as the largest party in the National Assembly and formed a government with parliamentary support from the Turkish ethnic party. The leader of the Green Party, Filip Dimitrov, became prime minister, although he was removed from office following the 1992 elections. This was followed by a period of political instability, interim government and the return to power of the BSP in 1994. Neither the coalition government under Professor Berov (1993-94), nor the BSP (1995-97) showed any serious interest in environmental matters.

The 1996-97 period saw much of the region, including Bulgaria, shaken by large-scale protests against governments who had made promises of social protection but whose record of misrule had deprived them of much of their remaining legitimacy (Gallagher, 1998: 54). As a consequence, the BSP lost power and the UDF was returned to power in the 1997 elections. However, the UDF remains weak and its room for manoeuvre is constrained by the political and economic vulnerability of the country, by external demands, including the re-structuring demands made by the IMF, and the constraints imposed by its Association Agreement with the EU (Baker and Baumgartl, 1998).

Despite the shift to parliamentary politics and the holding of what are generally regarded as fair and free elections, reform in post-communist Bulgaria remains relatively superficial. A wide gap has opened up between the procedural changes that have occurred since the end of communist rule and the substantive changes that are required for transformation (Baker and Jehlička, 1998: 24). There is a great deal of corruption in political life, reform has not penetrated to the local level, the civil service remains partial and discretionary and the rule of law has not yet been fully established. The police often ignore legal proceedings and there have been numerous claims of abuse in police custody and serious corruption allegations. This has contributed to the weak public faith in state institutions. In

1994, for example, a gun fight in Sofia between rival police factions forced the then Minister of the Interior, Viktor Mihailov to admit that the police were both corrupt and incompetent (Crampton, 1997: 230). The secret service is considered to be insufficiently controlled and are not accountable to parliament for their activities (CEC, 1997: 1.1). The process of dispensing justice is also marred by "serious short comings" (CEC, 1997: 1.1), with a shortage of qualified judges and legal proceedings which are both very slow and highly complex. There are also growing levels of organised crime.

Economically, the country has gone through deep crisis since 1989. Prior to 1989 Bulgaria was heavily dependent on trade with the Soviet Union, which accounted for almost 60 per cent of foreign trade. The collapse of CMEA trade badly effected Bulgaria. Furthermore, the USSR has provided Bulgaria with access to cheap energy sources and, as a consequence, it never implemented any fuel efficiency measures, typically found in the West after the oil shocks of 1973 and 1979. The post-1989 economic problems were deepened by the war in former Yugoslavia, which resulted in the loss of Bulgaria's primary overland route through Serbia-Montenegro and a USA imposed embargo on trade with or through Yugoslavia. This resulted in severe shortages of almost every commodity (Hutchings, 1997: 235). Bulgaria also suffered from the loss of trade with Iraq following the Gulf War. Iraq was Bulgaria's biggest foreign-currency debtor, owing some US$1,200m that was to be paid in the form of crude petroleum (Crampton, 1997: 227). Added to this was a severe banking crisis, currency collapse and periods of high inflation that followed the end of communist rule. The European Commission has called Bulgaria's financial system "dysfunctional" (CEC, 1997). Since 1989 industrial output has fallen, the country has experienced a severe decline in GDP, and large scale unemployment. In January 1995 almost 65 per cent of Bulgarians were reported to be living in poverty, with more than one-third of households suffering from malnutrition, though 70 per cent of their incomes is spend on food (Hutchings 1997: 235). However, since 1998 there has been some economic improvement, with a growth in real GDP, a reduction in inflation rates and a slight rise in industrial production, but with unemployment settling at around 12.7% of the labour force *(Financial Times*, Special Report, 8 March 1999).

The lack of political will, particularly on the part of the BSP, combined with the weak state of the economy, has blocked the creation of a Western-style market economy. This has not stopped, however, the privatisation of the most profitable sectors of state enterprises into the hands of the old nomenklatura. Managers and former government and party officials bought the profitable parts of state enterprise. As a result, many of the old elites have been able to convert their previous political monopoly into a new economic hegemony (Baker and Baumgartl, 1998: 183). The lack of independence of the media, judiciary, public administration and of the secret police all facilitated the old elites in taking control over important areas of the economy. Efforts have been made since the return to power of the UDF in 1997 to deal more effectively with the economic reform, particularly within the banking sector and to win international support for economic rebuilding *(Financial Times*, Special Report, 8 March 1999).

However, the consequence of the country's political and economic weakness is that structural reforms are still at a very early stage. There are also specific barriers that have proved difficult to overcome, including the lack of transparent and enforceable property rights and the problematic nature of land restitution policy – the law on ownership and use of farmland has been changed more than ten times since 1990 (Soveroski, 1997: 33). Underlying this, however, is the more serious problem of Bulgaria's inability to find a direction to its transformation: the old communist party elites reject the Western model of transition, while encouraging their own aggrandisement through Mafia style illegal behaviour; the anti-communists are weak, and while promoting Western-orientated marketisation and democratisation models, have been severely hampered by internal party divisions, economic weakness, short periods in office and a high turn over rate of government, low levels of foreign aid, limited international political support and the uncertainty surrounding EU membership. The enthusiasm with which the UDF has embraced Western models has not been met by corresponding, enthusiastic responses from Western agencies and governments. As such, the direction of future change remains uncertain: as Bulgaria struggles to adopt an increasingly unsuitable Western transition model large sections of its population, including as many of its political elites, hanker after the security provided by the previous, but now disintegrated, communist regime.

The development of civil society has also been restricted and it has only limited capacity to stimulate more effective environmental protection policies. The development of civil society is blocked by the widespread belief that engagement in politics is a tainted business – a belief strengthened by the self-enriching behaviour of many elites since the collapse of communism. Political instability, the lack of an independent media, and general apathy among the population in the face of food shortages and impoverishment have all reduced the terrain available for the development of civil society. Despite the central role that the environmental movement played in mobilising opposition to communist rule in Bulgaria, the post 1989 period has seen environmental considerations displaced by the harsh realities of economic decline and political instability (Baker and Baumgartl, 1998). In this context it is not surprising to find that environmental considerations have but a tenuous position on Bulgaria's policy agenda. Nevertheless, despite the numerous constraints, some attempts have been made since 1989 to deal with the worst aspects of environmental degradation and pollution inherited from the old regime and to develop more adequate structures for and capacity to engage with environmental protection, particularly at the legislative level.

2 Main Environmental Problems

2.1 Main Environmental Problems

Like all Balkan countries, Bulgaria has a high proportion of forests and woodlands (35% in 1993) (Eurostat, 1995: 8, 5).

Bulgaria is one of the most biologically diverse countries in Europe (Biodiversity Support Programme, 1994). However, as in other Eastern and Central European countries, Bulgaria experienced severe environmental degradation during the period of communist rule. After the Second World War its industrialisation process prioritised economic over environmental considerations. This resulted in

Table 1. Selected Statistics for Bulgaria

Total Area 1994 (km^2)	Agricultural Land 1994 (km^2)	Wooded Area 1994 (km^2)	Population 1994 (1,000)	Population Density 1994 (per km^2)	GDP 1994 (Mio US $)	GDP per capita 1994 (US $)
111,910	60,180	39,130	8,549	77	16,985	1,106

Source: Eurostat (1998)

wasteful resource use. However, it is difficult to gain an accurate picture of the extent of environmental pollution in Bulgaria. During the period of communist rule data gathering was limited and access to environmental information severely restricted. Since 1989 there have been numerous but often conflicting estimates available (Baker and Baumgartl, 1998: 189-199).

As in other Eastern and Central European countries, Bulgaria has experienced degradation in air, water and soil quality. It is generally accepted that air pollution is the most significant environmental problem that arose during the period of communist rule, a problem that has remained since the collapse of the old regime. The energy sector is a main cause of this problem, especially the burning of low-grade, high-sulphur lignite for both electricity production and for domestic use. Air quality is particularly poor in the Maritsa-Istok agglomerations in the eastern Upper Thracian plain, where large deposits of opencast strip-mined lignite supplies three lignite-burning power stations (Carter, 1996: 45). The pollution in Pernik, one of Bulgaria's major industrial areas, comes largely from the thermal power plant which burns lignite, although the cement and metallurgical industries have also contributed to making this Bulgaria's most polluted town. The burning of lignite for domestic purposes has also contributed to poor urban air quality, especially in the capital Sofia. The city is located in a valley surrounded by high mountains, which act as a pollution trap. Sofia is also exposed to other sources of pollution, in particular from the Kremikovtsi metallurgical complex, located in Sofia-North industrial zone. Prevailing winds carry this pollution towards the city centre and eastern residential areas (see Carter, 1996: 54-57). There is also pollution from the chemical complex at Vratsa, west of Sofia. The metallurgical industry has made a significant contribution to SO_2 and NO_x air pollution. Industrial sources of air pollution are also associated with cement production, especially at the infamous "Vulkan" plant (Dimitrovgrad) and the Zlatna Panega and Beli Iskur cement plants and the non-ferrous metal smelters in the southern cities of Plovdiv and Kardzhali.

Bulgaria suffers from water pollution problems. The country has very limited water resources and water quality and availability has become an issue in national elections. It has been estimated that all major rivers in Bulgaria are either organically dead or becoming so in their middle or lower reaches (Russell, 1991: 18).

Table 2. Detailed Inventory of Emissions by Source (CORINAIR 1990) for Bulgaria

	SO_2	NO_x	NMVOC	CH_4	CO	CO_2	N_2O	NH_3
1. Public Power, etc.	1,452.5	86.3	1	1	13	31,822	7,444	0
2. Combustion commercial, residential etc.	120.6	5.6	17	18	273	7,103	253	0
3. Industrial combustion	326.4	71.5	9	4	105	22,529	3,125	0
4. Production processes	80.7	29.2	30	3	48	5,142	2,732	9,595
5. Extraction & distribution of fossil fuels	1.0	0.2	4	261	0	10,736	7	0
6. Solvent use	0.0	0.0	41	0	0	0	0	0
7. Road transport	10.3	136.6	74	2	372	7,597	286	52
8. Other mobile sources & machinery	16.3	29.3	6	0	15	2,118	291	3
9. Waste treatment & disposal	0.4	1.4	7	62	64	2,673	76	25,671
10. Agriculture	0.0	1.0	30	236	1	1,154	21,109	288,425
11. Nature	0.0	0.2	176	2	10	180	12,869	0
Total (for estimates listed)	2,008.2	361.4	394	589	901	91,054	48,192	323,746

Source: Eurostats (1995)

During the 1970s efforts were made by government to clean up the country's waterways, but to little effect (Pickles et al, 1998: 10). Drinking water contamination is a major problem. Much of this pollution stems from livestock effluent, although there are notable industrial sources, including from the opencast lignite mines and from the lead and zinc mining complex in southern Bulgaria. Bulgaria has numerous major chemical plants that have contributed to water pollution, including in rivers, ground water and in the Black Sea area. The Black Sea coast is exposed to air pollution from the Neftochim petrochemical complex in Burgas and various chemical plants in the Varna-Devnya region.

The Kozloduy nuclear power plant presents a particular environmental hazard and has been the focus of considerable international attention. The plant began operations in 1970 and produces about half of Bulgaria's electricity. Kozloduy has a history of accidents, a poor safety culture and its equipment is outdated. International attention has also focused on poor safety standards, a problem made worse by the departure of the skilled workforce of Russian engineers after 1989. Following international concern, Unit 1 of the plant was shut down for repair in 1996 and international funding, largely motivated by self-interest, has been used to improve safety and training at the remaining units of the plant (*Financial Times*, 5 May 1993). However, very serious concerns remain about the plant's safety. Despite this, a continued role is projected for nuclear power in Bulgaria, given that it has very limited energy resources and insignificant oil and gas deposits.

Bulgaria is a contributor to transboundary pollution, particularly SO_2, ranking twelfth among European dumpers in the 1980s. However, in 1981, Bulgaria signed the Convention on Long-Range Transboundary Air Pollution, making it one of the first East-block countries to do so. In November 1988 it also signed the European protocol on nitrogen oxides (NO_X). In turn, Bulgaria suffers from imported transboundary air pollution. The case of the transboundary air pollution in the town of Ruse was particularly significant. Ruse is located across the Danube river from a Romanian chlorine and sodium plant in the town of Giurgiu. In Ruse chlorine levels were often fourteen times above the nationally recognised safety levels (Carter, 1996: 46). Local protest about this plant played a formative role in the development of an independent environmental movement in Bulgaria (Baker, 1996; Carter, 1996: 53-54). In 1991, foreign aid was made available for environmental improvement and, partly under pressure from the EU, Romania and Bulgaria agreed to co-operate in monitoring pollution in the Ruse-Giurgiu area (Hutchings, 1997: 232).

2.2 Developments Since 1989

There is no simple, direct relationship between the collapse of communism and improvements in environmental ambient quality in Bulgaria. The complexity of the relationship between environmental quality and the transformation process can be illustrated by developments in the transport sector. Here there has been a shift from the old, highly polluting cars to Western models, reducing air pollution. However, public transport has also declined and there has been a general increase in car ownership, resulting in a doubling of the contribution of car emissions to air pollution between 1991 and 1994 (Pickles, et al., 1998: 19). Furthermore, international funding for road building programmes, in particular those aimed at improving trans-national road linkages in the region and funded through EU programmes, are also contributing to increased pollution.

Similar complexity can be seen in the industrial sector. Bulgaria has seen a decline in levels of industrial pollution since 1989, especially with regard to air emission levels. These are mainly due to the slump in production that accompanied the end of communism. The exposure of Bulgaria's industry to market competition resulted in the closure of numerous industrial plans, especially those that were most inefficient in resource use. Industrial output in 1994, for example, was 7 per cent lower than the previous year, with a decline of up to 15 per cent in some sectors (Hutchings, 1997: 233). Overall, industrial output fell by 40 per cent and agricultural output by 33 per cent between 1989-1994. During this period GDP fell by more than 24 per cent (UNDP: 5).

The achievement of improvement in environmental quality through production reductions, however, has done little to enhance Bulgaria's environmental management capacity. Indeed it may even diminish such capacity by, first, giving the appearance of progress where in fact there has been none. This makes it difficult to ensure continued mobilisation of popular support for environmentally sound programmes and policies. Second, it can mean that, "without substantial restructuring of practices and technologies ... economic growth, – if and when it occurs – may lead to new rounds of environmental degradation in a context in which there has been little intervening change in public attitude" (Pickles, et al., 1998: 4).

Table 3. Trend in air emission levels (1980-1995) for Bulgaria

	1980	1985	1986	1987	1988	1989
Total carbon dioxide emissions (Mio t CO_2)				100	102	98
Total methane emissions (1,000 t)				646	660	638
Total sulphur oxide emissions (1,000 t SO_2)	2,050			2,420	2,228	2,180
Total nitrogen oxide emissions (1,000 t NO_x)				416	415	411
Total ammonia emissions (1,000 t NH_3)						
Total carbon monoxide emissions (1,000 t CO)				997	995	985
Total nitrous oxide emissions (1,000 t)						
	1990	**1991**	**1992**	**1993**	**1994**	**1995**
Total carbon dioxide emissions (Mio t CO_2)	91	68	68	72	70	73
Total methane emissions (1,000 t)	759	580	621	665	528	507
Total sulphur oxide emissions (1,000 t SO_2)	2,020	1,678	1,128	1,426	1,479	1,477
Total nitrogen oxide emissions (1,000 t NO_x)	376	266	239	242	230	264
Total ammonia emissions (1,000 t NH_3)	323	280	230	220	146	136
Total carbon monoxide emissions (1,000 t CO)	900	608	768	830	858	819
Total nitrous oxide emissions (1,000 t)	48				49	48

Source: Compiled by author from Eurostats (1998)

However, there have been some environmental improvements that are not merely attributable to production collapse. The improvements at the Kozloduy nuclear plant are a case in point. Similar improvements have also taken place in the industrial sector, for example at the lead smelter at Plovdiv. Prior to 1989, both local residents and Ekoglasnost activists campaigned for the closure of this plant. Following new investment in pollution abatement technology in 1992, SO_2 emissions from the plans have been considerably reduced, although reductions in production levels also contributed to this (Manser, 1993: 79).

The development of a market economy in Bulgaria, albeit limited, has resulted in a complex array of environmental consequences. Water pollution, for example, has increased as a result of increased numbers of largely unregulated small private food producers dumping waste in rivers (Georgieva and Moore, 1987, quoted in Pickles, et al, 1998: 19). But, as we see from the case of privatisation, the picture is complex. On the positive side, the exposure of firms to the rigours of the market has led to more efficient resource use in surviving industry, particularly regarding energy use. Furthermore, privatisation deals for particular plants have also resulted in environmental clean-up: for example, the Belgium company, Union Miniére, which has acquired a controlling stake in the Pirdop metallurgical plant, is implementing a four year environmental refurbishment programme financed through international loans to the Bulgarian government *(Financial Times*, Special Report, March 8 1999). However, privatisation has also brought negative environmental consequences. Government has been slow to implement measures for dealing with the environmental liability of firms from the earlier years under state control: only since the UDF-led government came to power in 1997 have they accepted legal responsibility for past environmental damage *(Financial Times*, 8 March 1999). Further, privatisation has reduced state income, without corresponding compensation through increased tax revenues (Pickles, et al, 1998: 4), thus decreasing the financial resources available for environmental management. In addition, many firms have been privatised into the hands of the old nomenklatura, who have a history of eschewing environmental regulation (Baker and Baumgartl, 1998). This has allowed the "embeddedness" of Bulgaria's institutions of governance, production and environmental regulation, to continue in the post-1989 period and has done little to remedy the problem of unclear and overlapping environmental policy jurisdictions.

3 Development and Main Characterises of Environmental Policy

3.1 Main Laws

The 1991 Constitution of Bulgaria imposes the protection of the environment and natural resources as an obligation on all state authorities. It also proclaims the right of citizens to a favourable and healthy environment as a fundamental human right, although citizens have also a duty to protect the environment. Furthermore, the Constitution declares natural resources, with the exception of land, to be state property. As well as Constitutional protection, Bulgaria's environment is also guarded by a wide range of environmental and nature protection legislation.

The main laws on nature protection and environmental management in Bulgaria can be grouped into four categories, following Friedberg & Zaimov (1998: 92).

1. *General laws* deal with matters of general significance and may be regarded as framework legislation: the Law on Protection of Air, Water and Soil (1963), the Law on Nature Protection (1967) and the Environmental Protection Act of 1991.
2. *Specialised laws* regulate pollution from particular industries or sources, including the Law on Mines and Quarries (1957), the Law on Forests (1958), the Law on Water (1969), the Law on the Protection of Arable and Pasture Land (1973, revised 1996), the Law on National Resources (1982), the Law on the Fish Industry (1982); the Clean Air Law (1991), which suspended the 1963 Law on the Protection of Air, Water and Soil; the Statute on taxes on motor fuels and the Statute for a national fund for environmental protection (ecofund) *(State Gazette,* February 23, 1996, No, 16, 1996) and EIA (1995, No. 73, Regulation No. 1).
3. *Laws regulating other matters* contain rules on the environment, including the Law on Public Health (1973), the Law on Town and Country Planning (1973) and the Law on the Protection of Farm Property (1974), to name but a few.
4. *International conventions,* including The Convention on Transboundary Air Pollution, The Convention on the Protection of the Ozone Layer, the Montreal Protocol and the Convention on Trade in Endangered Species.

The post-1989 period has seen the development of new, legislative environmental capacity in Bulgaria. Of particular note is the 1991 environmental framework legislation. Nevertheless, from the point of view of regulation and legislation, there is a great deal of continuity between the period of communist rule and the post-1989 period. Most of the laws in existence prior to 1989 are still on the statute book. They include statutes, executive decrees, ministry regulations and norms promulgated by the still existing National Committee on Standardisation (Friedberg and Zaimov, 1998: 96). Furthermore, the problems encountered in the utilisation of this capacity, that is, the historic problems encountered in putting these laws into effect through environmental management and countryside protection policies, continue to have a negative influence on the country's ability to manage its environment effectively.

3.2 Capacity Development and the Implementation of Legislation

Bulgaria has a long history of countryside and nature protection. The Council for the Protection of the Countryside was founded in 1928 and Bulgaria passed its first nature conservation law in 1936. The 1930s saw the designation of the first nature reserves (Parangalitsa) and the first National Park (Vitosha) in the mountain region outside Sofia. However, much of the impetus of these decisions was lost during the Second World War and it was not until 1960 that the first state decree for protecting the countryside was passed *(Izvestiya,* 1960), with further amendments on nature protection added in 1967 *(D'rzhaven Vestnik,* 1967) (Carter, 1996: 38). The 1967 Law on Nature Protection led to the designation of a large number of nature reserves, national parks, places of special interest and beauty spots, and the drawing up of preservation lists for wild plants and animals. The Popular Committee for Nature Protection was also created during this period. In 1974 The Council for Environmental Protection and Reproduction was charged with combating the country's environmental problems. Environmental considerations were also integrated in the national environmental plans. In 1988, a new national programme for environmental conservation was launched, to cover the period to the year 2000 (Carter, 1996: 39). A new Ministry for Land, Forests and Environmental Protection was also established, with responsibility for the new environmental policy. During the period between 1960 and the 1980s, new laws setting standards on ambient quality were also passed, comparable to those passed in the West.

Despite these developments, however, the gap continued to widen between policy formulation and implementation, which can largely be attributed to the prioritisation of economic production over environmental protection goals. This meant that the country's capacity to implement environmental legislation lagged behind its industrial development process. This has resulted in deep, institutional lacunae, including shortage of expertise, weak monitoring capacity, underdeveloped institutional structures, and poor administrative co-ordination. These lacunae continue to hamper environmental management in the post-communist period. Furthermore, the lack of a tradition of advocacy against authority within the legal profession has meant that the legal profession has not been able to facilitate the implementation of environmental regulation nor capacity improvement. Furthermore, there were loopholes in many of the laws, in particular the 1963 Law on the Protection of Air, Water and Soil, which granted exemptions in the national interest to new projects. There was also a provision in the penal code making it a crime to reveal pollution information concerning state enterprises. Despite this, however, public concern about pollution levels grew during the 1980s, particularly as studies revealed alarming increases in childhood respiratory and circulatory diseases in urban areas (Carter, 1996: 51). This public concern was eventually to lead to the foundation of

Ekoglasnost in 1988, which was subsequently to play a significant role in toppling the communist regime (Baker, 1996).

After the collapse of the old regime, a new Environmental Protection Act was passed in 1991. This provides basic, but broad environmental framework legislation. It introduced the "polluter pays principle" and environmental impact assessment (EIA) for major projects. It also contained a number of provisions relating to the right to a favourable and healthy environment, to the role of the minister of the environment and to pollution taxes, which places a tax on pollution and natural resource use. The law also makes provision for an annual report on the state of the environment, a ban on the trade in hazardous waste and the harmonisation of Bulgaria's legislation with that of European Community law.

While the EPA of 1991 is a welcome development, it suffers from a number of weaknesses, especially a lack of blanket pre-emption of the old laws. Furthermore, some of the provisions are very general, particularly on nature conservation and forest management and cannot be implemented without further supporting regulation. However, there has been little regulatory progress made since 1991. Capacity development in relation to sustainable forest management policies, to take an example, is hampered by the failure of government to address the issue of land restitution adequately, which may involve returning some forest land to private hands. In addition, there are still whole sectors ignored by the 1991 EPA and which are thus still not subject to any fundamental controlling legislation. Bulgaria has, for example, no coherent national policy or legislation on waste management. Air and water standards are regarded as inadequate (CEC, 1997: 3.6) and regulation of chemical substances and agricultural pollution is lacking or needs updating.

A 1995 legislative amendment has also weakened the law. The amendment exempts projects of "vital importance" from the public participation requirements of the EIA process and authorises the conduct of EIAs one year after, rather than before, such projects are initiated (UNDP: 6). The law was amended when the BSP had a majority in the National Assembly and was prompted by a water project to deal with Sofia's water shortage problems. Both local residents of the rural region from which the water was to be drawn and environmental NGOs protested

Table 4. State of Waste Legislation and Policy in Bulgaria

			Environmental permitting of disposal and recovery facilities						
Waste Management	Priority to prevent & reduce waste harmfulness	Waste Eco-taxes	Producer Responsibility	Prevention	Recovery/ Recycling	Hazardous Waste Controls	Basel Convention	Hazardous Waste	Other Waste
×	×	×	✓	×	×	✓	✓	✓	✓

Source: EEA (1998)

against the amendment. The amendment is regarded as a setback both for environmental protection and for democratic progress (Friedberg & Zaimov, 1998: 91). The amendment has also restricted the participatory opportunities available to environmental NGOs, diminished their input into the policy-making process and reduced the range of activities available to them that may have capacity development potential.

3.3 Main Policy Instruments

Despite the collapse of the communist regime, regulation has remained the main instrument of environmental policy in Bulgaria. Most of this legislation is of a "command and control" type, with a particular emphasis on end-of-pipe regulation. An example of this is the Clean Air Law (*State Gazette*, May 28, 1991, No. 45) which sets standards for air quality, limits for particular emissions and empowers the Minister for the Environment to determine admissible concentration limits for the emission of harmful substances. The National Committee on Standardisation continues to be responsible for norms on ambient quality.

Reliance on legislative policy tools presents severe problems because the rule of law remains weak in Bulgaria. Improvements in Bulgaria's environmental protection legislation cannot have a multiplier effect as a capacity enhancer unless more deep-rooted reform is instigated. Equally problematic is the fact that the country lacks the administrative capacity to implement its environmental legislation, particularly at the local level. There are shortages of human and technical resources as well as, more problematically, a lack of political will to put the laws into practice. Environmental monitoring capacity is also weak, with little dependable environmental data (Pickles and Mikhova, 1998: 105). There are very few environmental lawyers in Bulgaria, making environmental lawsuits rare and retarding the drafting of laws and regulations, prosecution of violators, administrations of existing standards and the pursuit of civil remedies (Friedberg & Zaimov, 1998: 95).

Since 1989 there have been some attempts to make use of policy instruments other than legislation for environmental management purposes. However, without supporting structures, their usefulness is limited. A National Environmental Protection Fund was established in 1995, financed by fines and taxes for industrial pollution and dues. The Fund is supposed to provide grants and loans to municipalities and enterprises for environmental measures. However, there have been complaints that central government is failing to distribute this Fund adequately to local authorities. Here we find a good example of the limits on Bulgaria's capacity to make use of tools other than legislation for environmental management – in this case the limits are imposed by lack of local government reform and by the failure of post-communist governments to implement a culture of accountability, transparency and control within central government. Further, the use of fiscal and other non-regulatory tools requires an effective and operating market system, which, as yet, remains underdeveloped in Bulgaria.

3.4 Policy Style

Environmental legislation and policy in Bulgaria have adopted many of the general principles that underlie Western policy and that are found in international environmental management forums, such as those following from the UNCED process, including the Rio Declaration. These include the precautionary principle, the "public right-to-know" principle, EIA and the "polluter pays principle" (UNEP).

However, while these principles have been adopted at the formal, legal level, many barriers remain to their effective utilisation in policy formulation and implementation. This is because their usefulness as guides to policy is in turn dependent upon deeper, structural capacity within the political and administrative systems of the state. Their effective use requires, for example, a co-operative, consensual style of environmental policy making and co-operation between the state, industry, and the citizen. This in turn is premised upon open, accountable forms of governance. But, although Bulgaria has lost its traditional Communist Party institutions and its supporting external network, the old political culture still thrives: both political leaders and government officials continuing to treat opponents as enemies to be purged rather than as legitimate participants in public debate. Here the culture of favouritism and bureaucratic discretion continues – favouring private gains and partisan policy at the expense of the public good. The unstable political climate, corruption and the polarised nature of political discourse and party support has meant that a consensual style of policy-making has not developed. On the contrary, successive governments have prioritised their own interests, at the expense of either environmental management or developing a more open political climate and policy-making culture.

Consensus building is made more complex by the history of antagonism between the police, the state, and the environmental movement, especially during the final years of the communist regime (Baker and Baumgartl, 1998). Despite the central role played by environmental groups in the downfall of the communist regime, environmental NGOs are still outsiders in the Bulgarian policy-making process. While the first government elected under the new constitution contained a number of committed environmentalists, in general the capacity of environmental groups to participate in environmental decision-making via parliament or the courts has remained limited (Kodjabashev, 1995). Activity is confined to street protests and attempts to use the media to accomplish environmental objectives.

3.5 Major Policy Initiatives

Bulgaria adopted a major new environmental strategy in 1992, which was heavily influenced by the World Bank. The strategy emphasised three needs:

- to build upon the environmentally beneficial results of economic reform;
- to develop a suitable institutional framework for policy development and implementation;
- to address the most serious environmental "hot spots", which are characterised by high environmental risk to human health. This led to the production of environmental action plans (World Bank, 1992).

More specifically, the strategy emphasised the importance of economic reform and structural change in promoting more efficient use of natural resources and energy. It also argued for a shift towards less environmentally damaging economic activities and technologies. Moreover, it addressed how government might generate the revenue to finance environment expenditure. The strategy also outlined future developments in the country's institutional, legal and regulatory frameworks as well as identifying areas that require priority action. The latter included air emission levels and drinking water quality.

The Strategy was updated in 1994, following the criteria established in the Environmental Action Programme (EAP) for Central and Eastern Europe adopted by environmental ministries at the Lucerne *Environment for Europe* Conference in 1993. Its 1994 update is in effect Bulgaria's environmental action programme for implementing the EAP. To date, implementation of the strategy has not begun.

4 Main Actors

4.1 Environmental Institutions

4.1.1 Central Government

A Standing Committee on Environmental Protection exists within the Bulgarian National Assembly. Its task is to draft laws for the assembly's approval and to make recommendations to different state authorities for improving their environmental protection activities. During the first post-communist government, this committee contained a number of committed environmentalists and was responsible for drafting the landmark Environmental Protection Act of 1991. However, since 1991 the capacity of the committee has been severely reduced, not least by the defeat of many of the environmentalists in the subsequent elections and the shift in government priorities following the election of the BSP to power.

Overseeing the implementation of the laws passed by the National Assembly is the task of the Council of Ministers. While all ministries perform environmental protection activities within their specific areas of competence, the Ministry of the Environment has the most specific powers with respect to environmental management. They include: (1) imposing sanctions for non-observance of environmental legislation; (2) the supervision of the EIA process; (3) the issuing of permits and (4) the promulgation of environmental standards and norms of maximum admissible concentrations of harmful substances in the environment (Friedberg & Zaimov, 1998: 84-85).

There are a number of specific bodies within the ministry that help shape the institutional dimension of Bulgaria's environmental management capacity. There is a research centre charged with information provision, data gathering and environmental monitoring. The centre lacks resources and, as of 1998, is undergoing a process of reorganisation. There is also the National Centre for Environment and Sustainable Development in Sofia, which co-ordinates research, conducts analysis, supports the development of environmental protection programmes and helps set national standards. By and large, however, standard setting is the task of the long-established National Committee on Standardisation. In the Bulgarian context, "add-on" institutional developments, such as the establishment of the National Centre for Environment and Sustainable Development, are of only limited value. This is because they are examples of what we may term "formal institutionalisation", that is, the establishment of specific state institutions charged with, in this case, specific environmental responsibilities. This can be contrasted with a process of "substantive institutionalisation", that is, resource deployment, internalisation of environmental values, and the establishment of stable, effective implementation conduits – all of which have yet to be established in Bulgaria. The Commission of the European Communities, for example, found that the ministries are unable to exercise control over state enterprise managers, which has severe implications for environmental management, given that industrial production is a major source of Bulgaria's environmental degradation (Baker and Baumgartl, 1998)

Moreover, within the ministry as a whole there is a lack of technical expertise, especially with respect to monitoring, standard setting and enforcement, and there are financial and other resource shortages. The ministry is relatively small, with approximately 140 staff. Its institutional weaknesses are compounded by the fragmentation of administrative responsibility: control of industrial enterprises is carried out by sectors, such as the Committee on the Use of Atomic Energy for Peaceful Purposes in Bulgaria (CUAEPP), which monitors the Kozloduy nuclear plant (Nikolova, 1998: 49); but there is little co-ordination between sectors.

There is also a system of regional environmental inspectorates within the ministry. The 16 inspectorates represent the backbone of the Bulgarian environmental management system. The inspectorates' rights and responsibilities have not changed substantially in the post-communist period, nor has the territory they administer. However, their internal organisation has experienced major changes, which have helped increase management flexibility. Until 1992, the departments within each regional inspectorate were structured around the regulation of different sources of pollution. Since then new structures are build around a "component" principle – air, soil, water and radiation. Specialised groups also monitor forests and other natural ecosystems, solid waste and the Black Sea (Koulov, 1998: 224). Under the new system, different inspectorates are now able to design and staff their own departmental structures according to the environmental needs of their specific region.

However, the regional inspectorates suffer from poor financial resources and, by and large, there are only vertical lines of co-ordination and information dissemination, that is, upwards to the ministry and not horizontally to other inspectorates. Staff in the inspectorates suffer from low wages, and poor career prospects, leading them to need to engage in private work, which may give rise to conflicts of interest. Furthermore, high inflation rates have eroded the real value of pollution fines imposed by the inspectorates. Furthermore, the proliferation of small firms in post-communist Bulgaria has not only put pressure on the limited waste treatment facilities available and has also created a monitoring nightmare for the inspectorate (Koulov, 1998: 224).

More generally speaking, the inspectorates have to exist in a context where there continue to be problems with the functioning of the executive, in part due to the instability arising from the high turnover rate of government, making it difficult to ensure policy continuity. There are also problems with the impartiality of the civil service, given the failure of post-communist governments to introduce a civil service act, although the constitution calls for it. As a consequence, there is little regulation of public administration, divisions of responsibility are unclear and the exercise of discretionary power continues to lead to problems of corruption.

Post-1989 attempts to introduce changes in the administration of nature conservation exposed some of these problems. After 1989, the National Nature Protection Service was formed within the Ministry of the Environment, with responsibility for conservation. However, this has led to a dispute over the demarcation of administrative responsibility between the National Nature Protection Service and the older Committee on Forests. This latter body has traditionally enjoyed jurisdiction over much of the forestland in the country and has access to financial resources outside of central government. The failure to clarify the administrative responsibilities of each of these bodies has led to cases where parallel park administrations have been set up for the same park and to cases of implementation paralysis (Cellarius, 1998: 63).

Other ministries take responsibility for environmental protection matters, including the Ministry of Agriculture and Food Processing. In additions, there are a number of academic institutions, such as the Institute of Water Problems, and the Central Laboratory of General Ecology of the Bulgarian Academy of Sciences, the latter dealing with matters relating to EIAs. While individual ministries need to continue to take such responsibility, it has become increasingly clear that effective environmental management requires the establishment of inter-organisational co-operation mechanisms. These are by and large absent, although some progress has been made. Memoranda of understanding have been signed between the Ministry of Environment and the Ministry for Regional Development and Construction, the Ministry of Health, the Ministry of Transport and the Committee on Forests, clarifying tasks where environmental responsibilities overlap. This has helped, for example, in the elaboration of a National Environmental Health Action Plan by the Ministries of Environment and of Health (WHO, 1999). However, these integration efforts are hampered by the strong tradition of vertical organisation within the central administration, which is in part a legacy of central planning. Further, the major sectoral ministries, including those dealing with energy and industry, give low priority to environmental matters, and budget cuts have led to the reduction or elimination of environmental units in some ministries (UNEP: 7).

4.1.2 Regional and Local Government

The 1991 constitution provides for the division of Bulgaria into municipalities *(obshtina)* and regions. The country is divided into nine regions, in which the regional governor co-ordinates action by the state. There are plans for administrative reform, replacing these nine regions with between 12 and 21 districts with rights to self-government (CEC, 1997: 3.5). There are also 255 municipalities, whose governing bodies – the mayors and town councils – are elected for a four-year term. Free municipal elections were held in late 1995. The 1991 constitution gives municipalities responsibility for managing local public services, including some environmental services. Although the constitution provides for the autonomy of the local authorities, this remains, as yet, limited, particularly due to their financial dependence on state funding. There is also a lack of clear definition of jurisdictional boundaries between central and local authorities.

While the evolution of effective decentralisation is not a sufficient condition for effective environmental management, in the Bulgarian case there is little doubt that decentralisation is a necessary condition for a socially rooted environmental policy – not least because such democratisation can increase stakeholder involvement in policy-making and can contribute to a sense of shared responsibility for environmental management, especially between the citizen and the state.

As far as environmental management practice is concerned, most local governments have not fully tested the limits of their powers to regulate pollution against the backdrop of the prerogatives of the Ministry of Environment, exercised in part by the ministry's regional inspectorates – which can be regarded as a arm of central government. Further, local authorities feel too politically and economically weak to confront large economic enterprises on environmental issues, many of whom provide much needed local employment. There is also disagreement over entitlement to funds generated by environmental fines – with some localities claiming the central government was withholding its share of such money (Friedberu and Zaimov, 1998: 94).

4.2 Green Organisations

The last decade has seen the rise of both environmental political parties and environmental NGOs in Bulgaria. In the period leading to the collapse of the old regime and in the immediate post-communist period, the most influential environmental activists and environmental groups in Bulgaria became deeply involved in the electoral process and parliamentary politics. As a consequence, it is important to examine this electoral and parliamentary experience if we are to explore the capacity of Bulgarian environmental proponents to shape environmental awareness and ultimately environmental policy.

4.2.1 Environmental Political Parties

The most politically significant environmental group established in the period leading up to the collapse of the old regime was Ekoglasnost. Ekoglasnost had its origins in a local protest in the town of Ruse, on the Bulgarian-Romanian border. Here local mothers began to act on their concerns about the health effects on their children of a chlorine production plant located across the river in Romania (Baker and Baumgartl, 1998). This early case of civil opposition laid the foundation for the development of a broader-based environmental movement, which ultimately was to stimulate the mobilisation of the wider political opposition that brought down the communist regime (Baker, 1996). As a result of their role, there was widespread goodwill towards and respect for environmental activists in the immediate post-communist period in Bulgaria.

On the strength of this, the first free elections in June 1990 saw several environmental activists elected to parliament under the UDF coalition. Ekoglasnost won 17 seats and the Green Party obtained 14 seats in the National Assembly. Some Green Party and Ekoglasnost members were later promoted to key positions in government. These included the 1990 Minister of Environment, Valentin Vassilev from Ekoglasnost; the Prime Minister Filip Dimitrov (1991) who was former Vice Chairman of the Green Party and member of the Conservative Green Party; Minister of Environment Dimitar Vodenicharov from the Green Party in 1991 and the Minister of Finance (1990).

The presence of environmentalists in the early years of the National Assembly was important for environmental capacity building: their influence helped to keep environmental issues on the policy agenda and ensured that some environmentally sound decisions were made by the first post-communist governments. These include the decision to halt the construction of a new nuclear reactor near the city of Belene. Environmentalists in parliament also put intense pressure on the Bulgarian and Romanian governments, leading to action on pollution in Ruse and they also negotiated agreement for step-by-step closure of the most dangerous industrial source of pollution (Georgieva, 1992: 15). These were also instrumental in the development of the new environmental framework legislation in 1991.

Despite these positive influences, however, the wider political and economic situation in Bulgaria continued to limit the capacity of environmentalists in government to instigate effective environmental management initiatives. In the face of mounting economic problems, there has been a decline in the priority given to environmental considerations (Pickles et al., 1998a: 245-248). During the economic crisis that followed the March 1992 miners' strike, for example, the green Prime Minister Philip Dimitrov was forced to drop environmental issues to near the bottom of this government's priority list (Georgieva, 1992: 16). The prioritisation of economic over environmental considerations also contributed to the defeat of many environmentalists in the October 1991 democratic election. Here green representation was reduced from almost 9 per cent to 4 per cent. The position of environmentalists in government was weakened by the fact that several of the key figures who were members of the first post-communist government and the first post-communist National Assembly have either taken up more lucrative positions in the private sector or have become disillusioned by politics.

The Bulgarian electoral system has also contributed to the problem of green representation. On one hand, the system of proportional representation has allowed many small to accumulate a sizeable share of the vote parties in any one election (up to 25%) (Olson, 1998: 132). On the other hand, however, the imposition of a 4% hurdle, that is, the minimum proportion of vote that a party must obtain to place any of its candidates in parliament, has resulted in a sizeable proportion of voters been without direct representation in parliament. The fact that the party system is also characterised by "polarised pluralism", with few parties dominating parliament from the beginning, has also reduced the opportunities for environmental political parties to enter the political arena, leaving them heavily reliant on coalitions with the UDF, which at times have proved problematic.

Ekoglasnost has also been weakened by internal problems. One of the first new parties to be registered in post communist Bulgaria was the Green Party, which formed following a split with Ekoglasnost. Since then both groups have continued to suffer from internal divisions. As a consequence, by the election of October 1991, two separate Ekoglasnost groupings had been formed – the Movement Ekoglasnost *(Ekoglasnost Dvizhenie),* which continued to form coalitions with the UDF, and the Political Club *(Politicheski Klub),* which has worked in alliance with the BSP. There are also two green parties, the Green Party and the Conservative Ecological Party (Baker and Baumgartl, 1998). These splits within Ekoglasnost are in part due to the diverse and somewhat opportunistic nature of its membership during the last years of the communist regime. Many had joined the environmental opposition prior to 1989 not because of their commitment to environmental protection and conservation, but because it offered a relatively safer arena for dissent activity than did direct political protest. After 1989 their interest in the environment waned. Ekoglasnost was also weakened by the difficulties it encountered as it ceased to be a movement of mass opposition and sought instead to adopt the structure and form of a political party. Sharing political responsibility with other parties inside the UDF also exposed greens to public criticism, for which they were not sufficiently prepared (Georgieva, 1992: 16).

Examination of the fate of Bulgaria's environmentalists in government reveals that capacity for action depends not only on formal participation or the determination and competence of activists, but also on the structural, framework conditions within which they are obliged to act. Bulgarian ecologists were propelled into governmental positions by the force of circumstances and in office they were able to make a contribution, especially to the enhancement of Bulgaria's environmental legislative capacity. But their efforts were hampered by deep-rooted barriers operating at the political, economic, cultural and institutional levels. These blocked effective utilisation of the country's enhanced legislative capacity and severely restricted the scope for strategic and co-ordination action. Once in office, Bulgarian environmentalists found neither the opportunity structures nor the framework conditions for the policy action they desired.

4.2.2 Environmental NGOs

The first formally registered, independent, environmental NGO in Bulgaria was the Bulgarian Society for the Protection of Birds (1988). The years since then have seen a significant growth in the number of environmental NGOs in Bulgaria. Their growth has been helped by the high levels of urbanisation, education and scientific literacy, as well as deep-rooted popular concern about the negative health effects of environmental pollution, especially air pollution. There are now an estimated 120-150 environmental NGOs, most based in Sofia (see Cellarius 1998 for a listing). They include the Association of Bulgarian Ecologists, which describes itself as an educational and scientific organisation; Association Ecoforum, which is mainly composed of scientists and business people; the National Ecology Club and the Union for Nature Protection. The majority of these NGOs are concerned with nature protection. They range from small, local organisations engaged in place-specific issues to larger groups operating at the regional or local level and with a more diverse set of interests. Some of the organisations operate as single independent groups while others consist of networks of branches or clubs (Cellarius, 1998: 64). In 1992 an umbrella group, the Union of Bulgarian Foundations, was formed to foster co-ordination between NGOs (Baker and Baumgartl, 1998). However, there are few linkages between environmental NGOs, in particular between local campaign groups and national organisations.

Despite their numerical growth, the capacity of Bulgaria's environmental NGOs to facilitate the promotion of sustainable development and to build local capacity for resource protection remains limited. There are a number of reasons, including both "external constraints" and "internal weaknesses". A major external constraint is that environmental protection no longer has the priority support that it enjoyed during the early days of transformation, having been displaced by economic considerations. Furthermore, the polarisation characteristic of political life in post-communist Bulgaria also divided environmentalists – into what may be termed "red greens" (socialists) and "blue greens" (democrats) (Georgieva, 1992: 16). This has caused fragmentation and conflicts between and within NGOs and a consequent lowering of their potential to act as effective proponents of environmental protection. The underdeveloped nature of Bulgarian civil society, with its weak traditions of individual freedom, low political expectations and a long history of submissive mentality vis-à-vis the state (Gallagher, 1998: 44) combine to make it difficult to engage effectively in political activity, or for the public to be active in shaping the process of social change. Many individuals still believe in the protective role of the state and, although the state has relinquished this role, the culture of personal passivity remains. Furthermore, the history of oppression of independent thought also acts as a barrier. The current political instability make it all the more important to take this into account, especially because traditional elites, including the communist successor party the BSP, have been able to preserve much of their authority in the face of change. Indeed, the fate of the environmental movement, from the heady days of the late 1980s to their decline in the 1990s, reinforces the popular belief in the continued power of the corrupt and largely unreformed post-communist Bulgarian state. Nevertheless, we need to bear in mind that public awareness of environmental problems is high and that this laid the foundations for the movement that toppled the old regime. This potentially important resource for future change, albeit dormant under present circumstances, should not be ignored.

Environmental NGOs also suffer from internal weakness. In Bulgaria, and throughout Eastern and Central Europe, the post-communist period brought fundamental changes in how environmental concerns are expressed. After 1989 the movement evolved from a mobilising agent for populist protest into pragmatic, goal-orientated, professional organisations. It has been argued that the process of professionalisation has fragmented the movement and shifted their focus away from domestic issues towards issues deemed to be of greater concern to external funding agencies (Baker and Baumgartl, 1998). External environmental funding agencies, for example, tend to favour big not small projects, despite the fact that the latter have a greater potential to increase local capacity (Baker and Jehlička, 1998). The toppling of the old regime also robbed the environmental movement of its mobilising focus.

Many of Bulgaria's environmental organisations are, furthermore, dominated by Sofia based academic experts and many NGOs have a limited membership base. Their participation is often based on the belief that it can offer access to potential sources of research funding and employment opportunities. There is less interest in mobilisation, information dissemination, and hands-on engagement in environmental protection activity. The focus of many of the groups is on technical issues, stressing the need for technical approaches to environmental problems. Institutional capacity, including resources, are also limited. Public funding (local or national government ecofunds) only provide limited support, and environmental NGOs remain largely dependent on membership fees or donations, or on international funding, often through the Regional Environmental Centre. The legal basis for NGOs' existence is also weak, and NGOs are restricted in their activities by the 1949 Law on Persons and Family, that provides the legal basis for officially recognised non-profit organisations in Bulgaria. Under the law NGOs are forbidden to engage in political activity and, as non-profit organisations, they face restrictions on their ability to undertake "commercial" activities, for example selling campaign posters, T-shirts etc.

4.3 The Media

The impartiality of the media in Bulgaria has been questioned (CEC, 1997), and state interference in public broadcasting has continued despite the end of the communist regime. In the lead-up to the elections of 1994 and 1997, for example, the opposition parties were given less airtime than the governing majority (Soveroski, 1997: 31). More seriously, in June 1995 the heads of state television and radio services and of the official new agency were dismissed amid BSP claims that news reporting was not sufficiently objective and was distorting government policy. Further dismissals of a number of Bulgarian National Radio journalists later in the year did little to instil confidence in the democratic credentials of the BSP or to strengthen the independence of the media. In this climate of intimidation, the capacity of the media to raise public environmental consciousness is limited.

Current restrictions on dissemination of available knowledge act as another limit on the resources available to environmental proponent in Bulgaria. Unfortunately, the public dissemination of environmental data can still be politically dangerous and challenged as unpatriotic. Information can be used to illustrate the continued inability of government to address environmental problem, raise difficult legal questions about the citizen's right to know whether they are being harmed by pollution, and can threaten the privatisation of industry enterprises, which are still heavy polluters (Pickles and Mikhova, 1998: 109). In the past environmental knowledge has acted as a mobilising resource.

4.4 The Role of Economic Actors

Bulgaria has not reached a point where the business community has become an active proponent of environmental improvement. By and large, the industry that has survived the economic upheavals of the post-1989 period is not willing to jeopardise its competitiveness by increasing outlays on environmental protection. Similarly, new firms entering the market are not willing or able to assume the costs of environmental protection, and access to environmental infrastructure, such as waste disposal and recycling facilities, is limited. Furthermore, the concentration of wealth, also resulting from elite privatisation policies, in the hands of the least public-spirited segments of society, – black (illegal), red (former communist elites) and blue (UDF elites) – threatens environmental gains. These groups have already demonstrated their preference for private gain over the public good. Unless conditions change, they will continue to impede environmental progress because, first, their disproportionate influence on the legislative process will retard the formation of strong environmental norms; and second, their power will help shield their acts from rigorous oversight or environment scrutiny (Friedberg and Zaimov, 1998: 97). Thus post-communist Bulgaria has not seen any "actor constellation" changes that might come, for example, from business support for the environment or the strengthening of the relationship between environmental NGOs and the business community. There is no trend towards ecological modernisation or towards the emergence of a "green" business sector.

4.5 International Influence on Capacity Building

Bulgaria's increasing international links are potentially very important for enhancing the country's environmental management capacity. They have, for example, facilitated Bulgaria's participation in a debt-for-nature agreement with Switzerland. Foreign help has also been used in environmental clean-up operations, such as at UM Pirdop, Bulgaria's biggest copper producer, an operation that is currently been financed by the World Bank *(Financial Times*, Special Report, 8 March 1999). Similarly, Bulgaria's participation in international environmental management forums has exposed its policy-makers to new international norms, values and expectations, particularly with regard to the promotion of sustainable development. Bulgaria's participation in the UNCED CSD process is a case in point. The Bulgarian Deputy Prime Minister and Minister of Economic Development, Rumen Gechev chaired the 4th Session of the CSD in 1996-1997. This led Gechev to establish the Bulgarian National Commission of Sustainable Development, of which he is chairman. This Commission is engaged in the implementation of Agenda 21 in Bulgaria and is composed of representatives from key industries, national NGOs, and public and private enterprises, helping to facilitate the development of a more open and participatory policy style and political culture. The establishment of the commission, its broad-based membership, and its need to provide regular progress reports to the CSD are helping Bulgaria make the first steps towards the development of a more open and participatory policy style and political culture.

The development of regional environmental co-operation structures is also potentially important for capacity enhancement. During the 1990s Bulgaria improved environmental co-operation with Romania and with littoral states of the Black Sea. Bulgaria is now a member of the Danube Commission, the Joint Danube Fishery Commission and the Swiss based IIUCN and is involved in the Black Sea Economic Co-operation. However, given the deep-seated historical animosities in this area and the fragile domestic political situation of many of the countries in the region, the capacity of such organisations is limited (Hyde-Price, 1998: 262).

Bulgaria has also signed an Association Agreement with the EU (1993), although future membership of the EU remains uncertain (Baker and Baumgartl, 1998). Among the conditions laid down by this agreement were a number on the management of Bulgaria's environment, including the demand that development be guided by the principle of sustainable development and that steps be taken to facilitate the approximation of all Bulgaria's environmental legislation to that of the EU. Through the Phare programme, the EU has made available a range of funds and expertise to guide these reforms. While the environmental component of the Phare programme has been criticised for been too weak and under-funded, Bulgaria's association agreement nevertheless was a potential major stimulus for capacity development. But a recent Commission assessment of Bulgaria's adoption of the *acquis,* including in relation to environmental measures, found that progress, if at all, was slow. The Commission feels that Bulgaria's serious environmental problems have not been effectively addressed and that considerable investment and effort are needed to establish the administrative capacity the country requires for the full adoption, implementation, and enforcement of EC legislation. This, they believe, can be achieved only in the very long term (CEC, 1997). Neither the political nor the economic climate in post-1989 Bulgaria has been conducive to enabling the country to reap the potential benefits from its relationship with the EU. Equally problematic is the West's insistence on imposing their own economic and political model of transition, when this model may be neither appropriate nor acceptable for the region. In particular, we point to the EU's belief that trade and foreign investment rather than aid are the most appropriate instruments of change. On balance, the EU's focus on marketisation and privatisation impede progress on environmental matters and helps legitimise the long-standing belief that economic matters should take precedent over other social and environmental goals (Baker and Jehlička, 1998).

Consequently, international influence has had a somewhat contradictory impact on Bulgaria's environmental management capacity. On the one hand, it has helped establish new levels of commitment, has linked Bulgaria with international regimes with systems of accountability demanding that some progress be made at the domestic level, and has exposed the country's policy makers to new ideas, values and norms. This can be seen, for example, in Bulgaria's involvement in the CSD, which is helping to transmit the notion of "sustainable development", at least at the normative level, to Bulgarian environmental policy. However, it has had negative consequences, too. Pressure from international agencies, particularly for economic reform, has helped keep environmental policy low on the legislative and administrative agendas of governments, restricting Bulgaria's commitment to the promotion of environmental protection merely to the level of declaratory politics.

5 Prospects and Evaluation

There has been an increase in environmental capacity in post-1989 Bulgaria, particularly at the legislative level, where stronger environmental framework legislation is now in place following the 1991 Act. In addition, some enhancement of capacity has occurred within the Ministry of Environment, also from greater participation in international and regional environmental management regimes. Administrative developments have also occurred within the ministry, including the establishment of new bodies and committees, and first steps have been taken towards greater inter-organisation co-operation, especially at the sectoral level, including new forms of inter-ministry co-operation. Bulgaria has also seen the establishment of a National Commission of Sustainable Development, to facilitate the promotion of Agenda 21. Environmental information is more easily available to the public, and there has been a large increase in the number of environmental NGOs, which can ultimately contribute to the development of a more enhanced public sphere.

However, further progress in improving Bulgarian environmental management capacity is still severely hampered. During the period of communist rule, the capacity to deal effectively with environmental degradation was heavily restricted by the "embeddedness" of Bulgarian institutions of governance, production and environmental regulation. A close relationship existed between polluter and regulator, as government-owned enterprises were both, and there were no regulatory bodies independent of the state. This embeddedness weakened the country's environmental management capacity and has left a legacy of unclear and overlapping jurisdiction that weakens environmental policy (Pickles, et al., 1998: 11). This embeddedness has continued into the post-1989 period. Bulgaria's failure to implement the rule of law has meant that partisanship, corruption and bureaucratic arbitrariness are endemic to the political culture. But, while the "embeddedness" of institutions has continued, their effectiveness has been eroded during the post-1989 period by the leadership vacuum, the loss of traditional sources of economic and financial support, the legacy of servility among the population, the self-enriching behaviour of the nomenklatura and the impasse caused by rapid and intense change (see Meininger and Radoeva, 1996: 53). The economy of survival that now permeates Bulgaria has reinforced reliance upon the traditional vertical structures ("connections") and informal mechanisms (bribes), which bode ill for capacity enhancement.

In this context, the role of international agencies and economic actors becomes crucial. Already there is evidence that international linkages (such as through the EU), aid and grants are contributing to environmental clean-up and, most importantly, to the initiation of sound environmental management policies. While this has not yet led to a change in actor constellation, such as the greening of domestic business or new alliances between business and environmental NGOs, international involvement has the potential to influence general environmental norms, especially in Bulgaria's legal, commercial and business sectors. Further, international support for environmental NGOS, particularly if directed towards the development of an environmental agenda that meets the needs of Bulgaria as opposed to the wishes of international funding agencies, can also help capacity building. But these international routes of influence can only achieve their full potential if more profound changes are instigated.

The lack of consensus on the direction of Bulgaria's transformation, which has resulted in a lack of depth to its reform process, is at the root of the country's problems, hampering the development of increased political, administrative and institutional capacity to deal with Bulgaria's environment. Ultimately, the increase in Bulgaria's environmental management capacity continues to rely on Bulgaria's ability to find both a direction to, and political will for, change, and to move beyond the polarised politics of the post-1989 period. Here, and throughout Eastern and Central Europe, there is increased evidence that the transformation process does not provide a simple and guaranteed transition to liberal democracy and capitalism (Bryant and Mokrzycki, 1994: 11), as the West once hoped. The problem for Bulgaria is that alternative models of transformation have yet to be found.

The problems facing post-1989 Bulgaria provide us with a pertinent example of the ways in which environmental capacity building is intimately interrelated with and constrained by a country's historical legacies, the traditions that are rooted in its institutional and administrative cultures, the nature of its social relations, and the constraints imposed by its political and economic structures. The case of Bulgaria shows that successful environmental management requires more than legislative developments, "add-on" institutional and administrative arrangements and the utilisation of new policy tools for policy implementation. Capacity building has to reach and ultimately reform the deep, structural framework and the political and economic contexts within which policy is formulated and implemented. Bulgaria is still at a very early stage of environmental capacity building. Further improvement in that capacity is ultimately dependent upon the country finding a direction to its transformation and engaging in a process of consensus building to heal the polarisation that has beset the country since the end of the old communist regime.

References

Andreev, A., 1996. The Political Change and Political Parties, in: I. Zloch-Christy, ed., *Bulgaria in a Time of Change*. Aldershot: Avebury, 25-44.

Baker, S., 1996. The Scope for East-West Co-operation, in: A. Blowers, and P. Glasbergen (eds.), *Environmental Policy in an International Context: Prospects, Vol. 3*. London: Arnold, 135-65.

Baker, S., and P. Jehlička, 1998. Dilemmas of Transition: The Environment, Democracy and Economic Reform in East Central Europe: an Introduction, in: S. Baker and P. Jehlička (eds.), *Dilemmas of Transition: The Environment, Democracy and Economic Reform in East Central Europe*. London: Frank Cass, 1-26.

Baker., S and B. Baumgartl, 1998, Bulgaria: Managing the Environment in an Unstable Transition, in: S. Baker and P. Jehlička (eds.), *Dilemmas of Transition: The Environment, Democracy and Economic Reform in East Central Europe*. London: Frank Cass, 183-206.

Biodiversity Support Programme, 1994, *Conserving Biological Diversity in Bulgaria: The National Biological Diversity Conservation Strategy*. Washington: Biodiversity Support Programme.

Bryant C. G. A. and E. Mokrzycki, 1994, Introduction: Theorizing the Changes in East-Central Europe, in: C.G.A. Bryant and E. Mokrzycki (eds.), *The New Great Transformation? Change and Continuity in East-Central Europe*. London: Routledge, 1-13.

Carter, F.W., 1996. Bulgaria, in: F.W. Carter and D. Turnock (eds.) *Environmental Problems in Eastern Europe*. London: Routledge, updated ed., 38-62.

Cellarius, B., 1998, Linking Global Priorities and Local Realities: Nongovernmental Organisations and the Conservation of Nature in Bulgaria, in: K. Paskaleva, P. Shapira, J. Pickles, and B. Koulov (eds.), *Bulgaria in Transition; Environmental Consequences of Political and Economic Transformation*. Aldershot: Ashgate, 57-82.

CEC, 1997. Commission of the European Communities, Agenda 2000, *Commission Opinion on Bulgaria's Application for Membership of the European Union*. Brussels, Com (97), 2008 final.

Crampton, R.J., 1997. History, in: *Regional Surveys of the World, Eastern Europe and the Commonwealth of Independent States*. London: Europa Publications, 3rd eds., 225-231.

Eurostat, 1995, *Europe's Environment: Statistical Compendium for the Dobris Assessment*, comp. Eurostat, EEA task Force, DGXI and Phare, UN Economic Commission for Europe, OECD and WHO. Luxembourg: Office for Official Publications of the European Communities.

Eurostat 1998, *Europe's Environment: Statistical Compendium for the Second Assessment*, comp. Eurostat and the EEA. Luxembourg: Office for Official Publications of the European Communities.

EEA, 1998: *European Environment Agency, Europe's Environment: The Second Assessment*. Luxembourg: Office for Official Publications of the European Communities.

Friedberg, J and B. Zaimov, 1998. Politics, Environment and the Rule of Law in Bulgaria, in: K. Paskaleva, P. Shapira, J. Pickles, and B. Koulov (eds.), *Bulgaria in Transition; Environmental Consequences of Political and Economic Transformation*. Aldershot: Ashgate, 83-103.

Gallagher, T., 1998. The Balkans: Bulgaria, Romania, Albania and the Former Yugoslavia, in: S. White, J. Batt, and P.G. Lewis (eds.), *Developments in Central and East European Politics*. London: Macmillan, 2nd ed., 43-58.

Georgieva, K., 1992. Environmentalism in Bulgaria: What Colour are the Greens?, *European Environment*, 2 (October), 14-17.

Hyde-Price, A., 1998. Patterns of International Politics, in: S. White, J. Batt and P.G. Lewis (eds.), *Developments in Central and East European Politics*. London: Macmillan, 2nd ed., 255-275.

Hutchings, R., 1997. The Economy, in: *Regional Surveys of the World, Eastern Europe and the Commonwealth of Independent States*. London: Europa Publications, 3rd ed., 231-237.

Kodjabashev, A., 1995. Bulgaria, in: *Status of Public Participation Practices in Environmental Decisionmaking in Central and Eastern Europe*. Budapest: Regional Environmental Centre, 39-46.

Koulov, B., 1998. Post-Communist Change in Ecopolitics and Environmental Management: A Case Study of Bulgaria's Burgas Region, in: K. Paskaleva, P. Shapira, J. Pickles, and B. Koulov (eds.), *Bulgaria in Transition; Environmental Consequences of Political and Economic Transformation*. Aldershot: Ashgate, 201-228.

Manser, R., 1993, *The Squandered Dividend: The free market and the environment in Eastern Europe*. London: Earthscan.

Meininger, T.A. and D. Radoeva, 1996. Civil Society: The Current Situation and Problems, in: I. Zloch-Christy (ed.), *Bulgaria in a Time of Change*. Aldershot: Avebury, 45-76.

Nikolova, M., 1998, Technological Hazards in Bulgaria, in: K. Paskaleva, P. Shapira, J. Pickles, and B. Koulov (eds.), *Bulgaria in Transition; Environmental Consequences of Political and Economic Transformation*. Aldershot: Ashgate, 39-56.

Olson, D.M., 1998. The Parliaments of New Democracies and the Politics of Representation, in: S. White, J. Batt and P.G. Lewis (eds.), *Developments in Central and East European Politics*. London: Macmillan, 2nd ed., 126-146.

Pickles, J., P. Shapira, P. Yarnal, B. Koulov, and K. Paskaleva, 1998. Introduction: Development, Restructuring and the Environment in Bulgaria, in: K. Paskaleva, P. Shapira, J. Pickles, and B. Koulov (eds.), *Bulgaria in Transition; Environmental Consequences of Political and Economic Transformation*. Aldershot: Ashgate, 3-22.

Pickles, J. et al, 1998a. Who Cares about the Environment: Popular Attitudes to Pollution and Environmental Reconstruction: Two Case Studies in Bulgaria and the Czech Republic, in: K. Paskaleva, P. Shapira, J. Pickles, and B. Koulov (eds.), *Bulgaria in Transition; Environmental Consequences of Political and Economic Transformation*. Aldershot: Ashgate, 229-256.

Pickles, J., and D. Mikhova, 1998. The Political Economy of Environmental Data in the Bulgarian Transformation, in: K. Paskaleva, P. Shapira, J. Pickles, and B. Koulov (eds.), *Bulgaria in Transition; Environmental Consequences of Political and Economic Transformation*. Aldershot: Ashgate, 105-127.

Russell, J., 1991. *Environmental Issues in Eastern Europe: Setting and Agenda*. London: Royal Institute of International Affairs.

Smith, A. and J. Pickles, 1998. Introduction: theorising transition and the political economy of transformation, in: J. Pickles and A. Smith (eds.), *Theorising Transition: The Political Economy of Post-Communist Transformations*. London: Routledge, 1-24.

Soveroski. M., 1997. Bulgaria, in: M. Soveroski (ed.), *Agenda 2000: An Appraisal of the Commission's Blueprint for Enlargement*. Maastricht: IPA.

UNDP, United National Development Programme, *The Road from Rio: Bulgaria*, http://www.sdnp.undp.org/c21/rio/bulgaria.htm, accessed 04.02.99:5).

WHO, 1999. National Environmental Health Action Plan, *Bulgaria*, http://www.who.dk/nehap/bul/bul200.htm. accessed 04.02.99.

World Bank, 1992. *Bulgaria Environmental Strategy Study*, Report no. 10142. World Bank: Washington DC.

CANADA

Robert C. Paehlke

Canada has a very solid and positive reputation in international circles regarding its record on environmental protection. In many ways this reputation is generally warranted. In particular, many individual Canadians have played an important role in advancing the capacity for environmental protection, especially at the international level. Maurice Strong chaired both the Stockholm and Rio conferences and Jim MacNeill was a principle figure within the World Commission on Environment and Development and in developing the seminal publication: *Our Common Future*. Elizabeth Dowdeswell is the current head of the United Nations Environment Program (UNEP). Moreover, one of the most important international NGOs, Greenpeace, had its origins in Canada, and one of the most important and successful international environmental treaties, the Montréal Protocol (on ozone depleting substances) was signed in Canada. Clearly, Canada and Canadians have earned a positive international image on the environmental front. However, there are recent signs that Canada's environmental capacities, while still comparatively strong, are suffering an important decline as governments in Canada have downsized their public service and cut heavily into spending levels in all areas of expenditure, including environmental protection.

Four different forces are at work in this recent process of change. The first is a strong capacity for environmental protection built up over many decades. The second is the significant public commitment to strong environmental protection, consistently borne out in polling data. The third is a more recent national will to see through governmental budget cutting following on nearly twenty years of very high federal budgetary deficits and a very high and, until this year, rising debt-to-GDP ratio. This severe fiscal correction is now seen to be successful, with a balanced federal budget probably on the horizon within one or two more years. Environmental capacity has not been spared the axe and, as will be shown, has suffered considerable damage through these years of budget cutting. Fourth, as a nation whose economy is more dependent than most rich nations on the exports of its extractive industries, Canada has been under pressure from some industry sectors to permit and promote continuing resource exploitation even when conservationist wisdom and prudence would dictate otherwise. This combination of competing forces has led to very uneven outcomes in environmental protection. Capacity has at times been diminished on some fronts while advancing on others.

The second of these assertions is borne out by recent comparative opinion data (the other three will be elaborated in succeeding sections of this chapter). In a survey of opinion in 22 countries conducted in 1993 by the International Social

Survey Programme (ISSP) Canadians registered very high in environmental concern and efficacy, nearly as high as the nations with the highest levels (generally Germany, Australia and New Zealand). For example, in response to the assertion that "people worry too much about human progress harming the environment" 62% of Canadians disagreed or strongly disagreed. The next highest levels of pro-environmental opinion on this question were scored in West Germany (59%), Australia (56%), Japan (55%), and New Zealand (54%). In the United States 47% either disagreed or strongly disagreed with this assertion. In a variation on this question that looked more directly at the question of employment as an alternative priority, people were asked whether "we worry too much about the future of the environment and not enough about prices and jobs today." Here Canadians disagreed or strongly disagreed 58% of the time, equal to the level of Australians and slightly "ahead" of Norwegians and New Zealanders (57%), West Germans (54%), and all others surveyed. The total on this question in the United States was 41% (Frizzell, 1997).

However, when asked if they were willing to pay "much higher taxes in order to protect the environment", Canadians (37% as very or fairly willing) scored almost identically with Americans (38%), nearer (with the Australians, Germans, and New Zealanders) to the middle levels of willingness within the nations surveyed. Here the higher levels were recorded in the Netherlands (48%), Japan, Israel, and Spain (43%) and Russia (41%). In terms of self-reported environmental activism, Canadians again scored very high with 43% reporting that they had "signed a petition about an environmental issue in the past five years." This level was equaled by Australians and exceeded only by New Zealanders (52%). In the United States the level (29%) was the next highest but for Great Britain at 36% (Frizzell, 1997). Again, these high levels of public concern and activism make it all the more remarkable how sharply, for example, public expenditures on environmental protection have been reduced in Canada in recent years (though the question on taxation may indicate some softness in pro-environmental opinion).

It is possible that these levels of pro-environmental attitudes among Canadians were even higher in the late 1980s when, for a brief period (especially in 1987), environment was selected in open-ended surveys as the most important issue facing Canada. These levels have not been attained since in this type of survey. However, in surveys conducted by Environics International, Inc. (a Canadian firm), Canada again scored "toward the green end of the spectrum." (Knox, 1997) Specifically, on the question of climate warming 61% of Canadians agreed that "governments should act now to reduce human impact on the world's climate, even if there are major costs." In contrast, this view was subscribed to by 74% in France, 69% in Japan, 46% in the United States, 32% in China, and 27% in Russia. Despite this level of support for a strong action plan, the government of Canada – facing strong opposition from the government of the oil, natural gas, and coal producing province of Alberta – was extremely slow in coming to a position on the issue, and did not take a strong leadership role on the issue of climate change in Kyoto.

Despite strong levels of environmental concern, there has never been an effective Green Party in Canada. The party exists, but its vote has always been highly marginal. The principle institutional reason for this is the absence of proportional representation in the electoral system. Third parties in Canada are principally regional (Quebec-based or with origins in western Canada). Moreover, Canada's social democratic party, the New Democratic Party (NDP), already occupies seemingly permanent smaller-party status nationally and generally takes pro-environmental positions, thereby capturing much of the potential green vote. The NDP, however, has multiple constituencies and when in power (provincially) has a mixed environmental record.

1 Principal Environmental Problems and Improvements Since 1970

Canada's reputation for environmental quality has been achieved in part less through vigilance than through a massive gift of natural abundance combined with very low population density. Canada's forests, for example, are hundreds of kilometers deep and stretch nearly 10,000 kilometers from Newfoundland through to Vancouver Island. Within Canada's borders are lakes and rivers containing a significant proportion of the world's supply of fresh water, the largest proportion held within any nation in the world. Canada is a vast country, but more than 80% of its population resides within 180 kilometers of the country's southern border, with the greatest proportion of population and industry residing near the Great Lakes and the St. Lawrence river. The northern reaches of the country are sparsely populated and, overall, Canada is one of the least densely populated nations in the world. Canadians like to believe that environmental protection is high relative to the United States and elsewhere owing to high levels of vigilance in Canada, but it is as likely that the determinant cause of Canada's relatively high levels of environmental quality are due as much to the fact of less population pressure per unit area.

On this point Hoberg (1992) compared US and Canadian environmental policy regimes and concluded: "How does Canada's environmental performance compare with the US? Certainly, Canadians have nothing to be ashamed of; this comparison suggests that the Canadian environmental record compares favourably with the American record in some areas. But the analysis also reveals areas where the US performance exceeds Canada's. The pervasive Canadian perception of environmental superiority toward the US is clearly unjustified" (261).

Hoberg concludes that, in broad terms, the US has been more stringent and successful in controlling water pollution, and that Canada has been more successful as regards industrial air pollution. He also concludes that Canada has benefited historically from more stringent US standards on automobile emissions, which were then generalized to the continent via the North American Free Trade Agreement (NAFTA) and other trade arrangements.

The persistent Canadian belief in the "environmental superiority" of its governments and society is deeply rooted and derives in part from the very great significance of the land and nature in Canadian cultural life, from painting to literature and even music. In terms of environmental capacity, this reality, as well as the very fact of Canadian environmental abundance, has perhaps been a mixed blessing. A nation whose art is dominated by landscape and whose literature abounds in attention to wilderness and the land (and much of whose tourist industry is wilderness oriented) is not likely to overlook threats to the quality of its environment easily. On the other hand, it has also been a challenge to imagine that human intervention can or could pose a threat to anything so enormous, daunting, and even threatening as the Canadian environment (including its climate). Clearing forests for settlement and agriculture seemed an impossible challenge for centuries, how could one imagine that that forest habitat might one day be in short supply? Only in this century has that possibility dawned on many Canadians and, despite warnings dating to the early part of the century, it is still not fully appreciated how far down the road to forest (and wild habitat) depletion the nation has gone.

The late 1960s and early 1970s saw the origins of the modern Canadian environmental movement, and a corresponding rapid growth of environmental capacity, beginning with a new attention to the issue of pollution. One early issue that commanded public attention was an excess of water-borne phosphates, especially in the Great Lakes watershed. This was one of the issues which led to the formation of the important Canadian environmental NGO Pollution Probe, founded at the University of Toronto in 1969. Public concern with pollution in the Great Lakes, arising primarily out of the extensive report to the International Joint Commission (IJC), led to the signing by President Nixon and Prime Minister Trudeau of the first Great Lakes Water Quality Agreement (GLWQA) with the United States in 1972. The IJC itself dates back to a Boundary Waters Treaty of 1909, but was never previously able to command either significant public attention or governmental concern on either side of the border. The public mood regarding matters environmental had, by the late 1960s, changed considerably.

In the early 1970s new NGOs were created, and many of them grew rapidly. Environmental publications were launched, environmental bureaucracies reorganized and/or created, and legislation was rapidly passed. In 1969 the United States Congress passed the landmark National Environmental Protection Act (NEPA), and in 1970 established the US Environmental Protection Agency. This step smoothed the way for parallel actions in Canada. As Munton and Castle (1992), document: "Responding to ... public concern, the Canadian Parliament approved an array of environment-oriented legislation during 1969 and 1970, including the Arctic Waters Pollution Act, a new fisheries act, and the new Canada Water Act. To the latter, at the eleventh hour, was added a provision requiring reductions in the amounts of phosphates manufacturers included in household detergents. This provision was the direct result of the IJC studies on the Great Lakes" (203).

The effects of the first Earth Day in the US (April 22, 1970) spread over the border to perhaps accelerate the creation in Canada of numerous environmental organizations, university programmes, and journals (including *Alternatives: Perspectives on Society and Environment* still published, now at the University of Waterloo). In the US, new environmental NGOs originating at the time of Earth Day included Environmental Action, Natural Resources Defense Council, and the Union of Concerned Scientists. In Canada, in addition to Pollution Probe, there were the Ecology Action Center (Halifax) and the Society for Pollution and Environmental Control (SPEC) in Vancouver, and similar organizations in almost every major city in the country.

The most important environmental issues in Canada in the 1970s and early 1980s were mercury pollution from the pulp and paper industry, especially in Northwest Ontario, the first James Bay Hydroelectric Project, continuing and evolving Great Lakes pollution, and – by far the most dramatic and contested of all – acid rain. Television evoked wide concern for the health of people in the Native communities of White Dog and Grassy Narrows which were adversely affected by long term ingestion of fish caught in mercury contaminated waters. After considerable denial and resistance, the use of mercury was eventually eliminated from the industry, but the communities did not get a prompt resolution of their loss of the opportunity to fish, accepting a mediated settlement for damages in 1986 nearly twenty years after the issue arose. Furthermore, despite strong opposition (crucially stronger in English than in French Canada), the massive James Bay I project was completed following an agreement between the governments of Quebec and Canada and the Cree and Inuit peoples. Since that time there have been serious environmental problems associated with James Bay dams and water diversions, including the loss of thousands of caribou and the leaching of mercury from the soil into the impounded waters.

In the late 1970s the GLWQA was renegotiated to include an extended list of Great Lakes pollutants which both countries agreed should be reduced or eliminated. By the 1980s the IJC and state, provincial and national governments were focusing attention on what had come to be called Areas of Concern (AOCs) in the Great Lakes and the St. Lawrence. Following the 1980 revelations regarding the staggering contamination of the Niagara River, primarily by chemical industry toxic waste dumps on the US side, public concern regarding the chemical contamination of the Great Lakes reached new heights. Governments and citizens within each AOC cooperated to develop remedial action plans (RAPs) regarding pollution levels; fifty (50) RAPs in all. Over time some jurisdictions made solid gains, especially when and where industries were being modernized or phased out in any case. The problem of Great Lakes contamination is by no means, even at this point in time, resolved. Fish in Hamilton Harbor (near to Canada's steel industry) still develop tumors and other sources of hazardous chemicals, including nonpoint and airborne sources, continue to release chemicals into streams and rivers flowing into the lakes.

From a Canadian perspective, all of the above issues paled in comparison to 1970s and 1980s levels of concern regarding acid rain. No other single environmental issue has so aroused the attention of citizens and governments in Canada. Following Swedish discoveries regarding the long range transport of air pollutants (LRTAP) and acidification of Swedish lakes, Canada began, in 1973, a modest monitoring program. By 1977, then Minister of the Environment Romeo LeBlanc was speaking of acid rain as an "environmental time bomb" and the worst environmental problem Canada had ever had to face. (Munton and Castle, 1992; Carroll, 1995) By 1978 monitoring stations were established throughout eastern Canada and the eastern United States. However, the relevant legislation, the US Clean Air Act of 1977, failed to address, or even to acknowledge, the problem of acid rain or acid deposition. This is crucial to the acid rain story because the larger share of acid rain doing damage in Canada had (and has) its origins in the US. Damage has been most notable in some of the smaller lakes of Ontario (as well as to some of the fishing streams of New England, New York, and the eastern provinces of Canada. There has also been concern about the health of forests in some areas subject to acidic precipitation.

Throughout the term of office of US President Reagan (1980–1988), Canada pushed hard for a treaty to reduce US and Canadian acid emissions from factories, smelters, power plants and automobiles. The domestic political support for this uncommonly aggressive posture by Canada was widespread, and was at a very high level. For example, it is notable that many wealthy and influential Ontarians spend their weekends and vacations at lakeside summer homes and camps throughout central and northern Ontario.

Canadian NGOs lobbied directly, and with considerable effect, directly within the US political process in Washington, D. C. and elsewhere, often in concert with US environmental organizations. President Reagan, however, with strong support from electrical utilities and other coal-burning industries in Illinois, Indiana, and Ohio, resisted taking action. Reagan consistently denied that there was a problem, even in the face of massive scientific evidence. It was simply asserted over and over that sufficient evidence had not yet been assembled. In an attempt to shame the US into action, the Canadian government acted unilaterally to reduce emissions from Ontario Hydro's coal fired thermal-electric facilities and from the very large smelters of the International Nickel Company in Sudbury, Ontario. This latter initiative had previously been sought by environmental organizations, and some Ontario provincial officials in the Ontario Ministry of the Environment, without success for nearly two decades. By the time George Bush succeeded Ronald Reagan to office the United States was finally prepared to come to an agreement – what became the Canada-US Air Quality Agreement of 1991.

In many ways, before it was resolved (or merely removed from public attention as many environmental scientists currently argue), the acid rain issue was already a hangover from a previous era. As will be discussed below, environmental concerns and initiatives in Canada have come in two distinct waves. Acid rain arose as an issue at the time of the first wave (1968 - 1975), but was not resolved until the second (1985 - 1992) (See Paehlke, 1992). The second wave was characterized by two other sets of issues – global concerns such as climate warming, biodiversity, and ozone depletion and threats to renewable resource capacities and wildlife habitats within Canada. Canada's international role regarding global issues has been, as noted, very strong and generally positive. However, the second set of second wave issues suggests that Canada has not always paid adequate attention to more local matters within the Canadian domestic environment, especially regarding the overutilization and depletion of resources and habitat by forest, mining and fishing industries.

The early 1990s saw the collapse of the Grand Banks cod fishery off Newfoundland, a resource that had supported many east coast communities for five centuries or more, and had for a very long time helped to feed much of the world. This resource collapse is one of the great environmental tragedies of human history, one which could and should have been avoided. Inshore fishers and some government scientists had been sounding warnings for some years, but there was great reluctance on the part of governments to act, given the high levels of unemployment in the region and the delicate international nature of the initiatives that were, and are still, necessary to protect and restore the resource. In parallel with the collapse of the cod fishery two west coast resource concerns have also arisen: the depletion of the salmon fishery and the long-standing excess and careless removal of timber from the coastal forests of British Columbia. These two issues are closely related, as the salmon stocks are in part depleted by a loss of spawning streams to excessive clearcutting and to erosion induced by forest industry extractions. Also of very great recent concern is the overuse of such national parkland treasures as Banff National Park.

Strong steps have been taken in the case of the cod fishery, effectively shutting down the fishery for several years thus far, with a brief reopening this year in a cynical move not unrelated to an impending federal election. All along, on both coasts, both federal and provincial governments have been timid conservationists in the face of the overwhelmingly preponderant industries in these two locations. The government of British Columbia has now put forward a new forestry code of conduct, but already it has been criticized as inadequate especially given the levels of overcutting that have already gone on for decades. Nonetheless the code of conduct, if enforced, should promote significant reductions in erosion and habitat damage. Also important are the strong protective initiatives proposed to curb development within Banff National Park, an important part of crucial habitat for grizzly bears and numerous other wild species. (Banff-Bow Valley Task Force, 1996)

2 Legislation, Institutionalization, and Environmental Capacity Building

Canada is perhaps, besides Switzerland, the most decentralized federal system in the world. Canada's constitution, the British North America Act of 1867, provides little explicit guidance on environmental policy. In that document, responsibility for natural resources rests with the provinces, responsibility for fisheries is assigned to the federal government in Ottawa. That is as near as the constitution comes to being explicit in these matters; much is simply not mentioned. Governments and constitutional lawyers in Canada are frequently left grasping at legal straws. Broad residual, treaty, and taxing powers rest with the federal government, which also bears broad responsibility for "peace, order, and good government." However, as matters have evolved historically since 1867, the provinces have grown in relative strength through time owing in part to a strong sense of nationalism in Québec and a relentless decentralism in the resource rich western provinces, especially Alberta and British Columbia. Accordingly, environmental capacity in Canada has developed at both the federal and the provincial levels, and environmental and natural resources policy development has frequently been a point of contestation between levels of government.

Federal environmental policy legislation began in Canada with the Fisheries Act of 1868. This potentially powerful piece of legislation defines fish and fisheries very broadly, disallows the taking of fish without a license and prohibits "any work or undertaking that results in the harmful alteration, disruption or destruction of fish habitat" (Estrin and Swaigen, 1993). It also disallows adding to waters any substance that can be harmful to fish. Had this law been enforced literally and stringently throughout Canada there could be almost no water pollution and the logging industry would have been all but prohibited from operating in such a way that soil erosion or logging debris would enter any waterway habituated by fish. Needless to say, such enforcement has not occurred. Prosecutions under the Fisheries Act have been initiated, but only with caution, and in general the federal government has delegated enforcement of this and other laws to provincial governments. There, discretionary enforcement has been the norm (Schrecker, 1992).

Where US environmental law frequently provides for detailed objectives and timetables for achieving explicit effluent levels, Canadian environmental law has historically been characterized by administrative and judicial discretion. As Schrecker (1992, 89) put it: "... Ontario's Environmental Protection Act and ... the federal Fisheries Act, could be read as prohibiting the discharge of any form of pollution, anywhere, at any time ..." Needless to say these acts have not been so interpreted. On the contrary: "In 1987 the Canadian Environmental Advisory Council (CEAC), an advisory group to the federal Minister of the Environment, identified a pattern of almost two decades of systematic non-enforcement of the Fisheries Act in Québec ..." The Law Reform Commission of Canada has also noted that this pattern has been characteristic of the enforcement of Fisheries Act powers as and when delegated to the provinces; historically, inadequate enforcement has been notable, for example, against large logging firms in British Columbia.

There have been some differences in how the courts have ruled on the authority of provincial and national governments. Regarding parks policy, in *Green v. the Queen in Right of the Province of Ontario,* the courts ruled (in 1972) that the Ontario government could not be forced to prevent the commercial use of sand from Sandbanks Provincial Park, and that the provincial government could decide when it will and when it will not prevent or permit such activities. However, in a more recent case involving the federal government *(Canadian Parks and Wilderness society v. Queen),* it was allowed that the National Parks Act provides protections that were not being granted by the federal government and that logging in Wood Buffalo National Park was therefore illegal. It should be noted here that neither the provincial nor the federal governments were inclined to prevent commercial exploitation of resources within parks they had established. The provincial governments have been more often deemed in the courts to have the discretion to allow such actions.

In the noted Crown Zellerbach case (1988), the peace, order, and good government clause of the BNA Act was the basis for allowing that the federal government can act within a largely provincial area of jurisdiction (resources management), and that the federal Ocean Dumping Control Act applied in British Columbia. Furthermore, in a 1982 case, the Manitoba Queen's Bench upheld federal air pollution regulations, arguing in Harrison's (1996, 45) words: "... that because air contaminants do not respect provincial boundaries, control of air quality clearly is a matter beyond private or local concern. It is noteworthy that the court did not demand proof that individual sources affected by the regulations had interprovincial impacts; rather, the court found the entire subject of air pollution control to be within federal jurisdiction under "Peace, Order, and good Government". "Interestingly, Harrison also notes that the federal government had also "studiously avoided" relying on this clause when drafting the federal Clean Air Act. Such behavior is in keeping with a long-standing environmental policy hesitancy on the part of the Canadian federal government.

In the area of inland water pollution, the courts have also struck down provincial jurisdiction over pollution originating outside their borders in another province (in this case Manitoba legislation establishing liability for water pollution originating in Saskatchewan and Ontario). Again in Harrison's view (1996, 46) the federal government received jurisdiction with great reluctance: "The willingness of these ... judges to confer such broad authority on the federal government is striking; the argument of exclusive federal jurisdiction was not advanced by either side in the dispute, nor any of the intervenors, including the federal government itself. In fact, the federal government joined Ontario and Quebec in arguing in favour of provincial jurisdiction." It is difficult to imagine greater hesitation, greater reluctance to assume power or to accept responsibility. Ronald Reagan, who was as indifferent to environmental protection as any figure elected anywhere since the establishment of the environmental movement, sought to shift some federal environmental powers to the US states, but in this was starting from a clear position of federal dominance. Canada, with a government that at the time purported to be strongly in favor of environmental protection, seemingly sought to divest itself altogether of jurisdictional responsibilities.

Two recent cases have asserted further federal rights and duties in environmental matters, specifically with regard to environmental impact assessment. In the Rafferty-Alameda case the courts ruled that Federal Environmental Assessment and Review Process (EARP) guidelines do apply to a provincially approved dam in Saskatchewan because the federal government must license provincial irrigation projects. The federal government would rather not have been involved, as was the case with the Oldman dam in Alberta. Again there, the courts ruled that the federal government could have halted a provincial project under the Navigable Waters Protection Act (though it was too late to do so effectively by the time the courts made their ruling). Again it must be noted that the courts were forcing a federal government to enforce its own laws and procedures when it was loathe to do so (Estrin and Swaigen, 1993; Emond, 1995; VanderZwaag and Duncan, 1992).

Thus, it has been politically, not constitutionally or even legislatively, that the Canadian government is most restricted in terms of national environmental policy initiatives. As Harrison (1995) notes, when (as in the mid-1980s) there is a very strong surge of pro-environmental public opinion, both the federal and provincial governments increase their environmental activity level. When this happens more conflicts arise. However, I think it is fair to say that it is more normal for both levels to be hesitant and particularly for the Canadian federal government to be hesitant. It can be argued that the reason that the federal government tried to stay away from involvement in the dam projects of Alberta and Saskatchewan was partly reluctance to challenge the jurisdiction of touchy western provinces, but mostly worries regarding the potential for extreme conflict over the proposed massive second stage of the James Bay (Québec) hydroelectric project, which was then still alive, at a time when the second referendum on Québec sovereignty was looming.

James Bay II did not become a point of federal-provincial conflict, in part because the separatist government of Québec did not want a confrontation that placed them on the side of environmental damage and the federal government in the role of protector. The federal government was thus again more cautious than it need have been. The James Bay II issue was overridden when the Cree peoples of the region, and to a lesser extent environmentalists, successfully lobbied within New England and New York to block US purchases of additional James Bay power. The need for the electricity from this source had also been lessened by technical and policy successes in improving the efficiency of electricity use. The important point here is that a major issue was resolved at the international level which would likely have lost at the provincial level alone, and would also have seen the Canadian federal government yet again backing away from tough environmental action in the name of higher political priorities (in this case a combination of national unity and economic growth). Constitutional and legal authority are important, but do not determine outcomes if governments are unwilling to accept authority.

Several other observations are in order here. First, this discretionary provincial enforcement is sometimes effective, especially when public visibility and NGO pressure on governments is high. The pulp and paper industry has recently reduced its waterborne emissions in many locations. Several large polluting firms in a variety of industries have also ceased operations along the St. Lawrence river (moving to the southern United States on the Mississippi in one instance). Second, it should be clear that, since the principle responsibility for enforcement of both federal and provincial legislation has for the most part resided with provincial governments, that is where monitoring and legal capacities have primarily developed. Third, as will be seen shortly below, environmental policy capacities grew considerably through the 1970s and 1980s, but in the 1990s capacity has probably declined, one of the earliest casualties being CEAC, whose important report was cited above, and which was eliminated by the Progressive Conservative government of Brian Mulroney in 1992.

One measure of the growth of environmental policy capacity in the 1970s is the net total growth in federal and provincial spending on environmental protection. Federally and in most provinces, governmental expenditures on environmental protection, as well as staffing levels, increased sharply from 1970 through 1975. At the federal level, however, spending and staffing cuts were experienced by Environment Canada in the late 1970s (in four of those five years) and in the mid-1980s, especially 1985-6 (Doern and Conway, 1994). The early 1990s saw a restoration of funding, with the provision of $3 billion (total) in multi-year Green Plan funding. However, total provincial environmental spending has been substantially more than the amount expended by the federal government and, until recently, has been less prone to sharp fluctuations in many provinces.

Most striking are the cuts experienced in the late 1980s, because these alone do not conform to the general level of public and media concern. That is, other expansions and contractions can be explained as a function of rising and falling public interest in environmental issues. However, in the late 1980s concerns were at their highest levels ever, but federal expenditures fell (though they were then sharply increased from 1990). One explanation of this aberration might be that the Mulroney government was more concerned with budget cutting, and was convinced that environmental opinion and pressures were a fad that would pass and could be satisfied by the single and singular act of achieving an acid rain treaty with the United States. By 1989/1990 it became clear that much more would need to be done, and thus the Green Plan was finally approved after several failed attempts within the government and the federal cabinet.

In many ways Canada's Green Plan was a unique undertaking. It was multifaceted and very much in keeping with the concept of capacity building. It included large scale multi-disciplinary research projects based primarily in universities, emphasized the training of graduate level environmental scientists and policy analysts, and sought to establish connections between researchers and the communities and community organizations within which the funded research was conducted. It also involved the establishment of "green chairs" – senior faculty positions in major Canadian universities. The Green Plan, through several ministries (Natural Resources, Environment Canada, Fisheries and Oceans Canada and others), also funded extensive non-university technical research in many areas related to environmental protection. Moreover, the Green Plan pressed all corners of the government to modify their everyday operations in environment-friendly directions in relation to such diverse activities as building design and maintenance, paper use, and any number of other practices. Virtually every ministry in the government needed to develop environmental capabilities and behavior.

The principal problem with the Green Plan was that it was a "sunsetted" six year plan and was not continued; the program has been terminated. This despite the fact that (or perhaps because of the fact that) the targets and schedules of action set out in the Green Plan have by no means all been met. These targets included: (1) passage of regulations for 44 priority toxic substances; (2) a fifty percent reduction in solid waste generation by the year 2000; (3) stabilization of greenhouse gas emissions at 1990 levels by the year 2000; (4) setting aside 12% of the nation's land area as protected space; (5) a 50% reduction in sulphur dioxide emissions by 1994 and several others. (Doern and Conway, 1994) Progress toward some of these goals has been made and may continue to be made, but others, most notably the target on greenhouse gas emissions, are not even close to being achieved (indeed Canadian levels of carbon emissions have risen considerably from their 1990 levels). Moreover, much of the funded research program of the Green Plan was discontinued even before the full, previously announced, competitions were completed.

Canada's overall results regarding air pollution over the past thirty years are decidedly mixed. Urban air quality has improved with one notable exception, ground level ozone (cf. Table 1).

Overall air emissions have also seen mixed results. NO_x emissions rose 44% between 1970 and 1980 and have roughly stabilized since, but SO_2 saw a 30% reduction between 1970 and 1990 and has continued to improve (though in Ontario very recent replacement of nuclear power with increased coal burning is having an offsetting effect on gains in metal smelting industries). Airborne benzene declined from 34 kilotons per year in 1985 to 22 kilotons per year in 1991, with additional initiatives taken in 1995 (data, including figure 1, are from Government of Canada, 1996).

Table 1. Canadian Urban Air Quality (1979-1993)

CO	-56%
SO_2	-46%
NO_x	-28%
O3	+29%

The control of greenhouse gases in the 1990s has been a notable failure. Even though reductions were agreed to internationally in 1992, greenhouse gas outputs have not only not been stabilized and restored to 1990 levels as agreed, but have risen throughout the 1990s, by double digit amounts in both Canada and the United States. (Gibson, 1999, 134; Toner, 1996) New reductions were agreed to in Kyoto (6% below 1990 levels by 2012). The voluntary initiatives on climate warming have been a notable failure – the Voluntary Challenge and Registry Program, despite initiatives taken in a few industries, would seem to have had almost no overall effect (Government of Canada, 1996, 15-18; Gibson, 1999). This stands in contrast to other voluntary initiatives (within the chemical industry, for example), where some gains have been made (Gibson, 1999).

In other realms the late 1980s and 1990s have seen improved environmental policy making capacities achieved through new legislative initiatives. Most notable are: (1) a legislated federal environmental assessment program; (2) an Environmental Bill of Rights in Ontario, Canada's largest province; and (3) an attempt at passage of a federal-provincial initiative to protect endangered species. The latter has not yet gained approval and thus warrants only a brief comment on its importance at this point. The United States has long had endangered species legislation, which has been, of late, highly controversial as, for example, the legal basis for gaining an injunction against logging in old-growth forests. Canada nonetheless agreed to take an initiative of this sort as part of its obligations under the international Biodiversity Convention of 1993.

Canada's governments, and especially its western agricultural and resource extraction industries have long been hesitant about having such protections in place. The Endangered Species initiative has now finally come forward to a process of federal-provincial negotiations. The federal government, to its credit, took this initiative and has provided the chief impetus (however limited it might be in some ways). This Act, if passed, will codify and provide some outlet for the developing scientific claims regarding threats to a wide variety of wild plant and animal species in Canada. The initiative is unfortunately limited in some ways in the extent to which it can protect the habitat of species deemed to be at risk. Nonetheless, this legislation would encourage the development and maintenance of scientific and administrative capacities in the broad area of conservation biology.

Again, many years after the United States and many other countries had legislated national environmental assessment programs, Canada operated with only provincial programs and a federal "guidelines order" mandated only administratively (from 1973). The provinces frequently resisted federal initiatives in environmental assessment as an encroachment on provincial jurisdiction. However, several 1980s court rulings concluded that even under a guidelines order Environment Canada was obliged to conduct assessments where federal funding or federal jurisdiction is involved, as in the case of hydroelectric dams that alter fish habitat and thereby require a federal license. (Estrin and Swaigen, 1993) The Canadian Environmental Assessment Act and the Canadian Environmental Assessment Agency (CEAA) are now, more than twenty years after the establishment of the guidelines order, finally enshrined in federal law. CEAA is currently conducting a review of the concept (rather than a site) for the disposal of high-level nuclear wastes. Thus a long-standing gap in Canadian environmental policy capacities was finally filled.

The Ontario Environmental Bill of Rights is the very distinctive undertaking of the recent social democratic (New Democratic Party) government which served in Ontario until 1995. Faced with serious budgetary and economic problems from the day that it came into office, this government in some ways disappointed environmentalist organizations, but it managed to see through several interesting (low cost) initiatives such as this one. The Ontario Bill of Rights provides for full disclosure of environmental information of many varieties, for the possibility of private citizen administrative requests, for limited rights to initiate court actions in defense of the environment, and for the establishment of an Environmental Commissioner partially insulated from the powers of the government of the day. The Environmental Commissioner is responsible for reviewing the environmental performance of all ministries as well as the government as a whole.

Thus far, it has been the Office of the Environmental Commissioner that has proven to be the most significant part of the Environmental Bill of Rights, in that the incumbent has been openly critical of the current (arch-conservative) government of Ontario. In her 1996/97 Annual Report, Eva Legeti, the Commissioner since shortly after the act was passed, observed: "We have spent the last 20 years strengthening environmental protection in this province and in a very short time that strength is under stress" (Mallan, 1997). Clearly the Environmental Commissioner is commenting on reductions in environmental protection capacities, but in being there and doing so, she is herself very much a part of those enhanced capacities. The Environmental Bill of Rights also promotes environmental capabilities within the public as a whole and within the legal system. Uniquely, it establishes the important possibility of self-criticism within the government itself. Care was taken in the drafting of this legislation to avoid unduly judicializing environmental decision-making, and in this it appears to have been successful.

Needless to say, there have been numerous other developments within law and government which have not been discussed here. Worth mentioning briefly are the increasing presence in Canada of international environmental agencies and organizations, and the strong development of environmental capacities at the level of municipal governments in both large cities such as Toronto, Vancouver, and Ottawa and in smaller communities. Some of the international agencies present in Canada are the Commission for Environmental Cooperation (responsible for monitoring the environmental and social effects of the North American Free Trade Agreement), the International Institute for Sustainable Development, and the International Council for Local Environmental Initiatives, supported by municipal governments worldwide. In recent years, local governments in Canada have significantly expanded their environmental capabilities, especially in the areas of recycling, protection of wetlands and other significant environmental features, and even with regard to such large issues as climate warming. Cities have, since the late 1980s, been adding significant numbers of graduates of environmental science programs to their administrative structures, including their planning and engineering offices.

From the late 1960s through to 1995, funding for environmental protection has waxed and waned in Canada, but throughout this period, capacities have generally expanded and, as with the case of the Ontario Bill of Rights and environmental assessment public hearings, have spread from government to the NGO and educational and legal communities. Particularly notable in this regard have been several large-scale public inquiries, especially the 1970s Berger Commission on the MacKenzie Valley (natural gas) Pipeline and the Ontario Royal Commission of Electric Power Planning (the Porter Commission). Capacities have also moved from the wider society into governmental structures. The following section of this chapter discusses the societal developments that have paralleled those in law and government.

It is worth noting prior to that discussion that, until very recently, there has rarely been open political resistance to the growth of environmental policy capacity in Canada, even when governments have generally oriented to reductions in expenditures and public sector staffing cuts. The spread of environmental capacity out of environmental ministries and agencies and into other areas of government and from government to society has been fostered and advanced, especially by the environmental assessment process, by the Green Plan, and by Ontario's Environmental Bill of Rights. This spreading and the independent development of environmental capacities within the wider society have become a significant part of Canadian cultural and social history, but – as we will see in the final section of this chapter, following the discussion of societal capacities – reductions in environmental capacities at the governmental level remain a very real possibility.

3 The Canadian Environmental Movement and Non-Governmental Capacities

As noted above, the contemporary Canadian environmental movement was established in earnest in the late 1960s and early 1970s. The growth of environmental capacity outside government has paralleled the growth within. Environmental capacities have grown up within environmental organizations (NGOs), within educational and research institutions, within trade unions, and within the corporate sector. Development of NGO-, university- and research community-based capacity in particular has also continuously fostered, and has promoted capacity growth within governments at all levels. Considerable synergy has evolved through time among this particular configuration of sectors.

In most instances, environmental capacity begins outside government, within the environmental movement. Government attends to matters environmental when sufficient public attention is aroused by credible and concerned environmental activists and organizations. The environmental movement has been motivated all along by a complex blend of detailed scientific findings and a willingness to see through the application of that science to a critical questioning of the ways in which social, economic and political institutions (and everyday citizens) behave and conduct their affairs (Paehlke, 1989, chapter 5). Single spontaneous actions are ordinarily neither sustained nor sufficient without the evolution of at least a rudimentary organizational (NGO) presence. In Canada (and North America as a whole), NGO and environmental movement activity can be seen to have had peak periods of growth in the periods 1968-75 and 1985-92. These can be seen as two waves of environmental concern. During the two waves both the most important issues and certain characteristics of the movement were quite different.

The first wave in Canada saw the creation of leading environmental organizations, and was motivated especially by new findings, perspectives and concerns regarding pollution, the potential for future resource shortages (especially energy), nuclear power, and rapid human population growth. The movement had a significant apocalyptic and apolitical dimension. Industrial economies were seen by some to be on the brink of collapse, in danger of being suddenly overwhelmed by the pollution, diminishing resources (and the flagrant waste of those resources), and out-of-control population growth, especially in developing nations. Only some of the information advanced by early environmentalist organizations was carefully grounded in scientific findings, but they struck a public nerve nonetheless and immediately provoked governments, universities, and scientific organizations to expand their environmental research capabilities, reorganize their research priorities, and even to develop new interdisciplinary methodologies.

In the 1970s the study of matters environmental within Canadian universities increased very unevenly, with some of the largest and most established universities and departments (and public research funding agencies) openly resisting and even dismissing interdisciplinary approaches to environmental studies and science. Some newer universities, perhaps most notably York (in Toronto), Waterloo, Trent, and Calgary took bolder early steps. But also, in almost every university, sometimes in surprising places, individual faculty and students often turned their research and teaching attention to a wide range of environmental issues. The locus for consideration of matters environmental could be as varied as departments of geography, political science, chemistry, biology, history, economics, or urban planning. Some interdisciplinary environmental studies programs grew very rapidly only to suffer enrolment declines as public interest waned (between the "waves") in the early 1980s. The declines confirmed for many critics within the universities that these approaches were but a "flash in the pan" and not, as claimed, a new and yet enduring approach to scholarship.

Nonetheless, throughout the 1970s and early 1980s, graduates of Canadian environmental studies programs, as well as many who brought a sense of environmental concern to studies within existing disciplines, moved into ever-increasingly professionalized environmental NGOs and policy consulting firms. Government departments, previously often predominantly populated by planners and engineers, added ecologists and other environmental scientists to their staffs, and some engineering faculties added environmental components to their degrees. Individuals with extensive NGO experience were hired by government ministries, or returned to graduate study and/or joined professional consulting firms or university faculties. By the time of the second wave of public environmental concern (1985 -92), both the issues and the fundamental character of the Canadian environmental movement had changed profoundly. The movement itself had become highly professionalized (a change that was not without its critics from within and without the NGO community).

The second wave of environmentalism in Canada saw the re-emergence of "preservationist" and biodiversity concerns more central to the earlier conservation movement (dating back into the 19th century) and, as noted above, an increase in attention to issues of a global character. The second wave saw the far more professionalized environmental movement provide sophisticated ecological, economic, and legal analysis to government and to Canadian society at large. In addition, and in part as a result, environmentalism was much less "on the fringe" – environmental ideas were acceptable to, or at least acknowledged and seen to be accepted by economic and political elites from all political parties and viewpoints. A sophisticated array of technical options and policy tools were typically put forward rather than vague notions of withdrawal from society and/or advocacy of crude end-of-the-pipe pollution controls. Environmental NGOs and others discussed the environmental importance of various taxation regimes, environmental user fees, governmental procurement patterns, environmental audits of industries, state of the environment reporting; and there was much talk of economy-environment integration and sustainable development. By the late 1980s not only the environmental movement but Canadian society as a whole had become more environmentally sophisticated. Years of nature-, conservationist-, and ecologically oriented television – most notably the enormously popular work of geneticist/environmentalist David Suzuki – had educated a whole generation.

Environment-economy roundtables were established in the late 1980s at all levels of government (from municipal to federal) and brought together environmental activists, industry leaders, labor unions, and government officials. Some useful actions arose out of the roundtables, especially at the municipal level, but on the whole this approach resulted in few significant environmental policy initiatives. Some analysts who studied provincial and federal roundtables concluded that these bodies were "... designed not to enhance but rather to diffuse the spirit of environmental reform in Canada in the mid to late 1980s (Bruton and Howlett, 1992). Some of the provincial-level roundtables have been eliminated (e.g., in Ontario) and many of the municipal structures have languished during and since the recession of the early to mid-1990s, but the National Round Table on Environment and Economy (NRTEE), based in Ottawa, continues to operate (Howlett 1990; Roseland, 1992).

Less well-known, but equally important perhaps, have been the long-established links between labor unions and environmentalists dating back to the late 1960s and rooted in the initiatives of the United Auto Workers Union (later the Canadian Auto Workers) and the Oil, Chemical, and Atomic Workers' Union (Paehlke, 1981). The principle environment-labor linkage in the 1990s has been carried by the Canadian Labour Congress in Ottawa, and much mutual support has been engendered. However, there have also been quite severe labor-environmentalist tensions, particularly regarding resource extraction, most notably logging in British Columbia. Labor-environmentalist tensions have also been visible regarding other issues, including nuclear power (where the labor movement itself has been divided) and refillable beverage containers. In general, the conflicts have gained far more media attention than has the cooperation and thus it is the tension that tends to "frame" environment/economy debates in Canada.

More dramatic as a characteristic of the second wave of environmentalism was the introduction by industries and retailers of a wide variety of so-called green products. Many 1970s environmentalists would have been appalled at the very concept (and more than a few 1990s environmentalists had their doubts as well). Nonetheless, suddenly in the early 1990s products from less polluting household cleaning products to energy efficient lighting devices to road asphalt made from used tires to recycled paper and boxes were produced and marketed. One of Canada's largest supermarket chains geared much of its promotional effort to this initiative, and there was some attendant scandal as endorsements from one environmental organization were seen to be at less than arms' length. Whatever the limits of green product practices, the introduction and wide acceptance of green products is evidence of a considerable environmental capacity both within industry and within society as a whole. Change has taken place in the way some products are made and sold to environmentally sophisticated consumers. Green products also contribute to a shift in the mix of environmental actors, with some new businesses being added to the green "side" of the political equation (along with manufacturers of pollution abatement equipment and some insurance firms fearful of environmental liabilities).

One logical extension of this shift from regulation to marketing was the need for a new form of environmental responsibility for governments. That is, green products clearly require a neutral arbiter and some certification of authenticity. Canada, like many other nations, now has an appropriate governmental logo, in the form of three doves interlocked into the shape of a maple leaf. However, by the time the Environmental Choice Program was established, the mood of governments in Canada regarding the environment had shifted considerably in the face of what was seen as unmanageable levels of public debt and excessive levels of unemployment. Rather than establish a sophisticated public environmental testing capacity, government seemed more concerned to be sure that environmental certification of one manufacturer did no economic damage to other Canadian economic interests (Cohen, 1994). Since 1993 or 1994 the second wave has clearly subsided, this time with a considerable undertow that threatens to do more than just temporarily slow the expansion of public and private environmental capacities.

4 The 1990s: Governmental Deficits, the Global Economy and Capacity Shrinkage[1]

No trend in the 1990s has been more noted than the continuing and accelerated advance of globalization – the international integration of a global economy wherein a growing proportion of every national economy is devoted to imports and exports, and both the ownership and the organization of economic activity is structured internationally. This trend has enormous political implications, and creates pressures within nations to present a face that appeals to international investors. Canada has long been one of the most internationalized economies in the world, and accordingly is highly vulnerable to close scrutiny by the international business community. Of the G-7 group of industrialized countries, Canada's economy is the most trade intensive, and is becoming more so very rapidly. Exports were 25% of Canada's economy ten years ago and are 40% today, doubling as an absolute figure since 1989.

The preparations for the scrutiny of Canada by the international business community came to a recent peak in the early 1990s, and the twenty years of very large governmental budget deficits was rather suddenly no longer tolerable. Re-

1 Some of the paragraphs in this concluding section were based on materials developed in an earlier study on institutional trends (and co-authored by myself, Michael Kraft and Exequiel Ezcurra) for the Commission on Environmental Cooperation, the NAFTA environmental and social monitoring agency.

gardless of the ideological inclinations of any particular party in power federally or provincially, severe budget cutting at all levels of government has been the Canadian norm of the 1990s. When combined with the effects of the massive reorganization of the private sector economy associated with the North American Free Trade Agreement (NAFTA), the result has been the highest sustained levels of unemployment since the Great Depression, even in the face of economic growth.

This economic and social climate has also resulted indirectly in the rapid receding of the second wave of environmentalism. Other issues have simply pushed environmental concerns into the background for an indefinite period for all but the most dedicated of environmentalists. In Canada the high unemployment and public sector cutbacks have been compounded by the recent near-separation of the province of Québec and the continuing threat that it will yet become an independent nation. What is unknown at this point is whether high levels of unemployment and social dislocation are a temporary or permanent condition for Canada in a globalized economy. What is certain is that there has been enormous pressure on governments in Canada (and elsewhere) to reduce the size of the public sector generally, and there is also an added increment of pressure to soften environmental standards whenever those standards are sufficient to impose, or even appear to impose, short term pressures on economic activities or potential new investment.

The result of all this is that there is some question in Canada now as to whether the levels of environmental protection capacity built up over the past thirty years can be maintained. The federal government is in the process of reducing environmental spending by 32% over the period 1995 to 1998 (from \$737 to \$503 million). Environment Canada will lose in total 1,400 of its original 5,700 public service positions (Toner, 1996). Some duties and tasks will be downloaded to the provincial level, but some provinces, most dramatically Ontario, were at the time cutting environmental spending at an even faster rate. Interestingly Ontario transferred further downwards the testing of drinking water quality to the municipal level where there has been little experience of such tasks. Ontario has even reduced the capacity to monitor acid rain emissions and consequences – this but a few years after the enormous wave of public concern and the signing of international treaties.

Expenditure reductions for environmental production have been imposed despite the fact that there have been few open and explicit political attacks questioning the importance of, or even the extent of, environmental protection. Despite this absence of partisan conflict regarding environmental protection, sharp cuts in funding levels, in staffing, and in research capabilities have occurred and continue to occur. Environmental organizations object, but receive little press attention where, five years ago, whatever these groups asserted was reported extensively. The most widely noted federal environmental cutbacks were the cuts to funding for the \$3 billion 1990 Green Plan, the centerpiece of federal environmental policy in Canada for the decade. Immediately after the plan was approved, it suffered funding cuts – \$600 million in 1991 and \$600 million in 1992. Since that time the plan has been formally, but quietly, terminated. Provincially, in Ontario for example, funding cuts to environmentally relevant ministries have been particularly severe. Operating expenditures in the Ministry of Environment and Energy were cut by 37% between 1994-95 and 1996-97 and are now, overall, less than half of what they were in 1992-93. The Ministry of Natural Resources has also suffered reductions of 22% (1994-95 to 1996-97), falling from expenditure of $584 million in 1992-93 to $373 million in 1996-97. The cuts in environmental ministries in Ontario are more severe than cuts in many other ministries (Canadian Institute for Environmental Law and Policy, 1996).

Some, however, are now raising questions about whether the recent level of governmental spending cutbacks in Ontario will allow Canada to continue to honor its international environmental obligations. Globalization, it might be observed, cuts two ways – there are also international pressures which promote environmental protection. Moreover, firmly embedded environmental capacity is not easily dislodged. The Canadian Institute for Business and the Environment (an association that includes manufacturers of pollution abatement equipment) recently noted that: "Canadian environmental businesses are beginning to express concern about the 'free fall' in environmental protection in Canada. First the federal budget cuts and devolution of environmental responsibilities to the provinces. Then the 45 percent budget cuts to the CCME (Canadian Council of Ministers of the Environment), which is attempting to coordinate provincial matters. The severe cuts in environmental protection threaten the ability of the provinces to meet their commitment to protecting boundary waters between Canada and the United States. It may also jeopardize their ability to meet Canada's commitments to international environmental agreements" (Gallon, 1996).

Losses in environmental capacity are occurring despite objections, arguably because the most dramatic changes have come in the context of general spending reductions and strong assaults on government deficits and debts rather than as assaults on environmental protection *per se*. There has been widespread political support in Canada for budget cuts in general, though polls indicate that a majority of Canadians would defend environmental protection expenditures in particular. Deregulation and capacity reductions have taken place at both the federal and the provincial levels. However, the outcome of the cutbacks and deregulatory exercises has varied greatly from province to province, which in some provinces (notably British Columbia) have essentially not taken place. It is worth looking more closely, then, at three particular examples of provincial cutbacks and deregulation exercises: Newfoundland, Alberta, and Ontario.

In April 1994 the government of Newfoundland launched a Regulatory Review Commission for the identification of regulations unnecessary for the maintenance of public health, order or safety, or the environment, or which contribute to sustainable development. The commission report recommended the elimination of nearly half of all regulations on the books, and the legislature has since begun acting on the recommendations. Explicit losses to environmental capacity did not occur, and despite extreme economic problems, very high unemployment and general expenditure cuts, the budget of Newfoundland's environment ministry increased in both 1994/5 and 1995/6.

In Alberta the government passed the Government Organization Act, which allowed ministers to delegate "any power, duty or function" except the power to make regulations to any legal person, including corporations or business associations. Responsibility for the registration and inspection of underground chemical storage tanks was then delegated to the Petroleum Tank Management Association.

In April 1995 the Alberta government established a Regulatory Reform Task Force which was mandated, for example, to assure that all regulations are cost-effective, contain a sunset clause, enhance economic values, and are the only way to achieve their objective. It would appear that many environmental regulations will be transformed into non-binding codes of practice or guidelines, and/or be delegated to non-governmental enforcement. These actions in Alberta shaped a proposal for a federal Regulatory Efficiency Act, and other cutback and deregulatory initiatives in Alberta have influenced the new provincial government in Ontario. Overall funding for environmental protection in Alberta has declined modestly, especially in real dollars, but a prosperous economy has to some extent helped to protect environmental capacity in that province – while simultaneously raising many land use controversies.

Ontario is perhaps the Canadian jurisdiction where there has been the most open and consistent rollback on environmental protection in the late 1990s, and was probably also the province that suffered the greatest job losses in the economic transformations following the free trade (FTA) and NAFTA agreements. The Ontario government has launched a general assault on environmental and other regulations. An interim report, issued in June 1996 by the provinces Red Tape Review Commission, recommended the elimination of more than one thousand minor regulations in a wide variety of areas (predominantly non-environmental in nature). This relatively noncontroversial interim housecleaning helped to set a public tone for more significant actions less easily seen through. Many Ministry of Environment and Energy regulations and practices remain under review and, unlike both Alberta and Newfoundland, the Ontario review terms of reference do not contain a caution regarding either protection of the environment or the sustainable management of natural resources.

In particular, the environmental initiatives taken in the first year in office by the government of Mike Harris included: (1) elimination of provincial funding support for recycling; (2) repeal on an "emergency basis" of a law that required mine owners to "obtain government approval of mine closure plans" and to set aside funds in advance for cleaning up mine sites on closure; (3) reduction by 70% in provincial funding for local conservation authorities and by amounts which have led to 50% lay-offs by the Niagara Escarpment Commission, which protects this internationally significant land form; (4) elimination of the Environmental Assessment Advisory Committee, all intervenor funding for citizen participation in public hearings, and all provincial support for a previously established network of community-based energy efficiency and waste reduction organizations; (5) removal of a ban on the incineration of municipal solid waste; (6) radical reductions in provincial subsidies for public transit, and elimination of funds for the protection of sensitive agricultural lands; (7) severe cuts to participation in the RAPs associated with the GLWQA; and (8) approval of logging and mining in the Temagami region, the locus of the strongest environmental NGO actions of the past decade and strong objections from Native Canadians.

Given typical provincial performances both historic and contemporary, Canadian environmental organizations are wary of any further decline in the federal role in resource management and environmental protection. Though the federal role has always been limited, it would now appear likely to decline with the possible exception of a proposed new federal Endangered Species Act. The federal government is generally hesitant to duplicate provincial efforts of any kind. All governments, especially perhaps provincial and local governments, are highly fearful of jeopardizing economic development of any kind, particularly in the economically weak regions of the nation (which would certainly include Newfoundland, and usually everywhere east of the Ontario-Quebec border). The overall result of the changes on the international level and within Canada, some NGOs argue, is a danger of a double "race to the bottom" in environmental protection – one race between nations and another within Canada, with the least well-off nations and regions in greatest danger.

The best hope for avoiding such a negative outcome is the thirty year history of the environmental movement and environmental governance and the resultant level of environmental capacity within the public sector, within universities and NGOs, and within corporations. Moreover, Canada has international environmental obligations and an international environmental image to protect. The public is environmentally sophisticated and organizations dedicated to environmental protection exist throughout the society. There are thus real limits on how much can be cut from environmental standards and environmental spending. Even the very conservative government of Ontario (a government Europeans would call hyper-liberal) is cautious to say explicitly that it intends to weaken effective environmental protection (even when it is not afraid to disparage many other dimensions of governmental activity).

Furthermore, not all environmental capacity is immediately vulnerable to public sector spending cuts (a market for green products remains, as does general environmental literacy and networks of environmental NGOs and environmental studies programs at all levels of the educational system). More than that, it is also possible that almost as much environmental protection can be achieved even if and when governmental funding levels are reduced if levels of public concern for the environment remain and if the public service is able to add relatively inexpensive, recently-trained, younger employees as it sheds some of its older staff. Also notable is the response of NGOs to the shifting (less green) mood of Canadian governments in the 1990s. Some NGOs have become decidedly more pragmatic, and have even been willing to bargain with obvious enemies (as in the very recent negotiations over Ontario's Lands for Life allocation of wilderness lands to a mix of resource extraction and parks protection).

One important question at hand is what objectives these more conservative governments are pursuing: reduced costs or reduced impositions on polluters and environmentally insensitive developers. It could also be argued, for example, that in many instances environmental protection with high administrative costs is precisely that environmental protection that is hesitant in character. It is administratively expensive to set, monitor and enforce complex pollution standards in all media (air, land and water). It is less administratively expensive to impose outright bans on the worst of toxic substances or simply to reject the most potentially damaging of development proposals. In addition, it has been argued by some environmentalists that there is potential income to government in market-based environmental protection tools such as user fees on water or highway use or solid waste production. Thus, environmental protection and balanced governmental budgets could be mutually reinforcing rather than competing initiatives. If it is objected that *any* form of taxation inhibits economic growth, it could be replied that environmental taxes could replace job inhibiting forms of taxation such as payroll taxes, or taxes that are entrepreneurship-inhibiting, such as taxes on corporate profits. Such arguments have been made, but there have thus far been no major governmental initiatives on green taxation in Canada.

5 Environmental Policy in Canada: An Overall Evaluation

Environmental policy in Canada has at times responded quite effectively to problems perceived as urgent by the general public, especially during the two "waves" of environmental opinion and NGO activity. Basic legislative and administrative structures were established in the early 1970s. Significant action was taken on acid precipitation in the 1970s and 1980s, and SO_2 emissions have been reduced significantly. Great Lakes water quality has improved and some forms of air pollution reduced through this period as a whole. Solid waste reductions and recycling were widely advanced in the late 1980s and the 1990s. But, in other realms and in times of relatively low public and NGO attentiveness, actions have been ineffective. Ground level ozone has increased, the habitat of endangered and threatened species has not thus far been effectively protected, greenhouse gas emissions have not been slowed and Canada's massive forests have not been harvested on a sustainable basis. In the late 1990s, particularly, environmental protection expenditures by some governments have been sharply reduced.

The overall development of environmental capacity, however, has produced an environmental movement in Canada which has gone well beyond an apocalyptic rejection of industrial society and end-of-the-pipe pollution abatement. There has been a sophisticated political and policy response to the current drive for governmental cutbacks and other threats to environmental capacity (see, for example, Toner, 1996; Gale, 1995; Canadian Institute for Environmental Law and Policy, 1996; and Hoberg, 1996). Environmental protection is well established as part of governance in Canada, part of our legal structures and part of our educational system and society. Setbacks have occurred, but not without a mounting response rooted in an extended history and an entrenched and broad policy capacity.

References

Banff-Bow Valley Task Force, 1996. *Banff-Bow Valley: at the Crossroads*. Ottawa: Government of Canada.

Bruton, J., and M. Howlett, 1992. Differences of Opinion: Round Tables, Policy Networks, and the Failure of Canadian Environmental Strategy, *Alternatives Journal* 19, 25-33.

Canadian Institute for Environmental Law and Policy, 1996. *Ontario's Environment and the "Common Sense Revolution"*. Toronto: CIELAP.

Carroll, J. E., 1995. Acid Precipitation: Legislative Initiatives, in: Robert C. Paehlke (ed.), *Conservation and Environmentalism: an Encyclopedia*. New York: Garland Publishers, 5-6.

Cohen, D. S., 1994. Subtle Effects: Requiring Economic Assessments in the Environmental Choice Programme, *Alternatives: Perspectives on Society, Technology, and Environment* 20 (September/October), 22-27.

Doern, G. B., and T. Conway, 1994. *The Greening of Canada: Federal Institutions and Decisions*. Toronto: University of Toronto Press.

Emond, P., 1995. Environmental Case Law: Canada, in: R. Paehlke (ed.), *Conservation and Environmentalism: an Encyclopedia.* New York: Garland.
Estrin, D., and J. Swaigen, 1993. *Environment on Trial*, 3rd ed. Toronto: Emond Montgomery Publications Limited.
Frizzell, A., 1997 Environmental Attitudes Around the World, in: A. Frizzell and J. H. Hammett (eds.), *Shades of Green: Environmental Attitudes in Canada and Around the World.* Ottawa: Carleton University Press.
Gale, R., and S. Barg (eds.), 1995. *Green Budget Reform: An International Casebook of Leading Practices.* London: Earthscan.
Gallon, G. T., 1996. *Environmental Cuts Affect Canada.* Montreal: Canadian Institute for Business and the Environment, e-mail from cibe@web.net dated 05/07/96.
Gibson, R. B., 1999. *Voluntary Initiatives and the New Politics of Corporate Greening.* Peterborough: Broadview Press.
Government of Canada, 1996. *The State of Canada's Environment.* Ottawa: Government of Canada.
Harrison, K., 1996. *Passing the Buck: Federalism and Canadian Environmental Policy.* Vancouver: UBC Press.
Harrison, K., 1995. Federalism and Environmental Protection: Canada, in: R. Paehlke (ed.), *Conservation and Environmentalism: an Encyclopedia.* New York: Garland.
Hoberg, G., 1996. Governing the Environment: Comparing Canada and the United States, in: K. Banting, G. Hoberg, and R. Simeon (eds.), *Degrees of Freedom: Canada and the United States in a Changing World.* Montreal: McGill-Queen's University Press, 341-385.
Hoberg, G., 1992. Comparing Canadian Performance in Environmental Policy, in: R. Boardman (ed.), *Canadian Environmental Politics: Ecosystems, Politics, and Process.* Toronto: Oxford University Press, 246-262.
Howlett, M., 1990. The Round Table Experience: Representation and Legitimacy in Canadian Environmental Policy-Making, *Queen's Quarterly* 97, 580-601.
Knox, P., 1997. Environmental issues heating up, poll shows. *Globe & Mail* (November 10), A10.
Mallan, C., 1997. Report sees red over Tory visions for greens, *The Toronto Star* (April 23), A3.
Munton, D., and G. Castle, 1992. The Continental Dimension: Canada and the United States, in: R. Boardman (ed.), *Canadian Environmental Politics: Ecosystems, Politics, and Process.* Toronto: Oxford University Press 203-223.
Paehlke, R., 1992. Eco-history: Two Waves in the Evolution of Environmentalism, *Alternatives: Perspectives on Society, Technology and Environment* 19 (September-October), 18-23.
Paehlke, R., 1989. *Environmentalism and the Future of Progressive Politics.* New Haven: Yale University Press.
Paehlke, R., 1981. Environnementalisme et syndicalisme au Canada anglais et aux Etats-Unis, *Sociologie et Sociétés* XIII (April,), 161-179.
Roseland, M., 1992. *Toward Sustainable Communities.* Ottawa: National Roundtable on Environment and Economy.
Schrecker, T., 1992. Of Invisible Beasts and the Public Interest: Environmental Cases and the Judicial System, in: R. Boardman (ed.), *Canadian Environmental Politics: Ecosystems, Politics, and Process.* Toronto: Oxford University Press, 83-108.
Toner, G., 1996. Environment Canada's Continuing Rollercoaster Ride, in: G. Swimmer (ed.), *How Ottawa Spends, 1996-97: Life Under the Knife.* Ottawa: Carleton University Press.
VanderZwaag, D., and L. Duncan, 1992. Canada and Environmental protection: Confident Political Faces, Uncertain Legal Hands, in: R. Boardman (ed.), *Canadian Environmental Policy: Ecosystems, Politics, and Process.* Toronto: Oxford University Press.

NATIONAL ENVIRONMENTAL POLICIES: COSTA RICA

Eduardo Silva

Costa Rica, nestled between Panama and Nicaragua, enjoys a reputation as a haven for peace, democracy, and commitment to biodiversity conservation in Central America. Occupying 19,575 sq. miles (50,700 sq. km), it is slightly larger than Denmark. Costa Rica has an enviable history of democratic stability. Its democratic political system dates back to a bloody civil war that ended in 1948. That conflagration prompted the founders of the new regime to abolish the armed forces, leaving national peace-keeping duties to a militarized police force: the Guardia Civil. Beginning in the 1950s, Costa Rica's political leadership sought to diversify agricultural exports, industrialize, and to build a social democratic welfare-state (within the constraints of a developing country) that emphasized health, education, and fair labor relations.[1] Moreover, Costa Rica, still a heavily rural country, possesses a better land distribution pattern than most Latin American countries. In sum, Costa Rica, classified as a middle-income developing country with a per capita income of U.S.$3,000 in 1995, stands out because it has democracy and income distribution and quality of life figures superior to those in most of the region.

The nation's commitment to peace and democracy became evident during the terrible civil wars that ravaged the isthmus during the 1970s and 1980s. Costa Rica did not succumb to the violence. Instead, headed by its Nobel prize-winning president Oscar Arias, it initiated the diplomatic process that eventually brought an end to the violence. By the same token, the Latin America-wide fiscal crisis of the 1980s has been met with a gradual policy of free-market restructuring that has not gutted social welfare institutions.

These attributes drew the attention of the world to Costa Rica, and many international organizations and government aid agencies established regional headquarters and research institutions there. Thus it was that the international community and environmentalists around the world became aware of Costa Rica's commitment to biodiversity conservation. For example, by 1989 nearly 27% of the nation's territory was under some form of protection. Parks and national reserves (areas of absolute protection) make up about 12% of the nation's protected areas; forest reserves, protected zones, wildlife refuges, and Indian reserves account for the other 15%. According to one estimate, if Costa Rica succeeds in protecting these areas "it will preserve nearly 95% of its flora and fauna, which represents

---

1 For an up-to-date overview of Costa Rica's political, economic, and social systems see Booth (1998).

about 5% of all of the plant and animal species currently known to exist on the planet" (Umaña and Brandon, 1992:85).

Indeed, most of Costa Rica's environmental capacity-building has focused on biodiversity conservation and forest management, issue areas in which the nation has made impressive strides. Since its creation in the mid-1980s, the Ministry of Environment has steadily increased its ability to coordinate agencies and policy for biodiversity conservation. It has been relatively successful in raising funds and managing them efficiently and effectively. The ministry has worked well with international agencies, green organizations, and domestic consultants that make up a deep pool of highly sophisticated technical expertise. Between them, they have devised many innovative policy instruments for biodiversity conservation. Indeed, with the appearance of private reserves and ecotourism, one could argue that Costa Ricans are on the doorstep of the biodiversity conservation equivalent of ecological modernization.

The picture, however, is not so rosy when it comes to pollution control. Agricultural and industrial development and urbanization are generating waste streams that have outstripped the treatment capacity of existing waste disposal systems. Fortunately, Costa Rica's commitment to quality of health between the 1950s and 1970s built waste disposal and potable water systems that have kept the public health problems of pollution from reaching crisis proportions. But how long this will remain the case is a matter of some concern.

Building the institutional capacity and crafting the policy instruments to deal with rising pollution are more difficult challenges for Costa Rican policy makers than dealing with biodiversity conservation. To begin with, the coordination problems are much greater because the Ministry of Environment does not have direct control over most of the government agencies involved, as is the case in biodiversity conservation. Sharp rivalries exist between it and the Ministry of Health, which traditionally has had responsibility for the issue. Moreover the integration of environmental issues into key ministries such as agriculture and economy remains a problem. To date, those ministries only have small environmental departments whose policies only counter the worst effects of the environmental consequences generated by their ministry's broader policies. Moreover, those two ministries are beholden to economic interests not much interested in shouldering the costs of pollution control. Finally, it is doubtful that sufficient political power or financial resources exist to change this state of affairs. The one exception might be with respect to international firms, which generally are wealthier than domestic companies.

Under these circumstances, Costa Rica will continue to struggle to advance to the stage of dilution with end-of-pipe treatment, and to inch its way to the intensive end-of-pipe treatment stage. The greater coordinating power granted to the Ministry of Environment by the general environmental law of 1995 will help to reach those goals. A policy style of negotiation and compromise will ensure incremental gains in the context of low social mobilization and public conflict.

1 Main Environmental Problems and Proposed Action[2]

Costa Rica's main environmental problems are closely linked to its level of socio-economic development, the history of natural resource use, and settlement patterns. although agriculture accounts for only 17% of GDP, the sector is vital for Costa Rica's economy because agroexports earn the lion's share of foreign exchange (World Bank, 1997). Meanwhile, industrialization (24% of GDP) and the service sector (58% of GDP) have concentrated a significant percentage of the population in the Central Valley of San José. This area contains 57% of the population, 92% of the nation's industry, and 55% of the coffee business. Given this profile, the nation's environmental problems cluster around agricultural land use issues and environmental health in rural areas; whereas urban pollution is most serious in the metropolitan region of San José.

Biodiversity. The advance of the agricultural frontier (coffee, bananas, cotton, sugar, cattle and land distribution to peasants) threatens Costa Rica with high levels of biodiversity loss due to soil degradation, deforestation, and wetland deterioration.[3] Costa Rica has one of the greatest concentrations of biodiversity on the planet with a high degree of endemism. It is estimated that 500,000 species are represented in Costa Rica: 12,000 plant species, 850 bird species, more than 200 mammalian, and approximately 300,000 insect species. Since the 1970s, Costa Rica has emphasized the prevention of irreversible species decline and the maintenance of a reservoir of biodiversity. To this end, Costa Rican governments have steadily labored to protect forests, wetlands, and coastal areas by improving the administration of protected areas and the management of natural resources.

Air pollution is concentrated in the metropolitan region of San José (the Central Valley) which contains 88% of the vehicle park, 88% of industry, and 57% of the population. Air pollution affects town centers the most. The sugar growing regions of rural areas are also heavily afflicted due to particulate pollution from burning sugar cane fields during harvests. The metropolitan area of San José (the capital city) suffers most, especially from fine particulates suspended in the air (PM10), carbon monoxide (CO), Ozone (03) and lead. Vehicles account for 70% of the emissions that contaminate the metropolitan region. Indices have been rising rapidly since the early 1990s. Carcinogen air pollutants such as benzene and formaldehyde are also present. Although the authorities recognize the potential health hazards of these pollutants, they have not been quantified systematically.

Water pollution. The quality of river and coastal waters has been deteriorating steadily due to toxic wastes from industry, runoff from pesticides and other agribusiness byproducts in rural areas such as solid waste, and raw household sewage. Population growth and the absence of land use planning have contributed to the problem. The most affected areas include the basin of the river Tempisque and the fluvial systems of the banana plantation regions in the North Atlantic. The principal problem for both are solid wastes and pesticides. However, the basins of the river Tarcoles and Barranca in the Central Valley, where the metropolitan area of

---

2 Unless otherwise noted, the data are from Ministerio de Planificación (MIDEPLAN, 1996). For earlier assessments see Fundación Neotrópica (1988), José Gracia Bondía (1990); and MIRENEM (1990).

3 For wetland deterioration see Carriere (1997).

San José is located, are even more heavily affected by water pollution. Only three of the urban areas served by the Tarcoles have partial networks of sewage systems (Alajuela, Heredia, and the Metropolitan Area). Toxic effluents from the coffee industry are among the highest sources of pollution, followed by industrial waste, and then raw household sewage. So far, the potential health threats have been checked by the ample supply of potable water in Costa Rica. But, if these problems are not addressed, they have the potential for great harm.

Agrochemicals. Costa Rica suffers from the highest incidence of agrochemical use per cultivated hectare in Central America, and one of the highest in the developing world. For example, pesticide consumption stood at 16 kg per cultivated hectare in 1989, as compared to 10.8 kg/ha in Japan and 11.8 kg/ha in Central America. The world average stands at 2.7 kg/ha. Worse, many of the most frequently used agrochemicals have been banned, severely restricted, or retired from markets in the developed world. These include, Clorotalonil, 2.4D, Mancozeb, and Paraquat. Their use is especially high in the production of fruit, rice, coffee, sugar cane, vegetables, and nurseries. It is estimated that banana companies account for 50% to 60% of total pesticide use in Costa Rica.

This irresponsible use of pesticides accounts for high incidences of poisoning and chronic health disorders among the population exposed to toxic chemicals in the various phases of their use. These health issues are especially acute among workers who apply the chemicals. They are endangered by overuse, lack of protective equipment, and work shifts that do not respect proper time intervals between applications. Moreover, rural workers often are ignorant of basic safety rules involving protective equipment and proper application and disposal rules. According to one survey, only 23% receive any training, only 16% have taken technical training courses in agrochemical use, 33% stored pesticides in or near their homes, 42% leave discarded containers lying in the fields, 52% wash application equipment near water sources, and 22% do so near their homes.

The effects of such high levels of agrochemical use reach beyond the health of rural labor. Toxic agrochemicals also affect potable water sources and wildlife. Their runoff drains into rivers that service cities and protected areas. Moreover, pesticides frequently are applied close to the dwellings of agricultural workers.

Solid waste collection, transport, treatment, and disposal – especially in urban areas – are inadequate. According to the Plan Nacional del Manejo de Desechos (GTZ, 1993), the Central Valley generates 81% of solid wastes, of which only 71% are properly disposed of. The problem of solid waste disposal is especially acute in the metropolitan area of San José, where the growth rate of such wastes between 1978 and 1994 has been 8.1% per year. Moreover, neither hazardous medical nor industrial wastes receive proper treatment. The National Health Service is addressing the problem of medical wastes. However, little has been done to manage hazardous industrial wastes. Outside of the Central Valley the solid waste problem appears to be even more acute, only 19% of solid waste being properly disposed of.

Mining is not a highly developed economic activity in Costa Rica, contributing less than 1% of GDP (MIRENEM, 1990: 67). As a result, with a few exceptions, the sector does not yet contribute significantly to Costa Rica's environmental problems. The exceptions include siltation and erosion in some areas due to quarrying and placer gold mining in the Osa peninsula. Deposits of gold, bauxite, copper, zinc, lead, iron, and titanium, among other minerals, have been identified (MIRENEM 1990: 67-70). As of 1997, Costa Rican authorities have courted foreign and domestic investment in mining. Although no offers have materialized, the intention to develop large and medium scale mining demands a regulatory framework that minimizes the potential environmental impacts of mining. The current mining code requires environmental impact reporting at all stages of the mining cycle. What is needed is the regulatory and institutional capacity to implement the mining code (MIDEPLAN, 1996: 17). This requires strengthening the Minera Nacional, the state mining company, and the Refinadora Costaricence de Petróleo (MIRENEM, 1990: 67).

2 Development of Environmental Institutions and Policy

Institutionalization. Beginning in the administration of Oscar Arias (1986-1990), Costa Rican governments have steadily worked to build an institutional and legal framework capable of coordinating the government's role in environmental management. Before then, as was the case in many Latin American countries, Costa Rica only had a weak coordinating agency: the National Environmental Protection and Improvement System under the Ministry of Planning. The agency, created in 1981, was charged with reviewing, integrating and harmonizing environmental policy (Brañes 1991, 58-59).[4]

For all intents and purposes, this meant that, until the mid-1980s, environmental matters in Costa Rica were handled by a proliferation of separate institutions under many ministries, such as agriculture, health, and transportation. This hampered environmental control and management for two reasons. First, ministries had set up the agencies over many presidential administrations (beginning in the 1940s), each with different objectives and under pressure from different political lobbies. Second, ministries and autonomous agencies had established their environmental units to avoid outside pressures, a process that accelerated when the government began to take responsibility for environmental assessment and control after the mid-1970s. As a result, some 27 different environmental units frequently worked at cross-purposes, jealously guarding their bureaucratic turf (Carriere 1991, 197-194; Hopkins 1995, 56-57).

Since the mid-1980s, three successive presidential administrations have labored to establish institutions capable of coordinating the government's environmental management. The most notable advances occurred under the administrations of Oscar Arias (1986-1990) and José Figueres Olsen (1994-1998), both of the center-left Liberación Nacional party (PLN). Rafael Calderón's government (1990-1994), of the center-right Partido Unión Social Cristiano (PUSC), had a limited but earnest interest in environmental policy that focused on the administrative reform of the system of protected areas; an endeavor in which the Calderón administration ultimately failed.

---

4 The main exceptions to his pattern in the region were Venezuela and Mexico which had created environmental ministries during the 1970s; although several countries with coordinating committees (Brazil, Colombia, Ecuador, and Guatemala) managed to pass comprehensive environmental legislation in the 1980s (Brañes 1991, 119).

The Arias administration took the first major step forward in the institutionalization of environmental management when it established the Ministry of Natural Resources, Energy, and Mines (MIRENEM). With the MIRENEM, Costa Rica became the first Central American country that raised environmental protection to cabinet rank. The MIRENEM's principal function was to enhance coordination between agencies in the conservation area. To that end, a number of scattered agencies were brought under its wing. These were the Forestry Service (DGF), the National Parks Service (PN), the Wildlife Service (DVS), the National Metrological Institute, the Bureau of Mines, and the Sectoral Bureau of Energy.

Effective coordination, however, eluded the MIRENEM. Many of the agencies nominally under its control enjoyed significant independence from the minister's office.[5] For example, the DGF, the PN, and the DVS had been created by separate laws regulating the forest, protected areas, and wildlife sectors before the MIRENEM had been established. The administrative prerogatives they enjoyed under these laws often shielded them from the minister's directives in case of disagreement. The minister's hold over the mining and energy bureaus was even more tenuous because they were parapublic agencies.

The difficulties in coordination were illustrated by the fate of a key ministerial initiative: the establishment of a National System of Protected Areas (SINAC), which was decreed late in the Arias administration (July 1989). Since the end of the 1980s, MIRENEM ministers had pushed the SINAC as a means to enhance coordination by streamlining and decentralizing the administration of protected areas. The SINAC called for the division of Costa Rica into a number of independent Conservation Areas (ACs) under the overall management of a national directorate beholden to the minister of MIRENEM. In this system, regional ACs would be polyfunctional agencies that fused the DGF, the PN, and the DVS (MINAE, 1997). AC directors would be free to assign duties according to functional need, not agency prerogative. In effect, the SINAC abolished the administrative autonomy of the DGF, the PN, and the DVS. Jealous of their autonomy, and under great pressure from the executive office of the MIRENEM to conform, the DGF, the PN, and the DVS resisted attempts to implement the SINAC. Subsequently, their stiff opposition – from the DGF and its allies in particular – killed a legislative bill for a SINAC introduced during the Calderón government by MIRENEM minister Hernán Bravo and vice minister Mario Boza.[6]

Despite the Calderón administration's failure to establish the SINAC, The MIRENEM took two modest steps to enhance the ministry's efficiency. First, with an eye towards implementing Costa Rica's Tropical Forest Action Plan, it created a Subsectoral Secretariat for Natural Resources. The point was to weaken the DGF, which had so antagonized the heads of the MIRENEM. More significantly, in 1993, the MIRENEM founded the Directorate of Planning and International Cooperation as an advisory agency to the minister, which became the Directorate of Cooperation and International Relations in early 1994 (Segura, 1997: 109-110).

5 The Minister's office consisted of the administrative units of the MIRENEM. These included advisors to the minister and the offices of the vice-minister, legal council, bureau of planning, personnel, and public relations (Hopkins, 1995: 58).

6 Legislation was needed to annul the laws that had established the three separate directorates.

The Directorate of Planning and International Cooperation was a response to Costa Rica's reliance on external resources for its environmental programs, especially in the biodiversity issue area. It also addressed the need to connect international funding to national priorities. In the 1980s, Costa Rica had benefited from a plethora of bilaterally and multilaterally supported environmental projects. But Costa Rica failed to coordinate the search for external funding with a general strategy for environmental protection. Instead, donors offered Costa Rica a multitude of projects, which, interesting and useful as they were, remained a collection of scattered efforts, generating their own clienteles, and occasionally working at cross purposes. Equally pressing for Costa Rica, project cycles were on the wane by the early 1990s and fresh external resources were required.

During the Figueres administration, the institutional structure for environmental protection changed markedly. First, in 1995 minister of environment René Castro took matters in his own hands with respect to the SINAC. Since his attempts to push through the legislation fared no better than those of his predecessors, he simply established the SINAC through a bold act of administrative restructuring that skirted the edges of legality. Second, in November 1995 the new general environmental law increased the MIRENEM's coordinating powers, especially in the energy, mining, and health sectors; and the MIRENEM itself was renamed the Ministry of Environment and Energy (MINAE).

Main Laws and Regulations. Before Costa Rica enacted its general environmental law in November 1995, a number of laws, many of them still on the books, regulated environmental quality and the use, conservation, and preservation of natural resources. These included the Fishing and Maritime Hunting Act (1942), the Natural Resources Conservation Act (1953), the General Potable Water Act (1953), the Urban Planning Act (1968), the Forestry law (1969, 1986) the General Health Act (1974), the National Parks Law (1977), the Coastal Waters Act (1977), the Mining Code (1982), the Wildlife Conservation Act (1992) (MIDEPLAN, 1996: 58; Salazar Cambronero, 1991).

The Figueres government featured "sustainable development" as a centerpiece of its program. Accordingly, it vigorously shepherded a number of laws through the legislature (some of which had begun their life in the previous administration) and issued important decrees on environmental protection standards. Among the most significant legislation were attempts to regulate key urban and rural sources of pollution. These laws and decrees recognized that air pollution, toxic wastes, and pollution from agrochemicals posed serious environmental risks to Costa Ricans. They included the Transportation Act of 1994, which focused on emission control; the "Standards for Registry and Control of Hazardous Toxic Products" (1994); and the "Standards for the Registry, Use, and Control of Agricultural Pesticides" (1995).

In addition to these fresh initiatives, a new Forestry Law, which had begun a torturous journey through the policy process in 1990, was finally enacted in 1996 (Silva, forthcoming). The forestry law was significant for a number of reasons, not least of all because it eliminated the DGF. This was a victory in the MINAE's quest to institutionalize the SINAC, a goal that finally was reached in 1998 with passage of the long-awaited Biodiversity Law that gave legal standing to the SINAC. The Forestry Law and its regulations were also significant because they rationalized the administration of the forest sector and aggressively favored market instruments over control and regulation.

The General Environmental Law of 1995 codified the government's responsibilities in environmental protection and quality and specified the issues the government had to attend to: protected areas; maritime coastal and wetlands resources; forests, air, water, biodiversity, soils, energy, and pollution. To accomplish these ends, as will be seen below, the law strengthened the coordinating capabilities of the MINAE with respect to other ministries and agencies. This capacity was expressed largely through institutional restructuring of the former MIRENEM. Equally significant, the law also focused on citizen participation in environmental policy making and implementation, established the polluter-pays principle, ordered national land use planning, and mandated environmental impact reporting for new ventures likely to have environmental consequences.[7]

Main Policy Instruments. Regulation remains the principal policy instrument in Costa Rica. Laws tend to be general documents that define issues, and that assign the administrative functions, institutions, and powers necessary to implement the law. The regulatory statutes of the law, defined by the line ministry of autonomous institute, set the standards that processes, products, and companies must meet and assign punishment for infractions. Although Costa Rican ministries have significant autonomy in creating the regulatory body of laws, they also consult with organized civil society in particularly conflictual cases. Those organizations frequently include producer associations, nongovernmental organizations, professional associations, academic institutions, other government agencies, and municipal governments.

Under the Figueres administration, the MIRENEM/MINAE also encouraged the use of economic incentives and market mechanisms, especially where biodiversity protection was concerned. For example, it introduced a gasoline tax that destined one third of revenues to the MINAE. Unfortunately, the initiative backfired. The Ministry of Finance refused to release revenue from the gasoline tax to the MINAE. It argued that International Monetary Fund-induced fiscal austerity obligated the ministry to spend the much needed funds on higher-priority issues. This was perfectly legal, for under the provisions of the decree the funds first went to general revenue account of the Ministry of Finance. According to Costa Rican law the Finance Ministry has significant discretion over the disbursement of funds from that account.

A number of other economic instruments, mostly introduced or reinforced during the Figueres government, had a happier fate. Indeed, Costa Rica has become a leader among developing nations in experimenting with economic instruments for biodiversity protection, and it has gained some notoriety as a result. Virtually all projects are partnerships between the Costa Rican government, the private sector, NGOs, and international actors; or they include some mix of them. The programs are designed to tap international revenue sources, underscoring the nation's dependence on external sources of financing for its environmental programs. The most famous of the programs that emphasize economic instruments include debt-

7 The new environmental law strengthened the legal standing of environmental reporting, a requirement which had been established only a year earlier by decree in October 1994.

for-nature swaps, bioprospecting contracts with international companies, and joint implementation agreements for carbon-fixing.

In the 1980s, the Arias government pioneered debt-for-nature swaps (Hansen, 1989; Rainforest Alliance, 1991). These are complicated financial arrangements in which largely foreign investors (often NGOs) "buy" a portion of a developing country's foreign debt obligations at a deep discount in international secondary debt markets. The developing country agrees to use the financial resources to protect a specific habitat, biome, or ecosystem. In 1986, informal negotiations between the fledgling MIRENEM and the Central Bank led to the establishment of a debt-for-nature swap mechanism. By 1988, Costa Rica had entered into five such agreements yielding "$12 million in grants and donations [that] were used to purchase nearly $80 million of Costa Rica's debt and generate over $42 million in local currency bonds for conservation" (Umaña and Brandon, 1992: 97).

The Figueres administration aggressively promoted the idea of making environment profitable; profitability would ensure conservation. As a result, during the Figueres government the MIRENEM/MINAE strongly supported the National Biodiversity Institute (INBio) (Presidencia de la República, 1994: 48-50). The INBio began life in the waning months of the Arias administration in 1989 (S. Rodríguez, 1993: 128; Umaña and Brandon, 1992: 98-99). The INBio is charged with collecting, cataloguing and inventorying all the plant and animal species of Costa Rica. Inventories are carried out mostly in national parks. In addition to the intrinsic scientific value of the exercise, the INBio investigates how biodiversity can contribute to socioeconomic development (Umaña and Brandon, 1992: 99).[8] In practice this means that it sets prices for items of national biodiversity of interest to international companies (S. Rodríguez, 1993: 127). In other words, the INBio negotiates bioprospecting contracts with transnational corporations; most of the proceeds benefit the national parks system which are the site of the prospecting. INBio's contract with Merck & Company, Inc. in 1991, one of the world's largest pharmaceutical companies, was the first bioprospecting agreement. The contract was not as favorable to Costa Rica as INBio would have liked on the complicated issues of royalty payments, intellectual property rights, and technology transfer. But INBio directors are confident they have learned lessons on how to negotiate with international companies and are eager to enter into new agreements (S. Rodríguez, 1993: 137-138).

MIRENEM minister Castro and his team also aggressively pursued new international sources of financing for conservation based on market-oriented economic instruments. To this end, they established an office for joint implementation projects: the Costa Rican Office for Joint Implementation (OCIC). Since 1995, OCIC operates with the approval of the United States Office for Joint Implementation, and it is managed by an NGO established with USAID funds in 1986: the Fundación para el Desarrollo de la Cordillera Volcánica Central (FUNDECOR). Joint implementation ventures are complicated arrangements, but the central idea is this. At various stages of their development forests fix (or sequester) some greenhouse gases, especially carbon dioxide. Thus, in joint implementation ventures, foreign

8 INBio argues that the eventual training and employment of some 200 parataxonomists for the collection and identification of species covers its social obligation, for these are drawn from poor rural areas (Umaña and Brandon, 1992; Rodríguez, 1993: 133-34).

companies that emit greenhouse gases buy "pollution rights" by paying developing countries for the sustainable management of natural forests and for sowing tree plantations (Alfaro, 1997). The first such agreement was signed in 1996 between Norway and Costa Rica to fix some 200,000 tons of carbon; Norwegian utilities agreed to pay two million dollars for the conservation of thousands of ha of primary and secondary forest and to promote tree plantations in the Virilla watershed and Braulio Carrillo national park and surroundings (Vega, 1997; Araya, 1997).[9] Private property owners participating in the project will earn $50 per ha. per year; the number of years depends on the type of forest involved.

Policy Style. Cooperation and compromise among government agencies and domestic and international actors best characterizes Costa Rica's policy style. This cooperative style is deeply embedded in the political traditions and institutional arrangements of Costa Rican democracy. Policy formulation routinely involves consultation with the sectors of organized civil society affected by it. Such consultation is frequently institutionalized via advisory boards to ministries and government agencies. The laws that created those advisory boards generally mandate representation of the affected interests from the private, the non-profit, and academic sectors as well as from other government institutions. In addition to these formal mechanisms, once the organizations of the social and professional groups who are affected by a particular policy become aware of a policy initiative, they frequently lobby the government agencies and ministries directly. A similar process takes place in the policy process for legislative bills. Legislative committees hold numerous public hearings that involve all of the relevant actors. Intensive lobbying of deputies also occurs. The process is repeated in the ministries during drafting of secondary legislation. In the end, policies, laws, and statutory instruments all reflect compromises among the main actors.

The formulation of the new forest law illustrates the process well. The legislative committee appointed a technical advisory team that included timber industrialists, professional foresters, representatives from NGOs, and from the DGF (Silva 1997a). After that, the MIRENEM, which initiated the bill, formed a special working committee to fine-tune differences between the private sector and the government. Once this was accomplished, the bill went back to the legislative committee (Agriculture) for further elaboration. The committee then harmonized three different bills (the MIRENEM's and bills introduced by two deputies). During this process, deputies frequently consulted with the private sector, NGOs, and other government officials. In the formulation of statutory instruments after the bill had been adopted, the MIRENEM held extensive consultations with the private sector, NGOs, academic circles, professional associations (foresters), and municipal governments (Silva, forthcoming).

Naturally, the compromises resulting from these negotiations favor some groups more than others. Nevertheless, confrontation or open conflict is rare. By the time policies are decreed and laws and regulations appear most of the principal actors are satisfied they have achieved what they could. For the rest, highly dissatisfied actors look to the policy implementation stage of the policy process as the next place to negotiate changes favorable to them.

---

9 FUNDECOR's project area centers on Braulio Carrillo national park in central Costa Rica.

This is not to say that no confrontation takes place. Traditional issues, such as wage, price, and welfare policies cause unions to mobilize. The environmental issue area, however, rarely brings about such public action. In regions where peasants are organized by more militant unions, such as the Atlantic, demonstrations about land use policy and environmental protection (parks and forest use) may erupt. NGOs occasionally take highly public stances against the government as well. The struggle in the Osa peninsula against Ston Forestal (a large international pulp and paper company) was a high profile case (Hombergh, 1998). Environmental NGOs protested the use of an exotic species for monocrop plantation culture, the impact of the proposed port and ocean going ships on a delicate and unique maritime ecosystem, and the divisive effects on local communities.

Occasionally, sharp conflicts on environmental issues also erupt between the private and public sectors. Such was the case during drafting of the forestry law, which was more conflictual than many other legislative efforts. Sharp dissension over the degree of government control in policy-making and implementation generated strong private sector resistance. It was because of this conflict that the MIRENEM set up the aforementioned reconciliation committee between the two in 1994 (J. Rodríguez, 1997). The most public rift broke out over the MIRENEM's attempts to eliminate economic incentives for plantations and native forest management (CCF, 1997). The private sector initiated a media campaign against the MIRENEM and threatened to stage a demonstration to embarrass the government during U.S. president Bill Clinton's visit to Costa Rica (Al Día, 1997; La Nación, 1997; and Alfaro, 1997).

Rivalries between government institutions and among agencies within ministries occasionally are another source of friction. No love is lost between the MINAE and the Ministry of Health, especially because the latter (which has a longer history) sees itself overshadowed by the former. Conflicts within the MINAE have been strong also (Brockett and Gottfried, 1994). The newer, more technocratic and market-friendly tendency in the MINAE (as represented by minister Castro) used two tactics to emasculate the old-guard supporters of planning, government control, and greater emphasis on social forestry. First, top MINAE administrators succeeded in eliminating DGF. Second, after the establishment of the SINAC, Castro reshuffled personnel and demoted the legal department as a means to further weaken the old guard (Cubero, 1997).

3 The Main Actors

3.1 Environmental Policy Institutions

The Ministry of Environment (MINAE) is the main government proponent of environmental policy. Ever since it was founded it has labored to establish its authority and to take a central role in environmental policy-making. To that end, as previously seen, MINAE ministers constantly have sought to gain effective control over the agencies nominally under their management. They have also sought to fortify their coordinating role vis-à-vis other ministries and agencies with environmental responsibilities. (As will be seen below, the legislative assembly played a key role in that process.) The minister and vice minister are political appointees of the president of the republic; and, although the position is not among the most senior in the cabinet, upwardly mobile politicians seek the appointment because of the attention the issue area receives nationally and internationally. The minister is aided by a hand-picked advisory council. Under the minister, career bureaucrats head the various agencies and departments of the MINAE.

The general environmental law of 1995 defined the MINAE's administrative structure and mandate.[10] The administrative structure strengthens the MINAE's coordinating role in relation to other ministries and agencies with environmental duties. For example, inter-agency coordinating committees run by the MINAE and frequently chaired by MINAE officials were established. The ample scope of the MINAE's official responsibilities further underscored the drive to enhance its coordinating capability. These include administering the nation's forest resources and protected areas; controlling air, water, and soil pollution; sustainably developing the energy and mining sectors; conducting environmental impact assessments for mining, transportation infrastructure, electric utilities, oil and gas exploration and other large-scale industrial projects (MIDEPLAN, 1996: 57).

Citizen and local government participation are integral components of the MINAE's decision-making structure. To this end, the general environmental law mandated the creation of national and regional institutions. At the national level the MINAE has a National Advisory Council, with an executive secretariat housed in the MINAE. The National Advisory Council includes representatives from the presidency and the ministers of planning, environment, health, agriculture, education, and science and technology. Regional environmental councils, which the minister of environment must consult, include one representative from each provincial governorship and regional council, as well as one representative from the municipal league, environmental organizations, student government, and private business chambers (República de Costa Rica, 1995).

The MINAE's official connection to the Ministry of Planning extends its coordinating role to the planning stage. The capacity to coordinate the planning of sectoral policy with environmental relevance is a vital step given the intersectoral quality of environmental problems and the MINAE's new mandate to deal with the issue. The publication of the *National Plan for Environmental Policy,* under the auspices of the Ministry of Planning, was the MINAE's first coordinating effort (MIDEPLAN, 1996). The document diagnoses environmental problems in all of the sectors for which MINAE has responsibility, identifies all of the ministries and agencies linked to each issue area, and proposes general policy prescriptions.

The Ministry of Health, under the General Health Act, also has an important role in environmental policy. This role is performed by the Environmental Health Division (EHD) and its five departments: sanitary engineering, environmental control, industrial safety and hygiene, food safety, and toxic substances control.

10 The MINAE's institutional structure includes the National Environmental Council (advises the presidency on environmental issues); the Regional Environmental Councils (civil society participation in environmental policy making); the National Environmental Technical Secretariat (environmental impact reporting, key link to rural and urban industry); the Environmental Comptroller (oversees compliance with the environmental law); the Environmental Administrative Tribunal (attends to accusations of environmental degradation) (MIDEPLAN, 1996: 57).

The EHD recommends standards and technical procedures, and it designs and revises pollution control programs. For example, the HED sets water quality standards and regulates water treatment and industrial waste disposal plants. Moreover, the HED conducts field inspections, levies penalties for code violations, evaluates the condition of waste treatment plants, and collects data on industrial waste (MIDEPLAN, 1996: 57).

Many other ministries and government agencies have duties related to the environment. The Ministry of Agriculture protects soil and water and controls pesticide use. The Ministry of Public Works and Transport coordinates oil spill cleanups and other transportation-related pollution issues. The Costa Rican Aqueduct and Sewage Institute is in charge of potable water and waste disposal. The National Electrical Service sets utility rates. The National Groundwater and Irrigation Service develops agriculture by building and maintaining irrigation and flood control systems. The Costa Rican Electrical Service develops energy sources and protects natural resources. The Costa Rican Tourist Institute promotes and regulates tourism (including ecotourism). The Agrarian Development Institute implements land distribution policy, which is intimately linked to the conservation and sustainable use of natural resources. The National Housing and Urban Institute is in charge of urban land use planning. The Cantonal Agricultural Centers promote the participation of civil society in agricultural policy making and implementation, including agroforestry and agroindustry. Municipalities collect garbage and are involved in urban planning (MIDEPLAN, 1996: 57-58).

In addition to government institutions, legislative committees are playing an increasingly significant role in Costa Rican environmental policy. Traditionally, policy initiation rests with the executive. But legislative committees have aggressively pushed bills when they suspect the government of foot-dragging. For example, the Agriculture Committee, fed up with the impasse over a forestry bill, pushed alternative bills, forcing the government to react, and putting its stamp on much of the 1996 forest act (Silva, 1997a; forthcoming). Meanwhile, the Environmental Committee initiated, debated, and passed the 1995 general environmental law. By most accounts the law had very little input from the MIRENEM/ MINAE. The principal sponsor of the law, Deputy Luis Martínez of the Liberación party, maintained that it had been crafted and passed over the head of MIRENEM minister Castro (Martínez, 1997). In any case, relations between minister Castro and the two legislative committees were anything but cordial.[11]

3.2 Green Organizations

The environmental movement of Costa Rica boasts a host of non-profit private institutions and academic organizations, many of which have close links with international agencies (Thrupp, 1990). Some advocate policy prescriptions, others specialize in project implementation, and another subset dedicate themselves to research. A few combine these tasks.

---

11 Part of the problem may have been rivalry between Minister Castro and Deputy Martínez (both of the Liberación party), for the latter was the principal sponsor of both the general environmental law and the legislature's version of the forest bill.

Educational organizations carry out research and pilot programs in conservation and sustainable development. Most of these projects address conservation issues and the relationship between rural populations, development, and conservation. Legislative committees and blue-ribbon executive branch commissions often call on prominent scientists from these organizations to testify or to advise them. The most prominent of these institutions are the Centro Agronómico de Investigación y Enseñanza (CATIE) (a large research center that provides an institutional home for programs sponsored by the aid agencies of developed countries), the Organization for Tropical Studies (a consortium of largely U.S. universities that maintains two conservation biology research stations, of which La Selva is the best known), the Environmental Sciences Department at the National University-Heredia (which also has a masters program in natural resource economics), and the Biology Department of the University of Costa Rica (Hopkins, 1995: 98).

The implementation of large-scale projects for conservation and sustainable development funded by multilateral and bilateral international aid agencies generated another set of well-recognized non-profit organizations.[12] In principle, these NGOs mainly search for internationally funded projects and then implement them. In many cases, however, these organizations also take on policy advocacy roles, especially with respect to environmentally responsible natural resource use. This policy advocacy usually does not take a strident confrontational public tone. Rather, the NGO leadership stays in constant contact with policy makers in the government and in the legislature. They become the spokespersons for the ecosystems, social and ethnic groups, and scientific research interests that are the focus of their projects. Since the Costa Rican policy process involves negotiation with pressure groups, they try to influence policy by exercising their representative role.

The Centro Científico Tropical (CCT) occupies a special niche among this type of environmental organization. It is one of the most influential policy advocacy NGOs in Costa Rica and is well-connected to government circles. For example, the CCT took a leading role in formulating the SINAC in the 1990s, and one of its principal figures, who works well with either Liberación Nacional or Christian Democratic governments, became the first director of the SINAC. The CCT also took a very active role in drafting the current forest law. In addition to policy advocacy and formulation, the CCT runs projects as well, such as the Monteverde Biological Reserve.

Costa Rica is also home to many environmental membership organizations. They devote themselves to environmental consciousness-raising, education, research, and policy analysis. Many are focused on a specific issue, such as a particular habitat or endangered species; others concentrate on region-wide problems. A few have their eyes on the national scene. Among mainstream organizations, the Amigos de la Naturaleza (founded in 1973) successfully lobbied the Legislative

12 Some of the most prominent are the Fundación para el Desarrollo de la Cordillera Volcánica Central (FUNDECOR), linked to USAID projects; Cooperación en los Sectores Forestal y Maderero (COSEFORMA), linked to the German aid agency GTZ; the Asociación Guanacasteca de Desarrollo Forestal (AGUADEFOR), born from Dutch and Swedish aid; Comisión para el Desarrollo Forestal de San Carlos (CODEFORSA) linked to FAO, Great Britain's ODA, and the GTZ; Fundación Neotrópica, initially linked to USAID.

Assembly to ratify the International Treaty on Endangered Flora and Fauna in 1974 (Hopkins, 1995: 100). The Asociación Costaricence para la Conservación de la Naturaleza (ASCONA), founded in 1972, has been a very active policy-advocacy NGO for sustainable development. In the past (1979), ASCONA campaigned for broad policy and programmatic changes in Costa Rican environmental policy, including the creation of a ministry of the environment. In the early 1980s, it filed several lawsuits to protect some parks from economic development, including mining operations (Hopkins, 1995: 100-101). In the 1990s, ASCONA has concentrated on environmental education.

Several more radical environmental groups also exist, such as the Asociación Ecologista Costaricence (AECO). AECO, said to be allied with Greenpeace, led the struggle against Ston Forestal's project in the Osa Peninsula. Given its combativeness, some environmentalists criticize AECO for being too confrontational. Since the death of two of its leaders in a fire (suspected to be a case of arson and murder) AECO has become a bit more circumspect. By the same token, the Fundación Ambio and the Centro de Derecho Ambiental y Recursos Naturales (CEDARENA), which are rivals, provide legal aid to environmental organizations. More recently, CEDARENA has done some legal consulting for the government on the monitoring of illegal logging (CEDARENA, 1996).

For a number of reasons, many mainstream environmental NGOs have lost their status as central actors in environmental policy making (Hopkins, 1995: 101-102). They have been replaced by the non-profit organizations with close links to government and international donors. As will be seen in the next section, the line between private and public in some of these policy-relevant organizations is significantly blurred. Nevertheless, in search of greater political clout, environmental NGOs federated in 1988 in the FECON (Federación Costaricence para la Conservación del Ambiente). FECON has had a spotty history. Despite its good intentions, it became moribund in 1990 due to differences among its many member organizations. However, FECON revived in 1995 in an attempt to influence the forest law debate in the National Assembly. It was not successful. FECON concluded that it needed to learn more about the policy process to achieve greater effectiveness; for it realized it intervened too late in the process to have any impact (FECON, 1997).

It is important to keep in mind that these many and varied green organizations (and the semi-private, semi-public ones discussed below) do not all share the same approach to environmental problems. They cluster into two broad categories. One camp is mainly concerned with environmental protection in isolation from the social context. It includes conservative conservation organizations and their U.S. supporters. The other camp advocates a more radical green-left alternative that favors social development via redistributionist policies and is "deeply democratic in its political practices" (Carriere, 1991:198). Nevertheless, they share a commitment to sustainable development and a "nurturing attitude towards the environment" (Carriere, 1991:199).

3.3 Other Proponents of Environmental Protection and Target Groups

Some politically very influential private non-profit green organizations have such close relations with government that in practice they are quasi-public institutions. The two most important are the National Parks Foundation (Fundación Parques Nacionales, FPN) and INBio. The FPN is one of the most influential policy-advocacy NGOs for conservation issues in Costa Rica. It was founded in 1979 at the urging of the National Parks Service director and the presidential advisor on natural resources. Of its five executive board members, three are appointed by the attorney general's office, one by the presidency, and one by the canton (province) of San José. The FPN's mission is to seek international donor support for conservation and to promote internal private conservation efforts. In short, the FPN raises and manages funds for the national system of protected areas (Hopkins, 1995: 99-100). The network of professional relations between FPN directors and the heads of the National Parks System is very tight, as are their working relationships based on fund-raising and budget allocation. These connections give FPN directors enormous influence on conservation policy.

The INBio is officially a private non-profit organization (Umaña and Brandón, 1992: 98-99). But, as was the case with the FPN, the close connections to government of its prestigious high-level directors blur the line between private and public. In practice the INBio is a semi-private/semipublic institution. The intensity of government backing assures INBio of preferential political and material support (S. Rodgríguez, 1993). Because of INBio's focus on biogenetic prospecting, its quasi-governmental status has stirred controversy in Costa Rica. More alternative-oriented green activists chafe at the ease with which INBio flourishes due to government support while their own organizations are constantly hampered by red tape and discriminatory policies. Sovereignty issues are also frequently invoked (S. Rodríguez, 1993).

In a positive development, some target groups have taken more environmentally-friendly stances. For example, a number of producer associations in the forestry sector are trying hard to find ways to contribute to the sustainable use of forests. Of these, the Costa Rican Forestry Chamber (Cámara Costaricence Forestal, CCF) is the largest and it basically represents the interests of large-scale forest holders. CCF members own native forests, tree plantations, and industries that manufacture wood products (such as doors in the case of Portico). They are very active participants in the policy-making process. As of late, the CCF has pushed tree plantations as a means to generate timber for export (to reproduce the Chilean forestry model) and to relieve pressure on the native forest. The CCF has also worked hard to find a fit between producer interests and the government's new emphasis on subsidies for the environmental services of the forest (read conservation rather than production). The CFF found modest success when it convinced the government that several aspects of forest management contributed to enhancing the environmental services of the forest. Moreover, the CCF has designed a certification program for sustainably harvested timber.

Meanwhile, peasants and small-scale producers in forested areas formed their own organization, the National Forest Peasant Association (Junta Nacional Forestal Campesina, JUNAFORCA) (Solís, 1993). Their concerns and activities are similar to those of the CCF. The main difference being that the JUNAFORCA emphasizes peasant interests, which are often quite distinct from those of the large-scale forest owners (Solís, 1994). JUNAFORCA was once a member of the CCF, but left in 1997 after significant friction soured the initial uneasy alliance peasant organizations had struck with timber industrialists to free themselves of close government supervision. The relationship between the CCF and JUNAFORCA is antagonistic at present since the CCF has expended considerable energy lobbying government for policies that benefit large-scale timber interests (Salas, 1997). Moreover, the CCF has clearly plotted a strategy to subordinate smallholders to large-holders (Arias, 1997; Alfaro, 1997).

In other areas target groups are more resistant to the internalization of environmental discourse. These groups consist of the chambers of industry and commerce, the employers' federation, and producer organizations for coffee growers, cattle ranchers, banana plantation owners, and so on. They are mainly interested in expanding exports and internal production. Given sharp international competition, they are reluctant to bear the costs of end-of-pipe treatment, alternative growing methods, or any of the goals of ecological modernization. Environment only becomes an issue when faced with resource depletion (as occurred in the forest sector) (Artavia, 1997; Carriere, 1991: 196). If the economic sector is not very strong, government may become emboldened to regulate it more strongly, forcing a private sector response (as was also the case in forestry). In some areas, target groups are parapublic agencies (such as in electricity). Thus, regulating them becomes a matter of intrabureaucratic struggles between MINAE and the respective ministry.

3.4 The Media

The media cover environmental issues only sporadically, and then most of it focuses on biodiversity conservation issues. None of the media (newspapers, television, or radio) have correspondents who are specialists in environmental affairs on staff. A division of labor, of sorts, exists among the media. Television generally focuses on nature documentaries; the MINAE and the Ministry of Health also use television to air public service announcements intended to educate Costa Ricans about the environment. Meanwhile, the main newspapers such as La Nación, Tico Times, La República, Estrella del Sur, Al Día, and Anexión cover events. These events include park improvements, environmental policy pronouncements by government officials, and foreign investment with environmental impacts. Of these, the English-language Tico Times has the most consistent reporting, probably because its readership mainly consist of expatriates and tourists with environmental sensibilities.

The stories and issues reveal that reporters gravitate toward press releases by public and private agencies. Much environmental reporting also springs from the use of the media by environmental groups when they publicize a struggle to generate support for their cause. One of the two most widely publicized stories has been the campaign to stop Ston Forestal's forestry project on the Osa peninsula. The other one was the CCF/JUNAFORCA's conflict with the MINAE over the proposed elimination of government incentives for sustainable forestry. Interestingly, the Tico Times stays away from these stories and concentrates on ecotourism, parks management news, and issues related to biodiversity conservation in general.

4 Capacity-Building

4.1 Increase in Number and Professionalism of Proponents

Over the past 20 years, Costa Rica has seen an increase in the number and professionalism of the environmental policy proponents in both the public sector and in civil society. The public sector has a Ministry of Environment, and several administrations have labored to give it effective control over the various departments and parapublic agencies under its mandate. These administrations also strengthened the MINAE's coordinating role with respect to other key ministries, notably, the Ministry of Health. These actions bode well for the coordination of environmental policy in Costa Rica across government agencies. They are a definite advance over the situation to the mid-1980s.

The professionalization of MINAE's personnel (and that of the Ministry of Health) also bode well for the design of environmental policy. Administrators acquire degrees and specialized program certificates from universities in Costa Rica and abroad. Many of the top people in MINAE, whenever possible, apply the latest academic and professional innovations. This is especially strong in the area of biodiversity conservation, as we saw in the cases of joint implementation and bioprospecting. But the application of innovative policy is not limited to this sphere. For example, small-scale hydroelectric plants whose operation benefit local communities, and that have been designed not to harm the local environment, also have been encouraged. Some green taxes have been introduced, such as those on energy and on foreign tourists who pay substantially more for access to parks than Costa Ricans.

Expertise in policy making and project design and implementation is not limited to government agencies. The government draws heavily on the proliferation of environmental NGOs, think-tanks and consultancies, academic institutions, and international organizations. These institutions are full of highly trained, sophisticated professionals, many with advanced science degrees and successful relations with international funding agencies. The MINAE and other government agencies frequently call on them to carry out studies and to make policy recommendations. Collaboration between the MINAE and these other proponents of environmental issues, which have mushroomed since the 1980s, immeasurably strengthens the environmental institutions of the public sector. Without them, the MINAE would be a pale reflection of itself and probably would not generate as many innovative tendencies as it has.

International influences add to the synergy generated by collaboration between the public sector and civil society. Many of the better endowed NGOs were established to administer the projects of international organizations and the aid agencies of developed nation governments. Thus, foreign experts contributed to the professionalization of Costa Ricans.

Institutional capacity-building has even begun to flourish in a modest way in the legislature. True, environmental issues lack a specialized committee they can call home. Nor has the legislature hired environmental specialists to serve on technical staff of the National Assembly. But several individual legislators have begun to champion environmental issues and they have hired environmental specialists to serve on their staffs.

4.2 Policy Process and the Strengths and Limitations of Proponents

Several elements strengthened the proponents of environmental issues in Costa Rica between 1980 and 1998. In the first instance there are the many institutional toeholds environmentalists have carved out. Historically, line ministries initiated most policy in Costa Rica. Therefore, the creation of a Ministry of Environment was an important step forward and constitutes a significant source of political influence. The constant efforts to strengthen its agenda-setting and coordinating role add to their capabilities. The pool of expertise on which policy makers can draw in the myriad, vibrant environmental organizations (many of them directing large-scale projects and constantly applying for new project funds) further energizes environmental policy.

The relative openness of the Costa Rican policy process and its emphasis on negotiation – that is, the pluralistic nature of Costa Rican democracy – also contribute to the power of environmentalists, and to a generally non-confrontational policy style. When conflict over sharp differences of interest erupts during the policy process, all sides are brought together to negotiate a consensus. This emphasis on consensus permeates policy making in both the executive and the legislature. For example, both of these political institutions applied the technique during the policy debates over the forest law between 1990 and 1996 (Silva, 1997a and forthcoming). The low quorums required to block bills in the legislature are at the root of this non-confrontational style. Without consensus, inflexible minorities can appeal to their political party allies in the legislature to kill bills (González, 1997).

This open, consensus-seeking policy style benefits environmentalists because they can influence policy through their connections to government, connections which are reinforced by government's need to consult. In short, environmentalists must be heard and the more they participate in the policy process over time, the more effective they become. The broadness of participation ensures that points of view that otherwise might be excluded from consideration are heard. Thus the spectrum of policy proposals to solve environmental problems tends to be wider than in some other countries, such as Chile for example (Silva, 1997b).

As of late, the locus of policy initiation has shifted, rendering participation in the policy process more complex and opening opportunities for environmentalists. Until 1994, by custom the executive branch initiated most bills. Between 1994 and 1998, the legislature initiated bills more aggressively than it had in the past. This unlocked a new institutional realm for environmentalists to influence policy. If the executive appears reluctant to act on an issue, the legislature may force an administration to consider it. This occurred with the 1995 environmental law (Martínez, 1997).

Public opinion and the legal system are additional tools that environmentalists use to bolster their position. Environmentalists use both when confronted by target groups. To date, most legal action and public outcry have centered on land disputes in national parks. The creation of many national parks entailed the expropriation with compensation of private property. Naturally, controversy over which lands are expropriated, and private sector attempts to disaffect public lands, generate disputes. Courts may rule in favor of environmentalists, as was the case in Palo Verde National Park in 1981 (Hopkins, 1995: 121-122). Generally, in these controversies environmental organizations bolster their legal challenges with public opinion campaigns.

Costa Rica is a small country open to international influences from the United States, Europe, and multilateral development banks. These external actors empower Costa Rican environmentalists. They offer material and intellectual resources that otherwise would be missing. In their absence, Costa Rica's wealth of capable environmental organizations and expertise that contribute to the policy process would have been considerably less.

Despite these advances in capacity-building, many of these sources of strength for environmentalists are double-edged, they also yield restraints. The openness and consensus-seeking style of Costa Rican policy making provides opportunities for the target groups to resist. This was evident during deliberation over the forest law when timber interests sought to reduce the regulatory power of the government, and succeeded. Timber interests also used the legal system to have the 1986 forest law – which they did not like because it restricted property rights – declared unconstitutional in 1990, and therefore null and void. It was this act that opened the deliberations for the forest law of 1996, which began in 1990 (Silva, 1997a). Finally, dependence on international organizations holds environmental policy captive to the policy fads of external donors. Thus, the emphasis of national policy shifts as do the funding cycles of development agencies and multilateral development banks. Thus, roughly every five years, many programs, projects, and organizations face a dearth of resources that threatens their existence.

The most serious restriction, however, lies in the fact that environment is far from the most salient issue on the policy agenda. Free-market economic restructuring in the hopes of renewing economic growth, meeting IMF stabilization policy to ensure macroeconomic stability, promoting agroexports, reforming the government bureaucracy, these are all much higher on the list. Presidents and mainstream political parties must promote these first and are beholden to constituencies in the government, the private sector, and civil society interested in development above all else. These groups and agencies view environmentalists as obstacles, they just add costs to their operations or take away economic opportunities. So far, target groups have shown a remarkable capacity to keep most environmental issues off the political agenda (Carriere, 1991; Hopkins 1995).

Financiers, landowners, real estate developers, industrialists, and others have far more powerful connections to presidents, ministries, and political parties than do environmentalists. Indeed, the Ministry of Environment is not very high in the bureaucratic hierarchy. This explains, in part, why environmental policy has focused so much on conservation biology and parks, rather than pollution. The other reason is that the MINAE has more direct administrative control over these issues. Thus, with the exception of very high public health standards compared to other Latin American countries, Costa Rica has not advanced much on shifting to the intensive end-of-pipe stage, much less any movement towards ecological modernization. Policy emphases on free-market economy led growth and reducing the size of government translate into decreased tax-raising capability. The resulting lack of funding further reduces government's ability to influence the shift from one stage of environmental policy-making to another. Whether the government will be capable of improving the enforcement of regulation is also in question. (Costa Rica, like many other Latin American nations, is a land in which many laws are not properly enforced). Lack of fiscal resources forces policy makers to rely much more heavily on market instruments and external funding.

5 Evaluation: Policy Integration and Development of a Policy Paradigm

5.1 Policy Integration

Albert Weale (1997) distinguished between internal and external integration. Internal integration refers to clear-cut lines of responsibility among agencies. External integration involves the extent to which line ministries incorporate environmental considerations in their decisions.

Policy integration in the environmental issue area in Costa Rica is far more advanced in biodiversity protection than pollution control. The establishment of the SINAC has gone a long way toward solving the policy integration problems of the Ministry of Environment with respect to biodiversity conservation. The intra-ministerial battles between the forest, parks, and wildlife services are a thing of the past. Yet, although the formal intra-institutional conflicts are over, the SINAC has a long way to go before the individual conservation areas sort out how best to apportion their resources. Potentially more serious, however, is the increasing tension between the Ministry of Environment and the Ministry of Health over policy-making on pollution, an area traditionally handled by the latter. Conflicts between the Ministry of Environment and the Ministry of Agriculture have also been common. For example, the Ministry of Agriculture opposed one version of the forest bill during the Calderón administration because of a dispute over property rights (Bravo, 1997). Similar incidents have marred relationships with the Colonization Institute, the Energy Institute, and the Ministry of Economy.

Section 3.1 on environmental policy institutions suggests a relatively high level of external policy integration for a developing nation. The Ministries of Health, Agriculture, and Public Works as well as the parapublic agencies that deliver water, electricity, and tourism all have formal responsibilities to conserve natural resources and to control pollution. Municipalities and the Cantonal (provincial) Agricultural Centers also have policy-making and implementation responsibilities in relation to the environment.

The main problem is that, with the exception of the Electricity Institute, the Ministry of Health, and the Water Institute (water pollution) few other government institutions exercise their mandates. Or if they do, their meager efforts are overwhelmed by contradictory growth policies that overshadow their environmental efforts. This is particularly the case in the Ministries of Agriculture and Economy and Industry. In part, it is a problem of insufficient resources to deal with pollution. More serious, however, is the lack of effective external integration of environmental concerns into ministries and agencies. The point is to incorporate environmental considerations in the overall policy-making process of government agencies. But many ministries and institutes only exhibit limited "integration" to an underfunded – and often very minor – department that deals with the consequences of their broader policies, usually inadequately.

The strengthening of the coordinating role of the Ministry of Environment in the environmental framework law is a step in the right direction for more effective internal and external policy integration. The MINAE is now in a position to engineer integration through its mandate to organize coordinating committees. The MINAE's environmental watchdog role offers hope that in such committees – along with new environmental impact reporting rules for economic projects – it will force greater consideration of environmental concerns at the earlier stages of the policy process in other ministries. The MINAE's collaboration with the Ministry of Planning in the production of the National Environmental Policy Plan (MIDEPLAN, 1996) is an important first effort at diagnosis and policy prescription at the sectoral level.

The extent to which integration actually occurs, however, depends in great measure on the importance presidents give the environmental issue. The new PUSC administration of Miguel Angel Rodríguez (1998-2002) has been ambivalent on this point so far. Some evidence suggests its overriding commitment to rapid, free-market economic growth will lead to an emphasis on biodiversity protection (parks and wildlife reserves) over the increasingly pressing pollution problems.

5.2 Development of Policy Paradigm

Since the establishment of the MIRENEM in 1986, Costa Rica has drawn on the concept of sustainable development to guide its environmental policy making. To that end, three successive major planning documents (all under Liberación Nacional presidential administrations) have emphasized the need for environmentally conscious economic growth strategies (MIRENEM, 1990; Presidencia de la República, 1994; MIDEPLAN, 1996). The documents stress economic growth based on the sustainability of natural resource use, biodiversity conservation, pollution control for expanding industry and agriculture, meeting the basic needs of the poor, demographics, education, and participation by civil society in policy making and implementation. The documents also uniformly recognize the need for institutional capacity-building, stimulating internal and external policy coordination, and generating financial resources. In short, these planning tools reveal a sophisticated high-level policy making community that recognizes the holistic, intersectoral character of environmental problems.

Unfortunately, a large gap exists between the philosophy that inspired those studies and the actual policy paradigms that have emerged. The problems stem from two sources. First, limited political and fiscal resources force policy makers to consider trade-offs between policy objectives. How many fronts can be attacked at one time? Second, policy makers seem unclear over the relationship between the general concept of sustainable development and actual policy paradigms that operationalize environmental sustainability. Do policies emphasize biodiversity protection and if they deal with pollution only stimulate its dilution through waste control? Or are policies systematically coordinated to engender successively higher stages of pollution control such as the dilution of pollution plus end-of-pipe treatment to transform the waste generated by industrial processes, intense end-of-pipe treatment, or ecological modernization which requires new green technologies that create environmentally-friendly production processes capable of substituting polluting technologies (Jänicke and Weidner, 1997)? When the question is posed this way, a policy-making community struggling to define an operational policy paradigm becomes apparent behind the official rhetoric of sustainable development.

An operational policy paradigm has emerged most clearly in the area of biodiversity protection and natural resource conservation (especially with respect to the forest and watersheds). With the adoption of some functional equivalents for concepts developed for the pollution issue area, one could argue that Costa Rica has embarked on a strategy of ecological modernization for natural resource conservation. Following Jänicke and Weidner (1997), several elements of biodiversity conservation policy point in this direction.

Which policy paradigm emerges is closely related to the configuration of actors in the environmental issue area (Jänicke and Weidner, 1997: 147). Where only government puts pressure on industry, dilution is the most common paradigm. In this configuration industry is reactive and defends the status quo. When green organizations pressure government to demand larger-scale changes from industry, the dilution plus end-of-pipe treatment paradigm emerges. When green organizations and the media begin to act on industry directly, industry's perceptions of industrial processes and pollution begin to change. As a result, industry begins to negotiate pollution control in a more proactive, rather than strictly defensive, manner. In other words, a reciprocal relationship emerges between government and industry, while green organizations and the media pressure government and industry. This configuration supports the intensive end-of-pipe treatment policy paradigm. Ecological modernization is the product of an even more complex actor configuration. In this stage the relationship between the major actors – government, industry, and green organizations/media – is reciprocal (they all influence each other proactively). In addition, green enterprises begin to influence industry's perceptions of technological change, forcing it to consider substitution and changes in industrial processes.

What suggests that Costa Rica may be reaching the equivalent of ecological modernization in the biodiversity conservation issue area? First of all, government, the private sector, and green organizations have developed reciprocal relationships. The government has negotiated the creation of national parks with the private sector (landowners) through the purchase of private property within park boundaries. By the same token, the government has drawn on green organizations to help design and implement park policy. Green organizations and the media and the state all encouraged private sector conservation efforts ("green enterprises" in conservation). The number of these green enterprises in conservation has grown rapidly (Rovinski, 1991).

Where natural resource use is concerned, the government and green organizations have encouraged the private sector, through regulation and incentives, to consider alternatives to the use of the primary forest: tree plantations, innovative market-oriented measures to encourage forest conservation and reforestation (joint-implementation and tradable pollution permits payment of forest conservation). These measures have encouraged private conservation efforts. Meanwhile, a few green enterprises in the wood products sector only use plantation-grown timber or certified sustainably logged timber from natural forests. These practices are roughly the equivalent to the development of green technologies to substitute for polluting technologies in industry. With these practices and the establishment of "green enterprises" in conservation Costa Rica does seem headed towards ecological modernization in the area of biodiversity protection. The fact that policy integration has been relatively successful in this realm further advances the trend.

However, the emphasis on biodiversity conservation raises a significant question. Why are the issues of urban and rural pollution relatively neglected when all of the major environmental planning documents recognize the need to address them? It is plausible that deficiencies in political and fiscal resources rather than lack of awareness of the issues have forced policy makers to accept tradeoffs. As discussed above, private sector resistance to environmental policy in agricultural and urban industries, the pressing need for economic growth, and the lack of fiscal resources to create strong, capable governmental institutions militate against forceful, clear lines of action. The MINAE's mandate and administrative structure is also much stronger in biodiversity conservation. These conditions promote fractured, opportunistic responses to environmental problems; they hamper the well-integrated, cohesive development of policies designed to move Costa Rica toward more advanced policy paradigms.

This state of affairs raises two questions about pollution control policy in Costa Rica. Which environmental paradigm are actual policies of Costa Rican Policy makers promoting? Conversely, do Costa Rican planning documents point towards the development of higher policy paradigms in the future?

In answer to the first question, Costa Rican policy makers seem to be working squarely in the dilution paradigm, with some evidence that they are transitioning towards the dilution plus end-of-pipe treatment stage. Driven by a configuration of actor relationships where government puts pressure on urban and agricultural industry, most pollution policies have focused on waste disposal. Nevertheless, in some instances a few green organizations have pressured government for more effective pollution control. This has generated movement towards the dilution plus end-of-pipe treatment paradigm, as evidenced by several recent laws and decrees. These include the Energy Conservation and Rational Use Act (1994), the Transportation Act (1994) and decrees on the Registry, Use, and Control of Pesticides (1995); The Regulation, Registry, and Control of Hazardous Toxic Wastes (1995); and the Regulations for the Control of Emissions of Gases and Particulates Produced by Automotive Vehicles (1994).

Are Costa Rican policy makers thinking within the confines of more advanced pollution control paradigms? Costa Rica's latest planning document is ambivalent. It clearly targets more effective implementation of policies that either promote dilution or that mix dilution with simple end-of-pipe treatment (MIDEPLAN, 1996). The document stresses building more waste disposal treatment plants and effective implementation of existing waste disposal in rural and urban areas for run-off into water supplies and solid wastes. The same mixture applies to air pollution control in urban regions. Here, end-of-pipe treatment is largely relegated to acquiring a newer automotive park capable of processing fuel more cleanly and the introduction of cleaner-burning fuels.

Agribusiness is one industry in which dilution is accompanied by policy prescriptions rooted in the ecological modernization paradigm. There is a call to reduce agrochemical use through substitution by natural products. There are even a few experimental programs in banana and coffee production, largely financed by international donors. However, the current configuration of actor relationships makes rapid advances on this front unlikely.

Most unsettling, however, is the near-silence on urban industry in the 1996 National Environmental Policy Plan. It only specifically mentions urban industry in connection with water pollution. Yet here, and in references to waste disposal problems in urban areas, the plan only offers dilution measures (waste disposal) to tackle industrial pollution. Policies to stimulate end-of-pipe treatment in industry, to say nothing of its intensive phase or ecological modernization, are not explicitly mentioned. Perhaps these issues will come to the fore once environmental impact reports for new projects start coming in.

Costa Rica's economy and population distribution are heavily rural. Thus, the vision for ecological modernization in agriculture is an important step. But traditional urban industry is also a significant component of Costa Rica's economy and a large proportion of its citizens live in urban centers. Thus, the omission of the industrial sector from Costa Rica's latest environmental policy planning documents seems ominous; all the more so because the sector received more attention in the 1980s (MIRENEM, 1990). This suggests that further research is needed into the configuration of actors involving industrial interests and the orientation of green organizations/media.

5.3 Evaluation

Environmental capacity-building in Costa Rica has progressed most in biodiversity conservation. The nation has steadily built institutional capacity by strengthening the coordinating capabilities of the MINAE, especially in the conservation area. The MINAE has steadfastly professionalized the Parks Service, the forestry sector, and the Wildlife Service. These institutions, and their technical capacity, have been bolstered by the synergy between the public sector and NGO professionals, the private sector, and international agencies and universities. That relationship has resulted in the design and implementation of innovative policy instruments and the development of a policy paradigm approaching green modernization. Moreover, the MINAE, and the Parks Service in particular, have accomplished much with the funds they possess. In part this is due to the funding and management strategies used. These involved creating private foundations and institutes to raise and manage funds (Fundación Parques and INBio); they also involved relying on NGOs and their professional staffs to carry out many projects. Thus, by the standards of developing countries, Costa Rica has a relatively effective and efficiently managed system of protected areas.

Of course, as with any human endeavor, more must be done, especially in three areas. First, there is a need for greater coordination of international donor priorities with national priorities. Second, personnel cutbacks in the MINAE may adversely affect the SINAC's effectiveness. The cutbacks originate in Costa Rica's efforts to reduce the size of government in order to comply with International Monetary Fund-mandated economic stabilization measures. But they leave the conservation areas understaffed during a period of great uncertainty and stress caused by two developments: massive job reassignments driven by administrative decentralization and physical relocation from the capital (San José) to the provinces (cantons). The resulting low morale among many staff members in the conservation areas, then, is a third problem that may influence the effectiveness of the MINAE in conservation. All of these elements underscore a serious issue. The SINAC still needs to iron out how the conservation areas will implement policy directives from the central office, whose main function is now planning and fund raising. Compared to biodiversity conservation, gains have been much more modest in the area of pollution control. Strong efforts in public sanitation (potable water and waste disposal) that began in the mid-twentieth century have contained the eruption of widespread health problems related to increasing pollution. However, economic growth, industrialization, the renewed emphasis on bananas as the main export crop, and urbanization are generating more pollution than the existing waste disposal and water treatment systems were designed to handle.

Moreover, Costa Rica's environmental institutions seem ill-equipped to handle the new environmental challenges that increased pollution raise. For one, the MINAE cannot act as directly as it does in biodiversity conservation with the SINAC. Many other ministries and agencies are involved. Thus, friction between ministries and agencies tend to reduce efficiency of meager resources for pollution control through duplication of programs and administrative functions. Effectiveness and efficiency suffer further from resistance by many ministries and agencies to increasing the policy making role of their currently weak environmental units. This is especially true for the Ministry of Agriculture and Economy.

Of course, funding remains a crucial concern. Where will it come from? Pollution control is expensive for governments and the private sector alike, especially where end-of-pipe treatment and dilution technology is concerned; not to mention the technological transformation that ecological modernization demand. Who will pay for waste treatment plants, for incentives to agricultural and urban industry to adopt dilution and end-of-pipe treatment technology? The public sector can be held responsible for some of this but it does not have the funds. Private, domestically-owned corporations argue they cannot afford environmental controls. All look to the international sector to supply financing and technology. But Costa Rica is considered to be a mid-range developing country with good public health indicators and infrastructure; thus, it does not qualify in the eyes of international donors. Green taxes are a possibility. But the experience with the gasoline tax demonstrates the difficulties of ensuring that such taxes are actually used for environmental purposes.

Whether and how institutional capacity-building proceeds, how funds are raised, how regulation is enforced, and who will pay for what and how are political questions. The strengthening of the coordinating capacity of the ministry of environment is certainly a step in the right direction to solve this political problem. So is the legal requirement of environmental impact reporting for new ventures and the MINAE's mandate to oversee and enforce them. In the event funding becomes available, the MINAE may find itself in a better position to coordinate the implementation of the policy prescriptions it advocates in its latest planning document (MIDEPLAN, 1996). Current improvements in institutional coordination also potentially increase the effectiveness and efficiency of whatever resources become available to attack environmental pollution.

But there are some concrete measures of effectiveness that are not conditional on the influx of immense resources in the future. Putting teeth into some recent legislation – via regulation and enforcement in the first instance – will be a litmus test of that capacity. Likewise, effectively incorporating environmental concerns in policy making in the agricultural and industrial ministries would be a huge political victory. Proper implementation of environmental impact reporting is also a significant opportunity to improve environmental quality in Costa Rica over the long term, especially in its application to international companies that have more resources and public relations-driven motivation to comply with environmental regulations. These are necessary steps on the path towards sustainable development, and a yardstick against which Costa Ricans can measure their progress towards achieving sustainable development.

References

Al Día, 1997. Financiamiento forestal siembra discordia, Friday, July 4, 14.

Alfaro, M., 1997. Author interview with private sector forestry consultant, a member of the board of directors of the Cámara Costaricence Forestal, and director of FUNDECOR's planning department up to 1995.

Araya, M.V., 1997. Author interview with the director of the financial department of the MINAE's Fondo de Financiamiento Forestal.

Arias, G., 1997. Author interview with international forestry consultant, architect of the regulations for the 1996 forestry law, and high-level professional staff of COSEFORMA.

Artavia, R., 1997. Author interview with the president of the timber industrialists association.

Booth, J., 1998. *Costa Rica: Quest for Democracy*. Boulder: Westview.

Brañes, R., 1991. *Institutional and Legal Aspects of the Environment in Latin America, Including the Participation of Nongovernmental Organizations in Environmental Management*. Washington, D.C.: Inter-American Development Bank.

Bravo, H., 1997. Author interview with the minister of MIRENEM during the Calderón administration and PUSC deputy in the Legislative Assembley 1994-1998.

Brockett, C., and R. Gottfried, 1994. The Impact of the Public Policy Environment on the Adoption of Sustainable Forestry Practices in Costa Rica, paper presented at the XVIII International Congress of the Latin American Studies Association, Atlanta, March 10-12.

Cámara Costaricence Forestal (CCF), 1997. Apectos más relevantes de la negociación con las autoriades del MINAE para alcanzar la efectiva asignación de los recursos destinados al pago de los servicios ambientales, *Boletín Informativo Bimensual de la Cámara Costaricence Forestal* 5 (23) (May-June) 1.

Carriere, J., 1997. The Degradation of Central American Wetlands: In Search of Proximate and Remote Causes. *European Review of Latin American and Caribbean Studies* 63 (December): 100-110.

Carriere, J., 1991. The Crisis in Costa Rica: An Ecological Perspective, in: M. Redclift and D. Goodman (eds.) *Environment and Development in Latin America*. Manchester: Manchester University Press, 184-203.

Centro de Derecho Ambiental y de los Recursos Naturales (CEDARENA), 1996. *Monitoreo de denuncias por el delito de tala ilegal, enero 1994 a enero 1996*, CEDARENA (mimeo) (February).

Cubero, G., 1997. Author interview with head of the legal department of the ministry of environment.

Federación Costaricense Para La Conservación (FECON), 1997. *Sistematización sobre un caso de incidencia proyecto ley de reforma a la ley forestal.* San José, mimeo (February).
Fundación Neotrópica, 1988. *Desarrollo socioeconómico y el ambiente natural en Costa Rica.* San José: Fundación Neotrópica.
González, O., 1997. Author interview with the principapl oppositon party (PUSC) member of the forestry law commissions in the National Assembley.
Gracia Bondía, J., 1990. *El deterioro ambiental en Costa Rica: Memoria del primer congreso ambiental de Costa Rica.* San José: Editorial de la Universidad de Costa Rica.
GTZ, 1993. *Plan nacional de manejo de desechos de Costa Rica.* San José: GTZ-Gobierno de Costa Rica.
Hansen, S., 1989. Debt for Nature Swaps: Overview and Discussion of Key Issues, *Ecological Economics* 1: 77-93.
Hombergh, H. van den, 1998. Bargaining for Biodiversity and Citizens' Participation: A Political Process Perspective on a Costa Rican Environmental Campaign. Paper written for CERES Summerschool, Amsterdam, 7-11 September.
Hopkins, J., 1995. *Policymaking for Conservation in Latin America: National Parks, Reserves, and the Environment.* Westport: Praeger.
Jänicke, M. and H. Weidner, 1997. National Environmental Policies: Germany, in: M. Jänicke and H. Weidner (eds.) National Environmental Policies: *A Comparative Study of Capacity-Building.* Berlin: Springer.
Martínez, L., 1997. Author interview with the main deputy in charge of shepherding the forestry bill through commission in the National Assembly. He was from the Liberación Nacional party.
Ministerio del Ambiente y Energía (MINAE), 1997. *El sistema nacional de áreas de conservación de Costa Rica: concepto, funciones y avances en su implementación.* Mimeo (May).
Ministerio de Planificación Nacional (MIDEPLAN), 1996. *Plan nacional de política ambiental: Eco 2005.* San José: MIDEPLAN/MINAE.
Ministerio de Recursos Naturales, Energía y Minas (MIREMEM), 1990. *Estrategia para la conservación y el desarrollo sostenible.* San José: Servicios Litográficos.
Nación, La, 1997. Compromiso Forestal, June 20, n.p.
Presidencia de la República, 1994. *From Forest to Society: A new Costa Rican Model of Development in Alliance with Nature.* San José: UNED.
Rainforest Alliance, 1991. Unique Debt Swap to Protect Forest, *The Canopy* (Spring): 1-2.
Repúblic a de Costa Rica, 1995. Ley orgánica del ambiente. *La Gaceta* CXVII, 215: 1-9.
Rodríguez, J., 1997. Author interview with a former director general of the DGF during the Arias administration, high-level consultant to the MINAE on forestry, and international forestry consultant for Central America.
Rodríguez, S., 1993. *Conservation, Contradiction, and Sovereignty Erosion: The Costa Rican State and the Natural Protected Areas* (1970-1992). Ph.D. Dissertation, University of Wisconsin-Madison.
Rovinski, Y., 1991. Private Reserves, Parks, and Ecotourism in Costa Rica, in: T. Whelan (ed.), *Nature Tourism: Managing for the Environment.* Washington, D.C.: Island Press.
Salas, F., 1997. Author interview with the executive director of JUNAFORCA.
Salazar Cambronero, R., 1991. Legislación y ecología en Costa Rica. San José: Libro Libre.
Segura, O., 1997. Políticas forestales en Costa Rica, in: O. Segura, D. Kaimowitz, and J. Rodríguez, eds. *Políticas forestales en Centro América: Análisis de las restricciones para el desarrollo del sector forestal.* San Salvador: EDICPSA.
Silva, E., 1997a. The Politics of Sustainable Development: Native Forest policy in Chile, Venezuela, Costa Rica, and Mexico, *Journal of Latin American Studies* 29, 2: 457-95.
Silva, E., 1997b. National Environmental Policies: Chile, in: M. Jänicke and H. Weidner (eds.) *National Environmental Policies: A Comparative Study of Capacity-Building.* Berlin: Springer.
Silva, E., (forthcoming). *Technocrats, Peasants, and Livelihood: Lessons in Sustainable Development from the Forests of Costa Rica.* North-South Center Issues, North-South Center, University of Miami.

Solís, M., 1994. Author interviews with executive director of JUNAFORCA.
Solis, M. 1993. Junta Nacional Campesina: Una experiencia de organización campesina. San José: JUNAFORCA.
Thrupp, L., 1990. Environmental Initiatives in Costa Rica: A Political Ecology Perspective. *Society and Natural Resources* 3: 243-256.
Umaña, A. and K. Brandon, 1992. Inventing Institutions for Conservation: Lessons from Costa Rica, in: S. Annis (ed.),Poverty, *Natural Resources, and Public Policy in Central America*. New Brunswick: Transaction Publishers.
Vega, L. 1997. Venta de oxígeno entra en la bolsa de Chicago, *La República*, 26 June.
Weale, A., 1997. National Environmental Policies: United Kingdom, in: M. Jänicke and H. Weidner (eds.) *National Environmental Policies: A Comparative Study of Capacity-Building*. Berlin: Springer.

THE CZECH REPUBLIC

Adam Fagin

1 Environmental Problems

Though political and economic reforms in the Czech Republic over the past decade have certainly had an impact on the environmental devastation of the late 1980s, much of the improvement in air and water pollution can be attributed to the decline in industrial production as a consequence of the country's integration into the global economy rather than through conservation measures or environmental regeneration. Such a reduction is therefore potentially temporary and the trend may be reversed as economic growth increases. The environmental dividends associated with democratisation and marketisation are also offset by the deleterious impact on the environment of the growth in consumerism and increased car ownership and usage.

1.1 Air Pollution

Air pollution remains the most pressing environmental problem in the Czech Republic today despite some improvement in the post-communist period. In 1989 the emission of dangerously high levels of sulphur dioxide from brown coal-burning power stations in Northern Bohemia and Moravia received a great deal of attention. A combination of out-dated energy-intensive technology, an inefficient economic model, and a natural endowment of brown coal with a low calorific value intensified levels of industrial pollution.[1] Though emissions of sulphur dioxide in the Czech Republic are still higher than the OECD average,[2] the decline in the country's heavy industrial output, plus investment in filters and other technology since 1990 has reduced emission levels of sulphur dioxide by approximately 50%, and of particulate matter by approximately 71% *(Table 1)*. However, it ought to be noted that the 25% decrease in emissions of sulphur dioxide between 1990-93 was significantly less than the corresponding 33% decrease in industrial production for the same period (Fagin and Jehlička, 1998: 116). This suggests that pollution per unit of industrial output remains high and may even have worsened.

---

1 The specific environmental problems of the former East European communist states have been well documented. See in particular, Carter and Turnock, 1993; Moldan et al. 1990.

2 Emissions of sulphur dioxide in the Czech Republic in 1996 equalled 92 kg/person p.a., compared with the European OECD average of 36 kg/person p.a. (MŽP ČR, 1997a: 8).

Control of nitrogen oxides has been less successful during this period due largely to the increase in car ownership since 1991.[3] Although aggregate data suggests a significant overall decline ins emissions of nitrogen oxides from 920 000 tonnes per annum in 1989 to 432 000 tonnes per annum in 1996 (MŽP ČR, 1997a: 8), this conceals the true extent of the problem; levels of nitrogen oxides in urban areas such as Prague, where the increase in car ownership and use is most acute, actually increased between 1991 and 1995, and remain significantly higher than the level for 1989 (ibid.: 17). The reduction in annual average emission levels for nitrogen oxide can be explained by the regulation of stationary sources and as a consequence of industrial decline during the early 1990s. However, the overall aggregate data for nitrogen oxides emissions does not distinguish between stationary sources and mobile sources. Thus, the true extent of the increase in emissions as a direct result of higher levels of car ownership is therefore hidden. Indeed, the impact of this increase in private transport on ambient air quality is perhaps the most serious environmental problem facing the Czech Republic today. The long term forecast issued by the Ministry of Environment is that there will be a gradual increase in the frequency of occurrence of above-limit concentrations of ozone at ground level and a continued rise in levels of nitrogen oxides despite the increased use of catalytic converters and obligatory vehicle emission tests (ibid.: 15). Between 1990-94 there was a 75% reduction in lead emissions which has been attributed to the increased use of lead-free petrol and the reduction of the lead content in other fuel (Moldan, 1997: 118).

Though there has been an *overall* reduction in emissions of carbon dioxide since 1990, it is interesting to note that the level for 1996 (886 000 tonnes per annum) was similar to the level for 1985 (889,000 tonnes), and that there was in fact an increase in emissions in 1991, 1994 and 1996 (MŽP ČR, 1997a: 8 [*Table II.2*]). The data suggests that there is by no means a significant declining trend with regard to carbon dioxide emissions. As regards the data on nitrogen oxides, the decrease in emissions is largely a consequence of climatic conditions and industrial decline rather than the success of regulatory measures introduced since 1989.

Table 1. Trends in emissions of principal air pollutants – 1990-1996 (kt p.a.)

Year	SO_2	NO_x	CO_2	Solid substances
1990	1876	742	1055	631
1991	1776	725	1102	592
1992	1538	698	1045	501
1993	1419	574	967	441
1994	1278	434	1026	355
1995	1091	412	874	201
1996	946	432	886	179

Source: MŽP ČR (1997, 1997a)

3 The number of cars, including vans, in the Czech Republic increased by over 30% to nearly 1 million between 1991 and 1997 (MŽP ČR, 1997: 34; Moldan, 1997: 118).

1.2 Water Pollution

Water pollution is a significant problem in the Czech Republic. On a positive note, overall surface water pollution decreased significantly between 1989-1994.[4] Moreover, the proportion of water discharged into public sewers that receives some form of treatment has increased by 16% since 1990 and now equals 90% of the total discharged. However, such apparent improvements are offset by the fact that, despite a reduction in the use of manufactured fertilisers, there was not a parallel decrease in concentrations of nitrates and phosphates, and over a third of water flows remained severely polluted (Moldan, 1997: 112). This is alarming, as it seems to confirm that the ecological benefits of industrial decline must not be permitted to obscure the fact that levels of air pollution per unit of production remain high or may even have increased (Moldan, 1993; Fagin and Jehlička, 1998: 116). Despite an overall reduction in consumption and waste water discharge, there is also worrying evidence that the quality of drinking water has in fact worsened in some regions since 1990 (Moldan, 1997: 112). The water system of the Czech Republic requires significant improvement. Over 70% of the country's drinking water comes from surface sources and there has been no substantial improvement in the quality of ground water during the past decade (MŽP ČR, 1997a: 6-8). Most waste water treatment facilities still lack the technology to filter nitrogen and phosphorus. Nearly a fifth of the population continue to rely on individual wells, most of which supply water containing unacceptably high levels of nitrates and bacterial pollution (ibid.: 9). Any improvement in levels of organic pollution, petroleum products, and apparent acidity/alkalinity can be attributed to a combination of industrial decline (particularly sugar mills), a reduction in the use of agricultural fertilisers, plus greater precipitation in recent years rather than an improvement in water treatment.

1.3 Waste

The dimensions of the waste management problem in the Czech Republic have altered since 1989 in that the regulation of hazardous waste emissions from inefficient state enterprises has been somewhat overshadowed by the substantial increase in municipal waste produced by the public and by private households. Despite its various inefficiencies, the supply-led command economy did not generate the amounts of consumer waste seen in Western market economies; there tended to be less packaging and the shortage of consumer goods dissuaded the unnecessary disposal of items. Increased consumerism since 1990 has led to a sustained rise in all kinds of waste. Since 1995 the amount of total waste produced in the country has increased considerably. In 1995 74 million tons of waste was produced (according to OECD methodology) compared with 93 million tons the following year, representing an increase of nearly 25% (MŽP ČR, 1997a: 9). It should be noted that the steep rise between 1995 and 1996 may be explained in part by the introduction of more stringent regulation and registration of waste (ibid.: 55). The bulk of all waste, 56% of which is hazardous waste, is still dis-

4 In this 5 year period BOD5 was reduced by 55%, insoluble substances by 43%, crude oil substances by 64%, inorganic salts by 25%, acidity/alkalinity by 72% (Moldan, 1997: 112).

posed of in landfills. Though the total number of landfills in operation has declined quite dramatically since 1991 from 10,000 to 380 in 1996 (ibid: 9), the estimated 7000 waste dumps and landfills including those no longer used pose a serious environmental hazard in terms of water and soil contamination (Moldan, 1997: 112).

1.4 Nuclear Energy

The commitment of successive Czech governments to expand nuclear energy production as an alternative to brown-coal power plants raises obvious environmental concerns regarding safety, levels of radiation and waste disposal. At present there is one nuclear power plant in operation in the Czech Republic situated at Dukovany in southern Moravia. The plant equipment is of Soviet design and approximately fifteen years old. The construction of a new plant at Temelín in South Bohemia was first proposed by the communist regime in the early 1980s. Though post-communist governments have maintained a commitment to completing the project, the development of the plant has been beset by financial difficulties and a great deal of political opposition from environmental NGOs. Although the plant is in fact state-owned, there has been substantial investment by the US company Westinghouse. At the time of writing the future of Temelín lies in the hands of an international commission, consisting of representatives from neighbouring countries, the OECD and the EU, set up to evaluate the feasibility of further investment in the project. Though the size of the proposed plant has been cut from four reactors to two, once completed the additional output would be significant in terms of overall energy production in the Czech Republic. At present nuclear energy accounts for 8% of total energy production compared to 55% for coal (MŽP ČR, 1997a: 105).

The issue of nuclear waste storage has provoked concern amongst environmentalists and the public. The government's proposal in 1995 to construct eleven waste storage sites across the country was eventually overturned after a lengthy campaign organised by Greenpeace which brought together local activists, politicians and sections of the general public. Though the protests were largely of a NIMBY nature, the incident highlighted a level of public concern regarding nuclear waste.

There are various other environmental problems in the Czech Republic, most more or less familiar to industrialised countries across Europe. Heavy industrial production has led to a reduction in biodiversity, and the excessive use of fertilisers and other chemicals in agriculture has inflicted a burden on soil. Since 1990 the damage to forests from air pollution has improved and there has been a reduction in the percentage of agricultural land under cultivation. However, the proportion of cultivated land remains high compared to the rest of Europe. A specific environmental problem for the Czech Republic and indeed other former Soviet satellite states is the damage inflicted on the environment by the Soviet army in their 73 locations across the country. The nature of the problem is essentially cleaning up contaminated land, the main pollutants being petroleum substances, toxic metals and chlorinated hydrocarbons. The process of cleaning up these sites began in 1992 and the cost so far has, in fact, been significantly lower than initial estimates (MŽP ČR, 1997a: 75).

2 Development of Environmental Policy and Regulation

2.1 Main Environmental Policies and Regulations

What can be identified as the first wave of post-communist environmental legislation (1990-1992) established a firm foundation for the legal regulation and protection of the natural environment.[5] In addition to the command-and-control mechanisms that had existed during the communist period, economic instruments such as charges and licences were introduced in an attempt to improve regulation and implementation. Emphasis was placed on the co-operation and participation of local communities in environmental regeneration, environmental education, and the active involvement of NGOs in the policy-making process (MŽP, 1991). Expenditure on the environment increased during this period from less than 1% of GDP in 1990 to 2.1% in 1992 and has subsequently increased to 2.5% (MŽP ČR, 1997a: 76; Moldan, 1997: 18). Though the regulatory framework now requires further amendment and improvement, particularly with regard to the disposal of waste, water pollution and the Environmental Impact Assessment (EIA) process, the Czech Republic has a comprehensive set of laws and regulations governing the main sources of pollution comparable to those to be found throughout much of Europe.

The General Act on the Environment (No. 17/1992) establishes the principle of sustainable development as a policy objective and decrees that the polluter should pay for damage inflicted on the natural environment. The charges to be levied, the fines for non-compliance, and the time frame for reducing emissions are all laid down. Two further important legislative tools were adopted in 1992, the Protection of Nature and the Landscape Act (No. 114) and Environmental Impact Assessment Act (No. 244).

With regard to air pollution, the specific act on air pollution (No. 389/1992) plus the Clean Air Act (No. 309/1991) establish a comprehensive list of the pollutants to be controlled and emission limits. Other acts include the requirement for keeping emission records for medium and large scale polluters (No. 205/1993), protection of the ozone layer (No. 86/1995), regulation of fuels used in combustion (No. 206/1993) and various regulations relating to specific pollution sources. Though a great number of new regulations were enacted during the 1990-93 period, they tended to focus on controlling air pollution, which was seen as the most urgent environmental task. No further legislation was forthcoming, so that the current legislation on water management and forestry dates almost entirely from the 1970s.

Whereas the communist regime had enacted legislation, however deficient, seeking to regulate air and water pollution, no such legal framework existed on waste. Since 1991 a number of legal norms have been introduced and the development of a coherent regulatory framework is in process (Moldan, 1997: 112). A new law on Waste Management (No. 125/1997) was hastily drafted during the second half of 1996 to update the categorisation of waste in line with EU and

5 Indeed, it is particularly notable that with certain exceptions (regulation of the ozone layer) the majority of the environmental legislation in existence in the Czech Republic originates from the 1990-92 period

OECD recommendations. The new law includes a controversial clause to ban the importation of any product containing PVC by the year 2001.

2.2 Evaluation of Environmental Policy

When evaluating the regulation of the environment and the development of policy in the Czech Republic it is often assumed that all initiatives stem from the post-1989 period. In fact, a concerted attempt to regulate air and water pollution began during the last years of the communist regime. Though investment in the environment had decreased in the previous fifteen years, from 1986 the share of total investment allocated to dealing with pollution increased to nearly 3% (Černá et al., 1995).

It is also easy to assume that since the collapse of communism there has been a coherent agenda and a sustained momentum to regulate pollution and regenerate the environment. In fact, after an initial period of legislative enactment from 1990 until 1992, virtually no new environmental laws were prepared or enacted between 1992 and 1996, coinciding with Klaus's first term as prime minister and the tenure as minister of the environment of Frantisek Benda. Indeed, for much of this period there was no long-term environmental plan or a set of future objectives; the "State Environmental Policy" bill was rejected by the government three times and was only finally passed in August 1995 after the deletion of any reference to sustainable development (Slocock, 1996; Fagin & Jehlička, 1998: 118).

Though the departure of Frantisek Benda as minister of the environment after the 1996 election and the collapse of the Klaus government in late 1996 restored momentum to environmental regulation, the absence of policy initiatives between 1992-96 has seriously thwarted environmental capacity building. Practically all environmental laws now require updating; for instance, when originally drafted in 1992, Act 114 on The Protection of Nature and the Landscape did not include a provision for the creation of new protected areas. Such an amendment was vital to the success of an otherwise effective legislative tool and would have enabled the Ministry of the Environment to prevent proposed developments by declaring an area to be a protected landscape. The necessary additional clause should have been enacted in 1993 but, under pressure from opponents from within the ruling ODS (Občanská demokratická strana – Civic Democratic Party), was not added until 1998. Similarly, it became evident quite early on that Act 244 on Environmental Impact Assessment required amendments to include public participation in the process at an early stage, to widen the scoping process and to ensure greater regulation of experts carrying out the assessments (Branis and Kružíková, 1994; Fagin and Jehlička, 1998). Such amendments are only now being drafted by the Ministry of the Environment, nearly seven years after the original law was enacted. The inertia of the Benda period also meant that the process of replacing antiquated communist legislation (for example, on water pollution), begun in the early 1990s, stalled and must now belatedly be resumed.

There is also an urgent need to pass legislation on new environmental problems that have emerged since 1990, such as waste and transportation issues. Though the new Waste Management Law (No. 125/1997) is certainly an improvement on the 1991 legislation, it is still not entirely compatible with EU norms. At one level the Czech act actually exceeds EU norms in banning the importation of products containing PVC by the year 2001. However, this is seen as a flaw insofar as there is likely to be a high degree of non-compliance with this clause which will weaken the overall impact of the legislation. In general, apart from the classification of chemical substances, the law is not compatible with EU regulations in terms of the definition of waste and secondary materials. In line with the recommendations of the Basle Convention the 1997 Act places restrictions on the importation of waste products from abroad. However, it does not prohibit the importation of waste for energy combustion or recycling. In both instances there is potential for the exploitation of such loopholes: under the current legislation old tyres from Germany could be brought legally into the Czech Republic supposedly for recycling or energy production; in fact they may be dealt with inappropriately as waste, remoulded or burnt in antiquated incinerators. There is particular concern amongst environmentalists that the OECD restriction shortly to come into effect banning the export of waste from member states to non-member states may result in an increase in the importation of waste to poorer OECD countries such as the Czech Republic.

This is perhaps a difficult point in time for evaluating environmental policy in the Czech Republic, as the country is potentially poised on the threshold of a new era in this field. Whilst the party politics and political scandals of the past two years ended the inertia of the 1992-96 period, this further delayed the much needed development of environmental policy. There have been three ministers of the environment since July 1996, each from a different political background and each with a different agenda and approach. In legislation, the main innovation so far has been a law on access to information relating to the environment which was initiated during Jiri Skalický's tenure (July 1996-December 1997). This enables the public to obtain all information on the environment, which has been available on the Internet since January 1998. The real momentum for establishing new legislation started during Martin Bursík's brief tenure (February-June 1998), and although he was unable to push through new legislation – the interim administration of which he was a member held office for only 120 days, the third day of which was the last for tabling new bills – he began the process of legislative review and drafted bills. Since taking office in July 1998, Miloš Kužvart, the new Social Democratic minister of the environment has continued with the drive for new legislation; amended laws on waste and the EIA process are expected in 1999.

2.3 Policy Implementation and the Success of Regulation

Any serious evaluation of environmental policy and capacity building rests upon an assessment of policy implementation and a consideration of whether legislation has improved the state of the environment.

As reducing air pollution was a priority on the post-communist environmental policy agenda, it seems appropriate to use this as a basis for evaluation. It would appear that the Clean Air Act and the Act on Air Pollution have been well implemented. Indeed, there is a general level of surprise amongst inspectors, environmentalists and industrialists at the extent to which the emission targets set by legislation have been realised. A deadline of January 1999 was laid down for industry to meet emission levels, and it looks as though most companies will meet the targets for sulphur dioxide and other pollutants. According to the Czech air inspectorate, the sources unable meet the targets tend to be state organisations such as hospitals and schools that have not received adequate investment to modernise incinerators and heating systems. The worst offender is the Ministry of Defence, whose installations have generally not been modified. The state electricity generating company, ČEZ, previously a major polluter and key target of regulation, has already met the statutory targets for sulphur dioxide and nitrogen oxides emissions.[6]

However, such apparent success requires some qualification. Though the policy approach of setting rigid targets for the reduction of air emissions by January 1999 appears to have resulted in a reduction in pollution from industry, this has been achieved largely through expensive end-of-pipe measures rather than by a reduction in energy consumption and new technology (Moldan, 1996; Fagin & Jehlička, 1998: 116). Advocates of this approach claim that this is an important first stage in reducing the deleterious legacy of the communist period and will be followed by sustainable strategies in the future. However, it seems unlikely that, having invested heavily in fitting scrubbers and other end-of-pipe technology to lignite-burning power plants, the energy industry will in the near future be prepared to invest in alternative, more sustainable methods of production, thus consigning their expensive filters to obsolescence.

Furthermore, the reduction in emissions cannot be attributed entirely to the new legislation and its implementation. As noted above, the reduction in sulphur dioxide emissions has coincided with a period of manufacturing recession and industrial decline. Had the reduction in emissions been greater than the decline in production and energy consumption during this period, this would have suggested that the new legislation was indeed having a definite impact. However, the decreases in air pollution and energy production and use during the period 1990-1993 were in fact smaller than the decrease in industrial output (Černá et al., 1995). This reinforces the claim that the "success" in reducing emission levels has more to do with industrial decline, and that Czech industry remains highly inefficient with regard to energy use. It also suggests that the reduction in emissions may well prove to be temporary.

3 Capacity of Proponents of Environmental Policy

In seeking to offer an overall assessment of the capacity of proponents of environmental policy over the past decade, it is important to note that there has been a significant change in actor configuration over the period, and that there has been a notable fluctuation in public attitudes towards the environment. The political efficacy of what might broadly be identified as the environmental movement has fluctuated from close involvement of scientists, activists, and policy officials, to a period of hostility in which non-governmental actors were excluded from any formal or informal involvement, to the current situation in which a functioning – though still embryonic – policy network has emerged, incorporating activists,

6 Information on implementation was obtained from interviews with inspectors Dr Stremmer and Dr Bosak from the Inspection Agency of the Czech Republic, 11 November 1998.

scientists, conservationists, and newly elected politicians. In terms of public attitudes, the initial enthusiasm for environmental regeneration expressed before and immediately after the 1989 revolution was eclipsed by the impact of economic reforms (unemployment, housing, inflation, etc.) on the general public. Whereas an opinion poll conducted in January 1990 by the Sociology Department of Charles University had recorded that 80% of respondents considered environmental regeneration as the new government's most urgent task, concern about the environment during the mid-1990s seemed to decline and NGOs found it difficult to win public interest and support for even the mildest campaign.[7] However, the attitude of the public towards the environment does appear to be changing. A poll conducted in late 1996 found that 87% of respondents thought ecological groups were useful and important.[8] In recent months, Greenpeace has noted a distinct increase in public support for its campaigns both financially and in terms of active involvement.[9]

3.1 State Organisations

Though the attitudes of particular government ministers and officials towards the environment have arguably impeded environmental capacity building in the post-communist period, the state agencies established after 1990 must be included as proponents of environmental regulation and protection. The expansive institutional and administrative structure that now exists in the Czech Republic is impressive by any standards, and is evidence of the concerted effort by the first generation of post-communist officials and ministers to establish a coherent institutional framework for environmental management.

The central authority concerned with environmental policy and regulation in the Czech Republic is the Ministry of the Environment based in Prague. In addition to providing a comprehensive environmental information and monitoring system (the Ministry's annual report includes a detailed overview of problems, strategies, and the policy agenda), the Ministry is responsible for co-ordinating co-operation with international agencies and for the formulation of policy in accordance with established norms.

The Czech Environmental Inspection Agency was established in 1991 to carry out state inspection of air and water pollution, waste management and forest and landscape protection. The Inspection Agency operates under the auspices of the Ministry of the Environment and regional inspectorates have been established in České Budějovice, Plzen, Ústí nad Labem, Liberec, Hradec Králové, Havlíčkův Brod, Brno, Olomouc and Ostrava to assist in the inspection process. The operation and jurisdiction of the Inspection Agency is not dissimilar to what exists elsewhere in Europe: in total there are over 300 inspectors with responsibility for

7 An example of the lack of public interest in the environment is the case of a campaign run by Zelený Kruh (Green Circle) to save trees in the residential vicinity of its premises in central Prague. Very few residents were willing to lend their nominal support.

8 This information was obtained from an interview with Hana Pernicová in July 1997. The poll was conducted in November 1996 as part of an annual Czech national opinion poll survey.

9 Interview with the then director of Greenpeace CR, Čestmír Hrdinká, November 1998.

assessing compliance with regulations, administering particular legislative clauses, identifying polluters, and setting fines.

In 1993 regional offices of the Ministry of the Environment were established. These are essentially branches of the Ministry rather than federal agencies with devolved responsibility, and should be distinguished from District Offices and authorised Municipal Offices (which exist to carry out the functions of the Ministry and to pursue various delegated administrative functions at the district and municipal level). The purpose of such territorial divisions of the Ministry is described thus:

The jurisdiction of the territorial divisions of the Ministry of the Environment ... includes examination of decisions of ... the District Offices in appeal processes, review of petitions and complaints against procedures and measures of District Offices, supreme inspection of implementation ... and application of the appropriate environmental legislation, environmental impact assessment for construction work and provision for professional training of employees in the District Offices. (MŽP, 1997: 296)

There are a number of state or state-funded agencies with responsibility for various aspects of administration, research and development, and policy development. Some of these operate under the auspices of the Ministry of the Environment and include:

- Czech Hydrometeorological Institute (establishes and operates monitoring stations, processes results, draws up forecasts)
- Czech Environmental Institute (responsible for establishing documentation and procedure for implementation, environmental information, and generally acts as a policy think-tank)
- Agency for Protection of Nature and the Landscape of the Czech Republic
- Czech Geological Institute
- T.G. Masaryk Water Management Research Institute
- Research Institute for Ornamental Gardening.

State expenditure on environmental protection comes from a variety of sources. The most important of these is the *State Budget,* the *State Environmental Fund* (considered below), and the *National Property Fund.* The latter consists of the proceeds from privatisation, a proportion of which has been made available for environmental regeneration.

The *State Environmental Fund* was set up in 1990 to provide financial support for environmental NGOs, the endeavours of the Inspection Agency and the activities of the Ministry in general. The funds come from charges and payments for the discharge of waste and emissions into the environment, withdrawal of ground water. Other sources of state funding include donations and subsidies from the state budget. A number of points should be noted with regard to state funding of the environmental sector. The most important point to emphasise is that the provision of state financial support to NGOs has been directly affected by political developments. The hostility of the Klaus government towards the environmental movement plus the overriding climate of economic austerity led to a rationalisation of state funds and the targeting of resources away from the more politically-oriented NGOs. Second, since 1992 funding for NGOs has been available only on the basis of conservation projects. NGOs are prohibited from using state funds to pay salaries or to develop their organisation. Despite recent political changes, this restriction has not yet been altered. Much to the detriment of smaller environmental NGOs, there is still no fiscal framework to provide any tax exemption or allowance on donations made to NGOs.

3.2 Environmental NGOs

Most of the environmental NGOs in existence today were established during the late 1980s or immediately after the revolution. It was estimated in the early 1990s that there were between 500 and 800 groups (Šilhanová et al., 1994). The number of NGOs has subsequently declined, largely due to smaller groups having merged and because the initial enthusiasm to establish independent organisations after the revolution has waned somewhat. The types of organisation in existence today range from conservation groups established during the communist period (e.g., Czech Union of Nature Conservation – CSOP; Brontosaurus), to political NGOs of which some are small interest groups and others are large NGOs with international backing (e.g., Greenpeace). Other NGOs focus specifically on environmental education for young adults (e.g., Tereza) or seek to provide information and advice for activists and the public (e.g., Zelený Kruh – Green Circle). Under the broad categorisation of NGO must also be included various non-profit organisations set up to offer professional, technical and policy expertise for government, business and industry. They include SEVEn (focusing on nuclear energy), the Centre for Energy Efficiency, and the Institute for Environmental Policy.

The efficacy of the more politically-oriented environmental NGOs that seek to extend environmental regulation and challenge policy decisions and approaches has changed quite considerably since 1989.[10] Three quite distinct stages are discernible: an initial phase of political prominence (1989-1992) in which environmental NGOs enjoyed an unusual degree of influence in shaping environmental policy; a period of marginalisation and demise (1992-96) during which NGOs were entirely excluded from what remained of the environmental policy process; and a current phase of gradual improvement (1996 onwards) in which a cluster of larger Prague-based environmental NGOs have regained a degree of prominence and influence within a newly invigorated policy process.

In the initial period following the velvet revolution, the environmental organisations that had emerged during the late 1980s enjoyed a great deal of influence. The new officials and political leaders with responsibility for formulating environmental policy had themselves until recently been involved in environmental protest, and there was thus a somewhat unusual overlap between activist and policy maker. The first post-communist Czech environmental minister, Bedřich Moldan, sought close co-operation with environmental NGOs and this sentiment found institutional expression in the formation of the Green Parliament in early

10 I would like to acknowledge the help of Hana Pernicová, formally Executive Director of Greenpeace CR, who has during lengthy interviews provided me with information and data on development in the environmental NGO sector over the past five years. I would also like to acknowledge the particular help of Dan Vondrous (Duha), Jindrich Petrlik (Détí Země), Prof. Bedřich Moldan (CSOP), Marie Hasiová and Zuzana Drhová (Zelený Kruh).

1990. The objective of the Parliament was to create a discursive forum in which a wide array of environmental NGOs and officials could co-operate in the formation of a programme for environmental regeneration. Activists who had decided after 1989 to remain outside the formal political machine and to continue their activities within civil society were nevertheless closely involved in the policy process; groups such as Greenpeace, Détí Žemě (Children of the Earth), Hnutí Duha (Rainbow Movement) all enjoyed a close relationship with ministry officials.

However, NGOs needed to adjust to the changed political circumstances in which they were now operating. Having emerged as illegal or semi-legal opposition movements under authoritarian rule, they were now required to co-operate and negotiate with the new democratic state. They were being invited to sit round a table and help draw up a concrete policy framework; they needed to deal in facts and realistic strategies to ameliorate the ecological degradation of the preceding decades. That their unity as a movement had been fractured by the collapse of communism is hardly surprising; they had now to confront ideological differences and degrees of "greenness" that had previously been eclipsed by their unifying opposition to Soviet-style communism. They also needed to alter their strategies and internal organisation, both of which tended to reflect the political climate and socialisation of communism. Such adaptation and internal changes were by no means easy to enact. Bitter rifts within the environmental movement – particularly over the nuclear energy issue and opposition to the Temelín plant in southern Bohemia – led to the demise of the Green Parliament as NGOs fragmented, reformed, argued and, in many cases, folded. There was something extraordinary and unique in the fact that during this short period after the revolution the new ministers and state officials were more radical and were moving far faster than the NGOs. In a sense the spirit of co-operation of the new political elite had come too early for NGOs, which could not take advantage of the political climate until they had reformed internally and adapted to the changed circumstances of post-communist politics.

The political demise of the first generation of post-communist ministers and officials and the rise to power of Václáv Klaus and his right-of-centre coalition in June 1992 was reflected within the realm of environmental politics. Instead of dealing with an environmental minister and set of officials who had themselves been activists and who sought to establish a progressive policy framework, the NGOs had now to face a political elite whose main objective was economic development and the opening up of the Czech Republic to external investment. This was to be achieved by neo-liberal economic reform and the deregulation of the economy and society. Environmental protection was seen as "the icing on the cake" rather than an objective in itself – something to be dealt with in the future once a prosperous market economy had been established (Klaus, 1994). The new minister of the environment, Frantisek Benda, refused to meet NGOs and to set up any formal dialogue. In general this was not an era of consensus politics, but about an elected government pursuing its radical mandate unobstructed by opposition forces. As opponents of untempered, unregulated growth, environmental NGOs were immediately identified as political enemies. Such antipathy towards the environmental sector reached a nadir in early 1995 when three environmental NGOs were included on a security services' list of "subversive organisations" that were to be the target of surveillance.[11] To argue that during this period NGOs were excluded from the policy process implies that an environmental policy process actually existed. In fact, as described in the section on Environmental Policy, apart from a few amendments already enacted in the previous period, no new legislation emerged during this time.

This period witnessed significant changes within the environmental movement, occurring in response to both the changed political climate and the worsening financial situation in which all NGOs found themselves. As described above, the ideological diversity of the environmental movement had narrowed after 1989. This process intensified after 1992 as NGOs desperately sought a measure of political influence in what became an increasingly adversarial political climate. Environmental associations rejected direct action, placed increased emphasis on local and domestic issues, and were reluctant to identify themselves as political organisations. Whilst the focus on local and domestic issues was a positive development, the rejection of radical ideas and strategies was a reaction to the absence of resources and a political climate in which the public apparently disliked radical action and the government were openly hostile towards the environmental movement.[12] Forced to compete for scarce resources, NGOs were reluctant to co-operate with each other and thus the movement became increasingly fragmented.[13] The economic and political context depicted above resulted in an ideological constriction of the environmental movement. Whilst Prague-based NGOs struggled to exist and were unable to develop during this period, local and regional activism all but disappeared. Small groups operating in places such as Ústí nad Labem were politically disenfranchised or simply folded through lack of any resources.

The loss of an overall majority for Klaus's ODS-led coalition in the 1996 election marked the tentative first steps towards a new era for environmental politics. Whilst NGOs were by no means suddenly invited to discuss policy, the replacement of Frantisek Benda as minister of the environment by Jiri Skalický after the

11 That the three environmental associations included on this list (Greenpeace, Déti Žemě, and Hnutí Duha) were committed to non-violent protest and to working within the democratic process, plus the fact that the other groups on the list were far-right neo-fascist organisations provoked political outrage. President Havel intervened as did a number of journalists and prominent figures in Czech society. As a result, 3 out of the 4 groups were removed from the list, though no apology was offered. For a more detailed discussion on the incident see Fagin and Jehlička, 1998, 122.

12 The Klaus government's hostility to political associations and environmental NGOs in particular found its most blatant expression in the reluctance of Frantisek Benda (Minister of the Environment 1992-96) to even meet ENGOs. In a number of interviews Klaus criticised notions of civil society and argued that the environment was the "icing on the cake", an issue to be dealt with after the free market had been established in the Czech Republic (Klaus, 1994).

13 The umbrella organisation Zelený Kruh (Green Circle), which had come into existence shortly before the velvet revolution to co-ordinate the plethora of small organisations that had emerged, existed throughout this period on the brink of financial crisis. They were continually threatened by the loss of their premises and the loss of staff who could not survive on the negligible salaries that the association was able to offer. Despite concerted efforts by the director (Marie Haisová) to bring cohesion to the sector and strengthen NGO participation in the political process, Zelený Kruh was unable to realise any significant results within the ostensibly hostile climate of the period. The experience of Zelený Kruh typified the situation faced by all NGOs during this time.

election saw a normalisation of relations between NGOs and the Ministry of the Environment. The climate of outright hostility was replaced by a period of tentative co-operation between the more prominent Prague-based NGOs, and a tactical politician who became increasingly distracted by corruption scandals and the politics of the fragile post-election coalition.[14]

The crisis which finally engulfed Klaus's embattled coalition towards the end of 1997 and led to the formation of a caretaker government, finally drew a curtain on the political isolation of environmental NGOs. Indeed, under the new minister of the environment, Martin Bursík, NGOs enjoyed an unprecedented influence and were granted unrestricted access to the process of beginning to draft and update existing environmental legislation.[15]

Since the election in June 1998 of a Social Democratic government, the political prominence of environmental NGOs established during the Bursík period has been maintained. The new minister, Milos Kužvart, is a former member of the pre-1989 dissident environmental association, the Ecological Society. More recently he has been involved with the non-governmental sector and is still a member of the highly influential STUŽ (Society for Sustainable Living). The Green Parliament has just been re-established as a forum for consultation between the ministry and NGOs. The first meeting (November 1998) was well-attended. Once again there is overlap between the non-governmental and government sectors within the realm of environmental politics. The new minister, his officials and deputies are working closely with their former colleagues in the more prominent NGOs.[16]

Despite the recent improvement in the political efficacy of certain larger environmental NGOs (Greenpeace, Duha, Détí Země, STUŽ), the political impact of NGOs generally within the policy process remains constrained. The most serious threat to the future expansion of NGOs is the absence of sufficient levels of funding and, in particular, the prospect of current sources being curtailed. The larger environmental NGOs have always depended heavily on foreign donations from European and American agencies and philanthropic foundations. Organisations have also received funding as part of the EU PHARE programme. In light of the relative success of the political and economic transition in the Czech Republic, many foreign donors have re-targeted their assistance. The American Ford Foundation has completely withdrawn from the Czech Republic, C.S. Mott (US) has rationalised its assistance, and USAID has also ceased. Competition is fierce for

---

14 Whether as a result of improved relations between the ministry and NGOs, or due to the fact that Skalický's attentions were diverted elsewhere during this time, Czech Greenpeace was able to influence the drafting of a new Waste Law to such a degree that a clause was inserted banning the importation of any product into the Czech Republic containing PVC from the year 2002. This certainly extends beyond the recommendations of the EU.

15 In terms of actual new legislation this period produced very little. This is largely because Bursík's four month tenure made it impossible for him to draft new laws and get them through parliament – indeed, his first day in office was the last opportunity for the submission of draft legislation to parliament. However, the foundations of a new policy agenda were laid during this brief period

16 The Society for Sustainable Living, of which Ku‡vart is a member, is perhaps the most influential and was recently able to halt the proposed amendments to the Waste Law which were deemed to be incompatible with the EU directive and open to manipulation by those intent on importing waste from abroad. Hnutí Duha and Dìtí †emì are also doing well and work closely with the ministry over new legislative amendments on waste, the EIA process, and water.

the small amount of money allocated by the state for environmental associations as part of the State Environmental Fund. Grants are given to conservation projects rather than political activity and it is forbidden to use state funding for the payment of salaries.[17]

The impact of the funding crisis on the ideological diversity of environmental activism is illustrated by the case of the České Budejovice-based NGO Jíhoceské matký (South Bohemian Mothers). This organisation was one of the few remaining NGOs in the Czech Republic actively to oppose the completion of the Temelín power plant in Southern Bohemia. After failing to attract funding from any of the foreign foundations or (unsurprisingly) from the state, the organisation has now folded. As a result there is now no organised opposition to the Temelín plant at a critical time in the development of nuclear energy in the Czech Republic. The new government must make a decision on whether to continue with the ill-fated plant, and a new feasibility study is currently taking place. The demise of Jíhočeské matký is to be lamented, as it was the one NGO to has campaigned continually against the Temelín plant and to acquired detailed information on its potentially deleterious impact on the local community.

The fact that environmental activists in the larger associations are now committed to working within the policy process and to articulating local and national issues is certainly a positive development. There is also a far greater degree of professionalism amongst NGOs, which has certainly helped to raise their public profile. There has been an informal division of responsibility on certain environmental issues amongst the larger groups and this has encouraged a higher level of specialism and expertise. The political profile of the environmental movement has certainly altered: Greenpeace played a high-profile role in the aftermath of the Moravian floods, whilst Duha's adept lobbying skills have raised the organisation from a group of marginalized activists to a professional NGO offering advice and information on ozone protection and waste management. Since 1996 it has become increasingly possible to identify an environmental network that combines experts, scientists, activists and political officials. The larger NGOs play a leading role within this informal network.

However, if NGOs are to develop as effective actors within the environmental policy process, channelling local and regional interests, offering professionalism and expertise, and acting as a counterweight to the power and influence of vested interests, they need sustainable and adequate sources of funding. The obvious long-term solution is for NGOs to be funded by supporters within society as occurs elsewhere in Europe. However, it is widely acknowledged that the problem NGOs face in raising sufficient funding from Czech citizens is that potential donors lack sufficient levels of disposable income to provide donations to charities and non-profit organisations, including the environmental movement. Whilst this will hopefully alter over time, at present it acts as a formidable constraint on the development of NGOs at a critical time in the development of environmental norms and strategies for the future. The recent experience of Greenpeace ČR, which undertook what turned out to be a very successful fund-raising campaign in

17 As a result, the co-ordinating NGO Zelený Kruh has received no state money at all in the past two years as it is not directly involved in environmental projects. It does however potentially contribute a great deal to the development of the environmental non-governmental sector.

Prague, does however suggest that the indigenous funding problem is not just about low levels of disposable income, but about the lack of fund-raising know-how amongst NGOs and perhaps in general. Though Greenpeace had an advantage over other NGOs insofar as it received advice and initial financial assistance to set up the fund raising venture, it is encouraging given that the organisation was able to quadruple its donations from amongst the Czech population. The problem for smaller NGOs that lack support from an international parent organisation is finding the initial funds to invest in fund-raising. Possible solutions to this situation include the introduction of legislation by the government that would allow for tax relief on corporate and payroll donations.

3.3 Other Proponents

It is important to note the role played by sections of the Czech media in promoting debate concerning the environment and in supporting the endeavours of those within the environmental movement. In particular, the media played an important part in highlighting the democratic (as well as environmental) implications of the inclusion of the four environmental NGOs included on the security services list in February 1995. The media uproar at the time of the incident served to embarrass the Klaus government and actually helped promote the environmental movement. The activities of the environmental movement have been widely reported, particularly in the right-leaning daily *Lidove noviny,* which ran lengthy interviews with Václáv Klaus on the environmental issue in March 1994 and offered his opponents, such as the former minister and dissident Ivan Dejmal the opportunity to respond. The proposals of Greenpeace to install solar energy panels in all new houses built after the Moravian floods in Autumn 1997 also received wide media coverage. The then director of Greenpeace, Hana Pernicová, was invited on to a high-profile television discussion programme with Professor Bedřich Moldan, the former environmental minister, academic and official environmental negotiator for the Czech Republic at the EU level.

3.4 External Agencies

The process of environmental regeneration in the Czech Republic has attracted both direct and indirect international involvement at a number of different levels. Direct involvement has tended to take the form of either the exchange of know how and technical assistance with regard to pollution control, or grants from international non-governmental organisations and foundations to indigenous Czech associations working on environmental issues and projects. This relationship between foreign donors and Czech environmental NGOs has already been described in some detail above.

In terms of indirect external influence, the role of the EU and, to a lesser extent, the OECD has been particularly important in shaping environmental capacity building. The desire of successive post-communist Czech governments to join the EU has encouraged the drafting of new environmental legislation in line with EU directives and norms (Fagin, 1994). The harmonisation of new laws with EU directives was formalised within the European Agreement between the EU and the Czech Republic in October 1993 (Fagin and Jehlička, 1998: 125). As the accession process has intensified, the Commission has placed increased emphasis on environmental management and emission targets as a condition of future entry (CEC, 1997). The Commission's Opinion on the Czech Republic, its most recent assessment of eligibility for membership, includes a section entitled "Quality of Life and the Environment" in which both state environmental policy and implementation are assessed (Slocock, 1998: 152-3). Since 1990 EU assistance for environmental regeneration in the Czech Republic has been delivered as part of the PHARE Programme. Funds have been made available for environmental education and other projects designed to improve the state of the environment.

However, the impact of EU influence on environmental capacity building is complicated and not as entirely positive as perhaps imagined. A first point to be noted is that the amount of EU-PHARE money that has been made available represents a modest sum compared to the Commission's estimated cost of environmental regeneration in the region. Of the total ECU 5417 million of PHARE funds made available between 1990-95, less than 10% (ECU 483 million) was directed towards environmental regeneration. The Commission has estimated the total cost of environmental regeneration in eastern Europe to be in the region of ECU 270 billion (Slocock, 1998: 153). It should also be noted that as a result of administrative problems only about half of the total money allocated to PHARE was successfully distributed.

There are further specific problems with regard to the PHARE Programme and the environment in the Czech Republic. Funds have been allocated towards ill-conceived projects that have not brought maximum benefit to the environment and, in certain cases, have led to other ecological problems. For instance, in the early 1990s, a PHARE-backed study recommended flue gas desulphurisation as the most appropriate method of reducing sulphur dioxide emissions. The approach was adopted by the Prunerov power plant. Yet this method of desulphurisation depends upon significant quantities of limestone to absorb sulphur dioxide from the flue gases. The quarrying of limestone in the protected landscape Český kras is already a serious environmental issue which the PHARE-backed study seemed to overlook (Fagin and Jehlička, 1998: 117).

International environmental aid programmes on the Czech Republic have generally tended to reflect the priorities of the individual donor country or the political and economic agendas of the institution involved, rather than necessarily the needs of the recipient country. This is also the case with PHARE funding. Environmental NGOs have hitherto benefited from the allocation of PHARE funding via the Czech organisation Civil Society Development Foundation (NROS – Nadace rozvoje obcanské spolecnosti). Associations receive grants for projects and to purchase equipment. Competition for such funding is fierce and the amount of money now available for this type of assistance is declining. This is partly owing to the relative success of economic reform in the Czech Republic, and also because of the changing focus of EU assistance in general. As the accession process gains pace, assistance to prospective entrants will be increasingly targeted towards their meeting convergence criteria. Whilst this does not necessarily mean less aid for the environment in general, priorities and funding will increasingly reflect the Commission's accession agenda rather than that of environmental activists within the Czech Republic.

Though the prospect of EU membership has hitherto acted as a pressure on successive Czech governments to enact environmental regulations, environmentalists now fear that, as accession gains pace, EU structural funds will be used to build motorways in order to facilitate investment. This, of course, will lead to an increase in traffic and levels of nitrogen oxides and carbon dioxide emissions. It will also cause concern about the protection of landscapes and habitats (Tickle, 1999). A further concern is that, once integrated into the EU, the Czech Republic will occupy a peripheral position, thus becoming a dumping ground for older, dirtier production technology for transnational firms (Slocock, 1998: 164-5). Though this is certainly less of a threat in the Czech Republic than in other former communist states, it is a dimension to the accession debate that is often overlooked.

It is also possible to identify a contradiction in the EU's approach to the environment that impacts directly on future capacity building in the Czech Republic. On one hand the Commission is pursuing a radical restructuring agenda in favour of sustainable development and against end-of-pipe technology. The most explicit expression of this dynamic agenda is the "Integrated Prevention Pollution Control" (IPPC) directive. On the other hand, the Commission will allow no flexibility with regard to transitional countries reaching emission targets. As argued above, such rigidity encourages an end-of-pipe approach rather than progressive restructuring. It is also particularly important in former communist states such as the Czech Republic for targets to be realistic to encourage respect for the legal process and to ensure successful implementation. One of the main obstacles to environmental capacity building is arguably the culture under communism of falsification to meet unrealistic plan targets. Assuming the Czech Republic is admitted to the EU early in the next century, the country will be expected to comply with the IPPC directive, which states that all investments and existing plants must apply best available technology by 2007. This will prove extremely difficult for Czech industry and could well provoke a reactionary response towards the EU and environmental protection in general.

The Czech Republic was admitted to the OECD in 1996 and the organisation's impact on the environment has since been quite significant. Like the EU, the OECD wants to harmonise standards to eliminate unfair economic advantage. During the summer of 1998 the OECD conducted an extensive Environmental Performance Review in the Czech Republic. All legislation and documentation was studied and a number of recommendations were made. The Czech Ministry of Environment is currently awaiting the final report and has intimated that it intends to act on the recommendations.[18] The most visible impact of the Czech Republic's membership of the OECD is the incorporation into Czech law of the principles of the Basel Convention. The convention, signed by all OECD countries, bans the export of all toxic waste to non-OECD countries. However, as with the environmental implications of EU membership, there is a fear that, being a poorer member of the OECD, the Czech Republic may well become a depository for toxic waste previously disposed of in non-OECD countries.

18 Information on the OECD Environmental Performance Review was obtained from an interview with Prof. Bedřich Moldan, currently official negotiator on behalf of the Ministry of Environment at the EU (11 November 1998).

4 Main Restrictions

4.1 Legacies of the Communist Era

One of the main restrictions on environmental capacity building in the Czech Republic is the legacy of four decades of Soviet-style communism. Though a great deal has been achieved over the past ten years in altering political behaviour, attitudes towards the state and the rule of law, notions of responsibility, and the role of citizens in the political process, the deleterious impact of the old regime is enduring and continues to act as an impediment on, for example, the implementation of environmental policy, the role of NGOs within the policy process, and attitudes and patterns of behaviour towards the natural environment in general.

In addition to being highly inefficient in its use of natural resources, Soviet-style communism had an adverse effect on the social contract between citizen and state. The perceived illegitimacy of the Soviet-backed regime, plus the tendency towards exaggerating the successes and hiding the failings of the economic system fostered a general disregard for the rule of law and respect for regulation. The emasculation of civil society and the ideological antipathy towards individual responsibility and independent activity fostered a passive political culture in which citizens were dissuaded from active participation.

Though there is evidence of recent positive changes in political culture and attitudes towards participation, Czech citizens remain reluctant to take assertive action to protect their environment or to influence the policy process. Particularly amongst the older generation, this emanates from a continued fear of political opposition and involvement despite democratic change. There is also a sense in which democratic participation is viewed narrowly in terms of voting at elections combined with deferring to and devolving responsibility to those in official positions. Moreover, the monumental economic reforms of the early and mid-1990s focused people's attention away from political participation towards the personal sphere, securing employment and housing.

Insofar as the enactment of appropriate environmental legislation and its subsequent implementation depend heavily upon public participation, eradicating the legacies of the communist era is critical.

4.2 The Political Opposition

As noted above, environmental capacity building in the Czech Republic during the 1993-1996 period was constrained by the prevailing political and economic climate. Václáv Klaus, as prime minister and leader of the main centre-right Civic Democratic Party (Obcanská demokratická strana – ODS), advocated the extension of private property as the only realistic "solution" to environmental problems (Klaus, 1994). He was scornful of environmental activists whom he portrayed as enemies of reform and a hangover from a bygone era of political opposition (Havel, 1996). Although his views on the environment were not endorsed by all within the ruling coalition, the pervading rhetoric of blatant individualism, economic liberalism and anti-regulation acted as a formidable constraint on environmental capacity building. The dearth of new environmental legislation during this period and the almost total marginalisation of environmental NGOs provide some indication of how powerful a constraint the political right was during this era. Unfortunately, despite the political demise of Klaus, the legacies of this period remain insofar as political precedents were established and vested interests pitched against the environment were able to consolidate their positions. For instance, in its attempts to update the Waste Law in line with OECD and EU standards, the Ministry of the Environment under Milos Kužvart currently faces a struggle with the so-called "incineration lobby", a group of local mayors from the ODS (such as Mr. Kokal, the mayor of Prague) who have sanctioned the construction of huge incinerators in their areas that far exceed the capacity of the locality.[19] Despite Klaus' departure from office and the fall of his ODS-led coalition, the legacy of this critical period is enduring and difficult to eradicate, particularly at the local level.

Moreover, disagreement on environmental protection transcends the left-right division of party politics and, as elsewhere in Europe, the opponents of tougher regulation are not confined to the political right in the Czech Republic. The new ČSSD (Česká strana sociálne demokratická – Czech Social Democratic Party) environmental minister, Milos Kužvart, faces opposition from within his own party and government, from a core of what might be termed "old" left ministers who reject the urgency of environmental action and argue in terms of a dichotomy between protecting Czech industry versus protecting the environment. Though there is certainly a pro-environment element within the party (including the vice-chair Mrs. Busková), both the transport minister (Mr. Peltran) and the industry and trade minister (Mr. Gregr) belong to the old left category. These key ministries are highly important in terms of affecting levels of pollution from transport and ensuring that Czech industry adapts to the IPPC (Integrated Prevention Pollution Control) requirements.

Yet, despite pockets of opposition from within both the left and right and the enduring legacy of the Klaus era, changes within the political parties have resulted in a more fluid and less ideologically constrained attitude towards the environment. Consensus appears to be emerging amongst political elements on both the right and the left on the need to regulate pollution and to up-date legislation. This is largely due to the break up of the right of centre coalition: the new Freedom Union (Mír slobode), which split from the ODS in January 1998, is pro-environment, though in parliamentary terms they are a marginal political force. Moreover, the Christian Democrats, former stablemates of the ODS, have also recently embraced environmental issues, though they broadly support the opposition ODS on transport issues.

5 Assessment and Prospects

It is a difficult point in time to assess environmental capacity in the Czech Republic. The inertia of the 1992-96 period is over and an array of legislative amendments and new policy initiatives are likely to be introduced in the coming months.

19 This information was obtained from an interview with Petr Stepanek, Director of Public Relations Department, Ministry of Environment, Prague, ČR (12 November 1998).

There is clearly renewed political momentum on the environment and a spirit of co-operation not witnessed since the early days of the velvet revolution. However, the new minister of the environment faces considerable tasks. The lack of legislation during the 1992-96 period has meant that the progressive framework of the early 1990s is now out-dated and regulations that should have been enacted during this period have to be quickly put in place. Although progress has been made in re-establishing a policy network involving NGOs, their political isolation and the economic hardship they suffered during the 1992-96 era has left an indelible mark on their development and capacity.

In assessing political capacity with regard to the environment, the most immediate issue is the relationship between the minister of environment and other members of the government. Regardless of how radical an agenda emerges from negotiations between the Ministry of the Environment and NGOs over the coming months, much depends on Minister Kužvart's ability to overcome opposition from within his own party at the parliamentary level and within the government. Though there is clearly a pro-environmental core of younger ministers within the Social Democratic Party, the old guard are well placed (industry, transport) to overturn policy initiatives. Additionally, political opposition from members of the former government and right-wing politicians at the local level cannot be overlooked. A decade in post-communist politics has been sufficient time for new vested interests to consolidate. Links between local politicians and the new private sector (for example, waste disposal companies and international mining companies) are likely to pose an increasing threat to the environmental lobby.[20]

Environmental capacity building implies both a long term dimension, a sense of putting structures in place to obtain future objectives, and the ability to deal with existing or short term problems. Whilst environmental management in the Czech Republic has certainly achieved a measure of short-term success, (attaining emission targets and improving the environmental devastation of the communist period), some approaches and solutions may in fact inhibit success (sustainable development) in the longer term.

The long term influence of external agencies in capacity building must also be evaluated with some caution. Since 1989 the Czech Republic has benefited from international aid and the transfer of know how. However, the objectives of the external agencies providing this support have not always been identical with the objectives of the environmental lobby or even those of the state. For instance, whilst foreign countries and other agencies have supported the Czech government's preference for nuclear energy as an alternative to the brown-coal power plants, which contribute heavily to transboundary pollution, issues of nuclear waste storage and the safety of the Dukovany power plant in South Moravia or the proposed power plant at Temelín have not been rigorously examined. Bilateral assistance from individual countries (such as Austria and Germany) has tended to be driven by a desire to arrest the effects of transboundary pollution. The OECD

20 An illustration of the pressure placed on local politicians by global corporations occurred recently when TVX Gold, a Canadian gold mining company intent on expanding operations in the Czech Republic, took a team of local politicians, officials and journalists on an all expenses paid trip to Central America to see how "successful" the company's operations were in other countries.

and the EU are ultimately concerned with integrating the Czech Republic into the global political and economic system and eliminating unfair advantage. The environmental agenda of the EU in particular reflects a desire to balance economic growth with environmental protection within existing member states. In some instances the Czech Republic has yet to experience the intensity of the problems EU member states are grappling with. For example, whilst the British government is currently attempting to construct an environmentally less damaging integrated transport policy around existing heavy dependence on private cars and decades of decline in public transport, the Czech Republic retains an integrated public transport system which, despite the recent increase in car ownership in and around large cities, is still widely used.

It cannot therefore simply be assumed that external influence and assistance will maximise capacity building. Though pressure from the EU to establish a legislative framework and to attain emission targets has undoubtedly been positive, and the inspections carried out by the OECD will similarly act as a pressure to strengthen capacity, such external involvement may well deflect from the Czech Republic dealing with other environmental problems, or encourage a government desperate to obtain entry into the EU to realise emission targets by a less than optimal route.

There is also the problem that such external assistance and the impact which it has exerted on capacity building may prove to be short term and in fact contradictory. It is already possible to detect a reduction in the amount of aid given to the Czech Republic by philanthropic agencies in light of more pressing issues elsewhere in the region and the relative success of the Czech transition. Once the Czech Republic has gained entry, the EU's impact on capacity building could arguably decline, particularly if the perceived benefits of membership are not apparent and the costs of accession result in a political backlash.

Perhaps the greatest cause for concern regarding environmental capacity building is the continued weakness of environmental NGOs. Although there is recent evidence of a cluster of NGOs occupying a stronger position within the policy process (Greenpeace, Hnutí Duha, Society for Sustainable Living, Děti Země), this should not be allowed to obscure the worsening financial situation faced by all NGOs. Whether lack of funding forces the closure of associations as in the case of Jíhočseké matký, or a dilution of their radicalism in response to corporate sponsorship, for example, the NGO *Tereza*'s decision to accept funding from Coca Cola.[21] The result is essentially that the position of environmental organisations vis-a-vis the economic lobby and opponents of environmental reform is becoming seriously weaker. In addition to a handful of large NGOs that have somehow managed to secure funding or that have links with international organisations, the arena of environmental policy in the Czech Republic requires the involvement of local grass-roots organisations.

The weak position of NGOs may well explain, or indeed be a consequence of, the narrow scope of the environmental policy agenda. This is particularly evident from the dimensions of the energy debate, which views nuclear energy as the only

21 *Tereza,* a leading environmental organisation, has recently received sizeable payments from both ÈEZ (the Czech electricity generating company and a major polluter) and Coca-Cola for contractual work (Interview with Jana Ledvunová, Director, *Tereza,* 17 July, 1997).

long term alternative to brown-coal power plants. Indeed, the future strategy for reducing air emissions is based heavily on the assumed completion of the Temelín plant (MŽP, 1997: 15). During his brief tenure as minister of the environment, Martin Bursik attempted to break this narrow approach by promoting a "third way" involving alternative sources of energy and a strategy based on reducing consumption. Though it remains to be seen how the new administration at the Ministry of Environment will proceed with regard to energy policy, the decision to place the future of the controversial and unfinished Temelín nuclear power plant in the hands of a commission of "experts" from neighbouring countries suggests that the energy debate may well have returned to its former parameters.

In conclusion, much remains in abeyance at the present time. There have certainly been some positive developments in recent months. The policy process has been reinvigorated and new actors have gained influence. Once amendments have been enacted the legislative framework is likely to resemble EU standards more closely; the administrative system established to ensure the implementation of policy is clearly achieving results. However, a decade since the collapse of communism, the more general problems of democratic consolidation in the Czech Republic are blatantly reflected in the problems facing environmental capacity building: associational activity requires strengthening, the assistance of external agencies in the reform process needs careful evaluation, and short term gains must not be permitted to obscure long term objectives.

References

Branis, M and E. Kružíková, 1994. The Environmental Impact Assessment Act in the Czech Republic: Origins, Introduction and Implementation Issues, *Environmental Impact Review* 14.

Carter F.W. and Turnock D. (eds.), 1993. *Environmental Problems in Eastern Europe,* London: Routledge.

Černá, A., Tosovská, E. and P. Četkovský, 1995. Economic Transformation and the Environment, in: J. Svejnar (ed.), *The Czech Republic and Economic Transition in Eastern Europe* San Diego, CA: Academic Press.

Commission of the European Communities (CEC), 1997. *Agenda 2000: For a Stronger and Wider Union.* Luxembourg: OOPEC

Fagin, A., 1994. Environment and Transition in the Czech Republic, *Environmental Politics*, 3 (3).

Fagin, A. and P. Jehlička, 1998. Sustainable Development in the Czech Republic: A Doomed Process? In Baker, S. and P. Jehlička (eds.), *Dilemmas of Transition: The Environment, Democracy and Economic Reform in East Central Europe*. London: Frank Cass.

Havel, V., Klaus, V. and P. Pithart, 1996. Civil Society after Communism: Rival Visions, *Journal of Democracy*, 7 (1) January.

Klaus, V., 1994. The Pollution Problem of the Natural Environment of Human Existence, *Lidové noviny* 14 February.

Moldan, Bedřich, 1990. *Zivotni prostředi České republiky* (Environment of the Czech Republic). Prague: Academia.

Moldan, B., 1996. Ekologie 1995 – Česká republika, *Lidové noviny,* 11 January 1996, 8.

Moldan, B., 1997. Czech Republic, in: Moldan, B. and J. Klarer (eds), *The Environmental Challenge for Central European Economies in Transition.* Chichester: John Wiley.

MŽP (Ministerstvo Životního prostředí ČR), 1997. Statisticka Ročenka Životního Prostředí České Republiky 1997 - (Statistical Environmental Yearbook of the Czech Republic 1997). Prague: MŽP.

MŽP (Ministerstvo Životního prostředí ČsR), 1997a. Report on the environment of the Czech Republic in 1996. Prague: MŽP.

Slocock, B., 1996. The Paradoxes of Environmental Policy in Eastern Europe: The Dynamics of Policy-Making in the Czech Republic. *Environmental Politics*, 5 (3) Autumn.

Slocock, B., 1998. Whatever Happened to the Environment?: Environmental Issues in the Eastern Enlargement of the European Union, in: Henderson, K. (ed.), *Back to Europe: Central and Eastern Europe and the European Union*. London: UCL Press.

Tickle, A., 1999. Political Transitions and Environmental Regulation in the Czech Republic (unpublished draft paper)

ENVIRONMENTAL CAPACITY BUILDING IN FRANCE

*Corinne Larrue**

1 Main Environmental Problems and Improvements since 1970

Compared with densely populated countries like Germany, the United Kingdom or the Netherlands, France has an environment that has remained relatively undamaged. The average population density in France is around 100 inhabitants per square kilometre, but with an extremely unequal distribution throughout the country. The population is concentrated within cities (eighty per cent of the population is urban) and mainly within the capital region, which accounts for approximately twenty per cent of the population and roughly thirty per cent of national GDP in less than seven per cent of the national territory. In contrast, some rural areas have been severely depopulated, some now having less than fifty inhabitants per square kilometre.

In France there are approximately two to three hectares of open space per inhabitant, compared with an average of less than one hectare per inhabitant in Germany and the United Kingdom. As a consequence, many parts of the country have been able to maintain a wide range of flora and fauna. France supports forty per cent of the flora species in Europe while occupying only twelve per cent of its territory.

However, this favourable environmental situation has been strongly damaged by the development of polluting and disturbing human activities. Such damage has played an important role in strengthening the environmental consciousness of the population since the mid sixties.

French environmental indicators show the same trends as in other European countries.[1]

In the 1970s and 1980s, the deterioration of natural habitats was obvious. The emission of pollutants in the air, water and the soils was considerable. Air quality in cities had deteriorated (13 zones exceeded the thresholds imposed by the EEC Directive for SO_2 and dust in 1982 for instance) and so had the quality of the main rivers (the first two national water quality inventories in 1971 and 1976 showed major bacteriological and toxic pollution). As far as soil was concerned, 44.2% of municipal waste in 1975 was disposed of by unauthorized means.

* I would like to thank Christiane Julien for the English editing of the text.

1 The following information on the environmental situation is from IFEN, 1994, IFEN, 1998, and OECD, 1997.

The main specific environmental threat in France remains the 57 nuclear power plants in operation, now account for about 75.9 % of French energy production. However, public acceptance of the nuclear programme remains very high.

Environmental policy as well as economic and energy consumption trends have helped to improve the environmental situation.

Since 1970 remarkable progress has been made in several areas. This is the main conclusion of the recent environmental performance review undertaken in 1996 by the Environmental Office of the OECD (OECD, 1997).

This is apparent in air quality. SO_2 and No_x emissions have decreased by 72% and 14% respectively since 1980. French emission of air pollutants per capita is now the lowest of the European countries. However, air pollution problems caused by the transportation sector are still substantial. Increased road transportation, especially in cities, has counterbalanced the important emission reduction efforts in other sectors (vehicle emissions of dust from transport have increased by 67% since 1980 and accounted for 77% of the emission of No_x in 1995).

Considerable progress has been made in combating water pollution. Investments in urban and industrial waste water treatment plants have improved water quality as regards classical parameters such as BOD or COD, and suspended matter. However, only 77% of the population is connected to a waste water treatment plant. Moreover, pollution from agricultural activities (i.e., nitrate, phosphate and pesticide pollution) are more and more noticeable. Similarly, the problem of water consumption, not considered a problem until the 1988-1992 drought, has now become an important issue.

As far as waste is concerned, major investments have been made since the 1975 law, and reinforced since the 1992 legislation. Waste produced by 99% of the population is now collected, and less than 5% is disposed of by unauthorized means. But more than 58% is still disposed of in landfills, and the recycling rate is very low (about 50% for paper, 42% for glass, 35% for packaging, and 8% for plastic, for instance). On the other hand, industrial waste treatment plants are numerous. They represent a major economic sector, importing waste from other European countries for treatment.

In nature protection, France still has a low proportion of protected land (9.7% of the national territory compared with 25% in Germany and 20% in the United Kingdom). And as the quality and the diversity of natural areas is high, many species of plants and animals remain threatened, especially in the southern part of the country. Agricultural and urban development, particularly in coastal areas, is responsible for this unfortunate situation (about 120 mammal species and 280 bird species are endangered).

The improvements in environmental quality have not yet been sufficient to impress public opinion on the environmental situation: according to a 1993 survey 50% of the population feel that the authorities should do more to protect the environment.

The 1990 National Plan for the Environment gave a cautious evaluation of the quality of the environment in France. According to the plan, the most obvious sources of environmental damage have been eradicated but pollution from less definable sources has increased. As a consequence, present and future policy are likely to be more difficult to implement.

2 Development of Environmental Policy and the Main Regulations[2]

2.1 Development

The following table presents the main French environmental laws. It shows that environmental regulation has been traditional in France.

2.1.1 The Roots of French Environmental Concerns

Three main concerns and traditions explain the development and the content of French environmental policy. They represent the main problem areas around which public environmental consciousness has developed. They also represent the principal fields of responsibility of the Ministry of the Environment, which was created in 1971.

The first is a tradition of health concern, which was an important issue in the enactment of the Industrial Pollution Control Act of 1810 and the Sanitary Regulations of 1902 (table 1). People like Villermé, Pasteur, and the architect Le Corbusier were prominent actors in this field. The aim was to improve public health by ameliorating environmental and living conditions, i.e., by combating smoke, dust, and other emissions, and by providing electricity, clean water and sewerage systems. These public health and hygiene concerns and this approach to environmental problems dominated administrative thinking throughout the 19th century and up until the mid-20th century.

The second tradition concerns natural science. It stems from a venerable French tradition devoted to the study of natural species and natural spaces, topics very similar to those dealt with by modern-day ecologists. A good example of the institutionalisation of this tradition is the National Museum for Natural History, created in 1791. Other examples are the local museums for Natural History and botanical gardens (Jardins des Plantes). The tradition was very strong during the French colonial period. At the beginning of the 20th century, voices were first heard in favour of nature conservation, the creation of natural parks and reserves, and the protection of endangered species. This movement supported the 1960 law on National Parks. In 1950 the Ministry of Agriculture took on responsibility for implementing nature protection policy. It was later transferred to the Ministry of the Environment after its creation in 1971.

The last tradition is an aesthetic one concerning sites and landscapes. It was represented by the Historic Buildings Service, created during the Revolution, which was quite prominent during the Second Empire with Vivet, Merimée and Viollet le Duc. The 1930 Natural Site Protection law introduced a system for the protection and preservation of certain sites. The sites and landscapes are listed in a survey and classified in terms of scenic, artistic or historical value. Once included in the list of protected sites, they cannot be altered without special authorisation

2 This chapter is drawn from an article written by L. Chabason and the author : "France: Fragmented policy and consensual implementation", in K. Hanf, Jansen, A.I., *Environmental Politics, policy and administration in Western Europe*, Addison Wesley Longmaan, 1998; 59-80.

from the minister. In 1971, a distinction was made between natural sites, the management of which was given to the Ministry of the Environment, and cultural sites, – produced by human activity – which continued to be managed by the Ministry of Culture.

The French environmental movement also reflects and is a product of those three traditions, which explains the diversity of the movement. Generally speaking, public environmental consciousness in France is intricately linked with the creation of the Ministry of the Environment in 1971. Before that period, the environmental "demand" was mainly generated by scientists, who were influenced by American literature, and, to a lesser extent, by NGOs involved in nature protection (Poujade, 1975, Charvolin, 1993). The story of the establishment of the Ministry of the Environment and its growing importance runs parallel to that of the development of public environmental awareness.

Table 1. The Main French Environmental Laws

Date of introduction	Date of reform	OBJECT
1810	1917 - 1976	Controlled industrial installations
1902	--- 1978 et 1984	Sanitary regulations
1913		Historical building protection
1930		Natural sites protection
1932	1961 - 1996	Air quality protection
1935	1992	Groundwater protection
1943	1972	Pesticides control
1943	--- 1979 - 1985	Advertising and shop signs
1948		Energy use
1960		National parks
1964		Hunting
1964	1992	Water protection
1964	1976 - 1983	Marine water pollution
1975	1992	Waste disposal
1976		Nature protection
1977		Chemicals control
1977	1981	Architecture
1980		Energy savings
1983	1993	Public inquiries
1983	1993	Transfer of competence to local authorities
1984		Fishing and continental water management
1985		Protection and management of mountain areas
1985		Forest management
1986		Coastal areas management
1987		Prevention of natural disasters
1991		Cities
1992		Noise
1992		Use of genetically modified organisms
1993		Landscape protection
1993		Quarries
1995		Strengthening environmental competence
1995	1999	Regional and planning policy

2.1.2 The Creation of the Ministry of the Environment

When it was created, the Ministry of the Environment was located within the Office of the Prime Minister. Such an arrangement was an easy way to create a new administrative unit without disturbing the existing overall administrative organisation. The Ministry of the Environment was given an inter-ministerial role of stimulating and co-ordinating action in the areas of pollution prevention, water management, nature conservation, management of hunting and fishing, and protection of sites and monuments, as well as the management of landscapes. The ministry was not given any executive power. It had no departmental or regional representative units. The new ministry was largely dependent on sectoral ministries and their local agencies for implementing its policy. It therefore did not threaten existing ministries.

The development of the new Ministry of the Environment has benefited from several structural changes. Some services were seconded from other ministries to form a new administrative unit within the new ministry. This was the case for the administration of nature reserves and parks, the hunting department (both of them previously under the jurisdiction of the Ministry of Agriculture), the administration of controlled hazardous installations ("service des installations classes"), previously under the jurisdiction of the Ministry of Industry), and the water agencies (under the jurisdiction of the Regional Policy and Planning Ministry). Those transfers constituted the organisational basis of the new ministry. It was also given authority over a number of secretariats, such as the secretariat of the Inter-Ministerial Commission for National Parks, of the High Commission of Sites (shared with the Ministry of Culture), and of the Inter-Ministerial Commission for Water. Lastly, for local implementation of its policy, the ministry was given the opportunity to use the local units of the Ministry of Agriculture, and the Ministry of Industry.

During the 1970s, two other areas of concern were introduced, as well as special administrative departments to deal with them, the rational management of natural resources and public participation, especially through environmental impact assessments (EIA) and environmental information.

With the creation of the ministry, French environmental policy assumed an obvious comprehensive dimension. By the mid-seventies, the main activities of the Ministry were to reform and complete existing environmental regulations. The laws on waste, nature protection and chemical products were enacted, and the law on industrial pollution control was reformed in 1976 (table 1).

In the early 1980s, the accumulation of several factors led to a decrease in the development of environmental policy making: the economic crisis, linked with the second fuel crisis, relegated environmental concerns to the background. The ecologist groups, who were mainly against the nuclear programme, were not followed by public opinion, and thereby lost their credibility. Leftwing parties, which came to power in 1981, attracted public attention, and the proponents of environmental policy were neutralised. France thus reacted during this period against the development of a European environmental policy that entailed environmental backwardness.

At that time, most of the enacted environmental laws did not set ambitious environmental objectives. For instance, the laws on the management of mountain areas (1985), forest areas (1985) and coastal areas (1986), proposed reconciling environmental concerns with the economic development of those areas, and did not really promote their environmental protection.

2.1.3 The Ecological Wave of the Nineties

After the Chernobyl accident and the return of economic growth, environmental policy and consciousness reappeared. This ecological "come back" appeared also in the political arena: the number of green votes increased, reaching about 15% in the 1992 regional election. Such political impetus helped the Ministry of the Environment to gain more responsibility within the field of environmental protection, especially in dealing with nuclear plant safety, infrastructure provision, landscape policy, and jurisdiction over the National Forestry Office, which manages 4 million hectares of land.

The ecological wave of the 90s found its main manifestation in 1990 with the formulation of the National Plan for the Environment. The plan may be seen as part of a new start in environmental policy (Chabason and Theys, 1990): new (ambitious) objectives and new instruments for national environmental policy were proposed. Enactment of the law on noise (1992), the reform of the water (1992) and air protection (1996) laws, as well as the law on the reinforcement of environmental responsibilities (1995) can be seen as outcomes of the Plan.

But the most important result was the creation of the DIRENs *(Direction Regionale de l'Environnement)* in 1991, which represents environmental state administration in each French region (see section 4).

After this period, with the right wing back in power and an increasingly severe economic crisis, environmental concerns were again relegated to the background. Public environmental awareness decreased, as the 1995 survey shows (IFEN, 1995): personal commitment slackened, and the number of people willing to accept tax increases for environmental purposes decreased by 9% between 1993 and 1994. Nevertheless, environmental concerns remained a priority in France, especially air pollution at the local level.

Just recently, environmental issues took a political turn. The 1997 electoral agreement between the Socialist and the Green Parties reinforced the legitimacy of environmental concerns, inducing the Government to pay greater attention to environmental policy. The enlargement of the portfolio of the new Minister of the Environment, who is in charge of both regional and environmental policy responsibilities, will undoubtedly increase the power of the Ministry and the coherence of French environmental policy. However, the merging of both central administrations will take time to be efficient.

2.2 Basic Instruments

The main instruments framing French environmental policy are command and control instruments and economic instruments. Moreover, the French administration of the environment has also developed informative regulations.

As regards command and control instruments, they provide the most important means of action for central government. Emission standards, planning tools, protection zones, controlled hazardous installations, environmental impact assessments have been gradually introduced by laws and regulations, and now constitute the main tools available to central and local government environmental agencies.

The command and control approach developed in France relies in the first place on individual constraints, usually imposed through a licensing system, applicable to industrial or agricultural emission sources, to water users, or to chemical products put on the market. The control system for industrial hazardous installations is a good example of such an approach. A permit is given to industrial or agricultural polluting plant through a procedure which includes the production of an environmental impact assessment and public participation. The plants are granted permits provided they respect individually defined technical standards and constraints based upon the best available technology which does not entail excessive costs.

The command and control approach is also based on collective constraints including collective standards applied to the whole national territory (usually for public health protection) and more severe collective constraints in some sensitive areas where environmental protection requires more attention. Such an approach applies to nature protection when special protection zones are designed in those areas where bio-diversity deserves strict protection. Such an approach also applies in air quality protection: some of the most polluted cities are designated as special protection zones, within which the emission of air pollutants is restricted. Such an approach also applies to protecting drinking water collection pipes.

Lastly, in the field of air pollution control, there is a semi-collective instrument, the so-called alert zone. In such zones only the more important emitters are submitted to restriction when a given threshold of pollution is reached.

French public authorities have also been developing economic instruments over the years. For instance, since 1967, taxes have been imposed on industrial water discharges. This type of tax is now imposed on air emissions, waste disposal, noise, etc. The recent OECD report on French environmental performance (OECD, 1997) underlines the existence of about 20 eco-taxes bringing about 30 billion francs in revenue. The 1990 National Environmental Plan has strengthened this economic approach.

Instruments based on the provision of information on the environmental impact of activities or products have recently been developed. Such an approach relies upon private sector environmental reporting and labelling.

We must mention that, at the implementation level, those basic instruments are usually involved in a joint implementation strategy, based on partnership between central public authorities and private companies (through sectoral agreements, for instance) or between central and local governments (through local environmental charters for instance).

The last characteristic of French environmental policy is the growing impact of EC environmental directives. As a member of the European Union, France must meet EU requirements in environmental policy as in other fields. A study evaluating the impact of EU environmental policy on French policy showed that, although the EU and France have similar policies, European norms and standards are stricter than French rules in several environmental areas (Larrue, Prud'homme, 1993). EU policy accordingly serves in France to reinforce regulatory pressure in favour of the environment. More importantly, many EU directives tend to reinforce the position of the Minister of the Environment vis-à-vis both industry and other – often competing – ministries.[3]

2.3 Policy Style

The main characteristic of French environmental policy style is flexibility. It relies upon a rather soft policy programme, and an implementation based on consensus, partnership and negotiation.

Because central government has implementation units at its direct disposal, the programme formulated at the central level is usually rather vague, leaving it to the implementation level to define many elements. This is also true for environmental policy despite the fact that implementation agencies are not always under the direct control of the ministry. For instance, as far as the programme for reducing SO_2 is concerned, until the implementation of the 80/779 EEC Directive, no objective was really set in terms of μg of SO_2 per m^3 of air, nor were national emission standards defined to be applied to individual firms. In many cases such soft programmes have provided a good basis for effective implementation, as extensive international comparison of national practice concerning air pollution shows (see Knoepfel, Weidner, 1986).

There is also flexibility at the implementation level. Implementation is based on consensus rather than on the imposition of constraints on the actors involved. The central environmental administration and its representatives at the regional and local levels try to involve target groups and/or local governments as their partners in the policy implementation process. A good example of such a partnership involving target groups and local government is provided by the water agencies and the charges they levy on polluting firms and water users. A financial aid system is closely linked to the charges system applied to water consumption and pollution discharges into water. The revenue raised by means of effluent charges are redistributed to those firms and local governments which bear them, through subsidies on investments in waste water treatment plants. Such a system of joint co-operation has also functioned as a strong argument for industrialists to accept the effluent charges system.

Two factors explain these characteristics of environmental policy implementation. The first concerns the number of actors involved in environmental policy. Environmental responsibilities are spread among many actors at the central government level as well as at the regional and local levels. In many cases, none of the actors is thus able to act alone in implementing a policy. Moreover, it is sometimes difficult to find any leadership at all within this implementation organisation. The administrative agency in charge of implementing an environmental programme usually needs to "borrow legitimacy" from other actors, and negotiation and consensus are therefore required in order to implement every single policy.

---

3 However, the above study also underlined that these benefits of the European environment policy are limited by the "compromises" produced by European decision-making processes. When policy negotiations at the European level do not incorporate the French government's position, the implementation of EU policy is made more difficult, because the position of the Minister of the Environment is weakened at home.

But even in the rare cases where a single agency is responsible for implementing a particular policy, as in the case of the DRIRE and industrial pollution policy for instance, the consensual aspect exists. In such cases a second explanatory factor can be pointed to, which is linked to the over-arching nature of environmental policy. Environmental policy is often implemented by administrative actors already involved in implementing other policies, and the way they do it is influenced by their habit of doing things. They usually have previous relationships with the target groups of environmental policy, which also influences the implementation of environmental policy. For instance, the DRIREs are responsible for both industrial pollution reduction and industrial development within each region. They are used to negotiating with companies to attract them into their regions. It is therefore not surprising for them to negotiate when implementing industrial pollution reduction policies.

3 Evaluation of Policy

The OECD Report on Environmental Performance in France, published in 1997, came to a relative positive view.

It states that the institutional framework of French environmental policy is now comprehensive, and that the combination of regulatory, economic, planning and voluntary instruments makes French environmental policy balanced and effective (OECD, 1997: 20). "Decoupling of economic growth and pressure over the environment has been achieved over the last 25 years" (OECD, 1997: 28). National spending on the environment is growing. It is estimated at 145.1 billion francs, which represented about 1.8 % of GDP in 1997, whereas it only represented 1.0% of GDP in 1978.

However, according to the report, policy implementation suffers from many problems. Polluters are often unknown, or benefit from exemptions. The most important efforts to be made are in waste water collection and treatment, flood prevention, and the management of urban transport pollution. The main environmental problems stem from the agricultural sector and from local governments. Particular care should be taken of the situation in coastal areas.

According to the 1990 National Environmental Plan, and as regards the situation in other European countries, the French environmental situation differs from domain to domain. The situation is good in water or energy consumption, open space available per capita, industrial waste treatment capacities, the quality of beaches or acid rain exposition. The situation is average for fertilizer and pesticide consumption, municipal waste treatment capacity, surface water quality, and air quality in cities. It is obviously worse for pollution stemming from transportation, noise exposure, municipal waste water treatment plants, forest fire prevention, and for the specific problems linked to nuclear energy.

In the OECD report, the positive environmental situation is linked to economic structural changes, to structural changes in energy consumption, and to environmental policy. French environmental policy can thus be considered "mature", with the main problems arising from administrative complexity (lack of internal integration) and the quality of implementation. A major effort is now required in more efficient policy implementation and better integration of environmental concerns in sectoral policy, especially in the transportation and agricultural sectors.

Finally, the OEDC report underlines the important effort made by the French government towards international cooperation in environmental protection with developing countries. Overall, the French government spends about 25,2 billion francs a year (1997) on the environment, mainly on water supply and waste water treatment.

4 Actors in Environmental Policy

4.1 Public Institutions

The French environmental politico-administrative system is comprehensive and complicated. It includes actors at the central, regional, and local levels.

4.1.1 The Central or National Level

The main actor at the central level is the Ministry of the Environment. Originally, the Ministry of the Environment was defined as a light structure designed to influence other more traditional politico-administrative actors. The ministry, with a staff of about 500, acted as a trans-sectoral agency. Its efficiency relied upon interministerial arrangements, like the High Committee for the Environment, the Interministerial Committee for the Quality of Life, the National Council for Noise etc., which were created to organize and coordinate contacts. All those committees consist of representatives of the ministries involved as well as independent experts, and in some cases, representatives of the actors concerned.

Since 1984, the Ministry of the Environment has included four main departments, dealing with nature protection, pollution, quality of life, and information and research. This organisational division corresponds to the main sectoral fields of competence given to the Ministry of the Environment, and also to the different administrative "corps" involved in the environmental field. The upper level of French administration is divided among a number of corps, defined with reference to the higher educational institutions offering advanced professional and technical training ("grandes écoles") where their members were educated. Three main "corps" are concerned with the environment: the "mining engineers corps", which deals with industrial pollution, the "water and forestry corps", mainly concerned with nature protection and management, and the "bridges and roads corps" in charge of the quality of life. The organisational structure of the Ministry of the Environment was thus divided into both functional and "socio-administrative" sections.

Since the early 1990s, and as a consequence of the implementation of the National Environmental Plan, the structure and efficiency of the Ministry of the Environment have been modified and enhanced.

A new organisational decree, promulgated in April 1992, added a fifth department exclusively devoted to water problems[4]. The jurisdiction of the ministry has also been expanded to include landscape protection. The ministry has a staff of over 2,500 (including regional agencies) and a budget of about 1.9 billion francs.

In addition, new inter-ministerial bodies were created in 1993: the High Committee for the Rights of Future Generations, the Inter-Ministerial Committee for the Environment, and the Commission for Sustainable Development (responsible for making recommendations for improving and integrating sustainability in environmental policy and, more generally, in all governmental policy).

Finally, the 1992 decree regulated how the ministry was to participate formally in the drafting transport, rural development and urban planning policy. The Ministry of the Environment had previously been excluded from the decision-making process. For instance, the National Highway Plan, enacted in 1987, was formulated without consulting the Ministry of the Environment; similarly, industrial and agricultural policies with a strong impact on the environment were often formulated and implemented without any dialogue between the ministries in charge and the Ministry of the Environment.

However, at the central government level four other ministries still play an important role in environmental policy, having retained some competence in the field. They are the most important "partners" of the Ministry of the Environment. The first is the Ministry of Public Health, whose "sanitation engineers" play an important role in defining and controlling the sanitary living conditions of the population. The second is the Ministry of Industry (the principal bastion for the "mining engineers corps"); it plays an active role in industrial modernisation policies as well as in energy policies. The third is the Ministry of Public Works (with staff from the "civil engineers corps" ("corps des ponts et chausses"). It deals with urban environmental problems. The last is the Ministry of Agriculture (whose staff are members of the "water and forestry corps"). It deals with most environmental problems in rural areas. Each important "partner" of the Ministry of the Environment has its corporate counterpart within the Ministry (except for the Ministry of Public Health).

Another important feature of the French environmental politico-administrative system is the existence of quasi independent technical agencies, responsible for one particular field. The most important is the ADEME *(Agence de l'Environnement et de la Maîtrise de l'Energie)*. It was created in 1991 by merging three existing agencies (one dealing with waste, another with air pollution, the third with energy). The ADEME has a staff of over 600.

The second type of agency corresponds to the six water agencies created a long time ago (1967), each managing one of the six great basins in France *(Seine Normandie, Loire Bretagne, Adour-Garonne, Rhine Meuse, Artois Picardie and Rhône Méditerranée Corse)*. The agencies are governed by quasi autonomous bodies called "basin committees". They act as real "water Parliaments", being composed of representatives of central and local governments, professional interests, and water users, and because – within limits set by the Ministry for the

4 Moreover, in 2000, a new department will be created entitled "environmental evaluation department" to strengthen the environmental economic approach.

Budget – they define the basis and level of water consumption and pollution charges.

Other agencies deal with hunting (Office national de la Chasse), fishing (Conseil supérieur de la Pêche), and special zones, such as coastal zones (Conservatoire de l'Espace Littoral et des Rivages Lacustres, since 1974), forests (Office National de la Forêt) or national parks.

All those agencies usually manage para-fiscal levies, which finance controls, promotion, and investment within their particular field. These agencies' aggregate budget is ten times higher than that of the Ministry of the Environment.

More recently, two other technical bodies were created: INERIS *(Institut National de l'Environnement Industriel et des Risques),* a technical institute created in 1990 with an extended expert capability in the field of industrial environment and hazards; and the IFEN *(Institut Français de l'Environnement)* set up in 1991, to provide the Ministry of the Environment and other actors with statistical and informational support in environmental problems and policy.

Under present institutional arrangements, in short, the Ministry makes rules whereas the agencies have technical, informational and financial functions.

4.1.2 The Regional and Local Level

France has a highly centralized administrative system. Most ministries have local agencies at their disposal, i.e., units located in each region and/or *département* able to implement their policy directly. These units are part of the central ministry, which they represent in opposition to or collaboration with local governments. They are usually under the joint control of their ministry and the local *préfet.*

Since 1991, the Ministry of the Environment has had such direct units at its disposal. They are referred to as regional departments of the environment *(Directions Régionales de l'Environnement* – DIREN), and are responsible for water quality management, nature conservation, and rural and urban landscape protection. However, the DIRENs are not in charge of industrial pollution, which remains the responsibility of the regional departments of industry and the environment *(Directions Régionales de l'Industrie, de la Recherche et de l'Environnement* – DRIRE), the local agencies of the Ministry of Industry (although the DRIREs are under the responsibility of both ministries).

More generally, at the *département* level, the Ministry of the Environment depends on local administrative units under the supervision of other ministries. Indeed, at this level, the ministry has no direct agency at its disposal and relies on *département* units attached to other ministries to implement its policy, such as infrastructure units (Ministry of Public Works), agriculture units (Ministry of Agriculture), and public health units (Ministry of Public Health).

At the implementation level, central and local responsibilities for environmental protection are still shared between different administrative units.

More recently, and as a consequence of the decentralisation movement started in 1982, new decentralised actors have also tended to assume some responsibility for environmental matters. Traditionally, some environmental jurisdiction has been granted to local, *département* and regional governments. Indeed, local governments have always been responsible for water supply, waste water disposal and treatment, and municipal waste disposal; and they usually perform their functions on an inter-authority basis. But more and more local governments are trying to organise a more comprehensive environmental policy on a local authority or inter-authority basis. However, local government powers are still limited in environmental policy. For instance, local power authorities have almost nothing to do with pollution caused by controlled hazardous plants, the so-called "classified installations", even if these plants have a strong impact on local environmental quality. The regulation of industrial pollution is the responsibility of central government.

Similarly, the government of the *département* (called *conseil général)* mainly followed central government initiatives until recently, because it partially finances the purchase of equipment decided upon by central government. However, since the reform of 1982, even though such competence was not explicitly transferred to them, regional, and above all *département* governments have tended to take an active interest in environmental and nature protection. They set up "environmental administrative units", define their own priorities; and try to formulate their own policy with – or without – central government participation.

To conclude, we can say that public support for environmental policy is increasing and diversifying. On the whole, public expenditure on the environment is about 158 billion francs, 15% of which is supplied by central government and 85% by decentralised governments.

4.2 Environmental Organisations

Since the early 1970s, nature protection policy in France has been explicitly co-managed by environmental NGOs. To compensate for its lack of resources and staff, the Ministry of the Environment has encouraged private organisations to become involved in the actual management of the local environment, as has been the case for social policy. Many environmental associations are, for instance, involved in the management of nature reserves, and receive subsidies for this purpose.

The 1976 law provides approval procedures for environmental associations, allowing the participation and representation of approved organisations in various environmental committees at the local, regional and national levels. The public recognition this procedure provides allows such organizations to take legal action on environmental issues.

The most important French environmental associations belong to "France Nature Environment", an umbrella organisation with 850,000 members and a budget of 4 million francs (50% in public subsidies). There are other, less important environment organisations, such as the Fondation Cousteau (80,000 members), Greenpeace France (30,000 members), World Wildlife Fund (100,000 paying members).

Ecologist parties are fairly recent in France. The ecologist vote first played a role in the 1989 European election (10% of the vote), and increased in the 1992 regional election (more than 14%).

Until recently the green political movement was mainly divided into "Les Verts" (the French Green Party) and "Generation Ecologie" a party formed by Brice Lalonde in 1990. After the failure of the 1993 "Entente ecologiste" (an electoral agreement between those two parties), Generation Ecologie disappeared, and

"Les Verts" succeeded in reaching an electoral agreement with the Socialist Party for the 1997 national election.

Until very recently ecologist parties were considered marginal, not really established within the party system. The situation changed after the electoral agreement of 1997, which made environmental concerns a direct priority in French politics. However, green politicians are still limited to environmental issues and have not yet managed to make headway in other fields of French politics.

4.3 Economic Support for Environmental Protection

The eco-industries represent an important lobby in France, especially those dealing with water treatment. Overall, the eco-sector directly employs 280,000 people with a turnover of about 125 billion francs, i.e., 2% of GDP (IFEN, 1995).

The sector is dominated by three important companies that are worldwide leaders in urban environmental services: Generale des Eaux, Lyonaise des Eaux and SAUR (from the Bouygue industrial group). The development of these companies is linked to the fact that many French cities and local governments have entrusted their water supply, waste water treatment and waste disposal services to private companies. However, the French eco-industry is neither really dominant nor organised to deal with clean industrial technologies: foreign companies (German, Swiss or Japanese ones) are more competent.

The environment is becoming an issue for more and more industrialists. More and more enterprises are tending to integrate environmental concerns in their industrial strategy by, for example, drawing up an environmental plan. However this is true only of certain large enterprises, small and medium size companies being largely unaffected by this trend. But with the standardization of environmental management (ISO 14.000 norm or European Eco-audit), this should become common practice.

5 The Changing Constellation of Actors

During the 50s and 60s, both the French politico-administrative system and the economic super-structure had a positive opinion of "productivism". This was particularly true for the nuclear energy production programme, which was intended to preserve "French energy independence". Generally speaking, French political parties as well as the administration agreed on the development concept implemented after the second World War, and saw the "environment" mainly as an obstacle to the modernization of the country.

At that time, environmental measures were taken only under pressure from local public opinion, but for a long time such pressure was inconstant, and mostly concerned with protecting nature.

At the beginning of the 70s, environmental problems were considered as a potential obstacle to economic development. The making and the implementation of environmental policy during the seventies relied on a common economic background: environmental externalities should not entail disturbance in economic growth; environmental problems were therefore to be solved and prevented as far as possible. Environmental policy was mainly the concern of central government and industry. Such a constellation was reflected in the so-called "contrats de branche" ("sectoral contracts") of the 70s and early 80s between the Ministry of the Environment and professional interests. These contracts stipulated the schedule for anti-pollution investment by companies in the particular sector. Central government subsidies aided such investments. Approximately 10 sectoral contracts were signed between 1970 and 1980. They involved the most polluting industries. The results have been positive, since polluting emissions into the air and into water decreased over the same period.

The environmental policy paradigm changed in the eighties. The economic crisis removed environmental issues from the economic and political agenda, and French public environmental actors could overcome their weakness only by resorting to European constraints and directives. There was no longer any consensus on environmental issues, and environmental policy was seen as a luxury to be taken into account once economic problems had been solved. The objective was to give a new impetus to the economy at whatever cost to the environment. During this period, the closeness of the economic and political arenas, which had previously been useful in implementing environmental policy hindered any development in the environmental field.

However, during this decade public environmental consciousness grew in parallel to an increasing lack of confidence in economic progress and development and in the capacity of the government to prevent environmental problems. "Environmental demand" grew, and became a political factor with the ecologist vote at the European election. Public opposition to national infrastructures became more and more common. At the same time, the decentralisation movement started in 1982, changing politico-administrative constellations and relations between central government and decentralised bodies.

This period without economic growth preceded a phase that began with a major "ecological electoral wave" in early 1990 (a 15% green vote at the 1992 regional elections). On the whole, the 90s brought real progress towards ecological modernisation as defined by Jänicke (1996). The trend has been formalised by two main changes in environmental actor constellations.

The first one has been a trend towards *internal environmental integration*. The National Environmental Plan, as well as public support has strengthened the legitimacy of environmental policy, and has led to the reorganization of public environmental institutions. The Ministry of the Environment has been given more human and financial resources, and has become less and less isolated. Environmental concern has again become an economic issue, and the enforcement of environmental laws is increasingly considered by the whole politico-administrative system as instrumental in preparing the French economy to meet international competition.

The second important trend has been *external environmental integration*. It is linked to the opening up of decision-making processes to environmental interest organisations. This trend has occurred at various levels. At the local and regional levels, it has mainly been associated with decentralisation. In formulating their own environmental policies, local, *département* and regional governments have paid increasing attention to local environmental organisations. They have supported them with subsidies and involved them in decision-making. At the national level, public opposition to important infrastructures such as the South East TGV or a number of major highways built in the late 80s has entailed various attempts to "organize the public debate". This was the case for the new participatory decision-making process on major infrastructure projects defined by a 1992 memorandum ("Circulaire Bianco") and for the newly created National Commission for Public Debate (set up by the 1995 law). In the near future, such arrangements should lead to more in-depth participation by the public and environmental organisations in the formulation of sectoral policy.

More recently, environmental issues have been linked to "territorial management". The specificity of the French situation in terms of open space availability and the unbalanced distribution of settlement, makes such a combination necessary. This has added a major spatial dimension to sustainable development concerns. This spatial approach to environmental issues may be considered as the new challenge for ecological modernisation.

6 Capacity Building

French environmental policy now has comprehensive capacities. As we have seen, these capacities were initiated during the 70s and reinforced during the 90s. However, they should now be used to the full.

Public institutions as well as target groups and environmental organisations are quite well structured and organised (see section 4). They have extensive technical know-how in the field, supported and nurtured by environmental research institutions. From early 1980, various research institutions in a range of technical disciplines and the natural sciences were created under the aegis of the National Research Centre (CNRS). An Interdisciplinary Research Programme on the Environment was set up in 1982, and a National Programme for Clean Technologies. Every year the ADEME provides 80 grants for doctoral studies in the environmental field. Masters degree courses and doctorates in environmental sciences are offered by most French universities.

However, because the "grandes écoles" training system for civil servants provides only technical know-how on environmental issues but very little knowledge of the natural and social sciences, the integration of environmental concerns into sectoral policy will remain very limited.

Moreover, as we have seen, the French green business sector is an important economic factor, but because it specialises in water and waste treatment it tends to favour end-of-pipe strategies, influencing public actors to that end. Environmental policy proponents usually lack political and social know-how.

Until recently, ecologist parties failed to gain a real foothold in the party system owing to second-ballot electoral system applied in general elections. However, green parties have had representatives in regional councils and local governments for about 10 years or more. They have learnt more about political behaviour and strategy. The recent alliance between "Les Verts" and the Socialist party is evidence. We can therefore expect the political weakness of environmental proponents to be progressively reduced. However, environmental issues are still outside traditional party concerns, and political integration is still lacking.

Until the mid 80s, the French participatory system was restricted to very local private interests. The 1983 reform broadened the basis for public participation but it was still restricted to the last stage of the decision-making process, without any possibility of influencing the whole project. The dialogue organised at the upper decision-making level involved only elected bodies to the exclusion of environmental interest organisations. The situation has recently changed. Some procedures and institutions have been set up to organise public participation at the policy formulation level. This will strengthen the environmental proponent's ability to influence decisions.

The French courts do not yet play an important role in the environmental field. Indeed, more and more public decisions are contested before the courts by, for instance, environmental organisations and some have been repealed; but the "pont de l'ile de Ré" syndrome explains why such judicial decisions are minimized. In this case, the decision to build a bridge linking the island with the mainland was taken to court, but the bridge finished before the court handed down its decision.

Informational capacity grew in the early 70s, but was fully organised only in the early 90s. The setting up of a Parliamentary Office for Technological Assessment in 1984 and the creation of IFEN in 1991 were the result of these organisational efforts. The Parliamentary Office for Technological Assessment produces interesting studies in the field of the environment. However, it has very limited resources and capacity to disseminate information. The IFEN now produces and publishes a lot of information on the environment: it observes social perception of and public opinion on the environment; it has published interesting studies on the state of the French environment at the national and regional levels; it has been able to establish a national environmental assessment, etc. However, its capacity to disseminate information as well as its analytic capacity remain limited. IFEN has so far been restricted to producing statistical data. In order to strengthen the environmental capacity building of environmental proponents, it should now develop its environmental policy evaluation capacities.

The media have played an important role in fostering public awareness of environmental issues. Many magazines specialising in environmental issues were created during the 70s only to disappear during 80s, but some reappeared during the 90s. Media attention for environmental issues varies. They usually concentrate on sensational news. However, most traditional magazines have an "environment" section and environmental specialists on the staff.

French environmental capacity-building can thus be seen as "mayonnaise ingredients" waiting for a chemical reaction. It could be produced by two combined stimuli, the first external, coming from European environmental policy, which will no doubt be soon reinforced because of the accession of environmentally more advanced countries like Sweden, Austria and Finland. The second is more internal. It will be the outcome of the two trends we have discussed, decentralization and participation, the results of which will be more and more evident in future.

References

Chabason, L., and J. Theys, 1990, Plan National pour l'Environnement, Rapport préliminaire en vue du débat d'orientation. Paris, Ministère de l'Environnement.

Charvolin, 1993. L'invention de l'environnement en France. Grenoble: IEP (PhD dissertation).
IFEN, 1994. L'environnement en France 1994-95. Paris: Dunod.
IFEN, 1995. L'opinion publique et environnement. Paris: Bordas.
Knoepfel, and Weidner, 1986. Explaining differences in the performance of clean air policies. An international and interregional comparative study. Policy and Politics 14 (1), 71-92.
Larrue, C. and L. Chabason, 1998. France: Fragmented policy and consensual implementation, in: K. Hanf, and A.I. Jansen, Governance and Environment in Western Europe. Oxford: Addison Wesley Longman; 59-80.
Larrue, C., R. Prud'homme, 1993. European Environmental Policies and Environmental Protection in France, in: Morizet, and Dreyfus (eds), France and EC Membership Evaluated. London: Pinter Publishers Ltd, 68-79.
OECD, 1997. Environmental Performance Reviews: France. Paris: OECD.
Poujade, R., 1975. Le ministère de l'impossible. Paris: Calmann-Levy.

HUNGARY

Joanne Caddy and Anna Vári*

1 Main Environmental Problems

Hungary faces a range of significant threats to its environment related to its geographical position in the heart of Europe, the legacy of over 40 years of resource-intensive command economy and the challenges of renewed growth in the current phase of consolidation.

Water quality represents the largest environmental issue for Hungary owing in part to the fact that 96% of its surface waters originate from abroad, and in part to the long-standing, intensive use of nitrogen fertilisers in agriculture, the limited reach of public sewage systems, and the lack of wastewater treatment plants. In 1996, 97% of the population were connected to the public water supply system while only 59% were served by a sewer system. Steady increases in water consumption have not been matched by equivalent growth in capacity to deal with the wastewater produced (see Table 1). Although the level of sewage purification attained has improved over the past 15 years, there is still a significant amount of raw sewage being released without treatment. Contamination of surface and groundwater has ensued which, in turn, pose a serious threat to the security of the drinking water supply (REC, 1994b: 39; KTM, 1994: 59).

Air pollution also constitutes a significant problem, especially in urban areas, and it is estimated that about 13% of the territory where 49% of the population lives is to be regarded as polluted (EHAP, 1997). The collapse in industrial production which marked the early transition period (1990-1993) did lead to a fall in

Table 1. Proportion of population with public utility service in Hungary, 1980-1996 (%)

	1980	1990	1992	1993	1994	1995	1996
Population with water supply by a public utility (%)	75	92	94	95	96	96	97
Population living in an area with sewers (%)	40	52	54	55	56	58	59

Source: KM (1998: 78)

* The opinions expressed and arguments employed in this publication are the sole responsibility of the author and do not reflect those of the OECD or of the governments of its Member countries.

emissions; however, these levels have since been climbing as economic recovery continues apace. Currently, the fastest growing source of air pollution is vehicle traffic. The boom in private car ownership (which increased by 27% in the period 1990-1996) has largely outbalanced the advantages brought by the modernisation of the existing vehicle stock. In Budapest, the nation's capital and most heavily urbanised city, vehicle traffic accounts for half the urban air pollution while the overall environmental pressure on residents has been estimated to be higher than that of Athens (EEA, 1995: 31).

The concentration levels of "traditional" ambient air pollutants (SO_2, NO_2, CO, settled dust and suspended particles) are relatively high. In addition, substantial quantities of toxic metals (such as lead, cadmium, nickel), aromatic hydrocarbons (benzene, toluene, xylene), products of incomplete combustion (polycyclic aromatic hydrocarbons), as well as asbestos enter the environment as a result of traffic and partly of industrial and communal heating, causing significant damage (EHAP, 1997). The repercussions on public health are reflected in the high incidence of allergic asthma in children, chronic bronchitis, and lung cancer among adults, and significant financial costs in terms of medical treatment and working days lost (KTM, 1994: 30).

Managing increasing amounts of waste – Hungary generated some 123 million tonnes in 1995 – is also a serious problem (EHAP, 1997). Waste disposal is thus another critical area, characterised by an estimated 2,000 illegal dumping sites, an almost exclusive reliance on landfills for municipal solid waste, and very limited facilities for hazardous waste treatment or disposal. It is likely that progressive economic recovery combined with the adoption of Western-style consumption patterns and packaging habits will soon lead over-burdened waste disposal systems to overflow. The disposal of high-level radioactive waste from Hungary's nuclear plant at Paks is a special case and, following a review by the plant's management of the available options (reprocessing abroad, temporary storage or permanent disposal in Hungary) in the early 1990s, it was decided to construct a temporary storage facility (Vári, 1996). As for nature protection, approximately 6.2% of the national territory may be classified as "major protected area" under the IUCN definition.

As regards the production of so-called "greenhouse gases", the greatest part of CO_2 emissions in Hungary (almost 97%) is generated by fuel combustion. In 1986 emissions levels were about 65 thousand Gg/year, but by 1994 had dropped to 57 thousand Gg; only to be followed by a slight increase that year. As far as the sectoral structure of CO_2 emissions is concerned, almost half of total emissions stems from transformation processes in the case of fuel combustion. The share of the residential sector in overall CO_2 emissions decreased from 24% to 20% over this period, mainly due to implementation of the country-wide household natural gas programme, while the share of emissions from transport increased, reaching 13% in 1994 (CSD, 1998).

The CH_4, N_2O, CO, NO_x emissions from several stationary sources show a substantial shift from 1991 to 1995 – during which period emissions from residential sources decreased by 25% to 60%. At the same time, emissions from commercial sources significantly increased. Thus in the case of CO, commercial-origin emissions doubled while total transport-related, non-CO_2 emissions show a certain decrease of 2% to 8% during this period (CSD, 1998).

Looking at energy intensity per capita reveals Hungarian energy consumption (103 GJ/person) to be some 30% lower than that of the average European OECD country (138 GJ/person). The high figure for energy consumption per GDP (17%, while the OECD European average is 7%) thus reflects the relatively low energy efficiency of the productive sectors of the Hungarian economy even after elimination of substantial segments of energy intensive heavy industry (CSD, 1998).

2 Development of Environmental Policy

2.1 Development

In line with the experience of many European countries, environmental policy in Hungary only gained recognition as a distinct policy field in the 1970s. Originally an offshoot of public health concerns, falling under the National Institute for Public Health established in the mid-1920s, the evolution of environmental policy may be traced through several key phases (KTM 1995; Emmott, 1996; Lehoczki, 1993).

During the first phase (mid-1970s to mid-1980s), awareness of the extent of environmental degradation and its impact on public health, which had initially been confined to expert circles (natural scientists, physicians), began to spread throughout society – often as a result of the unofficial dissemination of information by concerned experts themselves. Public pressure remained low, however, due to the severe constraints imposed by the political system then in place, while the main impetus for the creation of a distinct body of environmental law came from the scientific community and the increasing political salience of environmental policy issues on the international agenda following the 1972 UN Conference on the Human Environment at Stockholm.

The passage of the 1976 Act on the Protection of the Human Environment and the establishment in 1977 of the National Agency for Environmental Protection and Nature Conservation were the results of such efforts. Act n° II on the Protection of the Human Environment of 1976 contained important provisions such as: the prohibition of emissions over permitted levels, the classification of national territory in terms of air quality protection, the establishment of enforcement tools in the form of restrictions on certain polluting activities and the possibility of levying fines. A number of sector-specific pieces of secondary legislation (government decrees, orders) on protection of air quality from stationary and mobile sources, prevention of water pollution from organic wastes and management of municipal solid waste soon followed (Lehoczki, 1993; Emmott, 1996).

The second phase (mid-1980s to late-1980s) corresponded to the late socialist reform period, and witnessed the establishment of the Ministry for Environmental Protection and Water Management in December 1987 in an attempt to bring environmental administration under a single authority (REC, 1995b: 59; KTM, 1995: 7). Environmental policy, however, remained characterised by sector-specific regulation with a continuing lack of legal provisions for access to environmental information and public participation. Despite this, the regime was faced with increasingly vocal protests over the environmental health effects of local pollution sources or siting decisions. Of international importance was the conflict over the

Gabèikovo-Nagymaros dam, a project which had been jointly agreed by the sister socialist governments of the then-Czechoslovakia and Hungary in 1977. This aroused increasing public protest in Hungary from 1984 onwards and led to the foundation in the same year of the Danube Committee (later to become the renowned Duna Kör or Danube Circle) which at the time was an underground environmental movement. Public mobilisation was ostensibly over the environmental damage which would be caused by the project, yet the dam issue served many dissidents with an excellent pretext for protest against the government in power, and one relatively safe from retaliation. By the end of the 1980s, the dam had become, "a symbol of the old regime, against which all opposition forces were fighting" and it is widely acknowledged that this environmental dispute contributed to the collapse of communist power in Hungary (Galambos, 1993: 204).

The third phase (late 1980s - early 1990s) started off amidst the general euphoria of transition to democratically elected government, which was accompanied by the high hopes of many environmental activists that Hungary's future recovery could be based soundly upon the principles of sustainable development (Bowman and Hunter, 1992). Yet it ended with the effective sidelining of environmental issues within the government's agenda, in the face of severe macroeconomic crisis and its widespread social repercussions. That this would prove to be case was already clear, however, from the political campaign leading up to the 1990 general elections in which the environment remained a marginal issue (Galambos, 1993: 205). Such trends were not countered by pressure from the public who, despite the high political salience of environmental issues in the period leading up to the change in regime and the continued realities of environmental degradation, soon focused their attention on matters of more immediate concern – such as defending standards of living in the face of rising unemployment, high inflation and the erosion of social benefits.

Important legal and institutional innovations did take place during this period, however, mainly under the impetus of a relatively small group of committed ministerial and academic experts – efforts which were supported at a distance by those members of parliament elected as mainstream politicians but who had once been green activists. In the flurry of legislative activity which characterised the first post-communist legislature, many significant pieces of legislation with implications for environmental protection were approved. Foremost among these was the revised constitution of 1990 (approved by parliament in December of that year) which set out basic rights and obligations and introduced a specific "right to a healthy environment" set out in Chapter I, Article 18. The state, in its turn, was made responsible for ensuring the protection of the man-made and natural environment under Article 70/D (REC, 1994a: 198). Such developments were matched by the issue of a number of government decrees and orders which reflected a new policy style more in tune with the new domestic political system and international practice. However, this period of progressive legislation was not matched by serious efforts at implementation or enforcement, given the predominant belief that, "[o]nly after successful economic reforms will it be possible to begin implementing environmental protection measures" (Bochniarz and Kerekes, 1994: 19).

The current phase of Hungarian environmental policy development, from the early 1990s on, is characterised by a renewed commitment to environmental policy, which has arisen largely with the realisation that accelerated progress towards meeting international commitments would be needed in the run up to membership in such organisations as the OECD (which Hungary joined in 1996) and the EU (with whom accession negotiations have been underway since 1998).

2.2 Approximation to EU Environmental Policy

One of the pre-conditions for EU membership is successful integration of the *acquis communautaire,* including that on the environment, into the domestic legislation of each candidate country. The process of EU legislative approximation has had a significant impact on Hungarian environmental policy of recent adoption, a pattern common to many other candidate countries of the region, and is potentially a source of policy modernisation. The results of such efforts were recognised by the European Commission, in its July 1997 *Opinion on Hungary's Application for Membership to the EU,* which concluded that, "formal compliance to the EU environmental *acquis* is well advanced" (Commission of the European Communities, 1997a). However, in its Progress Report of November 1998 the Commission noted that the pace of legal harmonisation had slowed and concluded that implementation of those parts of the environmental *acquis* which, "require high levels of investment...and considerable administrative efforts...can be achieved in the long term" (Commission of the European Communities, 1998).

2.3 Principal Regulations and Policy Instruments

The passage of Hungary's new environmental framework law in 1995 (Act LIII on "General Rules of Environmental Protection"), following several years of drafting and debate, is a case in point. The law reflects many elements of EU environmental legislation and basic principles as a result of the close attention to the requirements of approximation. It provides an overall framework for Hungarian environmental policy, setting out several key principles including the polluter pays principle, rights of access to information on the environment, and participation in environmental decision-making (Government of Hungary, 1995).

Legislative approximation to the EU has also marked recent sectoral legislation (notably on waste), horizontal measures (such as environmental impact assessment and access to environmental information) and other bills currently being elaborated.

In 1995, a new law on water resources management was passed (Act LVII on "Water Resources Management') which, together with Act LIII of the same year, provides for the protection of both the quantity and quality of water resources.

New regulations on hazardous waste management were passed in 1996 (Government Decree No. 102), which define the rules for the registration, collection, storage, transport, and management of hazardous wastes. The import, export, and transit of hazardous wastes are regulated jointly by Government Decree 102/1996 and Government Decree 101/1996, where the latter is the adoption of the Basle Treaty.

Several important horizontal environmental policy measures have also been adopted with a view to achieving legislative approximation with the EU. The most important is the Decree on Environmental Impact Assessment (Government Decree n° 152) of 1995, which gave a more operational basis to the basic principles set out in the 1995 environmental framework law (Act LIII, articles 67-72), and constitutes one of the main sources of rights to public participation and access to information in environmental matters currently available in Hungary.

Other horizontal measures were also introduced by the 1995 environmental framework law such as the requirement of environmental audits (Act LIII, articles 73-76). While the potential value of integrated pollution prevention and control (IPPC) as an integrated policy tool for the future has been recognised by the Hungarian Ministry of Environment and Regional Policy, it seems unlikely to be introduced in the short term (KTM, 1995: 27; Emmott, 1996).

Recent legislation of direct relevance to the environment includes: Act LVIII. (1997) on Economic Advertisement Activities, Act I (1997) on Nuclear Safety (the second part of the Atomic Energy Act), Act CLV (1997) on Consumer Protection, Act CLIV (1997) on Public Health, Act XXVII (1998) on Gene Technology, and Act XXVIII. (1998) on Animal Protection. The use of economic instruments in pursuing environmental policy objectives was given further support by the passage of Act LXIX (1998) on Environmental (Product) Fees. Yet several important gaps in Hungary's legislative framework for the environment remain: most conspicuously on air pollution, wastewater treatment and waste management. The absence of secondary legislation setting out procedures, resources and sanctions has also hampered the effective implementation of several existing laws.

An important framework for the future development of environmental policy is provided by National Environmental Programme (1997-2002), which was adopted by the National Assembly on 16 September 1997. It offers a strategic overview of the state of the environment, the main environmental policy objectives and key areas for implementation. In its preamble, the programme recognises that implementation will require the concerted efforts of all ministries and sectors of society. The projected costs of implementing this programme are estimated to rise from 1.0 to 1.7% of GDP over the period of reference (KM, 1998). The 1998 Environmental Action Plan was based on the National Environmental Programme and incorporated into the National Budget.

The principal environmental policy instruments currently in use in Hungary are those of traditional command-and-control regulation accompanied by fines for non-compliance. Fines are imposed for: point source air pollution, sewage, waste water, hazardous waste, illegal waste imports, noise pollution and vibration, nature conservation, land protection. Although in theory fines serve as a tool for enforcement, the levels at which they are set are generally too low to serve as a deterrent for non-compliance, and their main practical purpose appears to be to generate revenues for environmental expenditures (REC, 1994d: 68).

However, some types of economic instruments have been introduced, and existing ones confirmed by the 1995 Act on the Environment, which allows for the imposition of charges for the use of the environment through environmental load charges, utilisation contributions, product charges and deposits (article 59(1)). User charges have been imposed for water abstraction and for sewage systems, product charges have been introduced for fuel (in 1992) and for car tyres, while a deposit system is in force for glass bottles (REC, 1994d: 68).

The judicial system has played a very limited role in shaping environmental policy in Hungary. Indeed, the administrative law system, "has in practice been growing hostile toward environmental NGOs, and ... tends to interpret the scope of environmental regulations in the narrowest sense" (REC, 1998, p. 236). Access to the courts by environmental NGOs has been facilitated by non-profit, environmental advocacy groups (such as the Environmental Management and Law Association based in Budapest) which provide legal support to cases brought by citizens and their representative associations.

3 Evaluation of Policy

Before the 1990s, environmental policy in Hungary was reactive with respect to environmental threats, closed to public input, based on command-and-control measures, and supported by a largely meaningless system of fines, which were collected from state owned enterprises (SOEs) by the state itself. The 1976 law on environmental protection, "is typical of such laws under the former regime, giving lip-service to strict environmental standards, while providing no mechanism for administration or enforcement" (REC, 1994a: 211). Among the factors which reduced the effectiveness of environmental policy were the imperatives of economic development under the command economy, the limited influence enjoyed by proponents of environmental protection within party circles, low public awareness, and limited capacity for "voice". The body of law developed under communism thus served, "more as a propaganda tool ... than as an effective instrument of environmental policy" (Bochniarz and Kerekes, 1994: 12).

Environmental policy-making under transition has suffered from a different set of constraints, foremost among which are the low political salience of environmental issues in the face of pressing socio-economic concerns and drastic budget cuts. Although more stable than many of its Central European neighbours, changes in government have also taken their toll, most notably by delaying the adoption of the environmental framework bill, debate on which was suspended with the elections of May 1994, the same bill being resubmitted by the new government and finally passed by the National Assembly in 1995. (Lehoczki, 1993).

Notwithstanding the introduction of concepts such as sustainable development and ecological modernisation in Hungarian environmental policy documents, the predominant policy paradigm remains one based firmly on a technological and end-of-pipe approach.

Policy integration remains elusive both within the sphere of environmental policy itself, which remains predominantly sectoral (with specific regulations on soil, water, air), and in other policy areas. Thus there has been only limited success in incorporating environmental protection in industrial, agricultural, and transport policy (as seen in the allocation of large infrastructure investments favouring road over rail, and the erosion of the previously dominant position of public transport systems through rising levels of private car ownership leading to growing urban congestion).

Yet the single main obstacle to environmental policy in Hungary today remains an abiding predilection for legislative drafting at the expense of its implementation, a characteristic of the previous regime. The sheer proliferation of new environmental laws has, of course, contributed greatly to problems of implementation, especially given frequent overlap with existing legislation, a lack of implementing legislation or regulations, and the absence, in many cases, of adequate training for those responsible for enforcing the law.

The emphasis on achieving policy convergence with the EU generated by the drive for future membership has further exacerbated this tendency. The consequence of such an approach has been the near-exclusive focus of attention, resources and expertise on the challenge of legislative approximation with the EU to the detriment of other equally urgent tasks – such as ensuring the effective implementation of the avalanche of new laws and decrees (whether of EU-origin or not) through specific institution-building measures (such as re-training of inspection officers) or the creation of new, purpose-built institutions (such as environmental agencies).

To date, Hungary's progress in approximation to EU environmental policy is impressive but must be balanced against its poor implementation record, as recognised by the EU Commission in the July 1997 Opinion on Hungary's application for membership, and reiterated in the Progress Report of November 1998:

> Full transposition of the environmental *acquis* can be achieved in Hungary in the medium term. However effective compliance with a number of pieces of legislation requiring a sustained high level of investment and considerable administrative effort (e.g. urban waste water treatment, drinking water, aspects of waste management and air pollution legislation) could be achieved only in the long to very long term. Substantive compliance will depend on implementing changes in the structure of the enforcement system and its instruments. (Commission of the European Communities, 1997a)

Today, the predominant policy style in Hungary is characterised by a reliance on command and control with only a marginal role for economic incentives, while voluntary measures (e.g. covenants, self-regulation) by industry are largely absent.

4 Capacity of Proponents of Environmental Policy

4.1 Public Institutions

The current institutional framework for environmental policy in Hungary includes, at the national level, a parliamentary committee, a council, a ministry, an environmental fund, an inspectorate and several research institutes, while at the subnational level it comprises both regional inspectorates and local governments.

National level institutions include the Committee on the Environment of the Hungarian National Assembly (composed of MPs from each parliamentary group), which holds expert hearings in the course of law drafting and presents legislation to the plenary session. Administrative support to the standing committees of parliament is very limited given their reduced budgets, with at most 3 or 4 persons to administer them and only 1 or 2 advisors. Although some funding is available from the State budget to hire external experts, the level of payment is such as to be merely symbolic.

The Hungarian National Council on the Environment, established in 1996, is a statutory advisory body to the government and its members are delegated by three social groups in equal proportions: environmental NGOs, professional and business associations, and the scientific community. The council provides opinions on acts and decrees and co-operates closely with the Committee on the Environment

Table 2. Environmental expenditures in Hungary, 1993 (million US$)

Source	Amount	% of total
State budget	122.8	49
Municipalities	75.0	30
Central Environmental Protection Fund	27.7	11
Enterprises	19.0	3
Other (including the Water Fund)	7.6	7
TOTAL	252.1	100.0

Source: REC (1994c: 45)

of the National Assembly and the Ministry of Environment. Although its effective influence on environmental policy has been relatively limited, it has certainly facilitated the exchange of information between government, business, academia and NGOs (National Council on the Environment, 1998).

In 1990, the Ministry of Environment and Regional Policy (Környezetvédelmi és Területfejlesztési Minisztérium – KTM) was established. Since May 1998, the ministry has been renamed Ministry of Environment (Környezetvédelmi Minisztérium – KM) and no longer has jurisdiction over regional policy. The ministry continues to be responsible for air quality, water quality, waste management and noise protection. Under the original mandate, the minister is responsible for developing the overall national environmental policy, co-ordinating and carrying out its implementation, monitoring and evaluating the state of the environment, and maintaining an environmental information network (Haaland Matláry, 1994).

In 1986, the Central Environmental Protection Fund was established by ministerial decree and operated under regulations which prevented the use of the revenues from fines being used for purposes other than environmental protection. In 1992, the legal provisions for the fund were updated by Act N° LXXXIII/1992 (REC, 1994c: 44) and the fund represents the third largest source of financing for environmental protection after that of the state budget and municipalities (see Table 2).

There are two main observations to be made about the Hungarian Ministry of Environment. First, despite the fact that during the first 8 years of its existence the same ministry was responsible for both environmental and regional policy (with the potential to promote the integration of environment and regional development) the reality of intra-ministerial coexistence had always been one of two distinct branches pursuing their respective objectives, to which the additional responsibility for the protection of historic monuments must be added (REC, 1995b: 59). Second, while the Ministry of Environment is responsible for the environmental aspects of water (quality), the Ministry of Transport, Telecommunications and Water Management (Közlekedési, Hírközlési és Vízügyi Minisztérium – KHVM) is responsible for water management issues (quantity). Other ministries also have environmental competences (Lehoczki, 1993):

- *Ministry of Public Welfare:* responsible for the public health aspects of environmental pollution, setting national ambient standards, and for the activities of the public health and medical services;
- *Ministry of Trade and Industry:* has a co-ordinating role for hazardous waste management, supervision of hazardous waste treatment plants, and responsibility for the national geological service;
- *Ministry of Agriculture:* oversees a network of plant and soil protection stations;
- *Ministry of Culture & Education:* is responsible for basic environmental education through its setting of the national basic curriculum and accreditation of environmental courses;
- *Ministry of Interior:* oversees the activities of municipalities in the field of municipal solid and liquid waste treatment and proposes the level of government subsidies for such activities.

The Ministry of Environment relies on two national level bodies for the oversight of policy implementation. The National Environmental Inspectorate (Környezetvédelmi Felügyelõség – Kfõ) with a staff of 20, is charged with the supervision of implementation and enforcement of environmental protection regulations and acts as a second level authority for appeals against regional environmental inspectorate decisions. The Nature Conservation Office, under whose responsibility lie 3 nature conservation directorates and 5 national park authorities, is the body responsible for the enforcement of nature conservation provisions.

Among the national bodies that produce policy-relevant research are the Institute for Environmental Protection and the Hungarian Academy of Sciences as well as the environmental sciences, law and economics departments of many leading national universities.

Regional level institutions for the enforcement of environmental regulations consist of the 12 Regional Environmental Inspectorates (Környezetvédelmi Felügyelõség – Kfõ) which represent the authorities of first instance in all environmental permitting procedures and have just over 1,000 staff. Created in 1990, at the same time as the ministry and national inspectorate, the regional inspectorates also bear the legacy of the historically dominant water management perspective, in that the territorial borders of the KFs have remained those of the regional water authorities, corresponding to river basin catchment areas rather than administrative boundaries, such as those of the counties (REC, 1996: 54).

The sphere of action of the regional environmental inspectorates (KFs) is wide and covers air quality, water quality, hazardous waste, noise and vibration pollution, protection against radiation as well as law enforcement activities within their jurisdictions. Tasks include (KTM, 1995: 14; Lehoczki, 1993) sampling and laboratory analysis; data-collection, monitoring and forecasting; collection, analysis and provision of information on the state of the environment; evaluation of environmental permit applications and organisation of EIA procedures; collection of environmental fines; preparation of proposals for environmental investments.

The regional environmental inspectorates face significant resource constraints in accomplishing their tasks. General cutbacks in the public sector have meant that an increasing proportion of their budget goes to cover the costs of personnel (salaries and social security payments), which may absorb up to 70% of the total. This clearly has consequences for spending on other items such as capital investments, equipment and the maintenance of existing stock.

Local level institutions with environmental competencies consist of the municipalities which, as the basic unit of local government in Hungary, are responsible for the issue of land-use permits, have the authority to set stricter local ambient air quality standards, and receive a proportion of the fines levied within their territory for violations of environmental regulations (Government of Hungary, 1994). Under Act LXV of 1990 on Local Governments (article 8(1)) the following are some of the environmentally-related local government responsibilities: settlement planning; protection of the built and of the natural environment; water resources planning and drainage (of rain water); sewerage systems; public sanitation. The level of investment needed to fulfil such obligations is currently beyond the reach of the majority of Hungary's over 3,000 villages and towns. Although local governments are the first point of call for citizens concerned with local environmental problems, they are also responsive to local electorate's demands that they deliver on promises of economic development. Attempts to satisfy such conflicting agendas often gives rise to ambivalent or contradictory positions and contributes to uncertainty about the implementation of environmental legislation.

4.2 Environmental Organisations

Environmental NGOs, public advocacy organisations and citizens' associations are all well-represented in contemporary Hungary. A part of today's national environmental movement has its roots in the mid-1980s, when activists took advantage of Hungary's "soft dictatorship" to organise action in favour of the environment despite official restrictions on public associations (Galambos, 1993: 203). The political leadership responded to such initiatives by organising "official" (i.e., top-down) environmental organisations (REC, 1997: 47). That such attempts to control the "grass-roots" were destined to fail was illustrated by the first mass demonstration (attended by some 30,000 people) against the Gabèikovo-Nagymaros dam which took place in late 1988 (Galambos, 1993: 182).

Since 1990, with the introduction of the 5 cardinal "freedoms" of expression, assembly, association, petition, and access to information, the number of environmental NGOs in Hungary has rocketed. Estimates of the total numbers of environmental NGOs active today range from 700 to 1,000, while precise numbers are difficult to obtain despite the fact that NGOs are required to register through the court system (REC, 1997).

At the same time, general public awareness remains low. The 1992 Health of the Planet Poll conducted by Gallup International revealed that less than 0.5% of respondents in Hungary ranked environmental problems as among the very important issues – placing Hungary last among the 22 nations covered.

Rights of access to information, as an important precondition for effective public participation (PP) in the implementation of environmental policy, have also been recognised and enshrined in the 1995 Act on Environmental Protection (Article 12(1)). Such rights have been given greater substance by the provision of certain "tools" for their use. They include the regular publication of state of the environment reports by the Ministry for Environment and Regional Policy and the services provided by its Public Information Office (established in June 1997):

published and on-line information on environmental indicators, data, current and proposed legislation as well as a toll-free “Green Telephone” handling inquiries from the public.

Despite the opportunities represented by the transition to democracy, most NGOs in Hungary (in keeping with the rest of the region) are struggling to cope with financial instability as they seek to define their profiles and enhance their capacities for action. In Hungary, as in all Central and Eastern European countries (CEECs), the potential for self-financing through membership dues, fund-raising and donations is severely limited by current economic conditions. As a recent regional survey of NGOs conducted by the REC in mid-1997 has found, one of the most commonly-shared characteristics of this highly diverse “population” was unstable or poor financial conditions – and Hungarian NGOs proved to be no exception (REC, 1997: 48).

As regards the tax regime for environmental NGOs, the general rules for non-profit organisations apply. According to recent regulations, only NGOs that can prove they are “especially serving the public interest” can apply for special exemptions. Most NGOs find the administrative burden of this process greater than the benefits conferred by this special status.

In 1998, a total of eight “green” political parties (i.e., parties whose name included the word “green”) were officially registered. However, the majority of these parties appear to be inactive. Only three of them – the Hungarian Green Party (with approximately 1,000 members), the Green Alternative (206 members), and the Hungarian Social Green Party (number of members not published) – even took part in the general elections of May 1998. None of them were able to muster the minimum 5% of votes cast necessary to win seats in parliament.

4.3 Economic Proponents of Environmental Protection

According to KTM estimates, there are currently some 1,500 environmental businesses active in Hungary, most established after 1990. However, in the absence of a precise definition of the term “environmental” as used in the private business context, it is difficult to establish the specific character of all firms in this category. A recent regional survey has shown that compared with their Central European counterparts, Hungarian environmental businesses earn more from technical services than from the sale of environmental technologies and products. What is clear is that the Hungarian market for environmental goods and services (estimated to be worth approximately $380 million in 1995), currently smaller than that of either Poland or the Czech Republic, will continue to expand as economic growth continues and environmental regulations of EU-origin are applied (REC, 1995a).

While on average multinational firms use stricter environmental standards, essentially those of their markets of origin, their impact as exemplars for domestic firms is likely to be limited unless imposed through subcontracting relations.

4.4 International Organisations

Although significant in historical terms, the amount of funding for the environment provided by international organisations falls far short of the estimated cost of environmental damages in the Central European region as a whole, which range from 8% to 15% of GDP (REC, 1995a: 54). Hungary is no exception in this regard and although most of the major international organisations with an "environmental mandate" are active in the country, their cumulative impact arguably lies more in their transfer of policy exemplars, information on best practice, and know-how rather than of funds *per se*.

The World Bank, the European Bank for Reconstruction and Development (EBRD), and the UN Development Programme (UNDP) are all present in Hungary with technical assistance and financing projects on a range of issues (from strengthening the management of the Lake Balaton catchment area to the implementation of local environmental action plans) and with varying scope (from local to multi-country initiatives, such as that regarding the Danube River Basin). Such multilateral initiatives are also flanked by numerous bilateral environmental programmes, notably those financed by the US, the Netherlands, and Canada. Finally, the participation of Hungary in European and international environmental data networks, such as those of the European Environment Agency (EEA) and the UN Environmental Programme (UNEP), are also significant in terms of raising data collection and processing standards to international levels.

One point at which multilateral, bilateral and national efforts to strengthen environmental capacity in Hungary meet is in the Regional Environmental Center for Central and Eastern Europe (REC), itself an international organisation based in Szentendre, Hungary and serving 15 countries throughout Central and Eastern Europe. Since its establishment in 1990, the REC's original remit of providing grants to NGOs, has been expanded to include the promotion of co-operation and exchange amongst governments, businesses, and NGOs within the countries of the region. The activities of the REC fall into 4 main programme areas: local and earmarked grants, information exchange, regional REC initiatives, fellowships and internships. Total contributions to the REC for the period 1990-1996 amounted to just under 21.5 million ECU. National programmes for Hungary are carried out under the auspices of the REC Local Office, based in Budapest.

The European Union (EU) is a multilateral actor of particular importance in terms of its impact on environmental policy-making in Hungary. This is so in the light of the country's 1994 application for EU membership under which it assumed the obligation to harmonise its national legislation with the body of EU law (the *acquis communautaire),* including that on the environment. Legislative approximation has gone on apace since then, and significant progress was recognised in the European Commission's Opinion on Hungary's application for membership, published in July 1997, which estimated that full transposition of the environmental *acquis* might be achieved "in the medium term" should current levels of commitment be maintained (Commission of the European Communities, 1997a). At the same time, the Commission's Opinion warns that substantive compliance (i.e., full implementation and enforcement) in Hungary will depend largely upon significant improvements in administrative structures and increased levels of investment in environmental infrastructure.

To aid such efforts, the EU provides technical assistance, training and investment for environmental policy-making and implementation. It does so through the PHARE Programme – the largest single grant assistance programme to the region – approximately 11% of whose budget during the period 1990–1995 was destined to single country or multi-country programmes for the environment and nuclear safety (PHARE, 1996: 3 and 32). Hungary's share of such environmental funding during this period amounted to 15.4%, comparing favourably with the highest single beneficiary, Poland (22.5%), and even more so with the Czech Republic and Slovakia (together amounting to 7.4%) (PHARE, 1996: 41).

Support to countries that have applied for EU membership will be further reinforced by the Accession Partnerships, introduced in 1997, which focus on strengthening public administration and institutions ("institution building") and increasing investment in business and infrastructure, respectively accounting for 30% and 70% of the PHARE budget (Commission of the European Communities, 1997b). Overall funding levels will also rise, bringing total pre-accession assistance to ECU 21 billion at constant 1997 prices for the period 2000-2006. Of this sum approximately a third, or ECU 7 billion, will be dedicated to the environment – roughly ECU 1 billion per year, a sum which is over twice the entire PHARE environment budget (ECU 483.4 million) for the period 1990-1995 (Commission of the European Communities, 1997c).

4.5 Interaction between Proponents

Today's policy context differs considerably from that of pre-transition Hungary, in which the main lines of communication on environmental issues were restricted to the state environmental bureaucracy, state-owned enterprises and the scientific community. In the absence of conditions for constructive criticism, let alone open conflict, such relations were clearly consensual – if of a constrained nature. Since 1990, many new actors, such as non-governmental organisations and political parties, have emerged to take their place at the table – together with several previously "dormant" actors, such as local governments.

Access to a free press and media coverage of environmental issues has undoubtedly contributed to promoting public awareness of the different issues, interests and positions at stake. Yet the limited number of qualified environmental journalists coupled with a distinct focus on the more sensationalist aspects of environmental disasters or controversies makes consistently high quality reporting rare.

Such developments have given "voice" to a wider range of environmental interests (those of citizens and environmental businesses, for example) while simultaneously exposing them to direct competition with many interests of more immediate and widespread concern (such as the pursuit of economic development and the reduction of unemployment) in the course of decision-making at both the national and local levels. As a result, policy-making and implementation have become far more open and yet far more unpredictable with regard to outcomes.

Government and NGOs: At the domestic level, interaction between government and non-governmental environmental organisations merits particular attention, given the relatively recent appearance of the latter. Two particular arenas of interaction are singled out: policy-making and funding.

Of only recent introduction, the procedure for public participation (PP) in legislative and regulatory drafting approved by ministerial order in September 1996 is an important innovation. It requires the Ministry of Environment and Regional Policy (KTM) to publish the list of bills and draft decrees on which it is working twice a year, while allowing environmental NGOs (who have registered their interest in being involved) to make suggestions for modifications to the drafts within a specified period.

The definition of "NGO" adopted by the KTM for this consultative exercise is of note, including as it does *all* forms of non-governmental groups – including specific interest representation (e.g., industrial associations). As of June 1997, a total of 29 NGOs were registered, of which only 12 were environmental NGOs. This low response rate (compared with the total "population" of Hungarian environmental NGOs) may in part be attributed to the nature of the draft legislation then under discussion, with interest expected to rise in time as this practice becomes more established. However, two important issues that may affect the implementation of this progressive measure in public participation have yet to be resolved: first, the stage of the drafting process at which a draft may be released for public comment; and second, guarantees that comments received will actually be taken into account.

A second element in Hungarian NGOs interaction with national government regards funding. A sum drawn from the Central Environmental Fund (CEF) is allocated to NGOs through yearly competitive tenders on the basis of project proposals presented. The total amount has increased sharply over the last few years, in 1997 reaching HUF 155 million with 795 applications for funding, of which roughly half were successful.

Given that the vast majority of Hungarian NGOs operate at the local level, it is not surprising that most report a level of co-operation with local governments that is several times higher than with national level authorities. Such relations may take on a variety of forms, ranging from mutual support to open conflict depending largely on the issue at stake and how it is framed. Most often, it is where the weight of public opinion falls that tips the balance, given local government concern to live up to the expectations of its electorate.

Intra-NGO relations: Although their diverse size, outlooks and specific interests are obstacles to effective co-operation, opportunities for networking amongst Hungarian environmental NGOs do exist. Zöld Pók (Green Spider) computer network and a recently established annual meeting of environmental and nature conservation organisations (REC, 1997: 48) are two such venues. Citizens' associations, characterised by their single-issue, local focus, are easily the most numerous of Hungarian non-environmental organisations. Although they often join forces with national NGOs in the pursuit of their goals, such alliances are not without tension, as the concrete goals of primary importance to local citizens' associations (e.g., mitigation measures and compensation for environmental damage) may become just one of a set of wider goals pursued by NGOs with a broader geographical base or long-term perspective.

National – international relations: Preparation for EU membership and the receipt of substantial amounts of technical and financial assistance for the environment add a further dimension to the issue of interaction between environmental proponents in Hungary. The first relationship to be affected is arguably that between the national government and the international donor organisation, which may be characterised by different "degrees of freedom" in terms of the ability of Hungary to shape the form, goals and terms under which foreign assistance is offered. Relations with the EU are a very specific case in point, given that the environmental assistance provided is part of the larger undertaking of preparing for accession. As such, EU-Hungary environmental relations tend to focus on only a specific sub-set of policy issues, namely those covered by the EU's environmental *acquis,* rather than the potentially wider range of problems to be addressed.

Prior to transition, the main axis of interaction between proponents of environmental policy was that between the government and the scientific community. Since 1990, the role of the latter as a privileged interlocutor of government has been flanked by a burgeoning community of environmental NGOs offering their perspectives and sources of expertise. Of more limited impact, at present, are green businesses, owing to their small numbers and limited capacity for self-organisation.

5 Main Restrictions

The distinction between "apparent capacity" and "effective capacity" is crucial to assessing the case of Hungary, where the contradictions of consolidation are ever-present and where environmental policy is no exception. As a result, the swift adoption of significant amounts of highly developed environmental legislation (including a large proportion of the EU's environmental *acquis)* collides with the limited capacity of environmental public administrations to implement such provisions effectively and of regulated entities to comply.

This difference between "formal and substantive institutionalisation" (Jänicke, 1997: 10) thus argues for a degree of caution when moving from the observation that certain institutional elements do exist (e.g., environmental legislation or inspectorate) to the assumption that they will work as expected. Of more concern is how they actually work in practice (e.g., existence of secondary legislation, number of inspections carried out, how violations are recorded and fines collected).

In terms of the structural framework conditions under which environmental policy is formulated and applied in Hungary, the following features may be highlighted:

- *Cognitive-informational:* although the restrictions that characterised the communist regime have given way to far greater freedom of information about the environment, there are still severe limits to the attention given to environmental issues by policy-makers and public alike, given overriding preoccupation with securing individual and social subsistence. This conflict is also reflected in trade unions' lack of interest in environmental health issues and consistent focus on defending jobs, salaries and employee rights. The use of new technologies (such as the Green Spider network) to exchange information and opinions on environmental issues is beginning to take place within the NGO community and in relations with government.
- *Political-institutional:* an indication of the participative capacity of the policy system is given by the presence of several measures for public participation in both policy-making (with provisions for NGO comments) and implementation (such as under environmental impact assessment). Yet the advantages these new opportunity structures bring have yet to be fully exploited by environmental NGOs and citizens' associations on the one hand, and recognised by en-

vironmental public administration officials on the other. In a similar fashion, integrative capacity remains limited – to date, little progress has been achieved in overcoming a long-standing sectoral approach to environmental policy or in incorporating environmental considerations into other policy fields, given the overwhelming emphasis placed on the pursuit of socio-economic development.
- *Economic-technological:* while the scientific knowledge and technological capacity to conduct environmental monitoring, introduce preventive measures and apply remedial measures exists in the majority of instances, effective action is often constrained by limited public and private financial resources. This gives rise to widespread non-implementation of environmental regulations and its tacit acceptance by public officials charged with enforcement, which ultimately undermines the rule of law. Furthermore, the economic recovery that followed the deep recession of 1990-1992 has produced few direct dividends for the environment. The commitment of public and private resources to the clean-up or prevention of environmental degradation has largely failed to keep pace with economic growth, thereby adding to the existing environmental burden (the steep and sustained growth in both car ownership and urban congestion being a case in point).

While the positive repercussions of approximation to EU environmental policy have been outlined above, mention must also be made of potential restrictions imposed by preparations for accession:

- *first,* even though EU policy advice and technical assistance activities undeniably represent an *opportunity* for Hungarian environmental policy, they must also be recognised as being a *constraint.* The narrow focus on EU legislation effectively restricts the variety of policy options on offer while, at the same time, substantially reducing the need for "policy initiative" on the part of national policy-makers, and curtailing debate among stakeholders;
- *second,* few calculations of the potential *cost* to Hungary of approximation for either the public sector (charged with its enforcement) or the private sector (which will bear the costs of compliance) have been undertaken. Given the realities of budgetary restrictions and limited capital investment for the foreseeable future, there is a strong likelihood that the implementation of newly-approximated legislation will be hampered by inadequate financial and technological resources.

6 Prospects for Future Capacity Building

In the decade since the transition to democratic governance and policy-making, Hungary has made great progress in strengthening the institutional foundations for the pursuit of environmental policy under conditions of a market economy and public accountability. Furthermore, the emergence of a network of actors representing a range of environmental interests (public administration, environmental businesses, citizens associations) promises ever greater use of new and existing opportunities for participation in the future.

If the "window of opportunity" for the introduction of sustainable development as the basis upon which to rebuild the national economy existed in the early

1990s, it is no longer among the policy options cited. Despite the best efforts of such bodies as the Commission on Sustainable Development, established in 1993 to disseminate information on Agenda 21 and the implications of Hungary's signature to international treaties on the environment, public awareness and commitment to treat economic and environmental goals as equal priorities remains very weak.

While the sources of capacity building are numerous, two main categories are highlighted here:

Interaction at the international level

- Participation in international environmental policy arenas itself constitutes an important source of policy learning for government, as do the preparatory steps necessary to participate in such events. Examples include the 1992 Rio Conference (for which Hungary prepared a "State of the Environment Report") and Hungary's participation in successive "Environment for Europe" ministerial meetings (with the drawing up of a National Environmental Action Programme). The premium placed on active participation in the international arena and the active pursuit of membership of international organisations (such as the OECD, NATO and the EU) has had significant repercussions on national policy development – in the environmental field no less than elsewhere. Whether policy innovations introduced by ratification of international environmental treaties or adoption of the EU's environmental *acquis* are consistently and fully implemented is another issue.
- Although subject to the limits outlined above, preparation for EU accession is also a key impetus for environmental policy modernisation through both legislative approximation and the receipt of substantial financial resources (destined to grow significantly under the proposed financing package for the period 2000-2006).
- Several Hungarian NGOs (especially the larger ones) are involved in international networks. The benefits of networking include: better access to information (on technical, legal and funding issues); involvement in joint actions (such as lobbying international organisations and the European Parliament); effective identification partners for joint international projects (e.g., EU Phare).

Institution building at the national level

- Programmes of public administration reform and efforts to improve policy implementation, whether of national or EU-origin, raise administrative capacity in all fields including that of environmental policy.
- Representatives of the public interest, such as environmental NGOs and citizens' associations, have demonstrated their willingness and ability to make a constructive contribution to environmental decision-making. This is supported by greater public awareness of environmental issues which proves, however, slow to mobilise given the conflicting interests and competing priorities which characterise contemporary Hungarian society.
- Improvements in environmental management in enterprises have been forthcoming and Hungary had, by mid-1998, reached the "basic capacity level" for

cleaner production as defined by the Task Force for the Environmental Action Programme for Central and Eastern Europe (EAP, 1998).

Capacity building for environmental policy in Hungary may best be described as the outcome of a set of mutual learning curves in which all of the main actors – from environmental inspectors, to NGOs and businesses – are involved. External support for improving the capacity of NGOs, businesses and government to address environmental problems jointly and proactively therefore continues to be of great importance. The nature of such learning over time, as witnessed by the case of Hungary, supports the proposition that capacity-building processes may be more important than their outcomes (Weidner, 1997). On the basis of progress to date, there would thus appear to be good prospects for the future despite the considerable distance that still has to be covered.

References

Bochniarz Z. and S. Kerekes, 1994. Deficiencies in the Existing System of Environmental Protection in Hungary, in: Z. Bochniarz., R. Bolan, S. Kerekes, and J. Kindler (eds.) *Designing Institutions for Sustainable Development in Hungary: Agenda for the Future*. Minneapolis-Budapest.

Bowman M. and D. Hunter, 1992. Environmental Reforms in Post-Communist Central Europe: from High Hopes to Hard Reality, *Michigan Journal of International Law*, 13 (4) 921-980, Summer.

Commission of the European Communities, 1997a. *Agenda 2000: Commission Opinions on Applications for Membership of the European Union*, Brussels: European Commission, 15 July.

Commission of the European Communities, 1997b. *Agenda 2000: Reinforcing the Pre-Accession Strategy*, Vol. II, DOC 97/7, Brussels: European Commission, 15 July.

Commission of the European Communities, 1997c. *Enlarging the Environment: A Newsletter from the European Commission on Environmental Approximation*, issue n° 5, Brussels: European Commission (DG XI), July.

Commission of the European Communities, 1998. *Regular Report from the Commission on Hungary's Progress towards Accession*, Brussels: European Commission, 4 November.

CSD, 1998. *Hungary: Inventory, Stabilisation and Scenarios of Greenhouse Gas Emissions and Removals*, Budapest: Hungarian Commission on Sustainable Development.

EHAP, 1997. *Hungarian Environmental Health Action Programme*, Budapest.

Emmott N., 1996. Requirements for the Implementation of Integrated Pollution Control in the Republic of Hungary, LLM essay, School of Oriental and African Studies, University of London.

Galambos J., 1993. An International Environmental Conflict on the Danube: The Gabèikovo-Nagymaros Dams, in: Vári A. and P. Tamas (eds.). *Environment and Democratic Transition: Policy and Politics in Central and Eastern Europe*. Dordrecht/Boston/London: Kluwer Academic Publishers.

Government of Hungary, 1994. Act LXV of 1990 On Local Governments, *Hungarian Rules of Law in Force*, no. V./24, Budapest, 15 December.

Government of Hungary, 1995. Act LIII on General Rules of Environmental Protection, *Hungarian Rules of Law in Force*, no. VI./17, Budapest, 1 September.

Haaland Matláry J., 1994. The European Union and the Visegrád Countries: The Case of Energy and Environmental Policies in Hungary. *Regional Politics and Policy, Special Issue: Protecting the Periphery: Environmental Policy in Peripheral Regions of the European Union*, 4 (1) 136-152.

Jänicke M., 1997. The Political System's Capacity for Environmental Policy, in: M. Jänicke and H. Weidner (eds.) *National Environmental Policies: A Comparative Study of Capacity-Building*. Berlin: Springer.

KM, 1998. *National Environmental Programme 1997-2002*. Budapest: Ministry of Environment (KM).

KTM, 1994. *Environmental Indicators of Hungary*. Budapest: Ministry of Environment and Regional Policy (KTM).

KTM, 1995. *Environmental Policy and Institutions of the Republic of Hungary: National Report for the OECD*. Budapest: Ministry of Environment and Regional Policy (KTM), March.

Lehoczki Z., 1993. *Environmental Administration in Hungary*. (Mimeo.). Budapest, January.

National Council on the Environment (1998). *Evaluation of the two-year activity of the Hungarian National Council on the Environment: April 1996 – February 1998*. (Mimeo.).

PHARE, 1996. *The Phare Programme Annual Report 1995*, COM (96) 360 final, Brussels: European Commission, 23 July.

REC, 1994a. *Manual on Public Participation in Environmental Decisionmaking: Current Practice and Future Possibilities in Central and Eastern Europe*. Budapest: Regional Environmental Center for Central and Eastern Europe.

REC, 1994b. *Strategic Environmental Issues in Central and Eastern Europe: Vol. 2 – Environmental Needs Assessment in Ten Countries*. Budapest: Regional Environmental Center for Central and Eastern Europe.

REC, 1994c. *National Environmental Protection Funds in Central and Eastern Europe*. Budapest: Regional Environmental Center for Central and Eastern Europe, November.

REC, 1994d. *Use of Economic Instruments in Environmental Policy in Central and Eastern Europe*. Budapest: Regional Environmental Center for Central and Eastern Europe, December.

REC, 1995a. *Emerging Environmental Market Survey – Vol. 1*. Budapest: Regional Environmental Center for Central and Eastern Europe.

REC, 1995b. *Government and Environment: A Directory of Governmental Organizations with Environmental Responsibilities for Central and Eastern Europe*. Budapest: Regional Environmental Center for Central and Eastern Europe, September.

REC, 1996. *Approximation of European Union Environmental Legislation: Case Studies of Bulgaria, Czech Republic, Estonia, Hungary, Latvia, Lithuania, Poland, Romania, Slovak Republic and Slovenia*. Budapest: Regional Environmental Center for Central and Eastern Europe, January.

REC, 1997. *Problems, Progress and Possibilities: A Needs Assessment of Environmental NGOs in Central and Eastern Europe*. Budapest: Regional Environmental Center for Central and Eastern Europe, April.

REC, 1998. *Doors to Democracy: Current Trends and Practices in Public Participation in Environmental Decisionmaking in Central and Eastern Europe*. Budapest: Regional Environmental Center for Central and Eastern Europe.

Stanners D. and P. Bourdeau (eds.), 1995. *Europe's Environment: The Dobrís Assessment*. Copenhagen: European Environment Agency (EEA).

Vári A., 1996. State-of-the-Art in Public Participation in Hungary: The Case of Siting a Spent Fuel Storage Facility, *International Conference on Environmental Pollution (ICEP.3) – The Proceedings: Volume 2*. London: European Centre for Pollution Research.

Weidner H., 1997. Implementation of EU environmental legislation – the theoretical perspective, presentation at the 1997 Spring Workshop Implementing EU Environmental Policy, Working Group on Environmental Studies, Robert Schuman Centre, European University Institute, 10-11 April, Florence.

ENVIRONMENTAL CAPACITY-BUILDING: INDIA'S DEMOCRATIC POLITICS AND ENVIRONMENTAL MANAGEMENT

Arun Agrawal and Noriko Yokozuka

The question of environmental capacity building and environmental movements in India cannot be separated from larger issues of India's democratic political framework and its recent political history. The stability and endurance of India's democratic government, it may be argued, present many observers with an intellectual puzzle: a multi-ethnic, caste-ridden, agrarian society with a rigid hierarchical structure, coupled with constitutional government, periodic elections, increasingly resilient political institutions, and freedom of expression. An ascendant regionalism, an unstable central government but still intrusive state, and international globalization pressures continue to raise doubts about the smooth functioning of Indian democracy. In the face of such threats, two aspects of the Indian nation state are perhaps the strongest guarantee of environmental rights and democratic freedoms in India. The first is a politically conscious citizenry whose literacy and awareness levels are constantly on the rise and which has proven itself capable of "throwing the rascals out" on at least four occasions in the past two decades. The second is the presence and development of a range of balancing institutional arrangements capable of engaging arbitrary actions of an intrusive state and an assertive market.

Politically conscious and increasingly participant Indian citizens have chosen to assert their demands on several occasions through mass movements that forcefully express farmers', women's, and civil rights, and in the past two decades, also the right to a safe environment and environmental products. These movements are significant because, although in part facilitated by the presence of a democratic political setup, their presence also simultaneously strengthens democracy by bringing to the fore a range of previously silent voices. In addition, the emergence of a vocal and independent media, the development of a vibrant NGO sector, and an activist judiciary have created a larger range and density of institutional avenues through which new actors can lobby for environmental protection, often by acting within the system. The presence of a vibrant community of non-governmental organizations concerned with citizen rights and subsistence has acted as a mechanism that can help focus the demands being made on the Indian polity. This paper argues that participatory and representative democratic political arrangements provide the best available option to address India's current environmental ills and secure sound protection of India's environmental future. This implies that at the grassroots levels, there should be a stronger network of participatory, semi-autonomous organizations that control local resources (pastures, forests, water bodies, irrigation facilities, village commons, or coastal fisheries), and that are effectively linked to exert political pressure at the macro level. In comparative terms, the lesson of India's efforts to protect its environment is that curbing arbitrary exercise of power in the interests of small minorities of privilege is most likely when citizens are mobilized against corrupt practices that threaten the environmental future of a nation.

1 India's Current Environmental Crisis

This paper, in assessing the capacity for environmental management in India, concedes that population and policy measures from central governments can be significant constraints on how humans live in their environments.[1] But it adopts a framework of analysis that emphasizes local capacities as the ultimate measure of the extent to which factors such as overpopulation and market pressures pose problems for the environment, and national policies and international environmental regimes can translate into actual capacities to manage natural resources (Agrawal and Yadama, 1997). Taking seriously Marx's admonition that humans make their own history, it examines those features of the Indian polity that influence the ability of the Indian population to live with its environment and to influence the Indian state to change environmental policies. As Redclift suggests, it is important to know how "human purposes can be married to environmental objectives, at both a local and international level" (Redclift, 1994: 56). Successful environmental protection cannot be achieved by the control of a single factor, but must be invented through a complex interaction of influences that involve the life-style patterns, geography, class structure and gender stratification in a given area.

Since India's first development efforts, issues of environmental degradation coupled with development have increased pressure on the state, and have challenged the state to accommodate new environmental demands while simultaneously guaranteeing economic growth. After independence in 1947, the Indian economy grew at a compounded rate of 3.6% between 1947 and 1981, but since 1981 at the much faster pace of 5% to 6%. This growth has been accompanied by a change in the structure of the economy as well so that manufacturing and industry account for a much larger proportion of the national product in 1991 compared to 1947, and the share of agriculture has fallen from a half to a third. The need to continue to grow has not abated, and therefore, the existing environmental policy making in India is best understood as a "politics of reconciliation" between the goals of development and environmental preservation.

But environmental problems are emerging as potentially one of the most pressing sources of conflict and crisis in India. This is evident from even the most cursory examination. The problems of underdevelopment, such as inequality, poverty and unemployment, seem to be never ending while environmental pressures stemming from rapid industrialization and subsistence pressures reaches critical

1 A large literature has focused on limiting population as the most important single factor in addressing environmental problems (Abernathy, 1993; Meadows, Meadows, and Ehrenfeld, 1992; Wilson, 1988). There is a similarly large set of writings that focus on government policies and institutions both as the culprits and as the potential solutions for environmental problems (Haas, Keohane, and Levy, 1993; Lipschutz and Conca, 1993). For a critique of environmental management in general, see (Sachs, 1993).

levels (Agarwal and Narain, 1997: 24). Lower levels of development reduce the room for maneuver available to address problems of environmental degradation through policy change, especially if powerful industrial or political interests see environment and development as involving tradeoffs.

India's population has increased from 336 millions in 1947 to an estimated 953 million in 1997. Although the increase has slowed down in the past few years, the four populous northern BiMaRU states (Bihar, Madhya Pradesh, Rajasthan, and Uttar Pradesh) continue to be characterized by high birth and population growth rates. In terms of urbanization, the five major cities of Bangalore, Bombay, Calcutta, Delhi, and Madras have witnessed more rapid population growth than the rural areas of smaller cities, but urbanization has also proceeded apace in other respects.[2]

Consider some basic indicators of environmental degradation and pollution in India.[3] Although the share of agriculture in the Indian economy has fallen, more than 60% of the Indian population continues to rely on agriculture, and income from land-based resources. As a result of the increasing population, each Indian today has less than 3500 sq. meters of space as compared to 9700 sq. meters in 1951. The total area of India is 328.6 million hectares of which the National Bureau of Soil Survey and Land Use Planning estimates that 187 million hectares suffers from some form of soil degradation. Soil erosion in India is 16.35 tons/ hectare by some recent estimates (Goel and Prasad, 1993). The area under water logging has increased from 6 million hectares in 1976 to 11.6 million hectares in 1994. Among the causes of waterlogging are higher levels of canal irrigation that also contributes to increased salinity of soils. As canal irrigation has increased from 8 million hectares in 1951 to 17 million hectares in 1991, large tracts of land have become waterlogged. Table 1 provides information on the estimated costs of soil degradation.

India has had one of the most professional and scientific forest services in the world that dates back to the colonial period (Fairfax and Fortmann, 1990). The Forest Department is in charge of nearly 77 million hectares of India's official forests, of which only 50% has a high density (40% or more crown cover). The total biomass on the forested lands is estimated to be around 90% of its level in

Table 1. Annual Costs of Soil Degradation in India (1 US Dollar = approximately Rs. 45.00)

Type of degradation	Cost (in millions of US Dollars)
Erosion	1,300 - 5,000
Salinity	480
Water Logging	200
Nutrient Depletion	120

Source: Based on TERI, 1998.

2 For example, the proportion of population living in cities with more than 500,000 people has grown from 55% in 1951 to 76% in 1991. Similarly, the number of cities that have more than a million people has grown to 23 in 1991 from 7 in 1951.

3 The figures on environmental degradation and pollution are taken, unless otherwise mentioned, from TERI (1998, 1999).

1947, or 4.7 cubic billion meters with an annual increment of around 88 million cubic meters. Part of the reason for the decline in forests has to do with domestic demand as the Indian population has grown by a factor of 3 over the same period. Even more important is the demand from livestock (increasing by a factor of 4), and industry (a 14-fold increase). Compared to many other countries such as Indonesia, Burma, Brazil, or Thailand, however, relatively little of the forested land has been diverted to other uses (about 7% since independence). One of the more significant converts of protection through community action, the Indian Forest Department is using the Joint Forest Management Program to build partnerships with the very people that it regarded earlier as the enemies of forests.

The statistics are somewhat more depressing where water resources are concerned. The availability of renewable fresh water fell from 6000 cubic meters per person per year in 1947 when India became independent to about 2300 cubic meters in 1997.[4] By 2017, no more than 1600 cubic meters will be the average amount of water available to a person.[5] Groundwater is overexploited so that the water table in cities such as Ahmedabad fell by 200 to 250 centimeters annually in the 1980s. The Central Groundwater Board estimates that in more than 90% of all observation wells that it monitored in the 1980s, the water table fell by more than 50 centimeters. The increasing scarcity of water lowers its quality and increases salinity, forces the time spent by households, mainly women, to collect water, and increases the expenditure necessary to raise the necessary amounts.

The 113 river basins that are another major source of water in India are becoming increasingly polluted as well. The quality of river water is monitored at 480 stations throughout the country under different programs such as MINARS (Monitoring of Indian National Aquatic Resources). The water in most of the rivers falls in use category C which means it is fit for drinking only after treatment for bacteria since the total coliform count has risen steeply since independence. The major sources of water pollution are flows from city sewage. Urban sewage flows have increased from 6 billion liters in 1947 to an estimated 30 billion liters today. Industrial waste and nutrient runoff from agriculture is now finding a place as an important source of water pollution. Finally, the industrial demand for water with increasing growth is becoming harder to meet.

Air pollution is fast emerging as a problem without easy solutions, especially in urban areas. A number of toxic waste disasters, the most prominent among them being the Bhopal disaster of 1984, focused the attention of the state on the need to create effective preventive legislation. But the everyday tragedy of air pollution is equally grim. The burden of many of the different types of air pollution, especially of domestic pollution, falls primarily on women and infants. Poverty and inadequate access to clean fuels has perpetuated the use of biomass-based fuels in rural as well as urban areas so that more than 90% of rural and 35% of urban households have high levels of unsafe indoor air pollution. But the problem of external

4 Regions where the per capita water availability is less than 1600 cubic meters annually are seen as water stressed. Although the figure of 2200 cubic meters is well above water-stress levels, there are wide regional variations, and many parts of India are already water stressed. The situation will only worsen in the years to come (TERI, 1998)

5 (Chitale, 1992; GOI, 1992). These projections have led some scholars to conclude that the effective constraint on agricultural production will be water, not land, in the years to come, especially as groundwater wells are deepened and attendant costs increase (Shah, 1993).

air pollution is far more severe in the urban areas. Indian cities suffer from high levels of major pollutants such as sulfur dioxide, nitrogen oxide, carbon monoxide, hydrocarbons, photochemical oxidants, metals, and other gases and vapors. The Central Pollution Control Board founded in 1975 launched the National Ambient Air Quality Monitoring Program in 1984 with 28 stations in 7 cities. Today, the program exists in 92 cities with more than 290 stations. An analysis of the data by TERI (Tata Energy Research Institute) for 62 cities reveals that the safe limits for particulate matter, sulfur dioxide, and nitrogen oxide were exceeded in 62, 3, and 3 cities. The air quality index established by the Board finds 20 cities in the "dangerous" and 14 in the "bad" zone.

Air pollution is a result of emissions from point sources as well as more dispersed locations. Industrial pollution in terms of particulate matter has increased enormously since independence. For example, emissions from power plants have increased nearly 50-fold, and those from major industries such as iron and steel, cement, fertilizers, and paper have increased more than 15 times. Similarly, motorized vehicles which are the largest source of non-point pollution have increased from 0.2 million at the time of independence to more than 36.3 millions in 1997.[6] Most of these vehicles are registered in the larger metropolitan cities so that Bombay, Calcutta and Delhi have 48% of all vehicles clogging the city roads, emitting noxious chemicals into the atmosphere.

These activities have a substantial economic cost. The economic loss owing to soil erosion alone is close to 11% to 26% of the agricultural output. According to a World Bank report, India loses close to USD 80 billion as a result of resource degradation and sickness and deaths owing to pollution.[7] The report attributes these losses to the familiar problems of poor sanitation, industrial effluents and chemical runoffs, lack of clean water, poor solid waste management, and air pollution (PTI, 1998). Considered at a more aggregate level, Table 2 provides an estimate of the major sources of economic losses to the country as a result of environmental pollution and degradation.

Current levels of environmental degradation and depletion are costly enough. But projection of current trends into the future is nothing short of frightening.

Table 2. Estimating Economic Losses owing to Environmental Degradation

Source	Annual Cost (in Millions of US Dollars)
Health Impact of Poor Drinking Water	2,711
Crop Production losses owing to soil losses	1,977 - 5,155
Forest wood and other non-timber losses	1,267
Air Pollution	19,666 - 94,444

Source: Based on TERI, 1998.

6 According to TERI, these figures represent all motorized vehicles. However, other sources estimate the total number of cars to reach 53 million by the year 2000 (CSE, 1998).

7 The figure of US dollars 80 billions is based on purchasing power parity exchange rates. Conventional exchange rate leads to a figure of US dollars 20 billions. For perspective, consider India's GNP in 1997 which was US Dollars 375 billions (Sirichanya, Robboy, and Mitchell, 1996).

According to TERI, if the Indian economy continues to grow at a rate of 5% (somewhat lower than current rates of growth), with the industrial sector moving much faster, the changes in the next 50 years will be staggering. Industry will account for nearly 40% of the national output and agriculture for just 5%. More than 50% of the population will be located in urban areas (compared to 25% in 1990), and city municipalities will produce 300 million tons of solid waste (compared to 48 million tons today). Emissions polluting the air will increase by 6 times unless new technologies are found and used to ensure cleaner fuels inside and outside the household. In trying to produce the necessary grains for the economy, Indian farmers will not only be faced with all the existing problems of soils, but also need to address the problem of nutrient replacement which will become far more severe. Indian forests will need much more careful management which would be harder to provide except through decentralization.

2 Environmental Issues and the Role of Democratic Politics

India's future ability to address problems related to the environment depend not only on the creation of the right information systems[8] and the adoption of better material and energy saving technologies (TERI, 1998), but also on the type of politics that emerges in relation to the environment. Politics always plays a central role in policy formation (Chatterjee, 1994). To understand environmental policy and capacity building, therefore, it is necessary to assess the forces that impact political negotiations in any context. Environmental politics in India is distinctive among developing countries precisely because of the availability of a longstanding democratic framework of decision-making that allows a variety of different interests to influence outcomes. India has already managed fifty years of competitive electoral politics, and has blended Western and modern forms of bureaucratic organization and participatory politics with indigenous practices and institutions. A number of factors lend a distinctive character to the Indian political scene, factors that also in turn impact environmental policy making and outcomes.

Owing to the diverse and socially fragmented nature of the Indian society, regional and other political parties have been imperfectly integrated into a "national party system" (Brass, 1995: 67). The presence of multiple competing interests, who are now often mobilized into electoral politics through populist, or regionally specific agendas (Bose and Jalal, 1997) implies grave difficulties in ignoring a mobilized group of people, whether they are mobilized around idioms of caste or region, or around development and environment. The high level of politicization pressures the state to accommodate groups, and forces it to open new avenues of participation. (Khator, 1991: 21). Specific groups of the Indian population are increasingly able to realize the potential of a democratic political system to represent demands and produce policies that reflect some aspects of demands that are forcefully expressed.

---

8 A number of efforts are under way to create better compliance and monitoring systems regarding environmental management and detection of pollution (Sirichanya, Robboy, and Mitchell, 1996).

This is because the normal accouterments of a democracy – elections, voting, lobbying, checks and balances through institutional arrangements – are not always available to the most disadvantaged groups in the population. Common men and women are seldom able to participate fully or equally in the decision-making processes that would help improve their socio-economic position, or safeguard their environment. As a result, popular demands are often expressed through the extra-constitutional methods of direct actions such as protest and mass mobilization. Although the ideas of Gandhi have receded into the background as India has moved further along its own developmental path, resonance of his methods are still to be found in the political practices adopted by those striving to give expression to their demands through direct actions. The existence of a democratic framework for the exercise of power makes it more difficult to brutally repress non-violent protesters, unlike the situation, for example, in Indonesia.

It can be argued that direct action is as vital a practice of participatory democracy as is voting on election day. In India, mass movements among the scheduled castes and tribes, agricultural laborers and poor peasants have increased considerably since the dominant Congress party has collapsed. The hierarchical social structure is gradually giving way to a more horizontal integration of class and caste groups. Such increasing political participation and popular protests may appear at first blush to result in greater problems of governance, but they also ultimately make Indian policymakers and leaders more accountable and the political system more representative.

Mass demonstrations, protests, strikes and other forms of mobilization are thus an important aspect of democracy, because they offer alternative means of articulating environmental issues. As Goodin argues, if we are committed to the notion of democracy at the very basic level, we must also understand the need for equal consideration of interests (Goodin, 1996). Naturally occurring objects have an objective value and we must equally consider nature's interests. Democracy is not just about "one person, one vote." If we assume that each citizen honestly reveals his/her preferences, and trust that the aggregation process (such as voting) is not being manipulated, then environmental interests can be represented in a democracy only if they are internalized by a large number of people with sufficient leverage in the political system. Environmental protest movements, expression of environmental interests through mobilization of people, and the emergence of leaders who can articulate environmentally-friendly positions is necessary, then, for long-term policies to be chosen.

Participatory democracy with direct action can thus provide the most efficient way to internalize environmental preferences, and to induce the political system to maximize responsiveness to the expression of environmental interests. In addition, participatory democracy breaks down concentrations of power, and the franchise is extended more widely and exercised more vigorously. Power cannot be insulated, and political actors need to assume responsibility for their actions (Goodin, 1996: 842). The political system will increasingly become sensitized to values of the environment, and the more others there are who are concerned about environmental issues, the more nature's interests will be expressed in policy programs.

A democratic structure also permits demands for an unobstructed flow of information about environmental costs and benefits. Coordination of environmental policies internationally also depends on the availability of information about national costs and benefits for particular environmental policies, something that will be harder to locate for countries that do not allow the full costs of environmental protection to be revealed. Non-democratic states are often slow in responding to pressure from civil society organizations, and address environmental problems usually only after they cross some critical point. The democratic process may be the only way to aggregate, even if imperfectly, the valuation of environmental goods (Chadwick, 1995).

3 Centralization in the Indian Polity: Environmental Capacity until the Early 1980s

Although a democratic political arrangement provides citizens a number of alternatives for articulating their preferences and seeking redress through the political system, a number of features of the Indian polity make it hard for environmental issues to be addressed through normal bureaucratic politics. Although democratic, the Indian political system has for long been highly centralized. The type of centralizing tendencies that were visible in the 1970s and the 1980s have now begun to give way to a wider dispersion of decision-making powers. But it is useful to understand some of the factors that led to centralization, since it is through changes in these factors that a more inclusive style of environmental politics might emerge. Tendencies toward centralization became most evident during the reign of Mrs. Gandhi who used her own personal charisma to make a direct appeal to voters and thereby bypass the Congress party political organizational machinery. Populist policies were the instruments she found most useful to lead her party to electoral victory and this mode of political mobilization continued under successor governments, initially headed by her son Rajeev Gandhi. However, challenges to power from regional leaders, and a fiscal crisis that made it impossible to follow populist policies, has meant that India now might be witnessing a reversal of the centralization that had taken place under the leadership of the Gandhi family.[9]

While it existed, the top-down exercise of political authority created many environmental problems whose legacy still continues. Flood control in the Indo-Gangetic basin provides a good example. During floods, the demands on the land, increased agricultural activity and greater irrigation led to loss of life and property, and devastation of crops. The government of India, with its centralized planning system, administered flood control measures through a series of dams in the upper embankments. This provided only a temporary remedy – with increased siltation loads, reservoirs behind the dams began to fill up and dam capacities lessened. Dams had to discharge more water than designed, which lead to further flooding. The silt began to settle in river beds, which slowly became higher than the surrounding area and finally, floods took on an even greater magnitude. The land beyond the embankments could no longer drain water into the river and thus became marshy and saline. The Gangetic basin example shows how the intervention of the government not only proved to be inefficient but also led to further degradation of the ecosystem (CSE, 1985).

9 See Kohli (1990, 1994) for a discussion of some of these trends in Indian politics.

Although there are many signs of decentralization in the Indian polity, a second factor that prevents rapid incorporation of environmental issues into the political arena is that the administrative structure remains quite centralized. The government owns, controls and develops all of the country's forests, dams, major irrigation systems, power stations and railways. The specific issues of the environment are left to the Department of Environment (DOE) created in 1980. Although this department is in charge of diverse activities related to the environment, such as preventing and controlling air and water pollution, conducting botanical and zoological surveys, monitoring and researching the environmental impact of new industrial units, its influence on other central governmental agencies and state governments remains minimal. There has been little change in the activities of the forest, industry, irrigation and public works departments as a result of the presence of the DOE (CSE, 1985: 328). Both horizontal and vertical diffusion of environmental awareness remains limited. Nor is it likely that the DOE would ever come to exercise the overriding authority needed to improve environmental conditions.

India's democratic machine, defined by Kohli as a "well defined network of distribution of spoils in exchange for support," however, has the potential to respond to pressures from important interest groups at different levels of the political system, and absorb dissent and co-opt leaders of the subordinate classes (Kohli, 1990: 219). As a result there are several venues in which social interests can be articulated and find representation in the policy process. Over the years, through the process of intense bargaining and apportionment of benefits among different partners of the dominant coalition, the Indian government has become increasingly resilient and open. It has also thereby created demands that seem to break the economic stalemate of the coalition of upper classes, and curb the power of a meddlesome bureaucracy.

In relation to the environment a large number of statutes seek to regulate resource use, pollution levels, emissions, and conservation of existing resources. Although even the Indian constitution recognizes the need for a safe environment, legislation since the 1970s has given teeth to the sentiments expressed in the constitution. The major statutes that empower the Indian state to take action in relation to the environment are the Water (Prevention and Control of Pollution) Act, 1974, the Air (Prevention and Control of Pollution) Act, 1981, the Forest Conservation Act, 1980, the Wildlife (Protection) Act, 1982, and the Environment (Protection) Act, 1986 (Bansal and Gupta, 1987; Mathur 1987). The Environment Protection Act of 1986, passed in part to give effect to the decisions taken after the Stockholm Conference of 1972, is the most comprehensive step taken by the Indian government. Together with the founding of the Department of Environment (DoE) in 1980 headed by a cabinet minister, it represents the integration of a number of piecemeal institutions and legislation about the environment.

The Act establishes the procedures through which the government can set standards for air, water, and soil quality. It also allows the government to impose limits on pollutants and emissions from different sources, establish procedures for the handling of hazardous and toxic substances, prohibit particular types of industrial operations, and take remedial measures in case of environmental disasters. While the Department of Environment with its state counterparts has the overall control of environmental matters, a range of regulatory institutions seek to manage levels of pollution, chief among them being the central and state pollution control boards. By 1986, the Central Pollution Control Board had identified more than 4,200 industries that caused high levels of pollution, but only 1400 of these had applied for ratification of the pollution they produced (Garg and Tiwana, 1987: 110). Few industries, however, have shown much willingness to comply with the limits imposed by the government. Indeed, industries at present see environmental regulation as an added complication in their operations. Many do not even attempt to apply for grants of consent from the Pollution Control Boards, preferring instead to remain undetected until government orders force them to cease operations. For instance, many industries continued to pollute the Ganges river until the Supreme Court issued closure orders in 1993 against 190 industrial units located along the banks of the river.[10]

The main environmental actors in the above discussion are the government and industry on the one hand, whose relationship is at best conflictual, and the citizenry on the other hand, who fall victim to environmental disasters. Action by the government itself was only limited until the early 1980s, after which a series of factors have combined to make the environment an important issue on the agenda. Until the mid-1980s, the government tried to manage the environment through command and control policies but without adequate enforcement. This strategy had a paradoxical effect. It emerged as a leader on environmental issues by taking the legislative initiative to pass new laws. At the same time, the lack of enforcement of these laws meant that economic activities continued to undermine the possibility of a clean environment. Further, institutional changes that could lead to environmental protection on a more decentralized basis were never pursued. The root causes lay in the low public awareness of environmental issues, the failure of the enforcement machinery of the government, and the unwillingness of major actors to comply with regulations.

4 New Environmental Actors and Capacity Building

It can fairly be said that until the mid-1980s, the major civil society actors in India – industrial capitalists, large and medium farmers, and white collar workers[11] – had little interest in ensuring environmental protection. The costs of environmental protection would fall on them, but the benefits would be shared by a large, dispersed group of poor people who bear a substantial part of the total costs of continuing environmental degradation. Although urban rich and middle classes would prefer a cleaner environment and more comfortable public spaces, they are unwilling to pay the taxes or support regulations that would support such an out-

10 The case under which these orders were issued demonstrates the limited effectiveness of the Pollution Control Boards and the environment ministry which had made few efforts to enforce the provisions of the Environment Protection Act (Aggarwal, 1993; Anon., 1993). The Act empowers the government to impose fines up to Rs. 100,000 (US dollars 2,200) and prison sentences up to 5 years.

11 (Bardhan, 1984). Because none of these were hegemonic, Bardhan also remarks that the Indian political scene remained one of constantly shifting alliances and mutual concessions. It is because of this characteristic of Indian politics that specific subsidies and concessions form such a large part of government policies. One estimate places them at nearly 15% of the GDP (Mundle and Rao, 1991).

come. Their willingness to act in the interests of the environment has changed only little since the mid-1980s except for greater awareness of looming environmental disasters.

But there are additional changes that have enhanced environmental capacity building in India. The role of international donor agencies and NGOs, Indian nongovernmental organizations, and the news media has been critical. With increasing public awareness of the issue and several social movements related to the environment,[12] judicial activism has also come to play a very important role. It is clear that, in the coming years, the model of command and control that has characterized government policy must change. Government legislation on the environment itself had focused mainly on issues of industrial pollution until the 1980s, but this legislation and related institutions (for example, the pollution control boards) address only a limited set of environmental problems, not broaching issues of water, forests, pastures, land, and fisheries depletion. In India, environmental issues are inextricably tied up with livelihood questions. Whereas government action focused on control of pollution activities by the creation of new agencies, environmental problems in India also demand attention to equity and deeper institutional change that involves citizens more widely. One of the characteristics of a new regime of environmental protection must be to decentralize planning and enforcement.[13] The case study on forests provides a sense of the kind of decentralized action that may give reason for hope.

In India, institutional agendas are not highly constrained. This has two effects. One, it allows any issue to be placed on the agenda once articulated societally. But the very multiplicity of issues that are placed on the social agenda means that not all of them can be addressed. As a result, only those that receive strong political backing from institutional or popular supporters gain public attention. Thus, although citizens are offered a wide range of issues, they must ably exploit methods of participation in a democratic setting, which include street demonstrations, protests and hunger strikes. These actions can alter the institutional agenda through a variety of factors. First, calamities and catastrophes augment the need for action and encourage the consolidation of political support. Examples of these catastrophes include India's serious pollution and erosion problems. Secondly, internal pressures created by public protests, such as in the Chipko movement, can encourage the government to take action, thus altering the agenda. Finally, popular literature suggests that agendas are altered through issue diffusion from one agenda to another.

---

12 Recent years have witnessed a number of environmental movements against logging and new dams. Forest-related environmental movements include the Chipko movement in Uttar Pradesh, and the Appiko movement in Karnatak. The Narmada Bachao Andolan (Save the Narmada movement), the Koel Karo protests, the Silent Valley movement, and the protests against the Inchampally, Bhopalpatnam, and Tehri dams are some of the protest movements that have questioned government efforts to create development on the backs of the more marginal sections of the population (Agarwal, 1997: 42).

13 Even a recent World Bank assistance project for strengthening environmental management capacity accepts the need for decentralization of selected environmental functions (Sirichanya, 1996).

4.1 International Influences

A number of environmental issues involve the global population and require sophisticated technology if changes are to occur. As Indira Gandhi pointed out in her keynote address at the 1978 Toronto Conference, "if we are to effectively deal with global environmental problems, we have to forge a partnership which seeks simultaneously to protect and serve the environment, taking care, at the same time, of the special requirements of developing countries ... " (Ramakrishna, 1992: 146). The 1992 United Nations Conference on the Environment and Development (UNCED) prompted 5 accords which limited the emissions of greenhouse gases, protected the diversity of the planet's animal and plant life, provided guidelines for forest preservation and produced the Rio Declaration on Environmental Development. The summit also created Agenda 21, a detailed blueprint for environmental actions of the future. This document elaborates strategies and programs that will reverse environmental protection and promote sustainable development. Such environmental regimes provide developing nations such as India with forums in which to discuss environmental issues, but at the same time put pressure on developing nations to undertake environmental management. Equally importantly, international regimes can also give focus and recognition to the environmental problems of citizens in the developing world.

International aid agencies have played an important role in influencing the adoption of environmental policies in developing countries. India is no exception. One of the main forms through which international organizations have affected environmental awareness in India is through funding programs and environmental activities of local grassroots NGOs. But they have also helped in shifts in environmental policy. The adoption of the Joint Forest Management programs in various Indian states, and the declaration by the Indian government to involve local communities in the use and management policies around local village woodlots was aided by the actions of the Ford Foundation, which helped gain recognition for the activities of innovative forest department officials and supported some of the initial experiments that have finally led to a sea-change in forest protection policy in India (see below).

4.2 Role of the Media in Environmental Issues

The Indian press has matured as a result of the democratic process. It is not only independent of the government, but holds a dominant position in the political game. The role of the media has been effective in the environmental policy arena, yet its efforts, like popular protests, have been local and limited in scope.

A content analysis of The Hindustan Times newspaper from 1976 to 1981, a period that is significant because of the numerous environmental laws passed, and new agencies and organizations established, reveals coverage of the disputes concerning the Silent Valley Hydroelectric Project and the Mathura Oil Refinery. The Silent Valley hydroelectric project was disputed owing to the danger of submerging hundreds of endemic species in the tropical rainforest that would have been covered by water behind the dam. A combination of coverage by the news media and the protests generated by an alliance of grassroots non-government organizations were responsible for the final decision to abandon this project. The Mathura

Oil Refinery came to national attention because the release of pollutants from the refinery was eroding the internationally famous marble monument to love, the Taj Mahal. Once again, although the refinery has not been closed down, a number of steps have been taken to reduce the impact of the pollution on the Taj. This was again partially a result of the attention the issue received in the media.

It can be argued that, over time, the "messenger" role of the media has evolved into the more aggressive role of "exposer" of government apathy and neglect. Once an environmental policy is in place, the media takes an active and aggressive role in criticizing the government's actions, providing suggestions in which people should react and behave toward these policy decisions. Today, environmental issues are covered in non-English local newspapers, and the tone of coverage is often critical of government policies, and increasingly critical of governmental interference.

The government itself also uses the media as a vehicle to promote interest in and approval of environmental issues. For example, the government, when publicizing its plan to clean the river Ganges, used the media as an important tool to arouse public interest. The Ministry of Environment and Forests makes efficient use of the media, and, in rural areas, environmentalists and government campaign agencies use other communication strategies more suited to the rural lifestyle, such as folk songs, slide shows, or open-air theaters, to promote their cause.

4.3 Non-Government Organizations and Judicial Activism

According to one estimate, there are more than 10,000 environmental NGOs active today in India (Dutt and Rao, 1996: 297). These NGOs operate from the International to the local grassroots levels, and are active in an enormous range of environment-related activities. Many of them, among them the Center for Science and Environment, the Tata Energy Research Institute, and WWF (India), and the Kerala Shastra Sahitya Parishad (KSSP) are active in urban areas and their main activities are in the form of research, networking, lobbying, and working as information clearing houses. Others undertake direct action such as mobilizing popular action against environmental ills, or afforestation. Overall, Indian environmental NGOs can be seen to fall into one of four categories: a) relief groups that concern themselves with environmental disasters such as famines or floods, b) sustainable development groups that focus on economic and environmental programs such as afforestation, or commons development, c) Environmental action groups who try to mobilize people into protests against environmentally unsound state or corporate policies such as large dam construction or nuclear testing, and d) support groups who try to network, disseminate information, or carry out research. The last are usually urban whereas the first three are overwhelmingly located in rural areas.

Mobilized groups in India have had long-standing concerns related to of rural and urban poverty, social justice, inequality, civil liberties and rural development. They have only recently taken up environmental issues that involve deforestation or afforestation, dam construction, air and water pollution, and soil erosion. Because of their relatively recent adoption of the environmental agenda, their protests are often spontaneous, sporadic, and against perceived injustices of diverse types. They employ a range of methods, from peaceful strikes to struggles involving violence. Their causes vary from economic and ethnic to political and environmental. Another feature is the usual lack of cooperation between individual, localized movements. For example, protests arising from environmental issues do not necessarily unite the groups across regional boundaries. Indeed, a number of different efforts may be going on at the same time to protest against various environmental ills, without the forging of a common front (Shah. 1988: 298).

The growth of groups that have direct links to environmental movements within a developing country is important in establishing environmental concerns as a legitimate political concern as noted by Enloe (Khator, 1991: 174). Group activity in India is significant. In 1983, there were only 250 NGOs that declared themselves as having an environmental interest. If the number today has risen to more than 10,000, part of the reason of the growth is that many existing organizations have simply begun to include environment as one of their concerns. Nonetheless, such rapid growth of environmental NGOs, even when they are not highly organized or united, creates multiple points of local pressure on the political system to consider environmental issues.

Most environmental groups in India are locally based. As such their efforts are seldom directed at raising mass environmental consciousness. Their relationship with the government has tended to vary, but in recent years they have come to gain recognition as important agents of social change, and find mention in the 5-year plan documents. They have also begun to use public interest litigation to achieve their environmental demands. As a result, the judiciary has handed down a series of landmark decision in public interest litigation cases that have shown its willingness to step in where the executive arm of the government has failed in protecting the environment.

The courts have acted in a range of cases to compel the state to undertake action for the environment. In a case with far-reaching consequences in 1980, the Supreme Court ordered the Ratlam Municipal Council to treat sewage, create a drainage system, stop industry from spilling effluents, and implement health safely measures (Varandani, 1987). More recently, it was through a case filed in the Indian Supreme Court that the NGOs were able to stop the creation of any new aquaculture shrimp farms in coastal regions. These farms were destroying local fisheries and agriculture through the release of excessive nitrogen into the ecosystem. While even the piecemeal efforts of the NGOs are producing some effect, recently there have been some attempts to establish a network of NGOs and voluntary groups to pursue environmental conservation more effectively. The voluntary sector is bound to become a lead promoter in pushing environmental concerns as key issues on the national political and development agenda (CSE, 1985: 342).

5 Environmental Capacity Case Study: Deforestation

To examine the above general statements in a more concrete context, the remainder of this paper is devoted to a discussion of a specific environmental policy arena: forest management. The shift in this sector, from protection by the forest department, to environmental movements and top-down efforts to involve citizens, to the recently created Joint Forest Management Program shows the capacity of the Indian state to address environmental issues by becoming more inclusive.

Forests provide vital resources to rural economies. Since the colonial period itself, Indian forests were managed on an exclusionary basis with the forest department asserting its control over more than 25% of Indian's land and the right to manage and control all Indian forests. The policies changed little with independence.[14] In the process, local rights to forests received little attention from the state despite the fact that village forests are important sources of direct benefits, such as timber, wood and panels for construction, fuelwood, fodder, fruits, and medicinal plants.

In the early 1980s, as satellite imagery revealed that India's forest cover was a scant 11%-13% of the total land area, government targeted deforestation as the most important environmental issue on the national agenda. But its strategy to address deforestation was to undertake top-down tree plantation programs under the auspices of a new agency called the National Wasteland Development Board. Social forestry programs were launched by several states to promote afforestation. But most of the new trees were planted on private farms rather than community land, and large tracts of degraded forest lands were assigned to industrial firms. By planting trees, these programs aim to meet the needs of the industrial markets while ameliorating the fuel and fodder crisis among the poor. The species of trees chosen for planting (eucalyptus, teak and pine), seldom benefited the local ecology, and the new biomass from the plantations often ended up in the hands of rich farmers and industries.

5.1 Social Forestry Programs

The term social forestry was used by the National Commission on Agriculture to indicate tree-raising programs that supply firewood, fodder, small timber and minor forest produce to the rural population. These programs comprised three aspects: 1) farm forestry, where the farmers were supplied with free or subsidized seedlings, 2) community woodlots, where village communities were encouraged to plant trees on common lands and 3) forestry woodlots, where trees were planted for the community by the government forest department on public lands (Gupta, 1988: 12). These social forestry programs have been far from successful; the wood produced from social forestry programs often ending up in urban and industrial India instead of in the hands of the rural poor. They also reduced rural employment, land holdings of the small farmers, and promoted absentee landlordism (Chandrashekhar, 1987). The failure of the programs reflects the absence of any important pressure groups that could force the government or the administration to apply mid-course corrections once it became evident that the benefits from the programs was not flowing toward either the environment or the poor.

Community forestry programs in Gujarat reflect the problems of India's social forestry programs. In 1980, the Gujarat social forestry program received a five year $67 million project from the World Bank that targeted 37,000 hectares of strip plantations, 68,440 hectares of village woodlots and 100 million seedlings for

14 Rangan (forthcoming) provides a thoughtful analysis of colonial and postcolonial forestry policy changes. She argues that although forestry policies have changed over time, the nature of control exercised by the forest department did not change much with independence, but that the variations in forest policy can be traced to other causes.

distribution. Farm forestry programs were reported to be the most successful. Between 1980 and 1983, the farm forestry program achieved 202 per cent of its target and by the third planting year, 95 million seedlings were being distributed (CSE, 1985, 32). Yet, a report by the World Bank reveals that this success may not have achieved its ultimate goal of improving living situations of the poor. Many of the farmers converted irrigated land to farm forestry because of uncertain prices of agricultural cash crops. Also, the tree farming process was less labor intensive and failed to increase employment opportunities. These programs have benefited the rich farmers the most and failed to improve the circumstances of small farmers and the poor.

5.2 The Chipko Movement in the UP Himalayas

In stark contrast is the experience of the Chipko movement that sought to alter the government's position in relation to the management of forests in the hill regions of Uttarakhand in the Indian Middle Himalayas. The world famous movement, pioneered by the Dashohli Gram Swarjya Mandal in Gopeshwar, was among the groups to show concern for development and environmental issues. The roots of the movement can be found the coming together of a few residents of Chamoli as a cooperative labor society in the 1960s to prevent exploitation by public works contractors. At this point the organization's main interests were in generating economic benefits and better employment conditions rather than in protecting the environment.

In 1964, the hill residents in Chamoli formed the Dohsoli Gram Swarajya Sangh (DGSS) to start cottage industries based on natural resources from local forests. A collection of village cooperatives then set up a factory based on the use of pine resin from Chir trees (Pinus roxburghii), but the forest department refused to supply these factories with enough pine resin. The DGSS were soon to realize they had no rights of control over their local forest resources, while the rich contractors exploited these resources because they had established a long standing coalition with the government.

The extraction of resin and turpentine was commenced in 1965 in the area of Badyargarh, and in 1979, big contracts to fell chir trees was given out. This led to widespread destruction and the taking away of all the wood, including the branches of trees. The Chipko leader Sunderlal Bahuguna was entrusted to hear the villagers' grievances. Leading the villagers, Bahuguna went from village to village, informing people of the proposed felling of trees and the ecological consequences that would follow.

Over three thousand men and women from all social groups participated in the struggle in which women especially played a prominent role. The leader, Bahuguna was jailed on January 22nd, where he continued to fast. The determined resistance from the villagers even after their leader was removed lead contractor and forest officials to abandon tree felling activities (Guha, 1989, 178). In April of 1981, Bahuguna went on an indefinite fast to urge a total ban on tree felling in the Himalaya above 1000 meters. The government responded by creating a eight-member committee to prepare a comprehensive report on forest policies and agreed to follow a fifteen-year moratorium on commercial felling.

The Chipko movement brought commercial felling to a standstill, but state forestry is only one of the problems that threaten ecological stability. The intensification of resource exploitation is furthered by large dams and increasing mining operations. The Chipko movement has been meeting these challenges through ongoing struggles and widening the movement's horizons. At present the most active wing of the Chipko movement is that led by Chandi Prasad Bhatt. This wing organizes afforestation programs in the villages, encourages greater environmental consciousness, and believes in the use of local resources for local needs. In summary, the grassroots Chipko movement has helped define the linkages of rapid industrialization, technological modernization, and the environment. All the wings of the Chipko movement challenged Indian intellectuals and political elites on the environmental impacts of rapid industrialization. For example, Sundarlal Bahuguna holds commercial forestry and the close links that exist between contractors and forest officials to be responsible for environmental degradation.

Although the Chipko movement successfully curbed deforestation in one region of the country, its success hinged on the coming together of the local populace in a concerted effort to influence government policies regarding the use of forests. The change in policies meant that local forests could become available to local populations more freely, and the subsistence needs of the villagers could be met more easily. What the Chipko movement demonstrates in relation to the environment management capacities of the Indian government is its ability to respond to social movements without repression, to accept articulated demands into the policy-making agenda, and to diffuse a tense political situation by conceding the more reasonable demands that are articulated.

5.3 Co-Management in Forestry

In recent years, a different mechanism has been used to reorient forestry policy toward the needs of the poor. This is the Joint Forest Management Program which was initiated in the late 1980s through a central government initiative. Soon after, 16 states had passed government orders by 1992 directing their forest departments to initiate community-based programs to protect forests. At this point, Joint Forest Management is being implemented in 22 of the 25 Indian states.[15] The Program seeks to involve local populations in the management and protection of local forest department held land in exchange for a modest level of regular benefits, plus a substantial payoff once timber from the protected trees is sold in the market. By 1992, more than 10,000 village level groups were active in the implementation of the JFM (Sarin, 1995) and by 1995, the number had increased to nearly 15,000 (Poffenberger and McGean, 1998).[16] The institutionalization of benefits for the rural population through the sharing of products from forest lands belonging to the

15 The best recent study of the Joint Forest Management Program is (Sivaramakrishnan, 1996). See also (Poffenberger, 1994; Poffenberger and McGean, 1998)

16 The JFM is not without historical analogs in the Indian case. There are more than 3000 forest councils in Uttar Pradesh hills, and collectively they manage more than 3000 sq. kilometers of forest land in the region. They came into being as a result of the forest council rules of 1931, passed by the British government (Agrawal, 1994).

Table 3. Changes in Environmental Management in India: The Case of Forests

Aspects of Environmental Management	Until Mid-1970s	Late 1970s to late 1980s	1990s
Problem perception	People responsible for deforestation	Crisis of wood fuel, timber and other forest products; Conventional measures ineffective	Lack of people's participation responsible for ineffectiveness of forest policy
Policy measures	Centralized forest department control	Community and social forestry	Joint Forest Management to involve communities in forestry
Major actors	Forest Department, industrial development	Forest Department, NGOs, International donors, Industry	Forest Department, NGOs, Village communities,
Evaluation of Environmental Capacity	Technically capable, limited capacity to enforce	New directions, limited success	Promising new avenues, participatory management

forest department is a novel experiment in community-based conservation that is becoming a hallmark of forestry policies around the developing world.

The example of the forestry sector exemplifies the direction of environmental capacity building in India. Table 3 summarizes the lessons of the case study with a view to drawing some general lessons. It indicates that in the early years, the Indian state saw forests as its own responsibility. It tried to protect them by excluding people, especially rural residents. When better scientific information revealed the extent of deforestation that had taken place, the first responses of the state were to initiate forestry programs that attempted a technically oriented managerial solution through the social forestry programs. The adverse distributive impact of these programs and the inability of the effort finally to address deforestation prompted a search for alternative solutions. The involvement of NGOs and international donor agencies, and some grassroots experiments in the arena of forest protection and management found expression as a new direction in forestry in the shape of the Joint Forest Management Program. The JFM has sought to decentralize forest management by involving village groups, and witnessed regeneration of forests in several areas (Agarwal, 1997, 43).

6 Conclusion

India's democratic state is flexible in incorporating new social issues into the political agenda, and stable as a functioning democracy. In the 50 years since independence, there have been several shifts in its environmental management style. The democratic framework of the Indian state has led to new issues coming on the agenda as a result of environmental crises, and encouraged the formation of new coalitions to lobby for new policy management styles. The most important new actors on the Indian environmental scene have been NGOs, and village level user groups in the issue areas of forestry and renewable resource management. The management capacity of the state has been enhanced by the inclusion of these new actors and their coalitions. At the same time, these actors have made environmental policy more participatory in at least some issue areas.

Among the problems from which environmental organizations and movements in India continue to suffer are that they are often locally based, lack coordination, and share an inability to cooperate over the long run. Their efforts at mobilization need to coalesce into large movements and they need to work more intensively and persistently with international actors to further influence the government in moving toward a participatory stance to an even greater extent.

As a result of developments that are already visible, it is likely that the Indian state will continue to seek greater involvement of grassroots groups, NGOs, and international donors in several environment-related policy arenas. In some areas, involvement is more characteristically advisory, or oriented toward research and information production. They include urban pollution, biodiversity protection, steps to ameliorate the emission of greenhouse gases, and ozone depletion. The research capacities of the official establishment in these arenas are usefully supplemented by public and private research carried out in educational institutions and non-governmental organizations. In other arenas, the newly emerging actors can play a more direct role, especially in planning and implementation. In forestry, protection of national parks, watershed management, grazing lands development, community-level groups, NGOs, have already been instrumental in helping achieve national objectives, and are likely to continue to be highly important. Research institutions such as the institutes of management, institutes of technology, and some other centers like the Indian Institute of Science in Bangalore and the Institute of Economic Growth in Delhi have produced research that has helped move public opinion more in the direction of the involvement of people and communities in achieving environmental goals.

At the same time, some of the actors, traditionally seen to be unfriendly to environmental goals, have remained set in their orientation. Industrial capital in India has scarcely become environmentally aware or active. For a shift to take place in the attitudes of industrial enterprises, the general populace will need to become far more vocal in its desire to protect the environment.

References

Abernathy, V., 1993. *Population Politics: The Choices that Shape Our Future.* New York: Plenum Press/Insight Books.

Aggarwal, R., 1993. Lid off Ministry's role in Ganga pollution. *Economic Times.* Delhi.

Agarwal, A., and S. Narain, 1997. A proposal for global environmental democracy. *The Earth Island Journal.* 7, 31.

Agarwal, B., 1997. Gender, environment, and poverty interlinks: Regional variations and temporal shifts in rural India, 1971-91. *World Development* 23, 23-49.

Agrawal, A., 1994. Rules, rule making, and rule breaking: Examining the fit between rule systems and resource use. In: E. Ostrom, R. Gardner, and J. Walker (eds.), *Rules, Games, and Common-Pool Resources.* Ann Arbor: Michigan University Press, 267-82.

Agrawal, A., and G. N. Yadama, 1997. How do local institutions mediate market and population pressures on resources? Forest panchayats in Kumaon, India. *Development and Change* 28, 435-65.

Anon., 1993. Closure of 190 units ordered: Ganga pollution. *The Hindustan Times*. Delhi.

Bansal, V. K., and N. K. Gupta, 1987. Environmental protection: A constitutional Obligation, in: R. K. Sapru (ed.), *Environment Management in India*. New Delhi: Ashish Publishing House, 72-85.

Bardhan, P. K., 1984. *The Political Economy of Development in India*. Oxford: Oxford University Press.

Bose, S., and A. Jalal, 1997. *Modern South Asia: History, Culture, Political Economy*. London: Routledge.

Brass, P., 1995. *The Politics of India since Independence*. Oxford: Oxford University Press.

Chadwick, B. P., 1995. Fisheries, sovereignties, and red herrings. *Journal of International Affairs* 48, 559-84.

Chandrashekhar, D., 1987. Social forestry in Karnatak: An impact analysis. *Economic and Political Weekly* 22, 935-41.

Chatterjee, P., 1994. Development planning and the Indian State, in: T. J. Byers (ed.) *The State and Development Planning in India*. Delhi: Oxford University Press, 51-72.

Chitale, M. A., 1992. Population and water resources of India, in: V. Gowarikar (ed.), *Science, Population, and Development*. Pune: Unmesh Communications.

CSE, 1985. *The State of India's Environment 1984-85: The Second Citizen's Report*. New Delhi: Center for Science and Environment.

CSE, 1998. *Slow Murder*. New Delhi: Center for Science and Environment.

Dutt, A. K., and J. M. Rao, 1996. Growth, distribution, and the environment: Sustainable development in India. *World Development* 24, 287-305.

Fairfax, S., and L. Fortmann, 1990. American forestry professionalism in the third world: Some preliminary observations. *Population and Environment* 11, 259-72.

Garg, M. R., and N. S. Tiwana, 1987. Enforcement of environmental law and management of pollution control, in: R. K. Sapru (ed.), *Environment Management in India*. New Delhi: Ashish Publishing House, 103-22..

Goel, R. S., and K. Prasad, 1993. An overview of environmental impact of Indian WRD projects, in: *Environmental Impacts of Water Resources Development: National Round Table Discussion, Vigyan Bhawan, New Delhi, June 4 - 5, 1993*. GOI, New Delhi: Tata McGraw-Hill.

GOI, 1992. *Environment and Development: Traditions, Concerns and Efforts in India, National Report Submitted to UNCED*. New Delhi: Ministry of Environment and Forests.

Goodin, R. E., 1996. Enfranchising the earth and its alternatives. *Political Studies* 36, 835-49.

Guha, R., 1989. *The Unquiet Woods: Ecological Change and Peasant Resistance in the Himalayas*. New Delhi: Oxford University Press.

Gupta, A., 1988. *Ecology and Development in the Third World*. London: Routledge.

Haas, P. M., R. O. Keohane, and M. A. Levy (eds.), 1993. *Institutions for the Earth: Sources of Effective Environmental Protection*. Cambridge, MA: MIT Press.

Khator, R., 1991. *Environment, Development and Politics in India*. New York: University Press of America.

Kohli, A., 1990. *Democracy and Discontent: India's Growing Crisis of Governability*. Cambridge: Cambridge University Press.

Kohli, A., 1994. Centralization and powerlessness: India's democracy in a comparative perspective, in: J. Migdal, V. Shue, and A. Kohli (eds.), *State Power and Social Forces: Domination and Transformation in the Third World*. Cambridge: Cambridge University Press.

Lipschutz, R. D., and K. Conca (eds.), 1993. *The State and Social Power in Global Environmental Politics. New Directions in World Politics*. New York: Columbia University Press.

Mathur, L. N., 1987. Environmental management of urban areas in India, in: R. K. Sapru (ed.), *Environment Management in India*. New Delhi: Ashish Publishing House, 1, 29-35.

Meadows, D., D. Meadows, and D. Ehrenfeld, 1992. *Beyond the Limits: Confronting Global Collapse, Envisioning a Sustainable Future*. Post Mills: Chelsea Green.

Mundle, S., and M. G. Rao, 1991. The volume and composition of government subsidies in India, 1987-88. *Economic and Political Weekly.*
Poffenberger, M, 1994. The resurgence of community forest management in eastern India, in: D. Western and R. M. Wright (eds.) *Natural Connections. Perspectives in Community-based Conservation.* Washington, D.C: Island Press, 53-79.
Poffenberger, M., and B. McGean, 1998. *Village Voices, Forest Choices: Joint Forest Management in India.* Oxford: Oxford University Press.
PTI, 1998. India loses $80 bn due to pollution: World Bank. *Indian Express*, 1.
Ramakrishna, K., 1992. North-South issues, the common heritage of mankind, and global environmental change, in: I. Rowlands and M. Greene (eds.), *Global Environmental Change and International Relations.* Hampshire: MacMillan.
Rangan, H., forthcoming. State economic policies and changing regional landscapes in the Uttarakhand Himalaya, in: A. Agrawal and K. Sivaramakrishnan (eds.), *Agrarian Environments: Resources, Representations, and Rule in India.* Durham, NC: Duke University Press.
Redclift, M., 1994. The environment and structural adjustment: Lessons for policy interventions in the 1990s. *Journal of Environmental Management* 14, 55-69.
Sachs, W. (ed.), 1993. *Global Ecology: A New Arena of Political Conflict.* London: Zed Books.
Sarin, M., 1995. Joint forest management in India: Achievements and unaddressed challenges. *Unasylva* 46, 30-6.
Shah, G., 1988. Grassroots mobilization in Indian politics, in: A. Kohli (ed.), *India's Democracy: An Analysis of Changing State-Society Relations.* Princeton: Princeton University Press.
Shah, T., 1993. *Groundwater Markets and Irrigation Development: Political Economy and Practical Policy.* Delhi: Oxford University Press.
Sirichanya, D., R. Robboy, and P. Mitchell, 1996. India to strengthen monitoring and compliance of environmental laws. News Release No. 97/1220 Sasia, 1. New Delhi.
Sivaramakrishnan, K., 1996. Forests, Politics, and Governance in Bengal, 1794-1994, Vols. 1 & 2. PhD, Yale University.
TERI, 1998. *Looking Back to Think Ahead.* New Delhi: Tata Energy Research Institute.
TERI, 1999. Union Budget 1999-2000: Energy and Sustainable Development. New Delhi: Tata Energy Research Institute.
Varandani, G, 1987. Judicial approach to environmental pollution, in: by R. K. Sapru (ed.), *Environment Managment in India.* New Delhi: Ashish Publishing House, 2, 116-30.
Wilson, E. O. (ed.), 1988. *Biodiversity.* Washington DC: National Academy Press.

ITALY

Bruno Dente and Rodolfo Lewanski

1 Main Environmental Problems and Improvements

Starting from the 1950s Italy underwent a process of accelerated economic growth in terms of both production and consumption that rapidly made it one of the world's most developed countries. However, this process also caused a sharp decline in environmental quality. The expansion of heavy industry (chemicals, steel, cement, refineries, paper mills) brought extensive pollution. Even the numerous small and medium size factories – perhaps the real secret of the Italian "economic miracle" – spread across northern and central regions, are responsible for serious pollution due to their concentration in relatively small territories. 14 areas have been classified as exposed to the risk of an "environmental crisis", and 10 more are considered as critical due the concentration of high impact industrial activities (Ministero dell'Ambiente 1997, XVII).

One well known feature of the Italian economy is the deep split between a highly developed North and a more backward South. Central government's traditional strategy to promote the southern economy through infrastructure projects and by attracting heavy industry has been largely unsuccessful in achieving widespread development, while contributing to the deterioration of assets of considerable natural and economic (e.g., tourism) value. Large industrial plants built in areas of great scenic beauty, emissions and vibrations produced by traffic in ancient cities and even the pressure of tourism exert serious negative impacts on fragile ancient city centres, and contribute to the deterioration of historical monuments and buildings.

With a population of 57 million living within a territory of 300,000 sq. km, Italy has an average density of 191 inhabitants per square kilometre; in urban areas – especially in the Po Valley and along the coasts – this figure soars to 2,000 inhabitants per square kilometre, making it among the most densely populated areas of Europe.

Construction of buildings and roads, excessive excavations, lack of care of mountain areas abandoned by farmers, and inadequate maintenance of water drainage systems cause widespread soil erosion, landslides and coastal erosion (affecting one third of the 3,250 km of beaches) in a land that is geologically fragile, especially in the southern part of the country, where – according to a recent survey carried out by the Environment Ministry – there are risks of landslides in one third of all municipalities. Built-up land occupies approximately 9% of total national territory; in the Po Valley, for example, "consumption" of land for construction purposes has been proceeding at an annual rate of 6% (Lega per l'Ambiente 1989, 227). Land-use planning policies have been widely ineffectual in managing urban expansion, that has destroyed rich farmland and areas of natural value.

As far as polluting emissions are concerned, available data show contradictory trends. In the field of air pollution, sharp improvements are registered in the emissions of some pollutants (CO, ammonia, lead and especially sulphur oxides thanks to the increase in the use of natural gas imported from Russia and northern Africa), whereas those of other substances (VOCs, nitrogen oxides, suspended particles, CO_2) continue to increase (OECD 1994, 26; Istat 1993, 44; Ministero dell' Ambiente 1997, 162). In any case, SO_x, NO_x and CO_2 emissions per unit of GDP and per capita are lower than the OECD average (OECD 1994, 47). Further research and greater awareness are pointing out the alarming levels of air pollution in urban areas both in winter and in summer periods by substances previously ignored, such as ozone (EEA 1998, 100-101), benzene and polycyclic aromatic hydrocarbons. Noise in urban areas is also normally well above maximum levels established by law (60-70 db; Ministero dell'Ambiente 1997, 196).

Thanks to its relatively insulated geographic position in the South of Europe, the Italian peninsula is somewhat less affected by "imported" pollution than countries in central and northern Europe. Italy, for example, is a net exporter of long-range airborne pollutants such as sulphur dioxide. Nevertheless, about one-third of sulphur depositions and half of oxidised nitrogen come from abroad (Unep 1991, 43-44). Depositions of such acidifying substances as sulphur and especially nitrogen are among the highest in Europe in northern Italy (EEA 1998, 78-80).

Air pollution has caused serious decline in forest health (along with more traditional forms of human intervention such as fires, which destroy approximately 0.6% of total forest area each year). In a 1989 survey, 15-17% of forests were classified as damaged by acid rains. Dozens of plant and animal species are threatened with extinction (30% of mammals; EEA 1995, 223), though positive trends can be identified in the case of several animal species (OECD 1994, 33-35). On the other hand, after decades of neglect, the creation of a number of new land and marine national parks has brought the quota of protected territory from 3.2% in 1984 up to currently 6.5% (for a total of more than 2 million hectares), and should soon reach the threshold of 10% of national territory. At the end of 1996, there

Table 1. Air pollution emissions

years	1980	1985	1990	1995
SO_2 (1,000 t)	3,800	1,733	1,678	1,437
NO_x (100 t)	1,480	1,589	2,025	2,157
NH_3 (1,000 t)	–	436	416	389*
CO_2 (mio. t)	–	359	444	436
CH_4 (1,000 t)	–	3,806	2,539	2,508

* *referred to 1994*

Source: Eurostat (1998, 21-23 and 45-47)

were 18 national land parks, 13 national marine parks, 74 regional parks, 174 State natural "reserves", and 94 protected "oases"; there were also 69 natural areas managed by environmental associations (WWF and LIPU; see below) (EURISPES, Rapporto Italia 1998).

Organic and inorganic (especially toxic metals and agrochemicals) substances cause high levels of pollution of internal surface waters from domestic household, agricultural and industrial sources. The quality of many lakes and rivers has improved over recent years thanks to the expansion of waste water treatment facilities and the limitation of effluent discharges since the early 1980s (serving 61% of the total population; OECD 1994, 84; Conte e Melandri 1994, 298-301; however many treatment plants in fact are not working for a number of reasons; Ministero dellte Ambiente 1997, XVI; Milanaccio 1998, 93); the lower reaches of most rivers are highly polluted (OECD 1994, 77). Data on a total of 118 rivers indicate that 31% are in good condition, 29% are judged to be average, 28% in poor condition, and the remaining 12% is in very bad condition (Ministero dell'Ambiente 1997, XVI). Intensive stock-breeding and cultivation methods relying on agrochemicals heavily concentrated in the regions of the Po Valley were causing pollution of the soil and underground water reserves by nitrates and pesticides. However, the use of agrochemicals has since declined (from 192,500 tonnes in 1990 to 158,000 in 1994; Ministero dell'Ambiente 1997, 261). Forty-one per cent of all Italian lakes are plagued by serious levels of eutrophication, and the same phenomenon is at work in the Adriatic Sea. The waters of some 2,500 km of coastline (about one third of the total) were declared off-limits for bathing during the first half of the 1990s, though there has been a sharp improvement (the above figure was 20% higher in 1992 than in 1994) (ISTAT 1993, 94, and 1996, 107).

Due to the difficulties of organising proper disposal of solid waste – especially in several Southern regions – an unknown number of illegal dumps contribute to further pollution of water resources. The provision of adequate treatment and disposal is insufficient for both municipal and industrial waste; in the latter case, for example, capacity is approximately 14 million tonnes, far below the official 34.6 million tonnes produced annually (OECD 1994, 66-71). According to such data, the amount of municipal waste per capita is within the Western European average, whereas the amount of industrial waste per unit of GDP is well below this average (OECD 1994, 63). Treatment and disposal facilities still rely excessively on landfills (88% of total urban waste; Ministero dell'Ambiente 1997, XVI), whereas incineration (5.1%) and recycling (6.9%) are less developed than in other European countries.

Many of the developments mentioned above, including the improvements in air and water pollution and nature protection, obviously are hardly typical of Italy alone; they can also be found in many other developed nations. However, a feature somewhat peculiar to Italy as compared to the more advanced countries is the fact of being consistently a "follower" and not a "leader" in adopting policy measures, and even more so in general public and private behaviour, in coping with the environmental issues.

2 Development of Environmental Policy and Principal Regulations

2.1 Development

The first measures aimed at protecting environmental goods date back to the beginning of the century; two national parks (Abruzzi e Gran Paradiso) were created in 1922, followed by others in the 1930s (Circeo and Stelvio). The aim of Act no. 1497 of 1939 was to protect landscape (though it was hardly implemented). Articles 216 and 217 of the Health Code of 1934 and decree no. 147/27 concerning the use of toxic gases were the first attempts at tackling industrial emissions.

The first explicit piece of anti-pollution legislation was passed in 1966. Act no.615 was rather limited in scope; it aimed primarily at reducing air pollution caused by combustion processes in large urban areas, especially in northern Italy, where frequent thermal inversions during the cold seasons, coupled with unfavourable topographical conditions hindering dispersion, represented a serious threat to public health. The target of the provision was clearly represented only by the pollutants responsible for smog – namely sulphur dioxide and particulate matter (hence the nickname "Antismog Law") – whereas the possibilities offered by the provision to reduce emissions of other pollutants, produced by industrial and mobile sources, were, instead, very limited.

The actual birth of a modern environmental policy took place only in the early 1970s, mainly as a result of a major reform of the institutional system represented by the creation of the 15 ordinary regions. The post-war constitution, though not mentioning the environment, attributed relevant legislative and administrative powers in other fields potentially related to it (tourism, agriculture, forests, small enterprises, excavations, hunting and fishing, inland navigation, regional roads and public works, health, land planning). Central government started transferring powers and resources in these fields over to the regions, rather sparingly, in 1972 and again in 1976; more importantly, as environmental policy grew, state legislation entrusted the regions with enacting it. For example, when the Health Ministry issued the implementing regulations on Act no. 615 in 1970 and 1971, the regions acquired responsibility in the field of air pollution and the Regional Committees for Air Pollution (CRIA, responsible for industrial permits and emission limits) were brought under their control, allowing regions to develop policies of their own, at least to some extent. The same happened in 1976 in the field of water pollution, as will be discussed below. In essence, the development of environmental policy occurred somewhat by chance, as a result of the creation of a new institutional actor, interested in increasing its influence, at the same time as opportunities for doing so were offered by growing legislation.

Legislative outputs tackled other problems as well. Act no. 125/71 established biodegradability requirements for detergents in order to reduce water pollution; Act no. 256/74 regulated dangerous substances; Acts no. 880/73 and 393/75 dealt with emissions from fuel-powered and nuclear power plants (in order to foster plans aimed at expanding energy generation capacity). Other provisions, such as Act no. 171/73 concerning Venice and Act no. 1097/71 regulating excavation activities threatening the Euganei hills, were aimed at protecting specific and geographically limited problem areas. A major piece of legislation – Act no. 319 concerning water pollution control – was finally passed in 1976, after a 10-year-long debate. Health Ministry Decree no. 3/77 indicated drinking water characteristics. Act no. 373/76 introduced the first measures in the field of energy saving.

Other relevant laws, dealing with such issues as industrial and household waste disposal (Decree no. 915/82; Acts no. 441/87 and 475/88), sea and coast protection (Act no. 979/82), drinking water (Decree no. 515/82), the protection of landscape (Act no. 431/85, called "Galasso" after the name of its promoter), air quality standards (decree no. 203/88), parks (Act no. 349/86), the biodegradability of detergents (Act no. 136/83), the prevention of accidents by hazardous industrial plants (Decree 175/88), impact assessment (Act no. 349/86, Decrees 377/88) were passed only in the 1980s. Act no.183/89, after some 20 or so years of discussions, dealt with soil protection, water management and protection against hydraulic risks.

In the first half of the following decade legislation focused on solid waste, noise pollution (Decree of March 1, 1991; Act no. 447/95), impact assessment (Act no. 146/94), parks (Act no. 394/91) and traffic pollution (calling for traffic bans when air quality thresholds are bypassed, and improving the characteristics of vehicle fuels). Act no. 36/94 ("Galli") on water management allocated resources for modernisation of infrastructures in this sector and contained indications on water saving and recycling.

The resources allocated to deal with environmental problems remained insufficient until the end of the 1980's, when total expenditure on environmental issues by central, regional and local authorities finally reached 1% of GNP (Ministero dell'Ambiente 1992, 332-337), a level considered by international experts to be the minimum for a developed nation, but still lower than the EU average of 1,2% (OECD 1994, 94). Consistent resources have been allocated for the construction of water treatment plants (Act no. 650/79), waste disposal facilities (Acts no. 441/87 and 475/88), energy saving (Acts no. 308/82 and 10/91), reducing eutrophication of the Adriatic sea (Acts no. 283/89 and 57/90), protection against floods and soil management (Act no. 183/89). Allocations for water treatment increased substantially between 1986 and 1993 (from 6.9% to 23.3% of total environmental expenditure), whereas air pollution suffered a decrease (from 38.6% to 12.8%). In 1995, out of a total of 2,164 billion lire allocated to central ministries for investments in the environmental field, 132 were targeted for water purification, 50 for air pollution, 941 for waterway management and protection against floods, 82 for waste, 160 for nature protection, 232 for areas with risks of environmental crisis, 5 for industrial risks, 34 for environmental instruments, 0 for environmental education, information and research, 50 for creating new employment in environmental protection, and 476 infrastructures connected with water management (Ministero dell'Ambiente 1997, 325). It should be pointed out, however, that a very small quota of such financial resources are actually used to deal with environmental problems, rather than for actions aimed at preventing further pollution.

At the same time, it is interesting to note that the actual capability of spending allocated funds proved to be rather low: the Environment Ministry was able to disburse only approximately one-third of its total funds in 1987 (this was the lowest rate among all central ministries, which, on the average, are able to spend 50%

of their budgets) and 21% in 1990 (Istat 1993, 251); the procedural difficulties encountered by the administration in actually spending allocated sums is still a major problem (Ministero dell'Ambiente 1997, 323-324). Only approximately 60% of environmental funds are included directly in the Environment Ministry's budget, the rest being controlled by other ministries (Public Works, Agriculture, Industry); furthermore, less than 40% of such sums are spent directly by the central administration, whereas the 60% is transferred to regional and local authorities and private firms (Cesaretti 1995, 14-16). The ministry has tried to co-ordinate expenditure for environmental purposes by all public authorities by promoting triennial plans capable of defining priorities (Ministero dell'Ambiente 1992, 416). However, the exercise has not been a success: fund allocation in fact has not reflected strategic choices by administrations and actual needs on the basis of levels of pollution, but rather pre-determined criteria (e.g., regional surface rather than level of industrialization), it has not accelerated the expenditure process (quite the contrary, the delays of some administrations have slowed down the entire process) and has not mobilized additional local resources (Malaman 1994), nor has it improved spending capacity (in 1993 only less than one third of the monies allocated in the 1988-91 Triennial Plan had been spent: Cesaretti 1995, 43). An actual national environmental plan has never even been considered. The most recent National Health Plan (Decree of July 23, 1998) for the period 1998-2000 also includes a number of goals concerning improvement of environmental conditions.

During the late 1960s and early 1970s, pollution problems were dealt with primarily by local authorities and lower level judges, who resorted to the available legislation dating back to the first decades of the century, such as the 1934 "health laws" or other provisions originally introduced to protect specific economic activities (navigation, fishing; Belligni and Robutti 1991, 118; Lewanski 1986, 45). These interventions were hardly effective, but at least contributed towards making policy makers and the public aware of the existence of pollution issues; also, economic interest groups began to understand that modern legislation offering juridical certainty for their activities and decisions was preferable to random interventions based on antiquated provisions. In more recent years, the role of the judiciary appears to consist mainly in the enforcement of legislation rather than in promoting the development of environmental policy, as in the initial phase of the policy. In this regard it should be noted that the government has decided to introduce harsher punishment of certain types of behaviour implying damage to the environment – until now subject only to administrative fines – by considering them as criminal offences.

Also at the central level, specialized institutional actors began to appear in the early 1970s as a result of the interest taken by the political system in the topic at this period, probably owing to developments in the United States. In 1971 a "Committee for Ecological Problems" was created within the Senate. As the environment started to emerge as a prominent issue, many ministries attempted to assert their hegemony over the topic. The Health Ministry promoted the Anti-Smog Act in the 1960s; Public Works tried to control water pollution policy (but was eventually pushed aside in the course of subsequent developments); the Ministry for Scientific Research tried at one point to gain control of the policy, by publishing a first national Report on the State of the Environment in 1973. In 1971 a first Environment Ministry – albeit without portfolio – was set up. However, the interest of the political system in the issue rapidly declined; in 1974 the ministry was transformed into a "Ministry of Environmental and Cultural Goods", with powers mainly concerning the conservation of cultural heritage (monuments, museums and similar), and no interest in pollution problems or nature protection. The Inter-Ministerial Committee for Environmental Protection, set up in 1979 with competences in the fields of water and sea pollution and solid waste, was an attempt to create a body intended to co-ordinate powers dispersed among 16 different ministries. An Ecology Minister was created in 1983, but without portfolio. The establishment of a full-fledged Ministry for Environment in 1986 (Act no. 349) "officially" marked the upgrading of the status of environmental policy vis-a-vis other sectoral policies. It was initially entrusted to representatives of the small moderate Liberal Party, close to traditional economic groups, but from 1988 to 1994 it became a "monopoly" of left-wing Socialists. A strong influence in determining the policies of the new ministry was exerted in the latter period by Mr. G. Ruffolo who was minister for almost 5 years, an exceptionally long period by Italian standards. The new ministry found itself in a difficult situation in which, with an administrative apparatus that was very weak owing mainly to a lack of expertise and feeble political support, it had to patch-up a number of emergencies due to a serious policy backlog in the previous period, and at the same time pursue more effective long-term measures. It was also engaged in staking out its area of influence against the resistance of other ministries (Industry, Health, Civil Protection), which tried to limit the loss of powers to the newcomer. Notwithstanding such a difficult situation, the new ministry was instrumental in promoting sectoral policies.

The law that set up the ministry permitted environmental associations to litigate in the name of collective interests, included them in national advisory bodies, and created opportunities to take part in siting decisions through impact assessment procedures.

The availability of information on environmental conditions, initially very poor, has increased considerably since the late 1980s: a "Report on the State of the Environment" was published in 1989, a second in 1992, and a third in 1997. The ministry, together with scientific institutions such as ENEA and CNR, promoted the creation of a national information system (SINA) integrating existing monitoring networks operated by a number of private and public actors, and encouraged major cities to set up or upgrade monitoring activities, especially in the field of air pollution generated by traffic (Herman and Lewanski 1992; Istat 1993, 48ff.). Citizens are entitled to have access to documents held by public authorities and to environmental information (Acts no. 349/86 and 241/90, decree 39/97).

Important institutional actors were created during the 1970s at the sub-national level, such as the above mentioned ordinary regions and the Local Health Units, responsible for enforcement and inspection activities in the environmental field as well. The environmental tasks of the latter are being transferred to the regional agencies that are in the process of being created. Furthermore, the basic responsibility for implementing environmental policies rests on local government, which, for different reasons and to varying degrees in different areas, have upgraded their interventions in this field in recent years. Such regional/local policy focus is also responsible for the most interesting advances towards environmental protection (for example in the field of waste recycling) and for the somewhat fragmented implementation of the policy.

The political visibility of environmental problems has increased considerably since the mid-1970s, as a result of both the emergence of the nuclear issue and serious industrial accidents that attracted media attention and contributed significantly to the growing public awareness of environmental problems. Particularly important in this respect was the accident that occurred in July 1976 at the ICMESA chemical plant in Seveso, just North of Milan. As a consequence of dioxin emissions, thousands of people were evacuated from the area and 220,000 people were placed under medical surveillance. The seriousness of the Seveso accident, together with the difficulties the responsible authorities met in handling this emergency, coupled with another chemical accident in Manfredonia in the same year, had a major impact on Italian public opinion. The following year, nuclear energy suddenly became a major public issue as the proposed construction of a new power plant at Montalto di Castro (just north of Rome) met strong resistance from the local population, backed by a lively student movement.

The level of environmental awareness among the general public probably reached its climax in the second half of the 1980s, also due to environmental emergencies and disasters occurring both in Italy and abroad (Bhopal, Chernobyl) and reported in the media, which played an important role in stimulating public demand for environmental quality. Various surveys indicated that environmental quality was one of the public's main concerns, though there seems to have been a decline since the early 1990s. A number of surveys carried out over the last decade agree that concern for the environment is quite strong in Italy, but is not a top priority. Typically it comes fourth, after such issues as crime, unemployment, and drug abuse (Liberatore and Lewanski 1990, 36; Istituto per l'Ambiente 1993; Ecoimpresa 1998, 8). Traffic, air pollution, and waste are considered the most serious specific environmental problems (Istituto per l'Ambiente 1993, 21). Interestingly, however, only a minority of the population selected the environment when asked to name the single most serious social problem (in 1991, only 3%, ranking seventh; Chianura 1991, 5). Whereas in 1990 people considered the government and industry to be responsible for environmental problems, now 84% of the sample interviewed stated that solutions depend on active involvement of the citizenry (Ecoimpresa 1998, 8). However, people seem to be hardly willing to accept sacrifices in their lifestyles, but try to improve environmental quality mainly on a personal basis through a change in consumption patterns, e.g., by buying health foods and natural cosmetics, rather than engaging in direct social activism (Istituto per l'Ambiente 1993).

Citizens have had the opportunity to express their opinion on environmental issues in a number of referenda. In 1987 more than 70% of voters voted against the use of nuclear energy (proposed referenda against hunting instead were judged not to be admissible by the Constitutional Court; in 1980 the court had taken a similar position against proposals concerning hunting and nuclear energy). During the 1980s, several local referenda also took place on topics such as traffic restrictions (in such cities as Bologna and Milan) and the closing down of local industrial plants. A referendum on hunting and another one concerning the use of pesticides in agriculture were held in 1990, but, though approved by the great majority, did not reach the required quorum (only 43% and 45%) (another referendum against hunting in 1996 attracted an even lower turnout). In April 1993, 82.5% voted against the health units carrying out environmental protection activities (which led to the creation of the regional agencies – ARPA – mentioned above). In terms of actual impact on policies, the referenda on hunting and pesticides, as well as the local referenda on traffic, were unsuccessful in inducing a substantial shift in public policies in these areas; the referendum on nuclear power clearly accounts for Italy's decision to abandon this sector once and for all, though it had never actually managed to establish a full scale programme comparable to that of other countries. Perhaps the referendum that has had the greatest impact on environmental policy was that against local health units, since it gave birth to the regional agencies, though this is largely an unintended consequence that voters were unaware of when casting their ballot.

The relevance of environmental issues clearly decreased during the 1990s. Besides a possible a decline in interest among the general public, this is the result of growing concern about worsening economic and employment conditions, on one hand, and the most serious crisis that the political system has experienced since the end of WW II. The break out of *"Tangentopoli"* (the disclosure of extensive corruption within the governing parties and the administration) combined with the end of the East-West confrontation brought the Italian "First Republic" to an end; many of the previously powerful personalities and parties have disappeared from the political scene. A new political and institutional order has yet to emerge from a transitional period of great uncertainty. In this situation, environmental issues have hardly been a priority.

Also because of the prevailing features of Italian political culture, that will be discussed below, political parties have shown little actual interest in environmental issues beyond paying lip-service to them. Politicians who have distinguished themselves as entrepreneurs in this field have not had very successful careers (perhaps with the exception of G. Ruffolo). It is, however, interesting to observe that the largest Italian party, the PDS – previously communist – has incorporated some environmental themes (such as a shift from a pro- to an anti-nuclear position), in contrast with its traditional "productivist" workers class culture, which perhaps can be interpreted as an attempt to develop a new ideological identity.

Though environmental policy has certainly grown since the mid-1960s, governments are never openly characterized by a pro- or anti-environmental stance. More recently, things have changed somewhat. The Berlusconi government (1994-95) was perhaps the first to hold an openly anti-environmental position; the ministry was headed by a representative of the extreme right, who showed no interest in environmental protection and openly stated that he was in favour of nuclear energy, the construction of new highways, and of hunting in national parks (but on the other hand it was the first government to propose the introduction of environmental taxation). In the following centre-left government, headed first by R. Prodi and, since the autumn of 1998, by M. D'Alema, leader of the PDS, the situation was somewhat reversed; the Greens, as part of the governing coalition, obtained the post of the Minister of Environment. Though it is too early to offer an overall evaluation of its performance in environmental terms, the Prodi and D'Alema governments have passed a number of significant measures:

- Institution of a considerable number of new land and marine parks, along with an increase of resources allocated for their management (art. 2 Act 426 of 1998).
- Promotion of recycling and reuse of urban and industrial solid waste (Decree no. 22/97 and 389/97); national legislation laid down that 30% of municipal waste should be recycled by the year 2002, though municipalities are complaining that they are unable to respect such a target.
- Protection of the stratospheric ozone layer (Act 179/97 has strengthened previous measures).
- Standards limiting exposition to electromagnetic emissions (radio, TV and cellular phone antennas, high voltage transmission lines) on the basis of a national plan to reduce the number of sites hosting TV and radio antennas (from 700 to 487) and the transmission power of such antennas by a factor of 30 (Act no. 381 of 1998).
- Use of environmental projects (water management, renewable energy sources, natural parks) to create new job opportunities.
- Clean up of highly polluted and risky land and water sites (such as industrial areas; funds have been allocated by Act no. 426 of 1998).
- Demolition of illegal construction in protected natural areas; the Ministry of Environment can take action if municipalities do not. Demolition of one of the "symbols" of this phenomenon, the Hotel Fuenti, located in a site of particular scenic beauty along the Amalfi riviera, specifically mentioned in Art. 2 of Act no. 426/98, started in April 1999 (Act no. 426 of 1998).
- Reduction of air pollution caused by vehicular emissions, especially in urban areas. The Prodi and D'Alema Governments have used economic incentives extensively to induce owners to replace older cars and motorcycles by new, less polluting models (one third of automobiles are still not equipped with catalytic converters – Ministero dell'Ambiente 1997, XVII). The Government has announced its intention to pass a similar measure to promote the sale of electric vehicles. The sum of 360 billion lire was allocated in the period 1997-99 to assist local authorities in buying more up-to-date vehicles for public transportation, including ones with low emission levels, and 100 billion for buying less polluting service vehicles. For the first time resources have been made available to build bicycle lanes (Act no 366/98). Further funds have been allocated to promote construction of urban rail public transport systems (Art. 50 of Act no. 448/98). Lower emission standards for motorcycles (a very widespread mode of transportation in Italian cities: some 8 million), will be enforced starting June 1999, as a result of the implementation of EU Directive 97/24. Further reductions are to be pursued also through compulsory installation of devices capable of avoiding the dispersion of vapours at gasoline pumps in major cities (Decree no. 76 of January 20 1999).
- Also, the Decree of October 23, 1998 has strengthened the powers of mayors to introduce limitations to traffic when air pollution limits are bypassed; the provision has also set tight limits (starting January 1 1999) for typical traffic pollutants such as benzene, aromatic polycyclic hydrocarbons and fine dust particles; recent measurements have shown that present levels of such pollutants are well above the new legal limits. Since motorcycles seem to contribute consistently to such emissions (a survey carried out in Florence attributed almost 50% of to-

tal emissions to such vehicles), mayors will be confronted by the – politically very sensitive – issue of limiting two wheel motorized traffic.

- Reduction of CO_2 emissions, implementing the Kyoto agreement. The main strategies for pursuing the 8% reduction of CO_2 emissions goal are indicated by a Decree of CIPE – a governmental inter-ministerial committee (n. 137 of November 19, 1998). Measures include introduction of a CO_2 tax (discussed below), incentives for consumers to replace household appliances by less energy consuming models, introduction of energy consumption labels for new domestic appliances (Decree of the Ministry of Industry of October 7, 1998), substitution of oil by natural gas for electricity production. Funds for such programmes have also been allocated (Act no. 426 of 1998). All in all, the Kyoto agreement appears to be used by the Ministry of Environment as an opportunity to integrate environmental issues with other sectoral policies, especially in the energy and transportation fields.
- Promotion of renewable energy sources: after the liberalization of the electricity production market, hitherto a public monopoly, each firm on the market is compelled to produce a minimum of 2% by means of renewable energy sources (Art. 11 Act no. 79 of 1999).

Economic issues (inflation, employment and public indebtedness) are still in the forefront of the government's concerns. To be sure, no government has yet given serious consideration to sustainability issues in programmes or policies, though the word "sustainable" has started to appear in legislation (as in the Public Works Ministry Decree of October 8 1998 aimed at promoting innovative programmes for upgrading urban areas, and Article 3 of Act no. 426 of 1998 allocating funds for "urban sustainability"). The ministry is also attempting to foster the Local Agenda 21 process. In this field, however, Italy's record is quite poor compared with other European countries: By the end of 1998 only some thirty local administrations (mainly municipalities, but also two provinces – Modena and Reggio Emilia) had signed the Aalborg Charter, and two regions (Emilia Romagna and Umbria) had signed the Göteborg Charter; furthermore, actual LA21 activities (such as drafting an Action plan) appear to be extremely limited (Ramieri and Fiorentini 1998, 123). However, local authorities in 1999 set up a coordinating committee, and have held several meetings to exchange experience and develop capabilities. The Agenda 21 process thus seems to have been mainly "bottom-up" so far.

2.2 Instruments

As far as policy tools are concerned, Italian policy in this field relies mainly on a regulatory "command and control" approach based on emission, process (e.g., combustion), and product standards (e.g., detergents or fuels, as discussed above), and, somewhat later, on ambient quality standards. The Ministry of the Environment has also resorted to less traditional regulatory approaches, such as the creation of compulsory consortia (Act no. 475/88 and decrees of 1990) set up to promote recycling of selected materials, such as glass, plastic, aluminium cans, mineral oil and car batteries). The consortia are non-profit organisations, whose members are producers, importers and user firms; they receive financial resources from charges levied on the products. Minimum collection targets have been set for specific years; deposit-refund systems and other product taxes will be introduced should these targets not be met (Malaman and Ranci 1991). Recycling of plastic containers from liquid substances passed from 2,000 tonnes in 1991 to more than 32,000 tonnes in 1995 (equivalent to approximately 10% of the total quantity produced; Ministero dell'Ambiente 1997, 299). The amount of aluminium from beverage cans that is recycled has grown from less than 2,000 tonnes in 1989 to 6,400 in 1995, equivalent to one forth of the total quantity produced (Ministero dell'Ambiente 1997, 302). Approximately 70% of residual lubricating oil is collected by the compulsory consortium for used oil; a large portion of it is regenerated, and the rest burned for energy production or incinerated (Ministero dell' Ambiente 1997, 300). COBAT – the consortium for used batteries – collected 154,000 used vehicle and industrial batteries in 1995 (mainly from large public and private firms), compared to 160,000 sold in the same year (Ministero dell' Ambiente 1997, 300).

Economic instruments have supplemented such regulatory approaches. Economic tools were used as far back as the mid-1960s, contributing to the success of Act no. 615/66 in curbing urban smog by promoting the use of light oil instead of the heavy oil widely used at the time for domestic heating, through the reduction of taxes levied on such fuel (Acts no. 608/66 and 9/70). In 1979 water pollution legislation provided for effluent taxes to be levied by municipalities, but the provision was in fact hardly applied (Lewanski 1986, 71). Yet ecological taxes are still less relevant than in other countries. Proposals by the Environment Ministry to impose taxes on certain activities (e.g., pig-breeding, pesticide use, and others) and products (pesticides like atrazine, and several plastic materials) have been opposed by the Ministries of Agriculture and Industry in defence of their respective constituencies (Gerelli 1990, 197). For many years the only actual example of an ecological tax on products was the charge imposed on plastic shopping bags (Act no. 475/88) to induce consumers not to dispose of them improperly; weak implementation by the administration has caused the scheme to have a low impact (Mariani 1991) and has been substituted by a tax on polyethylene foil. A tax on older types of aircraft responsible for high levels of acoustic emissions was introduced in 1990; a part of the sums deriving from such tax are to used to mitigate noise levels (Act 165, art. 10).

The turning point however is represented by a policy document produced by the Ministry of Finance of 1994 ("Libro Bianco sulla riforma fiscale"), which attributed a relevant role to eco-taxes within a broader reform of the taxation system. Financial acts of recent years have introduced or consistently increased charges on environmental services such as refuse collection and disposal; according to measures recently passed by government, taxes on urban waste in the future should be calculated on the base of weight rather than floor space, in order to encourage recycling. Charges for water services – supply, sewage and treatment – were been introduced by Act no. 36/94, also in response to the need to cut public expenditure. Taxes and deposits have also been introduced on such products as batteries, lubricating oil, glass and metal beverage containers, plastic containers for liquid substances, and CFCs; the resulting revenues go to the above mentioned consortia. But the most relevant development occurred in 1998 (Act no. 448, art. 8) with the introduction of a CO_2 tax, within the context of the policies that the government is introducing to meet the emission reduction obligations deriving from the Kyoto Conference. The tax has caused an increase in the price of all types of fuels as from January 1, 1999. Further incremental increases are planned until 2005. Revenues from the tax are to be used for a variety of purposes, notably to reduce the labour costs and promote emission reductions, energy efficiency and renewable energy sources. Besides reducing emissions, the tax will also be used to pursue social objectives, such as promoting the creation of additional jobs. The provision has been strongly opposed by the major electricity producer ENEL and the national oil holding ENI; the result is that tax levels applied to fuels used for electricity generation are, in fact, slightly lower than for other uses.

Up to now, voluntary agreements have played a minor role. The ministry concluded several such agreements with major firms (such as petroleum companies, FIAT, Enichem, Montedison and the public railway company) willing to assume obligations to adopt clean-up measures in exchange for specific decisions to be taken by the government. In the FIAT case, for example, the government pledged it would pass tax measures encouraging customers to buy vehicles with catalytic converters in exchange for FIAT's commitment to provide all its vehicles with such anti-pollution devices one year in advance of the EC deadline (in the end however, the deal did not work out because of the opposition of the Finance Ministry). FIAT also agreed to adopt other measures such as the elimination of asbestos from its vehicles, the recycling of old vehicles and research investments. More recently, FIAT has agreed to cut the petrol consumption of its vehicles by 20% starting from 2005 and to produce a 3 litres per 100 km "city cart" in 2000 (Ministero dell'Ambiente 1997, VIII).

Interest of firms in adopting ecolabels is slowly growing, as consumers appear to be interested in products that offer a guarantee of the their actual environmental features. *"Cartiera Lucchese"* – a paper mill – was the first to obtain such a label; several other firms have applied for it. Environmental auditing has not yet reached levels comparable to those of more advanced European countries: there are only 51 registered ISO 14001 sites (Ireland: 65; Spain: 81, whereas the most advanced European country – Great Britain – has 650), and only 8 EMAS registered sites (Great Britain: 56; Norway: 43, not to mention Germany: 1285) (data as of October 1998; source: Il Sole 24 Ore, EMAS Helpdesk).

2.3 Policy Style

The style of Italian environmental policy can be characterized as strongly legalistic and rigid in both contents and procedures, and as highly incremental, fragmentary and reactive.

As far as the first aspect is concerned, the regulatory approach features an exasperated formalism, uniform standards covering the entire national territory, rigid and complex procedures, and an absence of administrative discretion that strongly limit the scope of action and the chances of success of administrative action. This situation is caused by the, at least until recently prevailing culture in both government and administration featuring a legalistic orientation towards strict application of provisions, typical of last century's administrative model (Freddi 1989), rather than a pragmatic interest in the actual attainment of results (La Spina and Sciortino 1993, 217-20) and problem-solving attitudes and values (Aberbach, Putnam

Putnam and Rockman 1981, 52). Such traits are quite evident in the case of EIA procedures, which until now have proved useless in siting decisions precisely because of the formalistic approach to implementation. Furthermore, technical competence – particularly relevant for environmental policy – is rare and enjoys low status in the eyes of the political and administrative actors.

An explanation of the second trait lies, at least in part, in the configuration of national political culture featuring familism, localism, clientism, a fragmented political culture, lack of trust in others and in public institutions in particular, in a word, a very imperfect civic culture (Cavalli 1992, 393). In some parts of the country one could go so far as to speak of the predominance of pre-modern political values. For the purposes of this discussion, one particularly relevant consequence of this situation is that public goods – including environmental quality – enjoy low priority in the prevailing political culture; Italy therefore suffers from a "syndrome" typical of the Mediterranean countries as opposed to northern European nations, constituted by an inherent difficulty in providing collective goods owing to a culture that discourages cooperative behaviour (La Spina and Sciortino 1993, 217).

Such traits provided fertile soil for a clientistic style of policy making: the political system, at least until recently, has typically generated consensus by granting immediate and direct benefits to particularistic groups in exchange for support and votes, rather than paying attention to more general societal needs. Besides a tendency to pass legislation along such lines (Di Palma, 1977), political parties have been keen on putting their representatives in key positions (public agencies, state-owned industries, etc.) and preventing experts, technical bureaucracies or actors representing societal interests from monitoring such decisions. The political and administrative system is highly "permeable" to particular interest groups and clienteles, a feature that provided ample opportunity to hinder unwelcome measures.

The traits described above largely account for the difficulty collective issues encounter in getting on the policy agenda, for the lack of clear strategies for the development of environmental protection, and for the largely reactive nature of environmental policy, that in the Italian case appears to be especially pronounced. In fact, almost all policy measures in this field are directly triggered by two types of stimuli.

The first such stimulus is environmental emergencies and disasters; policies to deal with industrial waste, for example, were precipitated by the discovery in the late 1980s of illegal dumps of toxic waste both abroad and in Italy (the latter "managed" by criminal organisations). Such an emergency-based approach calls attention to problems and creates windows of opportunity for policy development; on the other hand, its effectiveness is certainly inferior to proactive approaches (as a comparison between Germany and Italy in promoting the diffusion of vehicles fitted with catalytic converters shows; Desideri 1993, 32, 40). Furthermore, once the immediate emergency is over, the issue is often quickly forgotten because of inertia or counter-interests (farmers for example have been successful in countering attempts to control the use of agro-chemicals that were proposed after the drinking emergency that, as mentioned, hit the Po Valley in the late 1980s).

The second type of stimulus is exogenous to the domestic polity, namely international environmental obligations, and especially policy measures taken at the EC level. There is little doubt that a very considerable portion of current national environmental legislation would not have come into existence without the pressure to comply with Community directives (though Italy has one of the worse records in enacting them; Capria 1988, 201), as shown, for example, by an analysis of air pollution policy (Lewanski 1996, 74). Notwithstanding the relevance of international co-operation in this field for the development of domestic policy, Italian governments, at least until the mid-1980s, adopted a rather passive posture in promoting it, though Italy has joined all projects for reasons of general international relations and a general orientation in favour of European integration (Skou Andersen 1994, 11). More recently, the Environment Minister has taken a stronger stance in international negotiations, as in the case of the Kyoto Conference, in which Italy proved to be amongst the most active and more pro-environmental countries. The reason for such increased "activism" probably lies in the intention of the Minister to use the international obligations deriving from the agreement to legitimise domestic policies that he is currently promoting.

3 Policy Evaluation

All in all, the performance of Italian environmental policy still appears to be inadequate in both absolute terms – with regard to the deterioration of environmental resources – and in relative terms – compared with the results obtained by similar industrialised countries. Sole consideration of the time factor – when institutions were set up and principle legislation passed – shows that, all in all, Italy is consistently 10-15 years behind other countries.

The more serious limitations of Italian environmental policy, however, are to sought not so much in delay, but rather in its lack of actual impact on the environmental problems at stake (Alexander 1990, 183).

The lack of adequate time series of comparable data on environmental conditions is in itself a significant factor in evaluating sectoral policy, and one that makes it difficult to single out the trends under way. The growing damage to forests from acid depositions, for example, could be ascribed to an upgrading of the survey activities over recent years, rather than to the actual worsening of the phenomenon (Conte e Melandri 1993, 224). Some data on air and water pollution and nature protection have been provided at the beginning of this chapter.

Another partial indicator is provided by recent data on the results of enforcement action, which shows that violation of environmental legislation is widespread. The specialized section of the Carabinieri (NOE) found violations in 68.1% of public waste water treatment plants (out of a total of 1,791 inspections), and in 57.6% of privately owned facilities (total: 203), in 88.3% of 1,763 sewage discharges inspected, in 98.9% of 903 water discharges, in 65.7% of 391 excavations inspected, in 77.8% of public waste disposal sites, and in 69% of private sites; almost half of 4,300 waste disposal sites existing in 1990 were operating without the required permits (Arma dei Carabinbieri 1995, 44, 50, 64). Such data certainly indicate that serious deficiencies do exist, but should not necessarily be interpreted as a deterioration of the situation. On the contrary, since inspections tend to focus on plants and sites that are suspected of not operating under required conditions; more importantly, the availability itself of such data indicates an upgrading of enforcement activities, which will presumably reduce violations over time.

In terms of environmental quality, the impression is that certain "classical" environmental problems have been tackled, with a certain measure of success. One field in which progress is evident is nature protection, with a sharp increase in the size and the quality of protected areas, and with the growing effectiveness of so-called "landscape" protection (OECD 1994, 38-9). Secondly, industrial pollution has been the primary object of the policy, with some success being achieved in reducing discharges to the air or to waters; this is partly an effect of the "natural" death of heavy industry, and partly of the specific structure of Italian industry, mainly constituted by small and medium size firms. This specific character of the industrial sector has implications both "in terms of the likely small mass emission of pollutants contributed by each individual plant" (OECD 1994, 48), as well as in the fact that plants are concentrated in industrial districts that allow relationships based on cooperation and flexibility between the firms themselves and between the firms and the (local) controlling authority to develop, thus overcoming the rigidities of the "official" policies. Considerable success has also been obtained in improving the urban environment, as in the case of urban smog caused by stationary sources, in the purification of domestic sewage and wastewater; there has also been a sharp increase in recycling of waste, especially in some cities (e.g., Milan).

In urban areas, however, serious environmental problems remain, such as air and noise pollution caused by traffic, also because of weakness of public transportation and the late adoption of measures such as catalytic converters and unleaded petrol. Emissions from vehicles (NO_x, CO, CO_2) have significantly increased over the years, which is not surprising if one considers that car ownership is among the highest in OECD countries (OECD 1994, 58 and 125). Transportation in general is a serious environmental degradation factor; road traffic in Italy has increased much faster than GDP over the last two decades (132 % against 83%; OECD 1994, 125). Freight traffic, in particular, increased by 150%; the share of freight transportation by road has risen from 65% in 1980 to 72% in 1993, whereas transport by rail and water has constantly lost ground (ISTAT 1996, 248). Agricultural soil and water pollution is another serious issue yet to be tackled.

4 Capacity of Proponents of Environmental Policy

4.1 Public Institutions

The 1946 Italian constitution contains no explicit reference to environmental protection. The structure of the state is essentially a four-tier system consisting of central government, regions (15 ordinary and 6 special), provinces (105) and municipalities (ca. 8,100). Initially, environmental policy was mainly local in character; subsequently central government attempted, with some success, to assume responsibility; in any case, policy expansion has attributed increasing powers to all levels of government. It should also be noted that at the end of the 1990s Italy is undergoing a process of "federalisation" and that substantial powers in fields of relevance for the environment have been passed over from central government to the regions and the other levels of local government.

The State's principal powers, generally speaking, concern the distribution of competences, procedures, pollution standards and measurement methods, national sectoral planning, and financial resources. Central government is organised on a traditional ministerial model. The main actor at the central level, since 1986, is the Ministry of Environment. The ministry has been given responsibility for air, water, soil and noise pollution, solid waste, mining, parks, as well as sea and coastal protection. The initial objective of concentrating all environmental policy powers in the hands of one ministry has not been fully attained, since pre-existing ministries – such as Merchant Marine, Transportation, Industry, Health, Cultural Heritage and Environmental Goods, Interior – have retained substantial jurisdiction in matters of environmental relevance. The Ministry of Public Works, for instance, controls very substantial powers and financial resources, especially in such fields as water management and soil protection. Furthermore, the Environment Ministry's capability for action has been severely hampered by the procedural requirement that it consult with and secure the agreement of other interested ministries in a number of matters. Finally, in other areas of critical importance for the environment – such as energy, industrial and agricultural policies – the Ministry for Environment has very limited powers for ensuring that adequate attention is paid to environmental aspects along with more "traditional" technical and economic considerations. The government recently presented a proposal to cut the number of ministries to twelve; in this event, the Ministry of Environment would also be given powers and financial resources currently belonging to Public Works, in the field of land use planning, water policy and river basin planning.

Furthermore, the operational capacity of the ministry remains far too weak due to lack of professional competence. Personnel transferred from other ministries, mainly Health and Public Works, brought along with them traditional attitudes, values and approaches.

The ministry was set up much along the traditional bureaucratic model, hardly suited to the technical and political nature of environmental policy. It is structured along six departments: 1) ARS (water, waste, soil); 2) VIA (impact assessment, citizen information, reports); 3) IAR (air and noise pollution, hazardous industries; 4) CN (nature conservation); the remaining two deal with personnel management and internal organisation of the ministry (Frey 1993, 11; Salvia 1989, 37). Two committees, one formed by experts and the other by representatives of public authorities, provide advice to the ministry.

A national environmental agency *(Agenzia Nazionale per la Protezione dell' Ambiente)* was instituted in 1994 (Act no. 64), but is still in the process of being set up; it will have a staff of approximately 600 (mainly coming from existing authorities such as ENEA). Its tasks cover the collection, elaboration and diffusion of information and data, cooperation with the European Agency, diffusion of low impact technologies and production methods, study activities connected with impact assessments; it can also provide advice on standards, measurement methodologies and new legislation. It can carry out inspection activities and provide advice to authorities on the licensing industrial plants.

The ministry can also resort to other authorities that do not depend on it, such as the ISPESL *(Istituto Superiore per la Prevenzione e la Sicurezza del Lavoro*; with a staff of about 1,300) and the national technical services (seas, hydrographic, sysmic, dams, geologic), with a total staff of approximately 1,100 (Conte e Melandri 1993, 254).

Scientific and technical know-how on environmental issues are provided by other agencies such as the National Health Institute *(Istituto Superiore di Sanità),* ENEA (the former nuclear energy agency, which was partially reorganised in the 1980s to broaden its field of interest to alternative energy sources and environmental issues; in 1999 ENEA – now headed by Nobel prize winner C. Rubbia – was again reorganised on the basis of Act no. 36 and assigned tasks in the field of research and innovation connected with sustainable development), specialized institutes of the National Research Council *(Consiglio Nazionale delle Ricerche),* and by universities. All in all, however, the in-house and external technical expertise available to the ministry and other public authorities remains largely below the required level.

A special branch of the Carabinieri, the *Nucleo Operativo Ecologico,* carries out inspection and enforcement activities on behalf of the ministry, which can also resort to police forces dependent on other ministries *(Corpo forestale, Guardia di finanza and Capitanerie di porto).*

After considerable indecision and confusion during the 1970s on whether environmental powers should be concentrated at the municipal or the provincial level, Act no. 142/90 made the provinces (beside the regions – the most important institutional actors) chiefly responsible for environmental policy at the local level, with substantial powers in such fields as water and soil protection, waste management, natural resources (flora, fauna, parks) protection, pollution monitoring, and public health. Municipalities remain responsible for the management of public services (waste collection, domestic water distribution, sewage, effluent water treatment, in some cities energy production), often carried out through controlled public enterprises *(aziende municipalizzate;* approximately 500, concentrated mainly in northern and central Italy).

Summing up, institutional capacity is formed by:

- A centre that is very weak, more in terms of technical and managerial skills than in terms of financial resources; the Environment Ministry was and still is more part of the problem than part of the solution.
- A very diversified local and regional administrative structure quite, but not very effective in some areas of the country, and not at all in others. Considerable variance exists in performance records between different local and regional authorities. Using legislative production as an indicator, several central and northern regions (Emilia Romagna, Umbria, Tuscany, Lombardy, Veneto) have been more active than others (Putnam 1993). A synthetic evaluation of substantive environmental measures ranked Trient and Sud Tirol (two small northern regions, with a special status) as the most active regions, followed by a group of central regions (Tuscany, Emilia R., Liguria and Abruzzo), and subsequently by the large Po Valley regions (Lombardy, Veneto and Piedmont). In both cases, the southern regions rank lowest (Legambiente 1995, 199).

The explanation for such variance is to sought in the fact that some regions, especially in the wealthiest parts of the country, have followed more flexible and cooperative approaches rather than conforming to the traits of formalism and legalism (and therefore of fragmentation and ineffectiveness) typical of the national policy style.

4.2 Environmental Organisations

The "green archipelago" features quite different genesis and cultural matrixes within itself. The first associations born in the post-war period were mainly concerned with nature conservation: Pronatura in 1953, Lipu (protection of birds) in 1965, WWF in 1966, Federnatura in 1970. *Italia Nostra,* focused on the conservation of the historical heritage, was founded in 1955. Starting the 1970s, a second generation of environmental associations, with a stronger political character, emerged, some with a marxist, anti-capitalist and in partly anti-industrialist matrix *(Lega per l'Ambiente* was set up in 1980), others influenced by the American single-issue political culture like Friends of the Earth (in Italy in 1977) and later (1986) Greenpeace. The environmental movement gained considerably in visibility thanks to the anti-nuclear opposition that arose in the mid-1970s. Total membership is presently estimated at 1.5 million, but this figure includes associations based primarily on recreational and tourist activities (Alpine Club and Touring Club with some 900,000 members), as well as multiple memberships. The largest associations are the WWF (270,000 members, 200 local offices, a budget of 27 billion lire in 1998 and a staff of 110, including surveillance personnel in the protected areas it owns and manages), LegAmbiente (110,000 and a budget of approx. 10 billion lire) and Greenpeace (40,000, a staff of 6 and a budget of 1.7 billion lire), but a number of smaller associations are also active, such as the LIPU (concerned with bird protection), FAI (the equivalent of the British National Trust), Friends of the Earth, EkoClub. Associations can resort to friendly experts, and have several research centres (e.g., *AmbientItalia).* Though the major associations often join forces on specific issues, there are also occasions when their positions can diverge considerably, especially when relationships with institutions and political parties (including the Greens) are at stake, because of the different ideological matrixes described above.

The Green Party, strongly influenced by the New Left culture of the 1970s, has been on the political scene for some 20 years now: a Green list appeared in a regional election (in Trient-Sud Tirol) for the first time in 1978. Since then it has been by and large present in local, regional and national elections. On average, it obtains about 3% of the vote in national elections. It entered parliament for the first time in 1987. In the last elections held in April 1996, somewhat paradoxically, due to the majoritarian electoral system and agreements among the parties belonging to the "Olive" (centre-left) coalition, the Greens, while obtaining one of their worst results yet (2.5%, i.e., 937,000 votes), were rewarded with the largest number of representatives in the lower chamber ever (rising from 18 to 28 seats).

The capability of the environmental movement actually to influence policy-making has been relevant in a number of specific issues, but all together has remained low as far as general policies are concerned, perhaps also because the Green Party, being part of the leftist coalition, has contributed to giving a partisan underpinning to the environmental topic.

4.3 Economic Proponents of Environmental Protection

Eco-business in Italy was born later than elsewhere and is more backward than in other industrialized countries (about one third the size of similar German industry and half the French figure). This is mainly due to the lack of pressure exerted by public policy, which has been too weak, slow and uncertain over time (anti-pollution expenditure being only 1.2% of corporate investment in 1986, as compared to 2% in the US, 3.6% in Japan and 7.4% in Germany; Malaman and Paba 1993, 76), and to the lack of demand from the public sector (which is three quarters of the total domestic market). Firms specialized in this sector, grouped in an association of their own – ANIDA – had a total turnover of 2,300 billion lire and 9,600 employees in 1990. Firms are usually small (only 3 have a turnover of more than 100 billion lire), which hinders their innovation capability and their willingness to invest in research, thus making dependent on foreign patents (the quota of Italian European patents in this sector is only 2%) (Malaman and Paba 1993; Drouet 1990, 50). Exports go mainly to less developed countries.

Finally, a "pathological" form of business connected with the environment (though it certainly does not contribute to upgrading institutional capacity), is constituted by the presence of organised crime, favoured by the inefficiency of many public authorities. The phenomenon is especially relevant in southern Italy and in the field of industrial and domestic waste management (60% of all non-authorized disposal sites has been found in this area (data referred to 1990; Arma dei Carabinieri 1995, 113 and 163). According to a cautious estimation, waste disposal, illegal building activities and betting on animal fights generate 12,000 billion lire yearly profits (4,500 only in the waste disposal business).

5 The Changing Constellation of Actors

Even if the "environmental policy community" is still quite weak (very notable is the almost total absence of the scientific community as an autonomous actor), the system of relationships between policy actors has undergone a deep evolution that is both cause and consequence of a development of belief systems, stimulated by policy itself, the impact of the media and participation in international networks where they have been exposed to the experience of more advanced countries.

As the environmental issue initially emerged in the late 1960s and in the 1970s, relationships between actors were highly conflictual. In its infancy, the ecological movement was strongly critical on a number of issues (nuclear energy, hunting, use of pesticides in agriculture, industrial pollution). The prevailing position in industry was simply to deny the existence, or at least the seriousness, of pollution problems, and to accuse environmentalists of being opposed to progress and of making unreasonable and economically unsound demands; firms resisted innovation and perceived environment merely as a cost and a constraint limiting their entrepreneurial freedom and flexibility. Firms exacerbated conflicts by threatening to close down polluting plants; thus it is not surprising that trade unions in this period saw environmentalists as a threat to employment (they were, for example, pro-nuclear). There was also strong competition among institutions, which tried to stake out their areas of influence in this new field, rather than coping with the substantive problems. This can be seen in the conflicts both among central ministries and between central government and local authorities. Public institutions were not, generally speaking, keen to promote policies, and even when legislation was passed, implementation and enforcement were very feeble.

Only very slowly did a network of actors, connected at least by a common language, though not by common values, start to take shape in the second half of the 1980s, allowing for a more co-operative system of relationships to develop. Industrial actors began to recognize their share of responsibility in environmental deterioration, and to accept the idea that the environment is a production factor like any other, to be accounted for; for some firms, it also represented an increasingly interesting business opportunity, especially as traditional production became obsolete or uncompetitive. A number of polluting firms undertook clean-up actions for a number of reasons, besides public regulation, such as safeguarding their image in the eyes of the public and customers (e.g., chemical firms; automobile industry), exploiting the "green consumer" market, and reducing costs by adopting more advanced and cleaner production technologies (Dente and Ranci 1992). The environmental movement became less ideological and more pragmatically open to cooperation to achieve tangible results. In a number of cases, ecological associations accepted to cooperate with firms in promoting activities in favour of the environment (e.g., *LegAmbiente* with Volkswagen in the field of environmental education) or of products (e.g., the WWF: a part of revenues go to the protection of specific areas or species). There are a number of cases of cooperation between industry and NGOs (such as WWF, LIPU and *Italia Nostra)* in the management of protected areas. Trade unions realised that growth based uniquely on mass industrialization has its limitations, that employment could be created also by means of environmental protection, and that the health of its members should be protected not only inside but also outside the factories. Public authorities, once they became interested in actually achieving results, realized that these could be obtained more easily through co-operative relationships (e.g., the voluntary agreements mentioned above) with industries rather than by means of authoritative approaches they were in fact unable to enforce.

All in all, a significant change in attitudes towards environmental issues has certainly taken place among policy actors; however, the shift in dominant paradigms has been much more limited, as demonstrated by the low degree of entrenchment that concepts such as "sustainability" have achieved in the policy discourse.

6 Capacity Building

The number, relevance, and resources – human and financial – of governmental actors, as shown above, has certainly increased over the past 30 years: ad hoc institutions have been set up at all levels of government, central, regional, and local. The production and dissemination of environmental information has consistently increased, and the public is certainly much more aware of such topics than it was in the past. Participation rights, though asserted quite late, are now ample. Yet, overall capacity still appears to be very low, for a number of reasons.

First of all, the institutional system is highly fragmented, causing widespread use of veto powers and time delays (Dente 1989,146); this is not unique, but the legalism of the administrative culture makes it very difficult to bring about intergovernmental cooperation.

Secondly, institutional capacity has so far been discussed with implicit reference to the policy formulation "phase"; in this respect, Italy has certainly made progress, however slow: legal provisions are passed, resources are allocated, specialized political and administrative institutions set up. Yet an evaluation of actual capacity cannot ignore implementation capability, which in turn impinges on actual effectiveness. And this is where Italy shows major weaknesses. Allocated funds remain unspent, as we have seen; and even when they are spent, e.g., on the construction of effluent water treatment plants, such plants in fact perform rather poorly or never become operational (Malaman 1994, 48). The low level of performance in policy implementation is to be explained in terms of the distinctive characteristics of Italian administration previously described in the discussion of national policy style. Once again, there are exceptions that prove the rule, cases of success that feature authorities capable of abandoning legalistic attitudes in favour of result-focused strategies and entering negotiation relationships with polluters (Dente, et al. 1984; Lewanski 1986). Apparently strong and stringent regulation is typically not enforced, to the extent that the Italian case has been characterized as "regulation without rules" (Giuliani 1992, 89). A relevant consequence is that target groups have learned to count on low levels of implementation and enforcement, inaction by the authorities, suspension of deadlines, remissions (e.g., for building permits), loosening up of parameters (e.g., when levels of pesticides above EC limits were found in drinking water, the limits were simply increased for several years) and other similar devices that in the long run damage policy credibility in the eyes of societal actors.

7 Catching up?

The picture that emerges from the elements outlined above may be summed up as follows: after a rather slow take-off in the 1960s and 1970s, environmental policy in Italy has gradually been gathering momentum; by now most of the relevant policy instruments appear to be in place; also, environmental issues have acquired both legitimacy in policy terms and a considerable public profile. All in all, environmental policy in Italy has caught up in these respects with other nations with similar levels of economic development. Although policy responses have come later and to some extent have been less effective than in the most advanced countries, overall Italy is not a special case: policies have developed according to the patterns found elsewhere, albeit somewhat later. As mentioned above, a number of successes can be recorded. On the other hand, they have been achieved in the issues that are easier to deal with, whereas the issues that remain to be tackled, being generated by a multitude of diffused sources and impinging on consumer behaviour, are much more complex, demanding and "costly" in political, institutional, administrative and economic terms: urban air and noise pollution caused by traffic, land use patterns, transportation (especially freight), agriculture.

If one looks for an explanation for such results, the obvious answers are that point sources are easier to clean than diffuse pollution; that cleaning up is easier than prevention; that zero sum games are more difficult to solve than positive sum games; that medium size economic interests (typically the Italian SMEs) are more willing to comply than very large or very small firms, and, finally, that the little expertise that has been collected by public authorities is more likely to be found at the regional/local level than at the national level.

As far as perspectives for the future are concerned, there is little evidence that the system has the capability required to produce innovation in targets, approaches, and instruments to improve effectiveness. More cooperative relationships between pragmatically oriented societal actors are certainly one favourable condition in this direction. The "missing link", however, is the unfitness of the political and administrative systems to act as a catalyst of innovation: the institutionalisation process has still been too "shallow" (Jänicke 1991, 12). As pointed out, Italy's policy in this field has lagged some 10-15 years behind other industrialized countries. Being a latecomer is not necessarily a disadvantage in itself; actually, it could be a wise strategy to avoid mistakes and take short-cuts. What causes concern is the inability of institutions to take advantage of the delay to learn from other countries experience and single out the most effective approaches. The weakness of the public system is probably the main reason why Italy has by and large already missed many of the opportunities of an "ecological modernisation" phase in environmental policy development (though there still are abundant challenges to be met, e.g., the updating of "sustainable" transport infrastructures – railways, transportation of goods by sea rather than by road, energy production from renewable sources).

The two basic strategic directions in which countries most advanced in this policy area have been working for some years now – integration and consensus building – are in essence still weak in Italy. There are hardly any examples of the use of conflict management based on consensus building in siting decisions (Lewanski, 1996), let alone more general policy measures. Inter-policy integration, a necessary pre-condition for sustainable development, is also still out of the picture, and the organisation model of the Environment Ministry itself, based on single media, shows that even intra-policy integration has not been pursued.

References

Aberbach, J., R. Putnam, B. Rockman, 1981. *Bureaucrats and Politicians in Western Democracies.* Harvard University Press, Cambridge MA.

Alexander, D., 1990. L'inquinamento e le politiche per l'ambiente, in: R. Catanzaro, and F. Sabetti (a cura di), *Politica in Italia. I fatti dell'anno e le interpretazioni.* Bologna: Edizione 90, Il Mulino, 163-189.

Arma dei Carabinieri, E., 1995. Legambiente-Osservatorio permanente su ambiente e legalità, Le "ecomafie", *Up & down*, 2, febbraio.

Belligni, S. and M. Robutti, 1991. *Acqua e democrazia. Attori, procedure e risultati di una politica pubblica.* Milan: F. Angeli,.

Capria, A., 1988. Direttive ambientali CEE. Stato di attuazione in Italia, *Quaderni della Rivista Giuridica dell'Ambiente.* Milan: Giuffré.

Cavalli, A., 1992. Un curioso tipo di italiano tra provincia ed Europa, *Il Mulino* 3.

Cesaretti, C., 1995. *La spesa per l'ambiente dello Stato. Aspetti metodologici e primi risultati 1986-93*, Rome: Ispe.

Chianura, C., 1991. Un paese 'incompiuto" che ha voglia di eroi, *la Repubblica*, (November 30), 5.

Conte, G., and G. Melandri (eds.), 1993. *Ambiente Italia '93*. Rome: Koiné.

Conte, G., and G. Melandri (eds.), 1994. *Ambiente Italia '94*. Rome: Koiné.

Dente, B., 1989. Il governo locale, in: Freddi, G. (ed.), *Scienza dell'amministrazione e politiche pubbliche*. Rome: NIS, 123-169.

Dente, B., P. Knoepfel, R. Lewanski, S. Mannozzi, S. Tozzi, 1984. *Il controllo dell'inquinamento atmosferico in Italia: analisi di una politica regolativa*, Rome: Officina Edizioni.

Dente, B., and P. Ranci (eds.), 1992. *L'industria e l'ambiente*. Bologna: Il Mulino.

Desideri, C., 1993. L'inquinamento atmosferico e acustico nelle città: cronaca di un'emergenza istituzionale, in: C. Desideri (ed.), *L'inquinamento atmosferico e acustico nelle città: dall'emergenza all'intervento ordinario, Istituto di Studi sulle Regioni*. Rome: Quaderni per la ricerca, serie studi 31, 5-45.

Di Palma, G., 1977. *Surviving without Governing. The Italian Parties in Parliament*. Berkeley: University of California Press.

Drouet, D., 1990. L'industria ambientale e le prospettive per il 1992, in: E. Gerelli (a cura di), *Ascesa e declino del business ambientale. Dal disinquinamento alle industrie pulite*. Bologna: Il Mulino.

Ecoimpresa News, *Gli italiani e l'ambiente* 3, 8.

EURISPES, 1998. *Rapporto Italia* (www.mix.it/Eurispes/rita98/cap27scheda19.htm)

European Environment Agency, 1995., *Europe's Environment: The Dobris Assessment*. (D. Stanners and Ph. Bourdeau (eds.). London: Earthscan publications.

European Environment Agency, 1998. *Europe's Environment: The Second Assessment*. Oxford: Elsevier.

EUROSTAT, European Commission and European Environment Agency, 1998. *Europe's Environment: Statistical Compendium for the Second Assessment*. Brussels: European Communities.

Fontana, E., and L. Miracle (eds), 1998. *Rapporto Ecomafia '98. L'illegalità ambientale in Italia e il ruolod ella criminalità organizzata*. Milan: Edizioni Ambiente.

Freddi, G., 1989. Burocrazia, democrazia e governabilità, in: G. Freddi (ed.), *Scienza tab dell'amministrazione e politiche pubbliche*. Rome: NIS, 19-65.

Frey, M., 1993. Il governo dell'ambiente e i possibili ruoli di un'agenzia ambientale in Italia. *b Quaderni Iefe*, Università Bocconi, June.

Galli, C., 1992. La cultura politica degli italiani, *Il Mulino* 3, 401-406.

Gerelli, E. (ed.), 1990. *Ascesa e declino del business ambientale*. Bologna: Il Mulino,.

Giuliani, M., 1992. *Giochi regolativi. La politica di protezione del paesaggio*, doctorate dissertation. Florence: European University Institute,.

Herman, A., and R. Lewanski, 1992. Environmental Monitoring and Reporting in Italy, in: H. Weidner, R. Zieschank, and P. Knoepfel (eds.), *Umwelt-Information*. Berlin: WZB-Edition Sigma.

ISTAT, 1993 and 1996. *Statistiche ambientali*, Rome.

Istituto per l'Ambiente, 1993. *Secondo natura*. Milan: Rizzoli.

Jänicke, M., 1991. *Institutional and other Framework Conditions for Environmental Policy Success*. Berlin: FFU Report 2.

La Spina, A., and G. Sciortino, 1993. Common agenda, southern rules: European integration and environmental change in the Mediterranean states, in: D. Liefferink, P. Lowe, and J. Mol (eds.), *European Integration & Environmental Policy*. London: Belhaven, 217-236.

Lega per l'Ambiente, 1989. *Ambiente Italia, Rapporto 1989: dati, tendenze, proposte*. Turin: ISEDI.

Lega per l'Ambiente, 1995. *Ambiente Italia, Rapporto sullo stato del paese a confronto con l'Europa*. Milan: Edizioni Ambiente.

Lewanski, R., 1986. *Il controllo degli inquinamenti delle acque: l'attuazione di una politica pubblica*. Milan: Giuffré.

Lewanski, R., 1996. *Governare l'ambiente.Attori e processi della politica ambientale*. Bologna: Il Mulino.

Lewanski, R., and R. Malaman, 1994. Dal FIO alla programmazione triennale: gli investimenti per la protezione dell'ambiente in Lombardia, in: A.Balducci, B. Dente, P. Fareri, G. Gario, G. Longhi, and R.Malaman, *Il nuovo ordinamento della autonomie locali in Lombardia*. Milan: F. Angeli, 35-74.

Malaman, R., and P. Ranci, 1991. *Italian Environmental Policy*, paper delivered at the International Conference on Economy and Environment in the '90s, Neuchatel.

Malaman, R., and S. Paba, (eds.), 1993. *L'industria verde*. Bologna: Il Mulino.

Mariani, S., 1991. *I processi regolativi in Italia: il caso di una politica regolativa*, dissertation, University of Florence.

Milanaccio, A., 1998. Lo sviluppo insostenibile: le politiche ambientali italiane, *Quaderni di sociologia* XLII (16), 87-95.

Ministero dell'Ambiente, 1992. *Relazione sullo stato dell'ambiente*. Rome: Istituto Poligrafico e Zecca dello Stato.

Ministero dell'Ambiente, 1997. *Relazione sullo stato dell'ambiente*. Rome. Istituto Poligrafico e Zecca dello Stato.

OECD, 1994. *OECD Environmental Performance Reviews. Italy*. Paris: OECD.

Putnam, R., 1993. *Making Democracy Work*. Princeton: Princeton University Press.

Ramieri, E., and F. Fiorentini, 1998. The status of Local Agenda 21 Implementation in Italy. In, *Implementing Local Agenda 21 in Europe. Varieties of Sustainable Community Development, report*. Oslo: ProSus.

Salvia, F., 1989. *Il ministero dell'ambiente*. Rome: NIS.

Skou Andersen, M., 1994a. *From Narvik to Naples: Environmental Policy in an Enlarged European Union*, paper presented at the Conference on "Governing Our Environment", Copenhagen, 17-18 November.

UNEP, 1991. *Environmental Data Report*, 3rd ed. 1991/92. Oxford: Blackwell.

MOROCCO

Marie Fauconnet and Peter Knoepfel[*]

1 The Main Environmental Problems and Deterioration

The principal types of environmental problems and deterioration and the main sources of pollution are identified in an official report as follows:[1]

Water, in terms both of quality and quantity, is a major problem in Morocco. Indeed, excessive pumping, imperfect sewage and draining systems in urban agglomerations, insufficient collection and treatment of industrial effluents, soil erosion, and the excessive and irrational use of pesticides in certain agricultural regions have led to the depletion and poor quality of water resources.

Given the importance and scarcity of water, this problem has attracted most attention and funds, both public and international, for many years. Morocco implemented a large-scale policy of dam construction, providing the country with drinking water and large irrigated areas. In terms of quantity, Morocco has a surplus of water, but there are considerable differences between regions, and during the drought in the early 90s, water was delivered to the city of Tangier by ship. One third of the resources used originate underground; 87% of the mobilized resources being used for irrigation and 13% for industrial purposes and as drinking water.

---

* The authors were in charge of the institutional analysis of the project "gestion de l'environnement – lot 1" of the Moroccan government, conducted in 1996-98 under the direction of Urbaplan (CH). The final report (March 1998), containing proposals for fundamental conceptual and organisational changes, is still confidential and will not be quoted in this article.

1 Morocco's main environmental problems were briefly identified in view of the National Strategy for environmental protection and sustainable development dated May 26, 1995. This document was prepared in the framework of the UNDP/UNESCO project: MOR/90/001. It is not a report on the state of the environment, such as can be found in some countries, but it provides information and data on the present state of natural resources, the main sources of pollution and the principal trends, as concerns the environment. Most of the data used in this article is quoted from this document.
Notice that the second most important Moroccan income producing sector just behind agriculture is tourism which, according to a large governmental program elaborated in 1995/96, is currently being further developed. However, this strategy does not consider the sector a source of environmental problems. The potential of tourism of Morocco is immense and figures show regular increases in earnings from the sector for about the past three years. We could not find any data concerning, e.g., marine water pollution caused by an increasing number of foreign tourists visiting seaside resorts, none of which have water treatment installations. The list of environmental problems above also ignores the nuclear issue; Morocco, at least in civil energy production, does not run any nuclear power station.

Despite efforts, it seems – according to the Ministry of Public Works[2] – that certain areas will suffer a lack of water because development and demographic growth will increase demand. Furthermore, the price of water, kept artificially low, does not encourage water-saving behaviour.

The poor quality of water is mainly due to inadequate purifying techniques. The quality of ground water is usually higher than that of surface water, but some ground water tables in coastal areas near the Atlantic Ocean are a little saline, due to excessive pumping. In certain areas, there is some pollution caused by pesticides and nitrates from fertilizers or landfill leachates.

The quality of surface water varies considerable from one basin to the next. Domestic waste water is the main source of organic pollution, although, in the Sebou region, sugar and oil factories also cause industrial pollution. Nitrate and phosphate pollution is due to agricultural and domestic practices. The main heavy metal pollution in the Sebou region is due to chrome from tanneries, whereas in the Tensift region, lead, zinc and copper mines cause other types of pollution.

Domestic effluents are for the most part discharged into the sea, since most of the large urban agglomerations are located on the coast (Casablanca, Rabat, Agadir, Tangier).

Another problem is the siltation of dams. The continued policy of building hill dams and large retention dams has ensured large supplies of water, but the dams deteriorate quickly due to the large amount of alluvium. Indeed, each year, dam siltation increases by 50 million cubic meters, or 0.5% of total storage capacity, which is 10 billion cubic meters. Dams built more than 20 years ago have all lost 10% of their initial capacity, due to clogging.

Air quality is poor mainly in large cities and industrial areas (Casablanca, Rabat, Safi, Marrakech), owing to traffic and industrial pollution. The main sources of industrial pollution are power plants and phosphate transformation plants. Other polluting industries include oil refining factories, chemical industries, iron, steel, rubber, glass, plastic, metal, asphalt, cement, paper, oil and grease factories, and food transformation factories.

In general, industrial pollution is due to insufficient control of facilities, their poor state of maintenance, and to the use of solvents and fuel with high sulphur levels.

Predictions concerning gas discharges show a clear increase. SO_2 discharges will probably have tripled by the year 2020[3].

Solid waste is an major source of pollution and environmental degradation, especially due to insufficient waste collection and municipal recycling services, and to the absence of waste discharge control as understood in Western countries. The absence of efforts to reduce hospital and industrial waste at source, insufficient recycling and waste treatment programmes, and the uncontrolled discharge of waste in the environment generate numerous problems in several regions.

In addition to the problems linked to landscape deterioration and to smell, unsanitary landfills also raise the problem of leachate infiltration, water pollution and localized air pollution.

2 National strategy for the protection of the environment and sustainable development of 26.05.95.

3 op. cit. 25ff.

Waste collection rates varies considerably from one urban centre to the next (100% in Casablanca and Rabat, 40% in Tangier), but the average is about 85%, whereas in rural areas the average waste collection rate is only 2%.

Recycling is carried out on an informal and traditional basis, and entire families actually earn their living from it. This type of recycling concerns mainly recyclable plastic and cardboard.

In 1994, about 800,000 tonnes of industrial waste was produced. It was mostly made up of mud, sand or metal scraps (72%), by-products (22%), raw material waste (5%) and other miscellaneous waste (1%). The collection rate is nearly 100%, but the waste is stored in unsanitary (municipal or industrial) landfills.

Other waste, such as hospital waste, is sometimes burned, but most of the time it is discharged in municipal dumps.

Between 100,000 and 200,000 tonnes of dangerous waste is estimated to be produced yearly in Morocco; it is mostly industrial[4], not counting the waste from mines and thermal power plants. This waste is managed and treated to a varying extent. In some cases, it is stored by the producing enterprises, but most of the time it is either discharged directly into rivers or the sea, without having been pretreated, or sent with normal waste to municipal dumps.

As far as toxic waste is concerned, such as pesticides, chemical weapons, asbestos and other toxic chemicals, the arrangements and security conditions for transport, storage and use are not always satisfactory. Usually, enterprises find solutions according to their needs and depending on common sense, but nevertheless the risk of accidents is high.

The urban environment. Morocco has 16 large cities (of over 100,000 inhabitants), representing 60% of the country's urban population. The urbanization rate has risen from 35% in 1970 to 47% in 1992[5]. The rapid and uncontrolled growth of cities, due to the extremely high rate of population growth and to the influx of rural populations over a very short period of time, has led to the development of unsanitary and often informal housing, to drinking water and electricity shortages and inadequate sanitation, as concerns both liquid and solid waste. This has resulted in an inefficient use of resources (land and water), difficult living conditions, disease and premature deaths among the poorer sections of the population.

The *land* deterioration varies in extent from region to region. It is caused by the overexploitation of pastureland[6] in many areas where cattle are the main economic resource. Owing to the settling of some nomads and to population increase, and in the absence of any other resource, the land tends to be used beyond its natural regeneration capacities. In the east, grass-sowing programmes have produced good results, and the population is increasingly favourable to them, despite the constraints linked to the necessity of letting the land rest. However, a period of drought can reduce these efforts to naught.

---

4 Bleaching, dying, tanning, metal industry, thermal treatment and coating, mineral and organic chemical industries, simple or complex fertilizers, insecticides and fungicides, synthetic and plastic resins, paint, oil refining.

5 World Bank, 1994, An Infrastructure for Development, report on Development in the World, 1994, 236.

6 Land which falls under traditional forms of administration and which is used by nomads and cattle raisers for pasture.

In zones where planting is no longer effective, soil protection and restoration programmes have been launched[7] to fight water erosion, but the cost and the difficulty of reaching agreement with the population has slowed the process. These programmes are not coordinated with the retention dam construction projects. The planning and construction of dams are usually carried out regardless of measures taken for soil preservation in the drainage basins, despite the serious siltation problems tied to erosion by water upstream of the retention dams.

In southern regions, wind erosion is a general problem. Dune consolidation and windbreaker construction programmes are regularly set up, but they represent only temporary solutions to the desertification problem.

In coastal areas, and in certain areas of the interior, the salinization of the soil, caused by the overexploitation of certain ground water tables, is also a serious problem.

The use of agricultural land for urbanization purposes, in particular near large cities, also causes problems of resource allocation; these problems are difficult to control owing to the rural exodus, which so far is unregulated. As a result, good land is becoming scarce.

The natural environment. Morocco's forests cover approximately 9 million hectares. They are natural forests, Esparto grass culture zones and artificial forests. The forests are put to heavy use, especially during droughts, since they serve as food reserves for cattle. Rural populations also use wood for fire, and forest fires are frequent. Every year, 33,000 ha[8] of forestland are lost. This, along with the accompanying erosion and desertification phenomena, seriously contributes to soil deterioration.

Some species, such as the atlas lion, have disappeared and several are threatened, including the monk seal and leopard. Certain vulnerable zones are threatened by human activity. The department in charge of water resources, forests and soil protection conducted a wide-ranging survey of protected areas[9]. The project involves the creation of 8 national parks (including those already existing), 19 biological reserves, and 127 natural reserves. Implementation is now under way.

The marine environment is also under high pressure, owing to Moroccan and foreign fishing in territorial waters. Biological recovery periods have been scheduled, but they are not sufficient to ensure the regeneration of the environment, and less fish is being caught in these waters.

The Moroccan *coast* is particularly long – nearly 3,500 km, from the Atlantic Ocean to the Mediterranean. This specific ecosystem is under a great deal of human pressure. Indeed, most of the country's population lives on the coast, and most large Moroccan industries are situated on the Atlantic coast, between Casablanca and Rabat. The marine environment is highly polluted, both by household and industrial waste, which is discharged directly into the sea. 99.2% of industrial waste and most of the waste from tourist resorts[10] is directly discharged into the

7 According to the *Plan national d'aménagement des bassins versants* (National plan for watersheds), 1993, Ministry of the Agriculture and Agricultural Revitalization.

8 According to the "Plan bleu, Statistiques et indicateurs environnementaux", a working document of the Department of the Environment, 1988.

9 "Plan directeur des aires protégées", 1994, Ministry of Agriculture.

10 Which do not have sewage treatment plants, only septic tanks or cesspools.

sea. Since 1994, beach water has been regularly monitored and very few beaches are considered in good condition for bathing.

Coastal erosion is another consequence of human activity. Although it is forbidden, sand is often collected from beaches for construction purposes.

Furthermore, the decrease in fishing resources near the coast is due mainly to industrial pollution, but also to excessive fishing, which is not offset by the regularly decreed biological recovery periods. Marine waters are polluted by industry, by coastal human activity, as mentioned above, and by port activities. Fishing waters are particularly threatened by the risk of accidents, due to the frequent passage of oil supertankers near the Moroccan coast. These risks are increased by the dense traffic of industrial and oil ships near the Mohammedia terminal. In 1996, the *Office d'exploitation des ports* (ODEP) obtained the necessary equipment (special fast boats, equipment for the emptying of ballast tanks, etc.) to act rapidly in case of accidents, even out in the open sea. Eutrophication phenomena have also been observed in some sites on the Mediterranean coast.

Noise. Noise disturbances are particularly severe in urban environments, due to the intensity of the traffic and the poor state of motor vehicle maintenance. However, the population seems to be used to it, and this problem is not considered a priority issue, nor even perceived as a disturbance. There is practically no data available on the subject. In this respect, Morocco differs from other countries, where noise is often considered a priority issue in the fight against pollution of all types.

In short, there is a reverse side to Morocco's rapid development, which began in the 1980s. Economic growth also plays a part in the deterioration of the environment and the depletion or contamination of natural resources. Since 1988, every year between December and March, the water purification plants in Kariat Ba Mohammed and Mkansa on the Oued Sebou (north) have been unable to provide drinking water for long periods. Every day, the population of cities (especially Casablanca, Rabat, Safi and Marrakech) is affected by air pollution. The destruction of the natural environment (considerable losses of forestland) and the degradation of soil (erosion and salinization) are also cause for concern. According to the national strategy for the protection of the environment and sustainable development, if nothing is done to counter this trend, environmental degradation could hamper or even halt the country's development.[11]

So far, the results of the efforts made in the different fields of environmental protection cannot be assessed, and for the time being there is practically no data available on the subject.

The implementation of the national strategy for environmental protection and sustainable development is now under way. It will probably take several years before its impact can be evaluated.

11 Decree n°2-93-1011 concerning the reorganization of organizations in charge of the protection and enhancement of the environment (18 chaabane 1415/20 January 1995), art. 9.
Decree n°2-93-809 on the powers and the organization of the State sub-secretariat in charge of the environment, under the responsibility of the State Minister in charge of the Interior (1" hija 1414/24 May 1994)

2 The Development of Environmental Policy and the Main Regulations

2.1 Development

The government's environmental policy includes, on the one hand, traditional elements, such as the protection of water, and the protection of forests (implemented during the protectorate period and still in force today). On the other hand, there are new problems to be dealt with, such as the management of dangerous waste. As a rule, the increasing pressure on the environment has not been followed by any legal measures to offset the damage.

The Department of Water and Forests was initially responsible for environmental policy-making. The office of hydraulics of the Department of Public Works also gradually began to play an important role in the field of water management. The National Drinking Water Office *(Office National de l'Eau Potable)* is also competent in matters of water quality. However, specific awareness of environmental issues began to emerge only in the early 90s, in particular at the Rio summit. In consequence, a state sub-secretariat in charge of the environment was created within the Ministry of the Interior; however, its status later changed.

Nevertheless, despite the royal letter of 30.10.96 to the State Minister of the Interior concerning the cleanliness of the cities of the kingdom, and other speeches delivered by the king, in particular at the National Colloquium on Forests or the First World Forum on Water in Marrakech, and despite the emphasis laid on the importance of protecting the environment, there is little awareness of the issues on the part of both public and private actors and the means and measures taken are rather limited.

The Ministry of the Environment has elaborated a national strategy for environmental protection and sustainable development. This strategy outlines the major orientations of environmental policy. It is based on the cost-objectives method, in other words, on the calculation of the cost of environmental degradation and on the identification of quality objectives for each type of environment. The costs and advantages of the implementation of quality objectives are evaluated, and a comparison is established between a development policy which would take into account quality objectives and the necessary investment, and a development policy that would ignore environmental problems. The medium and long term profitability of environmental protection measures is thus demonstrated. The following priority fields have been identified: water protection, air pollution reduction, improvement of waste management, and soil preservation.

This strategy is now being implemented through the National Programme of Action for the Environment (Programme d'action national pour l'environnement – PANE), which organizes workshops bringing together actors concerned with either sectoral or transversal themes (for instance, agriculture and the environment, or industry and the environment); the aim is to identify priority actions and available means. These workshops bring together representatives of administrations as well as representatives from the private sector and local residents.

Other departments have also developed policies targeting specific sectors of environmental protection, such as soil protection and restoration, water resource management and protection, forest protection, protected areas, and fishing. These policies usually represent a solid basis for action by management-oriented departments. However, coordination is often lacking between these policies and the development policies implemented by various departments or the strategy of the Department of the Environment. For example, the pollution of marine waters by industries is to a large extent responsible for decreasing catches. Fishery policies determine biological recovery periods, regulations concerning the size of the fish or the types of fishing nets allowed, and are not equipped to deal with this problem. Indeed, they are powerless to do anything about industrial waste, a problem that is supposed to be taken care of by other public policies.

2.2 Legislation

Existing legislation in Morocco includes numerous documents more or less related to environmental issues (about 400); over one hundred have been selected for the constitution of an analytical directory of Moroccan environmental law[12].

The main fields covered are: the physical environment, in particular water and aquatic environment, as well as coastal areas, soils, underground, quarries, nuclear energy and the protection against ionizing radiation, listed facilities, the protection of nature and heritage, and human settlement impacting environmental protection.

This legislation is for the most part outdated (going back to the French protectorate), and barely known by the various actors who would be in a position to apply it. It needs to be updated and this is now being done. There are many gaps to be filled, in particular as concerns environmental impact studies, waste discharge and environmental quality standards, air protection, waste management, environmental hazard prevention, etc. Legislation on different sectors of environmental protection is gradually being prepared, but the approval process is very slow.

For example, the bill on the protection and revitalization of the environment[13] was submitted for the first time to the General Secretariat of the Government in the 1980s, but it has yet to be approved, even though it was later revised. There is nonetheless a great need for such a law, and the fact that approval is still pending means that other important bills are blocked in the meantime. This is the case for the bill on the protection of the air and the atmosphere[14], the environmental impact study bill[15], the bill on listed facilities[16], which were also submitted to the General Secretariat of the Government but are still waiting to be examined by government and parliament. Numerous other projects are either in the process of elaboration or already quite advanced, such as the draft forest code[17], yet they remain stranded somewhere in the administration.

The main elements of the recently approved legislation are the following:

The most important legal document on environmental protection is the law on water N°10-95, approved on August 16, 1995. According to this law, water is a

12 Generally called "environment code", and soon to be published.
13 Submitted by the Department of the Environment (1997).
14 Submitted by the Department of the Environment (1997).
15 Submitted by the Department of the Environment (1998).
16 Submitted by the Department of Public Works (1997).
17 Submitted by the Department of Water, Forests and Soil Restoration (1998).

natural resource, which must be managed rationally. The aim is to set up a national water policy based on forecasts, taking into account both changes in the state of resources and the country's needs. It also includes regulations to rationalize water use, make water accessible to all, organize inter-regional solidarity, reduce inequalities between city and countryside, in the framework of programmes aiming to ensure safe water supplies throughout the kingdom. The law on water thus provides for the setting up of an institutional partnership framework between the administration and rural municipalities, with a view to reducing the inequalities of access to drinking water between city and country. This document applies the principles of "polluter pays" and "prevention" through the creation of protected areas surrounding water sources. Polluting waste – the list is established by decree – can be discharged only by special authorization and a special tax must be paid. Violations are punished by sanctions, from fines to prison sentences. Water quality norms and standards, depending on the type of use, must be determined; for several types of operations, preliminary impact studies are required.

Despite several elements typical of resource protection policy-making, water policy is essentially geared to satisfy demand, in terms both of quality and quantity, and not to protect the resource itself. The established norms for residual flows do not aim to protect ecosystems linked to water, but mainly to maintain hygiene standards.

The amendment to the law on traffic control *(police de la circulation et du roulage)*[18] to limit air pollution has been approved. Emissions are inspected as follows:

- each time the automobile undergoes a check-up as prescribed by the existing legislation;
- at the technical inspections of each new type of vehicle
- for imported cars, at the time they are marketed.

The traffic police can also inspect motor vehicles on the road at any time.

This document also enforces obligatory maintenance of the vehicle. Repairs and adjustments must be carried out within a month after the inspection, and a certificate must be submitted to the technical service.

In addition, the law specifies that it is forbidden to leave the engine running if the car is parked.

These measures are particularly called for in Moroccan cities, where cars are often poorly tuned, and where people often leave their stopped cars running for a long time.

This document marks significant progress in terms of clean air protection, but its implementation has been delayed for legal and technical reasons; it remains relatively isolated in the Moroccan legal apparatus. Indeed, immediate implementation is hampered by the lack of means for carrying out technical inspections, in terms both of garages and police staff. In addition, the law is in contradiction with an international contract signed with oil company Corral for the privatisation of oil refineries (1998), whereby Morocco has agreed not to modify the ecological re-

18 Decree no. 2-97-377 of 29 ramadan 1418 (January 28, 1998), completing the order of 8 joumada I 1372 (January 24, 1953) on traffic control.

quirements for the products for a period of 5 years[19]. This contract thus represents an obstacle to the implementation of quality standards for car fuels.

One of the main weaknesses of the legal system for environmental protection is the absence of compulsory standards for waste or for the quality of the natural environment. Since 1998 an interdepartmental work group is in the process of establishing norms and standards for polluting waste and for the different sources of pollution, according to the type of environment. An emission-based system will be set up. For the time being, no emission standards have been fixed. If need be, the administration uses the European or WHO standards as reference, adapting them to the Moroccan situation. However, the main problem lies in the fact that they are only for information and not compulsory.

According to Decree no. 2-95-717 concerning measures in case of accidental marine pollution, approved in 1996, coordination competence is to be entrusted to the Department of the Environment, in the framework of the emergency plan. This law concerns all emergency operations carried out at sea or on land to limit the consequences of the discharge of hydrocarbons or other toxic substances into the sea.

To this end, a national programme to fight accidental marine pollution is being set up. The national coordinator of this emergency plan is the Government Authority in Charge of Environmental Protection. Thus, there is an office at the Department of the Environment responsible for the emergency plan. The Royal Navy, in permanent contact with the national coordinator, is in charge of operations at sea. Operations on land are carried out by the Inspectorate for Civil Protection *(Inspection Générale de la Protection Civile),* also in contact with the national coordinator. If possible, international cooperation is also requested.

In addition to this programme, substantial material resources have also been devoted to the purpose; in particular; four surveillance ships were posted in the port of Agadir in November 1998. These patrol ships are equipped with an anti-pollution system.

To sum up, recent legislation on environmental protection is not very developed at present; it is made up of various isolated elements, which do not constitute a coherent whole. Only the law on water specifies the implementation of various mechanisms such as the elaboration of norms and standards or environmental impact studies, but these have not been put into practice. Since there is no blueprint law defining general principles for regulating environmental protection, each sector makes up its own laws according to need, but there is no coordination and no coherent legal framework. It seems unavoidable that, sooner or later, there will be conflicts between regulations concerning crucial elements such as environmental impact study procedures. Indeed, the draft project on listed facilities includes a provision subjecting certain objects to environmental impact studies, as does the law on water. If both laws are in force, a single object could be subjected to two separate environmental impact study procedures. A single document on EIS was submitted to the Secretariat of the Government in 1997, but approval is still pending, and in the meantime, each department tends to consider that its own procedures are the right ones.

---

19 La lettre Afrique Expansion, Morocco chapter, no. 559, September 1998.

2.3 The Basic Instruments

Given the difficulty in getting blueprint laws as well as waste discharge norms or environmental quality standards approved, the administration tends to act without referring to the legal framework. Classical permit, control or prohibition procedures in environmental matters practically do not exist, except in the law on water and in certain cases related to the import or use of toxic or dangerous products, or concerning listed facilities.

Many administrations, pressured by funding providers, have set up environmental impact study systems, but these differ from one department to the next. In many cases, the department responsible for developing objects which may be harmful to the environment, such as roads, carries out the environmental impact study itself However, impartiality cannot be guaranteed when the study is carried out by the administration that commissions the project. Another problem raised by environmental impact studies is the monitoring of the measures suggested in the studies to make the project acceptable. No follow-up studies have ever been actually carried out, and the risk is that environmental impact studies might have an alibi function. Generally speaking, the departments that do carry out environmental impact studies are convinced of their usefulness, but they would like to be in control of everything themselves without being constrained by law. In fact, the Department of the Environment is only rarely involved in EIS procedures, both at the elaboration level and as concerns the confirmation of compensation measures, and it has no means of verification once the object subjected to an EIS has been built.

The bill on environmental impact studies lists a certain number of objects that must be preceded by an EIS. However, this list is for the time being incomplete and unenforceable, since it does not provide a scale of environmental risk, according to the size of the object, for objects belonging to the same category. It suggests the setting up of a national committee for environmental impact studies, made up of representatives from different departments[20], in charge of examining specifications and impact studies according to a procedure defined by the law. This draft document serves at present as a guideline for Department of the Environment officials, but it is completely ignored by other departments, which are developing their own regulations (for instance the Ministry of Public Works).

The law on water is the only recent legislation suggesting effective regulations, such as waste discharge authorizations, the creation of protected areas around water sources, or fines and sanctions, measures which are certainly more relevant than those provided for in the obsolete documents still in use.

Furthermore, the Department of the Environment tries to encourage general public involvement. It has launched a dialogue process for the elaboration of branch contracts, led awareness campaigns among the actors concerned and has strengthened its collaboration with NGOs. However, the impact of these actions will remain limited as long as they are not supported by appropriate legislation.

20 In particular the Departments of the Environment, Agriculture, Public Works, Economic Affairs, etc.

It thus appears that the administration has very few legal means at its disposal to protect the environment, making it impossible to adopt a transversal approach to environmental issues.

2.4 Policy Style

As previously mentioned, environmental policy-making is rather recent in Morocco. There are no strict guidelines and no effective and complete legal arsenal. This situation partly corresponds to the wish of some departments, which are unwilling to increase the complexity of procedures, to avoid discouraging foreign investors. Furthermore, they tend to follow the European trend towards deregulation and negotiation in environmental matters.

However, this attitude shows that the authorities are going too fast. There has been no learning process ensuring that environmental aspects are systematically accounted for in development policies. In this context, it is difficult to see how an environmental policy can be set up without going through the preliminary stages common to most other countries which consist of elaborating waste discharge and environmental quality standards, and without the help of appropriate and strict legal tools.

Another important factor may also explain this trend: since competencies were distributed long before the creation of a specialized Department of the Environment, the new department was forced to define its role within the existing system. The arrival of this newcomer created strong conflicts, mainly due to the existing ministries' fear that their powers would be diminished. Indeed, another reason why they prefer to act outside a strict legal framework is to avoid having to take another ministry's viewpoint into account.

Morocco's administration is very rigid and sectoral, based on a strict hierarchy, which makes transversal policy-making very difficult. There is a great deal of reluctance to transmit files and share information. At present, despite efforts made by the Department of the Environment to elaborate a national environmental strategy with transversal actions, based on dialogue and negotiation, there is no real environmental policy-making in Morocco. The environmental aspects of public policies made by other departments usually concern the protection of a single type of environment.

Existing actions depend on the goodwill of the administration and are mainly tied to opportunities stemming from international bilateral or multilateral cooperation. In spite of the royal speeches on the subject[21], political determination and popular pressure are not strong enough to have a concrete impact on the issue. The only exception, perhaps, is the problem of waste, which has led, in some neighbourhoods of large cities, to spontaneous popular actions to compensate for the state's inaction.

At the judicial level, judges are not well informed about environmental regulations, and therefore there is little enforcement control. In some cases, Department

21 Royal letter to the State Ministry of the Interior concerning the safeguard of the cleanliness of the cities of the kingdom, of October 30, 1996; royal message to the participants of the national colloquium on forests, of March 21, 1996; Message of HM the King to the participants of the first world forum on water in Marrakech, March 22, 1997.

of the Environment officials have to help judges interpret certain environmental standards. A training and information project for the judiciary has been planned, but so far no schedule for its implementation has been set.

In this context, awareness and information campaigns and training programmes are extremely important. Indeed, the first step is to convince. Many Department of the Environment officials are devoting their efforts to this task. Awareness campaigns have been organized in schools, as well as information campaigns for target groups, and a partnership between the Department of the Environment and journalists has been set up to spread information on the environment in the media. The Department also collaborates with NGOs. Similarly, the Department of Agriculture has launched popularisation campaigns, and the Department of Public Works publishes an internal newsletter, which includes articles devoted to information and innovations in the field of environmental protection.

2.5 Policy Evaluation

As mentioned above, the existing environmental strategy drafts are only beginning to take shape and for the time being, their objectives are rather imprecise. They have not been at all evaluated[22]. No report exists on the actual state of the environment in Morocco, which could serve as a basis for evaluating the effectiveness of policies. According to the law[23], the Secretary-General of the National Council on the Environment must submit such a report annually, but so far this has not been done, owing to a lack of means and political will. The only subjects for which quality data, covering most of the country's territory, can be obtained are continental waters and forests.

The Department of the Environment is conducting a project to establish regional environmental monographs for the different economic regions of the kingdom. A geographic information system will be set up in each region with the specific indicators defined for the region concerned, which should permit the situation to be monitored. The question of data comparisons between regions has not yet been examined.

A team is now elaborating other indicators, in the framework of the Blue Plan for the Mediterranean, which includes several Mediterranean countries. Thirty-three indicators, grouped into nine themes (population, soil, waste, biodiversity, forests, agriculture, water, air and desertification) have been determined. These indicators should make it possible to carry out comparisons of the environmental situation in each of the participating countries.

At present, an important problem in Morocco is the unwillingness of administrations to disclose information. This is an obstacle both for the Department of the Environment – the reason being the conflicts of competence between ministries – and for private citizens who would like to obtain information on the state of the environment. Since competencies in this field are spread throughout the ad-

22 Within the above mentioned study ("analyse institutionnelle"), we did some evaluation work (limited to the production of administrative outputs). The results are more or less disappointing.

23 Decree no. 2-93-1011 concerning the restructuring of organizations in charge of the protection and enhancement of the environment (18 chaabane 1415/20 January 1995), art. 9.

ministrations, it is extremely difficult to obtain information. Once the data finally becomes available, it often appears that the results and periods taken into account are so different from one office to the next that it is impossible to carry out a decent evaluation of the situation.

Certain policies are now being re-examined, in particular that on the participation of the population in the management of forests, implemented in 1976. According to this policy, the funds produced by forest development are entirely repaid to the local authorities of the territory where the forest is situated. The authorities must then use 20% of the funds received for the forest (whether for development – construction of roads or water sources – or for protection – fire prevention or protection against overdevelopment). After twenty years, it appears that no more than 1 to 5% of the funds are actually used for the forest. The Department of Water and Forests responsible for this policy has decided (1998) to proceed to a re-examination and is preparing a project for a forest code which will modify this point and harmonize forest policies.

Morocco's environmental policy is extremely dependent on international aid, and for this reason, a strict evaluation of what is being done and what the country really needs is necessary. Normally, this task should be entrusted to the Department of the Environment, but in fact, the last update on projects underway in Morocco dates back to 1994. There are many overlaps and gaps. The administration is unable to provide a unified picture of the situation and to come up with a coherent decision on the course to take and on national priorities. There are many different environmental projects under way in different ministries, but there is no coordination between them. Although coordination is in theory the duty of the Department of the Environment, it is very difficult for the latter to carry it out. Despite the fact that the decree[24] creating the Department of the Environment specifies that the Department must be informed of the projects of other ministries, information is very seldom spontaneously passed on to the Department of the Environment.

Nevertheless, there is one factor which should pressure Morocco into improving its concern for environmental issues and its assessment system, and that is the prospect of a free trade agreement with the European Union. Such an agreement will require Morocco to make a serious effort to improve its environmental regulations. However, the deadline is still too far away (2010) for any significant improvement to be perceptible at present.

3 The Capacity of Proponents of Environmental Policy

3.1 Public Institutions

Competencies in matters of environmental protection are extremely dispersed. Indeed, they are shared by more than five main ministries (Environment, Public Works, Agriculture, Water and Forests, Maritime Fisheries) (see figure 1), and also by sectoral institutions whose activities can cause environmental damage,

24 Decree no. 2-93-809 on the competencies and organization of the State sub-secretariat in charge of the environment, under the responsibility of the State Minister in charge of the Interior (13 hija 1414/24 May 1994.

such as the Ministry of Trade, Industry and Crafts, the Ministry of Public Works (roads and dams), the Ministry of Agriculture, the Ministry of Maritime Fisheries, to name only the main ones. Other semi-public institutions also carry out important tasks, either through research projects or through daily activities *(Office National de l'Eau Potable* (National Office of Drinking Water), *Office National des Ports* (National Office of Ports), national research institutes).

Within the different administrative departments, competencies in matters of development and resource protection are often concentrated in the same units. This means that in situations of conflict, civil servants must make decisions which should normally be dealt with through political decision-making, with the risk that decisions will tend systematically to favour short-term development over resource protection.

There are no figures available on the number of people working in the environmental sector or in the fight against pollution. According to our estimates, 60 people over and above the 140 Department of the Environment officials are scattered throughout other departments, at the central level. There are no regional or local sections of the Department of the Environment, but the other ministries have decentralized units, in particular a forest guard unit. At the local authority level, regional and local councils in charge of environmental issues are gradually being set up in large cities. However, according to our calculations, the total number of employees does not exceed 600. For comparison purposes, it can be noted that the staff of the Ministry of Agriculture and the Ministry of Water and Forests together total 40,000 (including decentralized units).

Fig. 1. Competencies in Environmental Protection

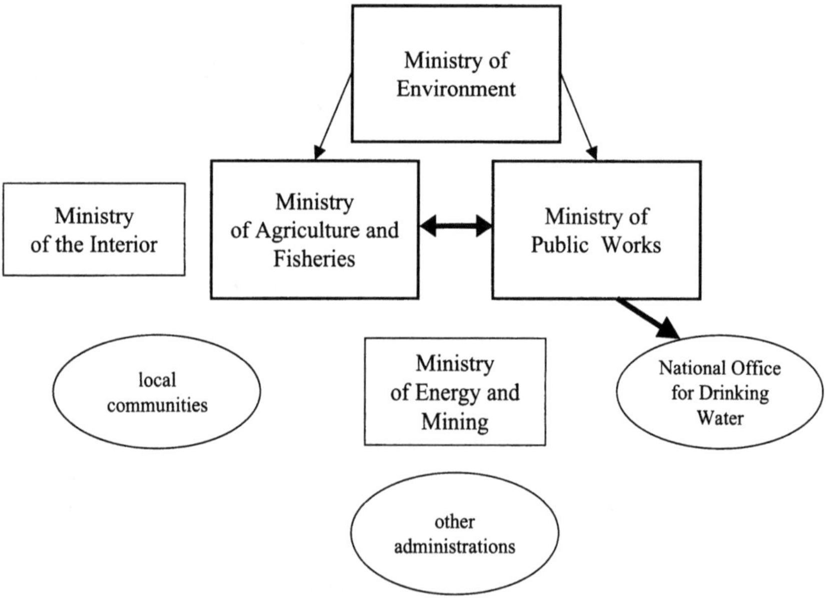

The Department of the Environment's financial resources for projects come mainly from international aid, whereas the running expenses are covered by the state. Once again, no figures are available.

The budget of the department only covers salaries, offices and some cars. The projects carried out by the department (such as educational and training programs, environmental impact studies and environmental quality observation) which represent the main part of its activities, are paid by bilateral cooperation programs (mainly with the German Gesellschaft für technische Zusammenarbeit and US Aid) or by multinational firms (e.g. World Bank, African Bank for Development, UNEP).

As mentioned above, the Department of the Environment first had the status of State Sub-Secretariat for the Environment, within the powerful Ministry of the Interior (as of 1992). In 1995, it became a fully-fledged ministry, but its organizational structure did not change, except for the addition of a general secretariat and a ministerial cabinet. During several transitional months, in 1997-98, the department became a State Secretariat for the Environment within a very large ministry covering public works and agriculture. With the arrival of the new government, in March 1998, the Department of the Environment was confirmed as a state secretariat but this time, it is part of the Ministry of Regional Planning, the Environment, Urban planning and Housing. Of course, there are good reasons for this regrouping, but since it is only recent, its efficiency cannot yet be judged. However, in the medium term, the stability of this administrative structure remains uncertain.

The tasks of the department have not changed since the first institutional set-up. The decree[25] creating the Department of the Environment is rather vague, which means it can be interpreted in many ways. The Department of the Environment has four types of competencies: monitoring, launching awareness and information campaigns, developing standards and rules, which are already partially elaborated. In addition, according to the decree, the Department of the Environment should also carry out tasks of a more interventionist nature. This type of task – promoting the protection of resources, preventing and fighting all forms of pollution, inspecting – has so far been more or less ignored. Indeed, the decree is interpreted in a very restrictive manner, emphasis being laid mainly on coordination tasks. The Department of the Environment is considered a "mission department", as opposed to "management departments" in charge of development tasks that are clearly defined and concrete. Owing to its inadequate organizational structure, the Department of the Environment has been unable to carry out its tasks in a satisfactory manner. The laws and procedures that should have supported the efforts of this young ministry were either not adopted or not implemented.

Various plans to reorganize the Department of the Environment and reinforce its competencies were launched in 1992 and 1996. The last project aims to strengthen the Department's competence in matters of environmental protection, both at the level of the Department of the Environment itself, but also at the level of other management departments. According to the project, the department would

25 Decree no. 2-93-809 on the competencies and organization of the State sub-secretariat in charge of the environment, under the responsibility of the State Minister in charge of the Interior (13 hija 1414/24 May 1994.

be organized on the basis of environment types, and separate units for each source of pollution would be created in order to establish a one-stop office system. The upper hierarchical level would be in charge of coordination and integrating the various problems tied to the different types of environment. In the "management ministries", the tasks relating to environmental protection would be separated from those to do with development, and separate units would be devoted to protection issues. For less important projects, a procedure for the transmission of files between units in charge of development, such as road construction, and those in charge of environmental protection, would ensure that they are drawn up in compliance with environmental protection requirements. For large-scale projects, the file would have to be submitted to the Department of the Environment, which would have the right to veto a project that does not meet environmental protection and pollution prevention requirements. So far, no decision had been made on implementing this system.

At present, the small Department of the Environment is struggling to hold its own among the large ministries with traditional competencies, well-established administrative practices and large budgets, and which are organized on a sectoral basis. These large ministries are the only ones to be really and systematically active in the field. Their methods are tried and tested and they have a wide range of competencies, especially as concerns water management and protection. It is thus understandable that these departments may resent a newcomer's attempts to "tell them what to do".

Files rarely circulate between the institutions in charge of protecting certain resources or the Department of the Environment, and those in charge of development. In general, files are transmitted when a decision to that effect is made at the upper decision-making levels, but there are no routine procedures ensuring the transmission of a file from one department to another for questions of conformity to environmental protection regulations.

The Department of the Environment does not have decentralized branches. Conversely, while most of the other departments whose activities have an environmental impact do. This is the case with the Department of Agriculture, which has provincial branches and offices for agricultural revitalization. The Ministry of Public Works also has provincial branches, regional hydraulics services (on the basis of watersheds), which will gradually be replaced by watershed agencies created by the 1995 law on water. The Ministry of Water and Forests also has decentralized units, and other departments, such as the fisheries departments are also represented at the local level.

Local authorities, which come under the Ministry of the Interior, are in charge of waste management[26]. When there is an urban agency regrouping several provinces, as in large cities, it is in charge of waste management. In addition, according to a recent law on regional organization[27], their number has increased from 9 to 16[28] and they have been granted competence in environmental matters.

26 Dahir n°1-76-583 of 30 September 1976 enacting law on municipal organizations.

27 Dahir no. 1-97-84 implementing law no. 47-96 on regional organization (23 kaada 1417/2 April 1997). The regions are local authorities, which are legal entities and are financially autonomous. Their mission is "to contribute to the economic, social and cultural development of the region, within the limits determined by the competence of the other local authorities ...".

Article 7 of this law specifies that: "within the limits of the regional territory, the regional council possesses the following competencies, in accordance with the existing legislation: approval of all measures concerning environmental protection; approval of measures aiming to rationalize the management of water resources at the regional level. To this end, the council participates in the elaboration of the master plan for the integrated planning of the water basins, if the region is totally or partially located in this basin, and participates in the elaboration of the national water policy each time its opinion is requested by the competent institutions and organizations". Furthermore, the regional council must set up permanent commissions, in particular a commission in charge of urban planning and environmental questions (article 36).

The first regional council elections took place in October 1997, and the final setting up of these entities should be completed within five years. As far as environmental issues are concerned, a few tasks still remain to be defined. Since there are already several regional councils for the environment, it will be necessary to define the competencies of each institution clearly. This will be all the more necessary if the Department of the Environment sets up decentralized branches in the future.

The judiciary has very little power in this respect, due to the judges' lack of training on environmental protection and anti-pollution issues. The few existing and relevant legal provisions are practically never enforced. Nevertheless, the implementation of Law 10-95 on water could change attitudes.

Besides the traditional administrations, Morocco has several councils, such as the Conseil National de l'Environnement (National Council for the Environment), the *Conseil Supérieur de l'Eau et du Climat* (Superior Council for Water and Climate)[29], the *Conseil National des Forêts* (National Forests Council)[30] or the future *Conseil Supérieur pour la Sauvegarde et L'Exploitation du Patrimoine Halieutique* (Superior Council for the Safeguarding and Development of Fisheries). These councils have variable official competencies, which are determined by decree. In addition to their official role, which consists most of the time in approving strategic orientations in various fields, these councils play an important role in promoting the circulation of information. The councils are often regarded as the showcases of the departments they come under. The decisions they make are semi-binding, in the sense that the departments concerned usually accept them, but

---

"They are governed by a democratically elected council, elected for a term of six years" (art. 1). They possess both their own specific powers as well as those delegated by the state (art. 6).

28 1. Oued Eddahab-Lagouira 2. Laâyoun-Boujdour-Skia El Hamra, 3. Guelmin-Es Smara, 4. Souss-Massa-Draâ, 5. Gharb-Chrarda-Ibn Hssen, 6. Chaouia-Ourdigha, 7. Marrakech-Tensift-Al Haouz, 8. Oriental, 9. Casablanca, 10. Rabat-Salé-Zemmour-Zaer, 11. Doukkala-Abda, 12. Tadla-Azilal, 13. Meknes-Tafilalet, 14. Fès-Boulmane, 15. Taza-Al Hoceima-Taounate, 16. Tanger-Tétouan.

29 The Superior Council for Water and Climate (Conseil Supérieur de l'Eau et du Climat) was created by the law on water 10-95 and the enforcement order 2-96-158

30 The National Council on Forests was created by the Dahir enacting law no. 1.76.350 of September 20, 1978, concerning the participation of the population in the development of forest economy.

they are not enforceable. At present, there are no contacts between the permanent secretariats of the different councils, despite common interests.

We thus observe that, in environmental matters, the actors concerned are very scattered and that there is very little systematic coordination between administrations.

3.2 Environmental Organizations

There are few active NGOs in the environmental field in Morocco and their influence is not very significant. For the most part they are national organizations working in specific fields, such as environmental rights, for instance, or neighbourhood associations whose aim is to ensure the well being of local residents. Some of them collaborate with the Department of the Environment to spread information about natural resource protection and the struggle against pollution.

Note that the big international environmental protection organizations such as WWF or Greenpeace are not present in Morocco. From these organizations only the International Union for the Conservation of Nature (IUCN) with its representative for all Maghreb States has an office in Rabat. This organization is welcomed by the Department of the Environment. Contrary to Western European traditions, the few non-governmental organizations in Morocco do not consider themselves as pressure groups. They rather cooperate with the government in order to create environmental consciousness amongst the population. This may explain why there has not yet been any important national or regional conflict about environmental issues in Morocco. Some local criticism is sometimes levelled at government nature protection measures, such as the creation of national parks or measures to combat water erosion that may hamper the economic development of the local community concerned. In most cases, such opposition can be overcome by demonstrating the success of the projects. A good example is the planned implantation campaign in arid zones against erosion which could convince the population that such measures will benefit local agriculture in only a few years even if extensification initially decreased yields. This "Oriental" program, implemented in the eastern regions of the country, has been extended in response to explicit demands from local inhabitants.

As regards professional associations, the ministries have found that it was much easier to collaborate with organized professions than with artisans, for instance, who are not organized in federations. Regular contacts have been established and agreements signed with some sectors to foster the use of clean technologies and the exchange of information between administrations and professional organizations. Similarly, the provincial branches of the Department of Agriculture try to encourage cattle-raising cooperatives; this would facilitate contacts with farmers, an important aspect in the fight against overgrazing.

The state of environmental consciousness among the Moroccan population has never been evaluated. In spite of the lack of systematic pool data, our experience shows that consciousness is generally very low. The only complaints articulated sometimes concern the bad condition of the urban sewage systems and the increasing number of private cars circulating in urban areas.

3.3 Economic Aspects

We found no economic actors, such as trade unions, economic interest organizations or larger individual firms interested in carrying out corporate environmental policies within their own realm or in playing an active part in national public policies in this field. This observation is surprising if we consider the data on the costs of environmental damage as presented in the national strategy for the protection of the environment. This document is based on an analysis of the costs of environmental deterioration and the advantages to be had from implementing a policy of quality objectives. In 1995, the cost of deterioration for different environments was evaluated as follows:[31]

- for water, the additional costs tied to drinking water purification, health care and the loss of income due to premature deaths: 13,953 million dirhams (5.7% of GDP)
- for air, the available data is not sufficient; an assessment on the basis of fuel consumption and a comparison with Malaysia yields a figure of 4.5 million dirhams (1.9% of GDP).
- for waste, the non-recycling of certain types of waste means a loss of income of 575 million dirhams (0.2% of GDP);
- for soils and natural habitat, the cost of dam siltation, which leads to a decrease in the irrigated surface, deforestation, the deterioration of pasture land and agricultural production losses, caused by the loss of arable land, are evaluated at 710 million dirhams (0.3% of GDP)

The total cost of the deterioration as evaluated in the strategy amounts to 19,738 million dirhams, or 8.2% of GDP.

1992 expenditure on environmental protection (water, air, waste, soils and natural habitat), according to the estimation of the national strategy for environmental protection, amounted to 800 million dirhams, or 0.33% of GDP. Since there is no separate accounting for the environment, it is impossible to identify all the expenses made by Morocco in this field at the time. The present situation is, to our knowledge, just about the same as concerns bookkeeping, but the amounts invested are probably higher now than in 1992, since international cooperation for environmental protection increased after that date. However, there are no more recent figures available.

Several types of *financial incentives* have been or are being provided. For example:

- Certain plants, soil analyses, the production of selected seed are subsidized.
- According to forest policy, the income generated by forest development goes to the local authorities of the area, on condition that the latter reinvest 20% of the sum in the forest. As mentioned previously, this system is not very satisfactory and is being revised.
- An industrial depollution fund has been recently set up to help enterprises purchase equipment that is more respectful of the environment.

31 UNPD/UNESCO 1995 (see note 1)

- Sectoral contracts are now (since spring 1999) being prepared; this has led to agreements with some professional associations working in chemistry and parachemistry, or with the Moroccan phosphate office, for information exchanges and increased collaboration between administrations and private enterprises.
- The law on water includes a tax system whereby the polluting enterprises will have to pay a tax serving to finance water purification.

In addition, a project to create an environmental fund is being discussed, but so far the financial question has not been resolved.

There are other examples, but for the time being, financial incentives to protect the environment are not very frequently given.

As concerns repressive measures, most of the sanctions and fines for violation of the law date back to the French protectorate period; they are so obsolete that they have no power of dissuasion whatsoever. The recently adopted law on water includes dissuasive sanctions (fines from 1,200 to 2,500 dirhams, and prison terms ranging from 1 to 12 months), but it has yet to be implemented.

4 Prevailing Strategies

No real strategy for action can be said to exist at present. Various methods are applied in different fields. For instance, the law on water applies both the principle of prevention (in particular through environmental impact studies or the setting up of protected areas around water sources) and the "end of pipe" principle limiting waste discharge levels, meaning that waste must be pre-treated before being flushed into rivers.

The monitoring of motor vehicle pollution, which is based on a recent regulation[32] (limiting authorized pollution levels), has been delayed for the present for the reasons mentioned above, but it represents a typical form of prevention practice. Indeed, the vehicles which have been inspected must be repaired to meet legal standards, at the car owner's expense (polluter pays principle).

The Industrial Depollution Fund is planning to finance enterprises to help them set up less polluting production systems, or install filters at the rejection end, but this represents only an incentive.

Given the importance of international funding in Morocco's environmental protection mechanisms, one of the "methods of intervention" is the financing of projects through bilateral or multilateral cooperation. For example, a pilot project in Fès plans to set up a lagooning plant for a group of tanneries, to purify water before it is flushed into the Oued Sebou. Such projects will enable officials of the departments concerned to increase their competence, and in the long run, they will be able to manage the projects on their own without external aid. However, very often, these pilot projects remain isolated experiments, since they are usually left without funds once the initial financial support has been withdrawn.

32 Decree no. 2-97-377 of 29 ramadan 1418 (28 January 1998), in complement to the order of 8 joumada I 1372 (24 January 1953) on traffic control (*police de la circulation et du roulage*).

5 Capacity Building

5.1 Actor Capacity

The Department of the Environment is a fledgling member of the group of concerned actors. The traditional ministries benefit from their status and have long-standing experience in the field. This is not at all the case for the Department of the Environment. For the time being, the latter is having trouble asserting its presence and legitimacy. However, given that, as recently as 1994, its status was only that of state sub-secretariat, and that since then it has become a ministry, with the status of state secretariat, it is clear that the state is truly determined to lay emphasis on this issue. Nevertheless, even if the staff were to be increased to 140, the budget allocated to the environment heavily relies on international cooperation, and judging by the amount of state funds allocated to the environment, the latter does not appear to be a real national priority. As concerns the "traditional ministries" active in environmental protection, actual protection does not seem to be a major concern for transversal policy-making.

As concerns capacity building, the specific competencies needed for environmental protection are more likely to be developed in the management departments. If development tasks were clearly separated from resource protection tasks, the different administrative units of a single department could control one another. For the time being, it seems that any attempt to transfer the competencies of management departments to the Department of the Environment will be unsuccessful.

5.2 System Capacity

The most obvious drawback in the present system is the lack of contact between the various departments in charge of environmental protection tasks, and even more so between the latter and departments responsible for development (agriculture, tourism, trade and industry, infrastructure, energy and mining, maritime fisheries, regional planning, urban planning and housing). In Morocco, environmental issues are not at all dealt with on a transversal basis. There are no procedures for the transmission of a file from one department to another, and even the environmental impact study procedure is generally carried out within a given department, without referral to the Department of the Environment.

To make up for the lack of transversal communication in the administration, which is organized according to a strict hierarchy and on a sectoral basis, interdepartmental committees or work groups are often set up. These committees and work groups are sometimes institutionalised, in particular when they are placed under the aegis of a council, such as the Superior Council for Water and Climate *(Conseil Supérieur de l'Eau et du Climat)*, or the National Council for the Environment *(Conseil National de l'Environnement)*. In general, these committees help accelerate administrative and decision-making processes and take care of spreading information on environmental issues. Some would like to see the competencies of the administration – which are blocked at that level owing to the lack of transversal contact – transferred to these committees, in the hope that the process will be more efficiently managed at the council level. This was done with the National

Council for the Environment. However, this system did not turn out to be satisfactory, since the councils have no power to enforce their decisions.

As concerns the judiciary and the inspection systems, almost everything remains to be done. Judges are barely aware of environmental protection laws. The inspection service of the Department of the Environment does not have the means to act. Other than forest guards, national park guards, inspectors of listed facilities, the surveillance of beaches by national guards (to prevent sand theft) and the future car pollution police, there is no pollution and environmental damage control system in Morocco to speak of.

The idea of creating training programmes on environmental issues is making its way; however, the administration has many very competent staff members. These are mostly people trained abroad, usually in France. As a rule, personnel is well trained and perfectly at ease in this field. However, there are not enough experienced people compared to the amount of effort needed according to the assessment made by the national strategy for the environment and sustainable development. The discrepancy between certain staff members' frequently very high level of competence and their actual position (e.g., chemist in the legal division) is quite obvious in some cases. This situation is due to many factors, such as nepotism or state programmes fostering the hiring of young graduates.

Universities increasingly create environmental science training and research programs. The most important scientific efforts are developed at Rabat and Casablanca (engineer's school Mohammedia, Veterinarian and Agronomic Institute Hassan II). Besides this university based scientific research, important studies are also conducted by international research and consultant corporations with local offices, which are often in charge of supervising internationally funded projects. Notice that the increasing scientific potential and the well trained human capital cannot be adequately exploited because of high graduate unemployment rates.

Various programmes conducted by the Department of the Environment (sometimes in collaboration with other departments) focus on awareness and information campaigns for such target groups as women, children, or the public at large. This is of crucial importance in Morocco today, since there is still very little awareness of environmental issues. The Department of Agriculture also organizes numerous popularisation and awareness campaigns. However, efforts have to be increased to raise public awareness of behaviours detrimental to the environment.

6 Capacity for What?

Much is to be expected of the future. Improving the situation is not just making things better – it means survival. According to the national strategy for environmental protection and sustainable development, "the environmental dangers threatening Morocco could hamper, or even halt its economic development, which so far has been quite satisfactory"[33].

The Free Trade Agreement to be signed with the European Union in 2010 is a concrete objective. Considerable effort is needed to raise the level of environmental protection in the country, but the means and energy spent will certainly be

33 Op cit. note 1.

worthwhile. Nevertheless, progress is slow. There is practically no popular pressure, and political will is sporadic. The change of government in February 1998 did not bring any significant changes as far as environmental issues are concerned.

Although systematic data on the political position of national and international enterprises on environmental policy issues are lacking, our investigations could not identify systematic opposition amongst these economic actors. Some of our interviews show that big industries actually welcome the very slow progress of environmental policies in Morocco, which reinforces their competitive position on world markets. Needless to say, these companies are not among the proponents of more consequent environmental policies. On the other hand, we did find some signs of concern among industrialists about this situation because they consider it an important factor of uncertainty that hampers solid investment decisions. These companies would rather prefer effective and internationally harmonized environmental policies. It should be noted that corporate policy induces these enterprises to apply much more rigid environmental standards than those laid down in (emerging) Moroccan legislation.

Management departments are strongly opposed to the development of environmental protection capacities, since they fear they might lose some of their powers to the State Secretariat for the Environment, considered a mission department devoted to coordination tasks. Given this situation, it might be easier and more profitable to strengthen competencies in environmental issues in management departments rather than to try at all costs to introduce a new actor into the complex Moroccan administrative system. Indeed, existing departments have many competent officials; their presence in the field is recognized and legitimate, and they are familiar with international cooperation.

However, what is really lacking is a genuine environmental policy, based on a firm determination to take environmental issues into account. The implementation of many intentions and projects is hampered by postponements or other technical, legal, political, or even economic obstacles. Unfortunately, the enthusiasm that followed the Rio conference seems to be waning; the quality of the environment is perceived less and less as a priority. National and international funds allocated to environmental protection are being gradually reduced.

Will efforts towards sustainable development re-ignite this enthusiasm?

The two major events in Morocco's politics in the last two years give no indication of rapid change. Neither the change of the government in 1998 nor the accession of the new king (Mohammed IV) was accompanied by programmatic statements on increasing efforts towards more sustainable Moroccan policies. The leftist Socialist Party which came to power following the elections of 1998 does not specifically address environmental issues in its party program, which is rather concerned with social issues, illiteracy, women's conditions, and the Sahara problem to be solved by popular vote in 2001, or future relations with other Maghreb States and Europe on the other. The new king has not yet given any sign of how willing he is to change his father's policy orientations. He will have to take decisions in this field because most of the mentioned legislation are still at the bill stage awaiting the royal assent.

ENVIRONMENTAL CAPACITY IN MEXICO: AN ASSESSMENT

Stephen P. Mumme and Donna Lybecker

Though a developing country, Mexico is well placed by many aggregate development measures. It is the second largest economy in Latin America; with a GNP per capita in 1995 dollars of $ 3,320 it is solidly ensconced among the world's upper-middle income countries, one of only six of thirty-six Latin American countries so ranked (World Bank, 1997: 214-215). Economic progress, however, has come at a high price in environmental degradation, a cost that it has begun to reckon only in the last decade.

While Mexico took formal measures to address the negative externalities of its rapid industrialization after Stockholm 1972, it began to take the environment seriously only a decade ago. In 1988, outgoing president Miguel de la Madrid, responding to rising public concern with environmental problems, undertook a revamping of national environmental laws. His successor, Carlos Salinas, undertook a broad reorganization of Mexico's environmental and natural resources administration and intensified Mexico's international cooperation for environmental protection spurred, in part, by economic liberalization and Mexico's bold initiative to enter into a North American free trade area,. These measures have certainly strengthened Mexico's capacity for environmental protection, placing it among the leaders for environmental protection in the Americas, though well behind the United States, Canada, and other OECD countries. Indeed much of the new architecture for environmental protection can accurately be described as incipient and still under trial. Moreover, these initiatives unfold in the context of both economic liberalization and a major democratic transformation of national political institutions, changes that at once harbor great potential but in the short run complicate both policy development and implementation and cloud efforts to assess the consequences of current policy reforms accurately.

1 Mexico's Main Environmental Problems

Mexico's environmental challenge is substantial whether the issue is pollution or the conservation of natural resources. In the half-century of rapid industrialization after World War II, Mexico's environment has been seriously degraded. Most efforts at regulation, remediation, or restoration are of recent vintage and have not yet significantly arrested or retarded basic trends as the following vignettes show.

1.1 Water Supply and Quality

Due to Mexico's remarkable geographical variation, its water endowment is unevenly distributed in relation to arable land resources. Precipitation is greatest in southern Mexico where measurable rainfall in tropical forests may exceed 400 inches annually. By contrast, rainfall is significantly diminished in the desertic northern zone, which averages under 24 inches annually. Infrastructural investments in the 20th century harnessed major rivers for human utilization, but water resources remain in short supply and by most measures are inefficiently utilized. Overexploitation of groundwater resources has increased rapidly in recent years. With respect to water quality, rapid industrialization and burgeoning human settlements have heavily stressed Mexico's surface and subsurface water resources with adverse effects on public health. By some estimates as much as 77% of Mexico's river basins were polluted to such a degree they threatened public health. According to a recent OECD study, of 24 major hydrological basins studied, most were described as polluted or strongly polluted (OECD, 1994a: 58). Mexico's 137 coastal lagoons all suffer from pollution. Nearly all national surface water requires pretreatment before it is safe for human consumption, and bacterial pollution of drinking water is a major contributor to gastrointestinal disease.

1.2 Air Quality

Urban air quality is certainly one of Mexico's most notorious environmental problems. Mexico City, in particular, has drawn considerable attention for its deteriorating air quality, which in the 1980s caused many residents to draw analogies to London's killer smog. Numerous air quality alerts, high rates of pulmonary disease, a cottage industry in street-corner oxygen bars, and flocks of starlings dead in the streets were some of the indicators drawing international attention to the City's worsening air pollution problem.

While various measures to monitor and assess Mexico City's air quality problem can be traced to the late 1970s, the metropolitan region's air quality problems were not seriously addressed until the Salinas administration took office in 1988. At that time a multidimensional set of air quality programs was set in place including a vehicle emissions control program, limits on vehicular circulation, areal circulation limitations, a shift to unleaded fuels, street paving in peripheral areas, improved garbage collection, pet control programs, and industrial emissions regulations coupled to closures of major industrial point sources, most notably, in 1992, the closure of the aging PEMEX refinery in the metropolitan zone (Mumme, 1992).

Most experts agree these programs have paid dividends though ozone, PM, and various other conditions still remain poor due to rising population and consumption in the metropolitan area. The OECD performance report noted that Mexico City exceeded its ozone safe standard 9 out of 10 days in 1997. PM levels have improved since the PEMEX plant was shut down in 1992 but the average daily concentration is still 200 on Mexico's IMECA index, making PM Mexico City's second worst air quality problem. Street paving, better refuse collection, and fecal control have contributed to PM/10 declines in the past decade. Even so, Mexico

City's air remains the number one environmental health threat in the nation's capitol.

Beyond Mexico City air pollution is a problem in most major urban areas, though only 11 of 32 of the largest urban areas were effectively monitored for a range of basic pollutants (SO_2, NO_x, CO, ozone, PM/10, VOCs and lead) in 1997 (OECD, 1998a: 82). Multidimensional air quality programs similar to those adopted by Mexico City have been put in place in Monterrey and Guadalajara, Mexico's second and third largest cities respectively. Programs in other cities have been adversely impacted by economic crisis in the 1990s and have been discontinued or temporarily suspended pending more favorable economic conditions.

1.3 Biodiversity

Mexico is classified as a megadiversity country, with 10% of the world's known species. Mexico's biodiversity is very much at risk, however, largely owing to deforestation, water and air pollution. At the present time 33.2% of its mammals, 16.8% of its birds, 18.1% of its reptiles, 16.9% of amphibians, and 4.5% of its fish are classified as threatened or at risk of extinction (OECD, 1998a: 98).

Mexico suffers one of the highest rates of deforestation in the Americas estimated between 1% and 2% annual loss of original forest cover – roughly 500,000 hectares annually. Degraded forest areas have increased from 17.8 million hectares in 1985 to 22.2 million hectares in 1994 (OECD, 1998a: 109). Important protected areas, such as the Lacandon forest, have been whittled down to less than 30% of their original forest. Typical of stresses on forests elsewhere in Latin America, much of the damage is from small-holding farmers employing slash and burn techniques for clearing lands and those clearing lands for grazing livestock and pasturage. Rural poverty and unresolved problems of rural land tenure are major contributors to this dynamic. Common lands associated with Mexico's historic practice of communal land ownership (ejidos) are particularly at risk, as these were excluded from constitutional reforms in 1992 enabling ejidos to sell or lease their property.

1.4 Waste and Hazardous Waste

Mexico suffers very serious deficiencies in waste disposal, with particular deficiencies in the area of hazardous and toxic waste disposal. By some estimates, about 30% of all municipal garbage is not properly collected or disposed of, and clandestine dumps have proliferated in urban areas, with potentially serious adverse impacts on groundwater, air quality, and public health – some 3,500 such dumps were recently identified in Mexico City. The country is also highly deficient in industrial and toxic waste disposal facilities. Only two such facilities currently exist nationwide, and these are located in the northern border area, distant from many major industrial zones. The OECD estimates that 5 million tons of hazardous waste are disposed of illegally on an annual basis (1998a: 119).

2 Environmental Administration and Policy Development

2.1 Institutional Evolution, 1972-1998

In general, the institutional apparatus for environmental protection in Mexico has remained centralized, with administrative powers concentrated at the federal level. At this level, the administrative apparatus for environmental protection in Mexico has evolved in three distinct stages. The first stage, associated with the adoption of Mexico's first general environmental law in 1972, loosely patterned after the US National Environmental Protection Act (NEPA), established environmental protection as a subsidiary function of national health administration, locating the subsecretariat for environment within the Ministry of Health. The second stage of administrative development came in 1982 with major reforms to the environmental law coupled to the integration of environmental protection with urban development in a new national ministry, the Secretariat of Ecology and Urban Development (SEDUE). SEDUE was given primary responsibility for implementing the environmental law but found itself highly dependent on other more powerful agencies with line jurisdiction in natural resources management, agriculture, and industry and hampered by a limited budget for environmental protection. The marriage of urban development and ecological functions was an unhappy one from the outset. The ministry's inherent functional asymmetry was exacerbated by Mexico's rapidly deteriorating economic conditions after 1982. SEDUE also suffered from a series of human and natural disasters in which the ministry was implicated – the 1985 Mexico City earthquake and the 1986 San Juanico explosion in Mexico City – as well as environmentally unpopular measures such as the initiation of Mexico's first nuclear power plant at Laguna Verde, Veracruz. It was the most troubled of the de la Madrid ministries with four different heads variously at the helm (Mumme, 1992).

The third and most recent stage of administrative development began in 1988 with a comprehensive overhaul of the national environmental law, replacing the old law with a new General Law of Ecological Equilibrium and Environmental Protection (LGEEPA). These reforms provided for a comprehensive approach to environmental protection, and laid a foundation for the further transformation of the federal administrative infrastructure for environmental protection. In 1992, the SEDUE was eliminated with its environmental functions folded into a new ministry for social development, the Secretariat of Social Development (SEDESOL). SEDESOL proved something of a way station for environmental protection as a new ministry, the Secretariat for Environment, Natural Resources, and Fisheries (SEMARNAP), was entrusted with principal oversight of environmental law in 1994. This new arrangement aimed to rationalize and coordinate the implementation of environmental law by placing all major environmental functions in the hands of a single ministry and reinforcing particular elements of environmental administration. The 1992-1994 administrative reforms also abstracted normative planning from the main ministry and placed these responsibilities in the hands of several semi-autonomous agencies, the National Ecology Institute (INE), the National Water Institute (INA), and the National Fisheries Institute (INP), all attached to SEMARNAP, and the National Research Institute for Forests, Agriculture, and Livestock (INIRFAG) attached to the Agriculture ministry. It further established a new office for enforcement of environmental law, the Federal Attorney General for Environmental Protection (PROFEPA). Of these agencies, INE and PROFEPA were respectively given the leading roles in policy development and enforcement. INE is now the leading agency for environmental policy development and the design of technical norms and standards for policy implementation and assessment. It also oversees the management of natural resources and protected areas and provides technical support to state and local agencies (OECD, 1998a: 102). PROFEPA has nearly exclusive authority for enforcing compliance with federal environmental laws dealing with industrial pollution and natural resources management, with the exception of water pollution, which falls under the jurisdiction of the National Water Commission (CNA). PROFEPA is also the leading agency for enforcing federal environmental impact assessment requirements, and is playing an important role in implementing the industrial eco-audit program. It also enforces international environmental agreements to which Mexico is a party (OECD, 1998a: 136-137). In 1996, the LEEGPA was extensively revised to bring the law up to date with intervening administrative changes, to emphasize sustainable development as a national priority, to clarify significant ambiguities in the law with respect to intergovernmental responsibilities, and to provide greater opportunities for public access to environmental information and public participation in environmental policy implementation (Alanis Ortega, 1996).

2.2 Environmental Policy Instruments

The 1996 LGEEPA relies on a mix of policy instruments to achieve environmental goals. Mexican environmental regulation to date has focused on command and control, end of the pipe measures, utilizing a system of governmental permits based on uniform national pollution standards. The LGEEPA as amended introduces a system of integrated pollution control based on the principle of ecosystem management and using integrated permitting in place of the old uniform emissions and permit approach to regulation. The new approach bases standards on the assimilative capacity of various media. This approach is being applied to water quality management first and is to be extended to other sectors over time. Since 1988 environmental impact assessments have also been required, and the 1996 reforms have clarified procedures for their application and provided greater public access to such documents.

Officially the Mexican government is committed to economic instruments for pollution control, and provision for such measures is strengthened in the 1996 reforms. In practice, however, the government has moved slowly to put such measures into place. Economic measures are best established in water management, where resource fees apply to non-agricultural water use, and pollution fees may be imposed when effluent is found to exceed permissible levels.

Air pollution has been tackled with a general fuels tax – in Mexico City a special surcharge is imposed to support introduction of VOC emissions reduction technology at filling stations (OECD, 1998a: 140). Since 1996 the government has been also trying to reduce the size of Mexico's aging vehicle fleet with sales tax reductions on new car purchases while increasing registration fees on vehicles more than ten years old. Another economic measure is found in EIA procedures that require affected developments to pay fees to support the assessment of environmental impact. Officials at various government levels are presently considering a range of economic instruments to achieve environmental goals, and more such instruments are likely in the near future. The federal government is also interested in using fiscal policy to achieve environmental objectives, and since 1991 has provided investment tax incentives to firms for acquisition of pollution control equipment. Taxes on natural gas used as vehicle fuel have also been lowered, and industries converting to natural gas also get tax breaks. Further, in accordance with the polluter pays principle, the 1996 reforms have strengthened Mexico's polluter liability standards. Any party that pollutes or impairs the environment is now subject to strict liability – it is "liable for and must repair the damage" (OECD, 1998a: 141).

Voluntary agreements with private sector organizations, or *convenios*, are also part of the government's formal strategy for achieving compliance with environmental norms. These are discussed further in Section 3.1 below.

Protecting biodiversity is emphasized in the 1995-2000 National Development Plan and the 1995-2000 National Environmental Program, with special emphasis on sustainable development of natural resources and arresting degrading trends. Mexico's recent shift to integrated economic and environmental planning has received perhaps its greatest emphasis in this issue area. Nationally, SEMARNAP plays the leading role in biodiversity planning and policy development, assisted by INE, INP, and INIRFAG. PROFEPA is the lead agency enforcing regulations, relying heavily on a system of permits for legal pollution, expropriation and use of natural resources, with inspections, fines, and in extreme cases, plant closures being imposed for noncompliance (Newman, 1994: 38).

Article 27 of the Mexican constitution endows the government with authority to regulate ecological balance, and a number of laws – particularly the LGEEPA – provide authority for wildlife management, the establishment and maintenance of protected areas, and natural resource conservation. A key feature of the LGEEPA approach is its provision for geophysical ecological planning, including the use of federal, state, and local plans detailing the ecological situation and indicating appropriate land uses, a program under INE's direction. These plans have legal status once approved by relevant authorities and are to direct local development. In 1997, a new program aimed at protecting coastal zones was adopted by SEMARNAP that utilizes integrated planning within a geophysical ecology approach (OECD, 1998a: 104).

In addition to these measures, a National System of Natural Protected Areas has expanded and now incorporates 117 officially protected areas for a total of 11.7 million hectares or 5.9% of the national territory. The management of these protected areas has been poor, however, and many diversity rich areas, including tropical rainforests, such as Las Chimalapas in Oaxaca and the Lancandon of Chiapas, as well as desert reserves like El Pinacate, are stressed by encroaching development. Only 20% of protected areas had a formal management plan in 1997 (OECD, 1998a: 104). In response the government is exploring various management and financial alternatives, including international assistance from such agencies as the Global Environmental Facility, eco-tourism, private sector sponsorship, NGO-state collaboration in management, and special funds like the newly established Mexican Fund for the Conservation of Nature. In the area of species protection, SEMARNAP has developed 21 special programs for particular groups and species of fauna and flora (OECD, 1998a: 107). An innovation in the 1996 national conservation program is the introduction of the "System of Units for the Conservation, Management and Sustainable Use of Wildlife" (SUMA). This program aims at balancing commercial needs of basic communities with sustainable development objectives, setting standards for the use of renewable resources based on distinctions between intensive resource units and extensive units. Once established, a conservation unit or SUMA is to be overseen by technical management committees including NGO representation and public participation by affected or interested parties. The protection of forest resources is associated with several laws and programs. Under the 1995 Forestry and Soil Program, forestry policy has been focused on defining property rights, arresting and reversing severe deforestation and land degradation, reforestation, and promoting sustainable forestry practices. A major loophole in national management practices was revealed as forest fires ravaged Mexico in 1998, demonstrating Mexico's lack of preparedness in fire control and prevention (Aridjis, 1998).

3 Environmental Policy Performance

3.1 Enforcement

Enforcement of environmental laws has long been a weak point in Mexico's environmental profile. This is so for several reasons. First, Mexico's continuing economic woes throughout the 1980s and 1990s led federal officials to place a high priority on voluntary compliance rather than enforcement, usually involving government-industry *convenios,* or agreements. Such *convenios* rely heavily on self-regulation, and; in the absence of sanctions, frequently fall short of objectives. Second, such sanctions as were imposed tended to be light, encouraging many violators to prefer paying the fines rather than modifying economic practices. Third, administrative resources were scarce, contributing to the difficulty of conducting inspections, analyzing results, and imposing sanctions. Administrative corruption, particularly in the patronage ridden environmental offices of the 1980s, was a serious problem. A contributing problem here has been the vagueness of many regulations and the lack of clear lines of administrative responsibility, contributing to the prevalence of negotiated solutions or, in many instances, procrastination in applying applicable laws. Moreover, citizen based accountability, including citizen initiated legal action, was weak due to statutory ambiguities and prevailing centralization of environmental administration throughout the country.

The reorganization of environmental administration and reforms in accordance with the LGEEPA have rectified this situation somewhat. Since 1992, PROFEPA, the new agency in charge of enforcement, has augmented the number and pace of its surveillance inspections, with 67,826 inspections conducted and 2,698 partial or total closures imposed through September 1997 (OECD, 1998a: 137). PROFEPA has also conducting environmental audits of large industries. It completed 795 such audits nationwide between 1992 and 1997, with 467 remedial action plans signed as a result during the same period (OECD, 1998a: 137). Since 1997 it has begun to focus its audits on medium to smaller sized firms, the so called micro-enterprises. PROFEPA is also responsible for monitoring phytosanitary standards of fish and agricultural products in conjunction with the NAFTA accords as well as enforcing Mexican compliance with other international environmental accords. Citizen complaints to the agency increased from 1,321 in 1992 to 6,249 in 1996, indicating a growing citizen use of administrative processes to hold polluters accountable.

3.2 Program Integration

While past environmental policy could be faulted for administrative balkanization and lack of integration with economic policy, recent reforms place a high priority on integrating environmental and economic policies. Formal articulation of sustainable development goals as relevant to economic planning dates to the 1990-1994 National Program for Environmental Protection, and is reinforced in the 1995-2000 National Development Plan and the 1996 LGEEPA reform. To improve coordination of private sector and intergovernmental planning for environmental protection, SEMARNAP in 1995 established a National Consultative Council for Sustainable Development and four regional consultative councils. These bodies review economic policy proposals for sustainable development concerns and make recommendations for their improvement. SEMARNAP's own environmental planning focuses on four areas, water management, fisheries, forestry and soil protection, and environmental protection in general. The 1995-2000 Environmental Program under SEMARNAP's direction details a wide range of environmental practices, including protected areas, land use planning, environmental assessment, hazardous waste, eco-audits, education, public participation, and research. At the inter-ministerial level, an inter-ministry commission on pesticides, fertilizers and toxic substances, and another dealing with Mexico City's environmental problems provide additional policy integration (OECD, 1998: 124-125). The results of recent policy integration are difficult to assess. Much recent innovation remains formal, and most observers agree that continuing economic difficulties reduce government incentives to incorporate environmental values in important economic decisions in the absence of international pressures.

3.3 Financing

Mexico's capacity for environmental protection is affected by its relatively low level of estimated total expenditure on pollution abatement and conservation. The OECD puts total pollution abatement expenditures from private and public sources at 0.8% of GDP annually, which places Mexico at the bottom of the OECD countries with Spain and Hungary (OECD, 1998a; 1998b). On the other hand, though hard data are wanting, it seems likely that Mexico ranks among the leading Latin American countries in per capita environmental expenditures. SEMARNAP's 1997 budget of over 14 billion new pesos (10 pesos = 1 US$) does represent a 33% increase since 1995 (OECD, 1998a: 128) but is largely based on new water and environmental infrastructure investments, much of this concentrated in Mexico City and the Mexico-US border area. SEMARNAP's actual operating budget –

not including the budgets of its specialized quasi-autonomous agencies – is much lower, and has fallen since 1997 to just over US$ 14 million (EcoAmericas, 1999: 2). Its budget for enforcement, research, and normative development has correspondingly shrunk during the same period. Mexico City's environmental expenditure in 1997 was 8% of its total budget, up from 5.8% in 1989. Recent budgetary trends suggest this, too, has fallen since 1997. It is worth noting that much of Mexico's budget for environmental expenditure is based on international assistance, a source of fragility in sustaining current commitments.

3.4 Intergovernmental Relations and Decentralization

Mexican environmental policy remains highly centralized with states and municipalities still strongly dependent on federal policy guidance and resources. Recognizing this, and with the aim of making environmental policy fit within the parameters of President Ernesto Zedillo's "new federalism", SEMARNAP has since 1994 sought to devolve or decentralize greater responsibility for environmental administration to Mexican states and municipalities, consistent with the constitutionally mandated responsibilities of these subsidiary governments (Newman, 1994: 28). Policy devolution is mainly achieved by transferring functions to SEMARNAP's and other federal agency's 31 state delegations and federal district office. SEMARNAP's state offices, for instance, are encouraged to develop particular priority areas reflecting regional needs, and to increase their coordinative activities with other levels of government, particularly with respect to Agenda 21, post-Rio concerns, and the development of new regional sustainable development plans. With respect to decentralization, the 1996 LGEEPA further extends the constitutional concept of *concurrencia* (concurrent and overlapping jurisdiction between government agencies at various levels) to environmental policy, and further specifies the division of labor between levels within various areas of environmental policy. The formal transfer of responsibilities to states and municipalities is achieved through consultation and bargaining and with the formal agreement of the affected governments (OECD, 1998a: 144). Affected policy areas include water, protected areas, forests and fisheries, environmental impact assessment, land use planning, air pollution, and solid waste management. Thus far SEMARNAP has signed agreements with 25 states.

Progress towards decentralization nevertheless remains limited. The decentralization initiative has not been accompanied by substantial downward transfers of fiscal resources or responsibilities. SEMARNAP's state offices and state and local environmental administrations remain underfunded, understaffed, and poorly trained. Continuing ambiguities in the LGEEPA on jurisdiction, the slow pace of reconciling the revised federal law with state environmental law, and the lack of formal regulatory ordinances at the state and municipal levels all hamper the exercise of local initiative. As yet, environmental organizations express little confidence in local institutions' capacity to cope with environmental problems lying within their formal competency. One potentially bright spot in the local capacity picture is found on the Mexico-US border, where North American Free Trade Agreement (NAFTA) generated environmental programs like Border 21 and the Border Environmental Cooperation Commission (BECC) have been directed at supporting local initiatives and strengthening local capacity for environmental administration.

3.5 International Cooperation and the NAFTA Institutions

Since 1983 Mexico has deepened its international ties, entering into a number of important international environmental agreements and other multilateral relationships that deepen its commitment to environmental protection. Already a member of a number of bilateral and multilateral conventions as well as the 1983 La Paz Border Environment Cooperation Agreement with the United States, Mexico made substantial new commitments in the 1990s. In 1991 it ratified the International Convention on Endangered Species. After Rio in 1994 it quickly moved to implement the Convention on Biological Diversity and the Montreal Protocol and the Convention on Climate – with respect to the Montreal Protocol it is now very close to attaining national objectives on consumption of ozone depleting substances (OCED, 1998a: 199). Its admission to the OECD in 1994 entailed a major commitment to moving towards a more comprehensive and transparent approach to environmental management. And Mexico has significantly broadened its environmental cooperation within the trinational area of North America as a result of NAFTA.

There is little doubt that Mexico's entry into NAFTA spurred its environmental policy effort. Mexico's 1992 administrative reform and NAFTA's environmental side agreements derive substantially from the Mexican government's interest in satisfying US congressional critics of its NAFTA initiative (Mumme, 1994). The NAFTA agreement, which took effect in 1994, made sustainable development a trinational objective of trade policy. Two subsidiary agreements respectively provide for trinational cooperation in environmental protection and special support for the development of environmental infrastructure on the US-Mexico border. Concurrent with these formal agreements, the two national governments have elaborated their preexisting agreement on border environmental cooperation, the 1983 La Paz Agreement, to embrace a wide range of binational inter-agency protocols and cooperative actions under the mantle of the Border 21 Program (Environmental Protection Agency, 1996). Together, these programs have channeled new funds for development of environmental infrastructure to border communities, and are contributing to much more effective communication and a wide range of cooperative activities for environmental protection between the two countries.

3.6 Public Participation, Education, and Environmental Information

In general, the level of public awareness of and public involvement in environmental concerns remains low. Since the early eighties, however, the government has sought to promote the development of environmental concern through public education, diffusion of environmental information through the electronic and print media, and through support of environmental organizations. These strategies have had rising success, particularly in large urban areas, where the media and public schools reinforce environmental messages. Since 1993, Mexico's Social Co-Investment Fund, a project sponsored by the Secretariat of Social Development, has supported Mexican non-profit agencies providing community services to include the diffusion of environmental and environmental health information and remedial activities (OECD, 1998a: 145). The 1996 reforms to the LGEEPA aim to strengthen SEMARNAP's commitment to deepening public participation in environmental policy-making, but public involvement is still weak. A major impediment to public participation has historically been the lack of access to government information and lack of transparency in public proceedings. The 1996 reforms aim to improve this situation by establishing a public right to access to environmental information. This initiative is offset, however, by the lack of a national open-records law and poor diffusion of basic information. Most environmental impact assessments, for instance, are either unavailable or very difficult for ordinary citizens to obtain. Mexico's administrative culture, steeped in centralism, has proven highly resistant to efforts to enhance access to information and transparency in public policy development (EcoAmericas, 1998: 11). On the positive side of the ledger, Mexico has recently made some progress in analyzing, compiling, and disseminating environmental information, stimulated by international obligations to OECD, the North American Commission on Environmental Cooperation, and other multilateral organizations. Mexico's national statistics institute (INEGI), for example, is now maintaining an inventory of 27 national environmental indicators in 11 topic areas (OECD, 1998a: 146). Mexico's overall research effort on environmentally related science and technology, while growing, is very small compared to other industrialized countries.

3.7 Summary: Evaluation of Policy Performance

Since 1988 Mexico has made significant strides towards developing its capacity for environmental protection. Its basic environmental law has been elaborated and a national regime of environmental standards based on command and control and end of the pipe regulatory approaches, and applicable across a range of media, has been promulgated, revised, and strengthened with the aim of moving towards an ecosystem approach to environmental management. National development and environmental planning now incorporate sustainable development principles and aims. Environmental enforcement has been strengthened, important international and regional environmental agreements have been signed, incipient initiatives towards strengthening the environmental capacity of state and local governments undertaken, and the government is committed to environmental education and raising public awareness of environmental concerns There is little doubt that Mexico has made real progress in the face of great odds.

Even so major deficiencies are evident. Expenditures on environmental protection remain low, and have been negatively affected by the post-1994 economic crisis. Key programs are heavily dependent on foreign assistance and insufficiently supported within the framework of national fiscal policy. Policy remains heavily centralized, with strong incentives to push critical decisions to the federal level in the absence of clear local authority or local capacity to address problems at subsidiary levels. Public participation in environmental affairs is low and, though public awareness of environmental issues is growing, administrative culture is resistant to facilitating access to environmental information or including the public in all but the most formal aspects of environmental policy-making.

4 Main Actors in Environmental Policy

4.1 Proponents and Their Capacities

Proponents for Mexico's environmental policy comprise three sets of actors, international actors, government institutions and political parties, and domestic, non-governmental actors including environmental non-governmental institutions (NGOs) and the so-called green enterprises.

International actors play a critical role in Mexico's environmental policy development. For example, in the early 1990s, criticism of Mexican environmental conditions was associated with the drive for a US-Mexican free trade agreement. This criticism spurred the then president Salinas de Gortari to new levels of environmental activism (Mumme, 1991: 10). Today NAFTA and the BECC are playing a major role in northern Mexico's environmental policy by developing regulations and enforcing greater compliance.

Other international organizations that are major actors in Mexico's environmental policy include the Interamerican Development Bank, which has been responsible for financing and influencing environmental conservation projects and policies *(El Financiero,* 1998) and international NGOs such as The Nature Conservancy, Conservation International, Greenpeace and The World Wildlife Fund. NGOs such as these have pressured the Mexican government to move forward with environmental policies concerning such issues as commercial exploitation of marine turtles *(Excelsior,* 1991) and prohibition of the importation of toxic wastes and use of certain pesticides (Kaimowitz, 1997: 23).

The strength of these international actors includes their access to the international media through which they have the ability to help make Mexican environmental issues more visible, and access to resources: money, people and information. On the other hand, international actors' capacity is often limited by their narrow focus on environmental issues that diminishes their ability to address the underlying causes of environmental degradations such as poverty and dependence on environmentally degrading activities for export revenues (Kaimowitz, 1997: 23).

Government agencies and political parties constitute a second set of environmental policy proponents. SEMARNAP is the major player here. Headed by Julia Carabias, a respected academic and ecologist with no party affiliation, SEMARNAP is respected among federal ministries for a certain institutional independence and its strong environmental advocacy. With its mandate to promote sustainable development, SEMARNAP maintains close connections with other government institutions that have interests in environmental issues: the Secretariat of Foreign Relations, Secretariat of Interior, Secretariat of Communication and Transportation, Secretariat of Energy, SEDESOL, and INEGI.

SEMARNAP has grown rapidly since the federal reorganization and consolidation of environmental protection functions in 1994. As seen in Section 3.7, its 1997 budget is equivalent to US$ 140 million. Its current staff totals nearly 14,000 (39,000 if the 25,000 employees at the CNA are included). Of these 14,000, over half are assigned to SEMARNAP's field delegations at the state level, the others divided among SEMARNAP's General Secretariat and Sub-Secretariats of Planning, Natural Resources, and Fisheries, the semi-independent institutes (CNA excepted), and other internal administrative units. Despite its growth, SEMARNAP suffers from severe resource shortages. In 1998 the agency saw its budget progressively reduced due to across-the-board budget cuts in Mexican federal government provoked by falling oil revenues. These rescissions affected most operations. State delegations, in particular, were forced to pare back their activities as they were hard pressed just to pay salaries and maintenance on basic facilities. The agency's 1999 operating budget has been cut a further 5%, an ominous development that some environmental critics believe will require major cutbacks in environmental enforcement (EcoAmericas, 1999: 2).

Political parties are increasingly important in pushing environmental policy. Nearly all parties now include environmental issues in their platform. Mexico's green party, the Partido Verde Ecologista Mexicana, has maintained its legal registry, and in 1997 won several seats in the Chamber of Deputies. It may be partially credited with pushing other political parties to strengthen their environmental platforms – although others believe it was created by the Institutional Revolutionary Party to split opposition votes (Umlas, 1996: 23).

Domestic, non-governmental actors are even more important at this point than the political parties, and comprise a third key group of proponents. The number and importance of environmental NGOs has grown steadily since the 1980s, now numbering over 400 accredited organizations. These include the Grupo de Cien, an agglomeration of Mexico's leading artists and intellectuals, the Pacto de Grupos Ecologistas (Pact of Environmental Groups), the Mexican Ecologist Movement, Union of Environmental Groups, and indigenous movements and networks such as the National Committee for the Defense of the Chimalapas (CNDCHIM). Additionally, companies and farmers who gain from the green markets through ecotourism, organic foods, recyclable containers, and products from sustainably managed forests also have impacted environmental policy (Umlas, 1996: 199). The national eco-industry, though still at a fledgling stage, is steadily advancing, and is modestly subsidized through credits from the government's National Finance Agency (NAFINSA). The sector has its own trade association, the National Council of Industrial Ecology, A.C. (CONIECO) which has organized and sponsored a number of international expositions promoting Mexican eco-industrial technologies and products (El Mercado de Valores, 1996: 16-19).

These groups see the emergence of environmental concerns as an opportunity to construct new alliances. The networks have been able to bring together the middle-class, urban-based environmental activists and grassroots indigenous and community-based groups. Likewise, the green companies and farmers have been able to argue that their activities can play a role in preserving diversified production systems and in conserving natural resources. By stressing their role as stewards for the environment, they have found a way to gain support from urban middle classes and foreign aid agencies.

4.2 Opponents and Their Capacities

Opponents are found among a number of groups. The World Bank and the Interamerican Development Bank have imposed adjustment programs designed to ensure that countries such as Mexico increase their foreign exchange earnings to service the debts they have accumulated. Thus, the environmental conditionality in project aid for countries such as Mexico may well be less significant than the ecological impact of the neo-liberal economic policies that these lending agencies have fostered. With the amount of resources accessible to these lending agencies, in addition to their ability to involve international governments and media, they can have a major impact on the development, or lack of environmental policy.

On the domestic side of the Mexican environmental policy issue, the so-called opponents include a large part of the Mexican government. Due to budgetary constraints stemming from the 1994 peso crisis and the government's primary goal of economic development, environmental policy has not so much been rejected as it has taken a backseat to economic development. Within the government, the opponents include such agencies as the Treasury Secretariat as well as the Secretariats of Commerce and Industrial Development and of Energy, Mines and Parastate Industry. Government opponents have frequently included state and municipal governments who see environmental policy as an obstacle to attracting industry and thus jobs to their areas.

Another set of opponents to environmental policy includes private and parastatal industries and companies. Industry favors self-regulation and supports the use of voluntary environmental audits. Consequently, though industries often pledge to comply with governmental objectives, they move slowly, resisting compliance with norms and regulations. Parastatal companies are, in fact, many of the worst polluters in Mexico. Yet because economic recovery and employment have top priority, the government has been loath to insist on compliance.

4.3 The Changing Constellation of Significant Actors

The relative novelty of environmental policy in Mexico insures a changing constellation of main actors as new groups and organizations come on the scene. In the 1990s the greater interaction of environmental networks such as CNDCHIM, the Movimiento Ecologista Mexicana, the Pacto de Ecologistas, as well as individual organizations such as Greenpeace, the Mexican Center for Environmental Law (Centro Mexicano de Derecho Ambiental), and the Grupo de Cien is leading to innovative strategies for advancing environmental agendas. An important development is an incipient resort to the courts and legal strategies for holding government accountable. The emergence of environmental law, and specialized legal organizations dedicated to litigation on behalf of environmental values is a brand new development in Mexico and is an important if unproven element in holding the government and the private sector to their legal obligations (Umlas, 1996: 288).

Another dynamic element is the increasing number of government sponsored state-society partnerships aimed at achieving cooperative solutions to environmental problems. One notable example is INE's multi-stakeholder council, the National Coordinating Group. This group is attempting to design a technical, administrative and legal framework for the implementation of a newly established Pollutant Release and Transfer Registry and is comprised of representatives from industry, government, academia and NGOs (OECD, 1998a: 146). SEMARNAP's National Consultative Council for Sustainable Development is another such example.

In the private sector, the growth of green industries provides an increasingly important impetus for strengthening environmental capacity. Mexico's green industry is small compared to other industrialized countries but growing rapidly – at present over 600 firms are active in environmental commerce, and the sector is among the most dynamic of the emergent economic sectors in Mexico. With growing impact on job creation and provision of needed products and services, these firms will play a larger role as an environmental lobby in the future.

5 Main Restrictions

5.1 The National Economy

While Mexico has made remarkable progress in developing a set of laws and regulations for environmental protection, incorporating Agenda 21 elements of sustainable development in its planning regime, and integrating vital elements of economic development policy with environmental programs, it remains a developing country with limited capacity for meeting environmental challenges in the context of pressing socio-economic needs. Allocating resources to government programs and building incentives for compliance with environmental rules and regulations is thus the critical challenge in building environmental capacity. The greatest immediate challenge to environmental policy is undoubtedly the continuing stress on the federal budget arising from its post-1994 economic crisis. The immediate effects of the devaluation were shrinking social spending in 1995-1996 and further retrenchment in implementing social programs (Review of the Economic Situation in Mexico, 1997: 369-373). Budgetary growth for environmental administration has been static or negative since 1994. In 1998, further constriction arising from falling petroleum prices led to three successive waves of federal and state budget cuts, diminishing hopes for strengthening Mexico's environmental policy effort in the near term.

5.2 Centralism and Democratic Transition

Mexico confronts the challenge of moving from a highly centralized political and administrative system to one that better balances the capacities of government at all levels of the system. As seen above, recent Mexican environmental policy reform promotes intergovernmental coordination, decentralization and empowerment of subnational authorities, but lags in transferring supporting fiscal authority to these governments. Mexico's democratic transition, which accelerated after opposition parties in 1997 captured a legislative majority in Mexico's Chamber of Deputies, promises to strengthen state and local capacity for environmental policy in the medium term, but in the short run has delayed progress on decentralizing reforms as opposition parties and the government focus on resolving important procedural questions associated with Mexico's newly competitive party system and the changing balance of executive-legislative relations.

6 Prospects for Mexican Environmental Policy

Compared to the more industrialized OECD countries, Mexico's environmental capacity remains low. Since 1988, however, Mexico's environmental policy capacity has grown substantially, this in spite of serious economic challenges arising from successive economic crises in the 1980s and 1990s and the stresses associated with trade liberalization and a shrinking public sector during the same period. Mexico's inclusion in a wider range of international economic and environmental agreements helped sustain and increase its environmental effort during this period, even as federal support for social programs in general declined. Its 1997 environmental regulatory reforms have unquestionably strengthened its formal capacity for addressing environmental problems on a comprehensive basis.

If Mexico, as a recent OECD report states, "has taken the measure of the challenge it must meet if it is to reverse the severe environmental degradation confronting it" (OECD, 1998: 148), it is also true that further progress is critically contingent on further structural reforms aimed at implementing these regulatory changes. Augmenting existing levels of funding for environmental protection is essential, as is supporting decentralizing initiatives with needed fiscal reforms. Mexico has a long way to go in incorporating market incentives in environmental protection. To supplement command and control regulation it must continue to strengthen its voluntary compliance programs. Mexico must further amplify its efforts to educate both professionals and the general public in environmental concerns. Mexico's gradual economic recovery augers favorably for gradual progress in these areas.

At the systemic level there should be little doubt that the means of strengthening Mexico's environmental capacity is integral to its democratic transition. Strengthening public participation in environmental policy is as much a function of strengthening political access to government generally as it is a strictly sectoral matter. Holding regulatory officials more accountable for performance depends in part on making legislatures more responsive to public concerns. Amplifying public access to information, establishing a right to information, involves a shift to a more democratic culture. Over the longer term Mexico must address fundamental problems of socio-economic dualism and poverty associated with its present stage of development if it is to strengthen public awareness and social concern with environmental issues.

References

Alanis Ortega, G., 1996. 1996, el ano de la reforma a la legislacion ambiental, *Teorema*, 8 (Marzo-Mayo), 48-49.

Aridjis, H., 1998. El Ano del Fuego, *Reforma*, (17 de Mayo), A26.

EcoAmericas, 1998. Around the Region: Mexico Group Building Toxic Waste Database, 1:1 (November).

EcoAmericas, 1999. Spare Mexican budget spells environmental spending cuts, 1:3 (January) 2.

Environmental Protection Agency, 1996. *US-Mexico Border XXI Program Framework Document*. Washington D.C.: EPA.

Excelsior, 1991. Anuncia SEPESCA un Program para la Proteccion del Delphin y el Aprovechamiento del Atun, 2 April, 11A.

Kaimowitz, D., 1997. Social Pressure for Environmental Reform in Latin America, in: H. Collinson (ed.). *Green Guerrillas.* New York: Black Rose Books, 20-33.
Mumme, S., 1991. Clearing the Air: Environmental Reform in Mexico, *Environment* 33:10 (December), 7-30.
Mumme, S., 1992. System Maintenance and Environmental Reform in Mexico: Salinas's Preemptive Strategy, *Latin American Perspectives* 19:1 (Winter 1992), 123-143.
Mumme, S., 1994. Mexican Environmental Reform and NAFTA, *North American Outlook*, 4:3 (March), 87-103.
El Mercado de Valores, 1998 La Proteccion del Medio Ambiente y el Programa de Nacional Financiero," 56:10 (October), 16-19.
El Financiero, 1998. Cuatro áreas al proyecto ecológico del Valle de México, May 8, 64-A.
Newman, G., 1994. *Managing Mexico's Environmental Challenge.* New York: The Economist Intelligence Unit.
OECD, 1998a. *Environmental Performance Reviews: Mexico.* Paris: Organization for Economic Cooperation and Development.
OECD, 1998b. *Towards Sustainable Development: Environmental Indicators.* Paris: Organization for Economic Cooperation and Development.
OECD, 1997. *Review of the Economic Situation in Mexico. The Anti-Poverty Program.* (September), 369-372.
Umlas, E., 1996. *Environmental Non-Governmental Networks: The Mexican Case in Theory and Practice.* Doctoral dissertation. Yale University, May 1996.
World Bank, 1997. *World Development Report, 1997.* Washington, D.C.: The World Bank.

NEW ZEALAND

Ton Bührs

1 Main Environmental Problems

New Zealanders pride themselves on the natural beauty of their country. Uninhabited by humans until about 700-800 years ago, with its majestic mountains and volcanoes, sparkling rivers and streams, splendid coasts, extensive natural forests, and scenic landscapes, the country is sometimes referred to as "God's own". This beautiful, almost pristine, and "clean and green" image is a carefully cultivated part of New Zealand's identity, and an important economic asset.

But however much one wants to believe in that image, there is no denying that New Zealand's environment has been drastically altered since the beginning of settlement by the Maori, the country's "first people", and particularly following the arrival of the first European settlers in the 1840s. Already prior to New Zealand's (Aotearoa's) European colonisation, fires lit by humans had caused a decline in the area covered by forests from 85 to 53 percent of New Zealand's total area, and a range of animals, among which the Moa, had become extinct (Ministry for the Environment, 1997a: 9.28-9.29).

With the increase in immigration from Britain, human impact on the environment intensified significantly. Forests were slashed and burnt to create space for large-scale agriculture (mainly sheep grazing), to the extent that the area covered by forests was reduced to 23 percent of the country, with 13 percent of the total area of New Zealand denuded in a single decade (Ministry for the Environment, 1997a: 8.30). The rate of animal extinction also went up, and it is estimated that since European settlement, 16 land birds and more than a dozen of other animals have disappeared (Ministry for the Environment, 1997a: 9.30).

Today, the threat to New Zealand's remaining biodiversity is considered by many as the most serious environmental problem facing the country. Despite the fact that most indigenous forests are now under some form of protection, these forests, and many animal species endemic to New Zealand with them, are still under threat, largely from introduced animals and plants. Also the kiwi, the flightless bird that became New Zealand's national icon, is in sharp decline and faces extinction if present trends continue (Ministry for the Environment, 1997a: 9.19).

The concern about New Zealand's natural environment may attract most attention, but it is not the only type of environmental problem affecting the country. Despite its small population (of about 3.6 million), low population density (of 17 people per square kilometre), and relatively low level of industrialisation (the economy is largely dependent on the export of primary products), New Zealand shares many of the environmental problems that prevail in much of the rest of the world, be it on a smaller scale. For instance, air pollution is a serious problem in some urban centres, such as Christchurch, where at times various pollutants reach levels higher than those recommended by the World Health Organisation. Water pollution is also a serious problem in some locations, where the treatment of industrial waste and/or sewage is inadequate, and because of agricultural run-off and the contamination of stormwater. Erosion and other forms of land degradation affect much of New Zealand. It has been estimated that 7,800 sites may be chemically contaminated, 1,500 of which seriously, from landfills, service stations, timber treatment, and other industrial practices (Ministry for the Environment, 1997a: 8.6).

Congruent with the image of New Zealand as a "clean and green" country is the perception that the quality of life in New Zealand is of a high standard. Yet, although defendable in some respects, also this view needs qualification. It is true that living in New Zealand offers advantages that many naturally less endowed, and more densely populated, countries can't offer, notably with regard to opportunities for outdoor recreational activities in beautiful and uncrowded natural surroundings. Yet, traffic congestion is a growing problem in some cities, notably in the Auckland region, where more than a quarter of the population of New Zealand is concentrated. Also, during the last fifteen years, income inequality, poverty, and many other social problems have been on the increase. On the social (quality of life) dimension of environmental quality, New Zealand is far from being the paradise that some people may (want to) believe.

In some respects, New Zealand's environmental performance or conditions are worse than those of many other countries. For instance, the level of energy intensity (the amount of primary energy used per unit of GDP), is high by OECD standards, and energy consumption per head of population is still increasing. In recent years, energy intensity has started to come down, but the decrease lags behind the decline observed in many other western countries (Ministry for the Environment, 1997a: 3.22). Ozone depletion in the upper atmosphere, for which of course New Zealand cannot be blamed alone, exposes (not only human) life in the country to increasing levels of damaging UV radiation. Also, although New Zealand's contribution to greenhouse gas emissions is a very small proportion of the global total, it is higher than that of many other countries on a *per capita* basis, in large part because of the methane produced by the large number of sheep and cattle (Ministry for the Environment, 1997a: 5.23).

Precisely how serious New Zealand's environmental problems are, and whether they are diminishing or worsening is hard to tell, mainly because the information is lacking or inadequate. The first state of the environment report was published only in 1997, and although environmental data have been gathered for some time, this has occurred without the benefit of a co-ordinated and systematic approach. Given the paucity of information on many aspects of New Zealand's environment, New Zealanders have been able, for quite some time, to hold on to the belief that their country was (or still is) one of the "cleanest and greenest" in the world. More recently, however, that belief has been shaken with the discovery that New Zealand is not free from the same kind of serious pollution problems that exist in the rest of the developed world. With more information becoming available, it is likely that the element of denial in the country's environmental psyche will be receding further.

2 Environmental Policy Evolution

Five stages can be identified in the evolution of environmental policy in New Zealand: the "pre-environmental" era until the early 1970s, the first wave of environmental reforms in 1972 and 1973, a period of stagnation from 1974 to 1984, a second period of reform from 1984 to 1991, and a period of consolidation since 1992. Environmental policy evolution can be characterised as a process of "punctuated equilibria", where periods of little and slow change are interrupted by relatively short bursts of rapid change.

In New Zealand, as in most other western countries, there was an upsurge in environmental concerns and demands during the early 1970s. Prior to that, concern about the environment was focused foremost on nature conservation issues, and had led to the establishment of a Nature Conservation Council with an advisory role in these matters in 1962. Water and soil legislation had a somewhat longer history, and was largely directed at the prevention or mitigation of erosion problems and conflicts over water use (by granting "water rights"). Since the 1950s, land use had been made subject to a planning regime akin to that in Britain (Town and Country Planning Act), based on district plans, zoning, and a permit system. Various other environmental problems were kept under control as "nuisances" under other legislation (Commission for the Environment, 1984; Milne, 1992; Williams, 1980).

1972 was a watershed year in the evolution of New Zealand's environmental policy. With environmental issues (many from outside New Zealand or of a general nature) receiving an upsurge in media attention, the first international conference on the human environment pending, and the environment becoming a political "bandwagon", the government appointed the first Minister for the Environment. Later in the year, a Commission for the Environment was established as a *de facto* government department with the responsibility to provide policy advice on environmental matters and to promote the co-ordination of environmental policy (Bührs, 1991a). Also in the same year, a Clean Air Act was passed, modelled on British legislation.

New Zealand was also one of the first countries to adopt a system of environmental impact assessment, introduced in 1973, influenced by the American and Canadian precedents. For all public and private development proposals that were considered to have significant environmental effects, or that required permits under environmental legislation, an environmental impact report had to be produced which was submitted to the Commission for the Environment for independent scrutiny ("auditing") and public perusal and comment.

After this first wave of environmental reform, not much change occurred for about a decade. During that time, the advocates of environmental protection were very much put on the backfoot by governments that assigned foremost priority to New Zealand's deteriorating economic situation, in particular by the promotion of capital intensive industrial and energy projects. During this "Think Big" era, in which the state itself was heavily involved in pushing projects that were perceived by environmentalists as environmentally damaging ("state sponsored vandalism"), life was made very difficult, and at times even precarious, for the Commission for the Environment. The principal reform that occurred during that period was the introduction of legislation (the National Development Act, repealed by a later government), that rationalised and sped up the process of granting permits for projects that were considered to be in the "national interest", with nefarious consequences for public participation and influence.

With the election of the Fourth Labour Government in 1984, New Zealand entered a period of radical reform, not only in all areas of policy, but also in the very foundations on which public decision making occurs. Virtually no stone remained unturned. Under the influence of "New Right" ideology regarding the role of the state, and neo-classical theories of economic management emanating from within Treasury and embraced by a "Troika" of key government ministers, the whole of the public service was "down-sized", reorganised and put on a "New Public Management" basis (Boston, 1991; Boston et al., 1996). Many state owned enterprises were first corporatised, then privatised. The speed and consistency with which this occurred was unprecedented, and turned New Zealand into a model for "New Right" protagonists in many other countries. But when the economic and social pain caused by the reforms started to bite, and their benefits appeared uncertain if not illusive, this led to also unprecedented levels of public cynicism towards politicians to an extent that the New Zealand political system can be said to suffer from a crisis of legitimacy.

Perhaps paradoxically, many aspects of these reforms, and the environmental reforms in particular, have been considered as major improvements from an environmental point of view. Financial support from the government for economic activities, such as subsidies for agriculture, were almost totally abolished, with positive environmental effects. In 1986, a Ministry for the Environment was established with a stronger mandate for environmental policy development to replace the rather weak (non-statutory) Commission for the Environment. A new Department of Conservation, amalgamating a range of fragmented agencies with responsibility for the protection of natural areas on publicly owned land, comprising almost twenty-three per cent of New Zealand, was created in 1987. A new office, of the Parliamentary Commissioner for the Environment, was also set up in 1986 as an independent "watchdog" over the whole system of environmental management (Bührs and Bartlett, 1993; Memon, 1993).

The Resource Management Act, passed in 1991 after more than three years of preparation and extensive public consultation, is the *pièce de résistance* among New Zealand's environmental legislation. The Act replaces a broad range of previously existing pieces of legislation, provides a framework for the development of environmental policy from the national to the local level, and establishes an integrated system for granting permits ("consents") for all types of development, based on an assessment of their environmental effects. Under the Act, Regional and Local Councils are required to develop and implement mutually consistent plans that address all environmental issues in their areas, and that provide a framework for decisions regarding the granting of consents (by the same councils). All plans and consent decisions must be based on the sole purpose of the legislation, the sustainable management of natural resources. The Act also allows for public input at all stages of plan formation and decision making, and provides opportunities for appealing decisions with an Environment Court, and higher courts.

The introduction of the Resource Management Act, with its emphasis on an assessment of outcomes instead of regulating activities, also marks a shift in the use of policy instruments. Traditionally, New Zealand's approach to environmental management, inspired by British influence, was dominated by prescriptive planning (zoning) and regulation, be it of a rather general and flexible kind (no legal standards, but guidelines, for water and air quality, waste management, no "technology forcing" regulations), which allowed affected parties (such as industries, farmers) to exert considerable influence over their implementation, even to the point that it can be argued that consent authorities were "captive" of vested interests. With the shift in emphasis towards the assessment of effects, and the imposition on consent granting authorities to consider the costs and benefits of alternative means towards achieving acceptable outcomes, the scope for flexibility in the selection of instruments to achieve objectives is formalised and extended. In principle, "anything goes" as long as it can be demonstrated that the environmental effects of proposed activities meet the sustainability criterion.

Congruent with the move away from the "command and control" approach in environmental policy (which was never strong anyway) has been an increased preference for a co-operative approach (voluntary agreements) and a growing interest in the potential use of economic instruments. On the national level, voluntary agreements between the Ministry for the Environment and industries are relied upon for the management of greenhouse gas emissions and waste reduction, even though there is growing doubt about the effectiveness of these agreements. And despite the initial enthusiasm for economic instruments among decision makers (in particular for the establishment of tradable property rights regimes), very few have been introduced, due to the complexities associated with their implementation, uncertainty and conflicting views about their effectiveness and appropriateness, and potential costs to the economy (in the case of taxes and charges) (Bührs and Bartlett, 1993: chapter 4).

Impressive as the Resource Management Act may be on paper, there are still many questions to be answered about its effectiveness and consequences. Given the sheer size and complexity of the legislation, many issues are as yet unresolved, even with regard to fundamental matters, such as what "sustainable management" means, the extent to which social (quality of life) concerns should be considered in plans and decisions, and how cumulative effects can/should be assessed. Concern has also arisen about the practical (in particular financial) obstacles to public participation, despite the generous "standing" provisions, inconsistencies between councils in the implementation of the Act, the capacity of councils to fulfil their tasks, the inadequacy of national guidance, and the long time it takes for plans and decisions to be made, and the costs associated with the whole process. Whether the reforms are actually achieving their stated purpose, the sustainable management of resources, remains to be seen, and is far from a foregone conclusion (Bührs, 1997; Bührs and Bartlett, 1993; Grundy, 1994; May, 1997; OECD, 1996).

After 1991, environmental policy developments have not been anything near as dramatic as during the previous stage. Since then, much energy and attention appears to have gone into implementing, fine-tuning, and consolidating the new environmental regime. On the national level, the most significant environmental policy initiative of recent years has been the adoption of the *Environment 2010 Strategy* in 1995. This document, hailed as the most comprehensive environmental policy statement ever produced by a government in New Zealand, is a variation on the move towards "green planning" or the development of sustainable development strategies which can be ascertained in many countries. New Zealand's *Strategy*, however, hardly deserves the "planning" label, as it contains not much more than an inventory of environmental problems and general principles and goals, but no definite priorities, few specific objectives or targets, no time-table, and hardly any sign of strategic thinking. In the broader context of New Zealand's other (in particular economic) strategic efforts, and present political and economic climate, its political weight and environmental significance is doubtful (Bührs and Bartlett, 1997).

Despite all the reforms, New Zealand's environmental policy style continues to be rather "laid back", influenced by the "clean and green syndrome", not characterised by the sense of urgency or crisis that is more apparent in some of the countries where environmental pressures are more intensive or apparent. The driving force behind the environmental reforms between 1984 and 1991 was not only, or even primarily, inspired by a concern for the environment. Foremost, these were an integral part of a broader agenda of change driven by "New Right" ideology and neo-classical economic theory. Their underlying rationale reflects a predominant concern about efficiency and the institutional facilitation of market-led economic growth. Much of the unease expressed about the Resource Management Act is that is does not seem to deliver the efficiency that was promised.

In some other respects, however, New Zealand's environmental policy style has changed significantly. Gone are the days when the government heavy-handedly bulldozed through (often literally) its favourite development projects, regardless of their environmental costs and consequences. Pulling itself back from direct involvement in development, the state seems to have become more accommodating of the demands for public input in environmental decision making. Now, the provisions for public participation are generous, even though in practice the extent and meaningfulness of that participation remain problematic. But the adversarial and confrontational relations between environmentalists and the government, common during the 1970s and early 1980, have been replaced by a more consensual style of decision making, although it is far from approaching the corporatist approach prevailing in some countries, such as Sweden (Lundqvist, 1997).

3 The Main Actors

On the national level, the main institutional agencies with regard to environmental policy development are the Ministry for the Environment, the Department of Conservation, the Parliamentary Commissioner for the Environment, and a range of government departments representing other sectoral interests, such as the Ministries of Agriculture, Forestry, Fisheries, the Ministry of Health, the Ministry of Commerce (which also has responsibility for energy policy), and, last but not least, Treasury.

The Ministry for the Environment is the principal environmental policy agency. Its prime responsibility is for environmental policy development and coordination, and it has almost no role in "hands-on" environmental management (such as pollution control). Given the broad definition of the "environment" concept in the *Environment Act 1986* on which it is based, the scope for the Ministry's policy initiatives, and its involvement in the development of policies of all kinds, is very wide. However, it is not a powerful agency, and relies primarily on a strategy of voluntary co-operation in its policy development and co-ordination efforts, although more recently it has been moving towards formulating a more substantive basis for integrating environmental policies with the adoption of the *Environment 2010 Strategy* (Bührs, 1991; Bührs and Bartlett, 1997). Also, given its role in "balancing" the broad spectrum of views and interests with regard to the environment, it is not perceived to be, and arguably cannot be, a strong or very "green" advocate for the environmental movement.

The Department of Conservation, according to its official mandate, is an advocate for conservation, combining policy and management responsibilities with regard to all protected areas on Crown land (the conservation estate). Its creation, under the *Conservation Act 1987*, can be seen as a victory for the conservation movement in New Zealand which, for a long time, battled for a more comprehensive and secure protection regime for New Zealand's remaining indigenous forests and other areas of high natural value under threat from continuing development. But although the Department still enjoys the support and sympathy from the mainstream conservation movement, it has become the subject of debate and critique regarding the effectiveness of its management, in particular following the collapse of a viewing platform that cost the lives of fourteen people (Chapple, 1995). In part, the critique of the Department is also fuelled by development interests who accuse it of "locking up" resources and thwarting desirable development, particularly in areas where unemployment is high.

The Parliamentary Commissioner for the Environment, also established under the *Environment Act 1986,* is a relatively rare breed of agency, even by international standards. Its main role is that of a "watchdog" over New Zealand's system of environmental administration (from the local to the national level), to identify gaps and weaknesses in the institutional framework, and to recommend measures to address these shortcomings. Although appointed by the Government, the Commissioner is independent in the execution of her/his role, and reports to Parliament. Recently, with the appointment of a new Commissioner, the role and performance of the Office has come under review, but there is much appreciation for the work that it has done and broad support for the continuation of its existence, despite some rumbling about its (almost inevitably controversial) evaluation of policy performance in some cases (Bührs, 1996; Hawke, 1997).

Although over the last two decades environmental awareness may have grown within most branches of the public service as much (or little) as among the *public* at large, there are still many obstacles to the integration of environmental values into other policy sectors, such as transport, economic policy, energy policy, and agricultural policy. Most of the government departments responsible for policy development in these sectors still treat environmental concerns as an "add-on" to what they see as their main business, which is predominantly oriented towards promoting economic growth. Although within some of these departments there are "pockets" of more environmentally oriented groupings or divisions (such as the Energy Efficiency and Conservation Agency within the Ministry of Commerce), they are usually not strong enough to put their stamp on the overall direction of policy development in their sector, with or without the involvement or support of the Ministry for the Environment.

In the context of New Zealand's government system, whereby Cabinet is not only part of but dominates Parliament, the kind and status of environmental policy is also strongly influenced the personal commitment, position and effectiveness of the Minister for the Environment and the Minister of Conservation. In large part, the introduction of the Resource Management Act can be attributed to the role and commitment of the Minister for the Environment at the time, who retained the portfolio when he became Prime Minister. In New Zealand's hierarchical Cabinet structure, the importance assigned to a policy area is reflected in the status of its Minister in the Cabinet ranking order. Since the creation of the environment portfolio in 1972, Ministers for the Environment have usually occupied a lower place in the government ranking order, with a few exceptions. More recently, the conservation portfolio was occupied by a minister who was not even in Cabinet. This mixture of personal interest, abilities, and the rank of Ministers continues to be a very important factor in the evolution of environmental policy in New Zealand, also after the changeover to proportional representation and the practice of coalition government in 1996.

Environmental organisations have also played a significant role in the development of environmental policy. Since the 1920s, the protection of nature has been pushed by a conservation movement that became stronger, more radical, and more professional over time. Although for a long time forced into fighting reactive battles (to halt one project or development after the other), it shifted its focus more and more towards changing institutions and policies, with considerable success, during the 1980s and 1990s. On a policy level, however, ideological divisions within the movement have also become more explicit and weakened its cohesiveness.

Although public support for nature conservation has been strong and growing since the 1920s, for a long time there has been a lower level of support, and weaker organisations as a result, for other dimensions of the environmental problematique, such as in the areas of pollution, waste management, and urban environmental issues. One notable exception is the extent of public support for the anti-nuclear movement, even though New Zealand has no nuclear industry or power stations. Much of the support for this movement can be attributed to the, now past, French testing of nuclear weapons in the South Pacific. Already in the early 1970s, indignation about this extended to the Government, which sent a warship to Moruroa to protest against nuclear testing. It also led to the introduction of a ban on nuclear ships from New Zealand ports. Continued French testing, and the bombing of *Greenpeace's* flagship the *Rainbow Warrior* by French secret agents in Auckland's harbour in December 1984, ignited public anger about France's apparent disdain for the concern about its nuclear tests in the Pacific to unprecedented levels, and led to a huge upsurge in support for Greenpeace, and for the Labour government's proposals to legislate New Zealand's nuclear free status, which eventuated in 1987.

With the increase in *Greenpeace's* support and resources, also other environmental issues, such as those associated with pesticides and hazardous waste, started to receive greater attention, thus strengthening the demands of other, much smaller, organisations that had been trying to raise the profile of such issues, such as the *Friends of the Earth,* and the *Soil and Health Association.* Although nature conservation issues still often dominate the agenda within the environmental movement as a whole (including the umbrella organisation *ECO* (Environmental and Conservation Organisations of New Zealand), the attention given to other environmental matters has increased significantly in recent years.

Another important actor behind the rise of concerns of a broader "green" nature is the Green Party of New Zealand. New Zealand was one of the first countries to give birth to a green party, the *Values Party*, in 1972. The emergence of *Values* can be attributed to a range of factors or influences, many of which associated with the rise of social movements, the globalisation of the economy, and the social and cultural poverty or crises ascribed to an increasingly materialistic world (Dann, 1997). *Values'* political agenda extended beyond environmental issues and, if anything, was directed at changing the economy and the socio-cultural infrastructure to reduce alienation and improve the *quality* of human life and society.

Despite a promising start in the elections in 1972, *Values* never gained enough electoral support (under the "First-Past-the-Post", FPP, system) to get representation in Parliament, and dwindled to virtual extinction in the late 1980s. In 1990, however, the remnants of *Values*, together with an influx of fresh blood, reconstituted themselves as the Green Party, and gained 6.9 per cent of the vote in the elections that year (but again no seat in Parliament). To enhance their chances of entering Parliament, the Greens joined forces with four other small parties and, in 1992, formed *The Alliance,* which obtained 18 per cent of the vote in 1993 (but only two seats, and no representative of the Greens due to the FPP electoral system). In 1996, finally, under the new MMP (Mixed Member Proportional representation) system, the Greens, for the first time, gained representation in Parliament, with three seats out of the twelve won by the Alliance.

In November 1997, however, the Greens decided to split off from the Alliance, in which they felt that the green identity could not be sufficiently expressed, and started to prepare themselves for the challenge of having to gain at least five per cent of the vote in the 1999 elections.

In the business sector, environmental awareness still seems to be lagging compared to that in many other western countries. Only a handful of New Zealand corporations has adopted environmental reporting as a practice, and even fewer have introduced environmental management systems. Also, very few companies have gone through the process of obtaining the "Environmental Choice" seal of approval, a New Zealand green product labelling scheme. Probably associated with that is the relatively small domestic market for green products. Often, such as in the case of much organically grown produce, production is destined for overseas markets to cash in on the premium prices fetched for these goods (Law, 1996; *The Press,* 30 April 1997).

The media in New Zealand frequently report on environmental matters, but mostly in a reactive way, and with a preference for the more dramatic events and issues, often from overseas. Serious environmental reporting of a more informative or educative kind is mostly to be found in some of the newspapers and on public radio, but hardly ever on television, which is totally commercially dominated. Apart from the newsletters and magazines published by environmental groups, there are no environmental publications with a readership on any significant scale. The role of the media in the development of environmental policy in New Zealand does not seem to go much beyond that of feeding snippets of more or less disconcerting news to the public, thus contributing to a general sense of environmental decline and creating a basis of support for the demands from environmental groups.

Regional Councils, established as recently as 1989, are the most significant environmental actors on the local government level. Under the Resource Management Act, they were given the task of taking an integrated approach to the management of the environment in their regions. Their responsibilities include long-term planning, the management of water and air (and land use in as much as it impinges on these), coastal management, the management of hazardous substances, pest control, and public transport (although the provision of the latter is contracted out to private companies). To a large extent, they are the lynchpin in the system set up with the Resource Management Act, expected to achieve a degree of environmental integration that the central government has never been able to approximate. And given the long-standing absence of regional government in New Zealand, the dislike of local governments for what they regard as another layer of bureaucracy and control above them, and the shaky support basis of Regional Councils (many people have little appreciation of their role or importance), their existence is still precarious and the need for them regularly challenged.

Increasingly, environmental policy development in New Zealand is influenced by actors and developments from outside the country. Australia has become a major actor, co-operating with New Zealand in the formulation of a range of issues and policies, such as food standards and hazardous substances, in part because of the increasing level of economic integration between the two countries. International agencies, such as the OECD, the European Union, the World Trade Organisation, and various UN organisations, also have a growing impact on the direction and shape of New Zealand's environmental policies. Increasingly, environmental policy development in New Zealand is driven by external forces and events, and influenced by examples set by other countries. On the other hand, New Zealand's highly integrated framework of environmental management has also aroused considerable international interest.

4 The Development of Environmental Capacity

Over the last decade, New Zealand's institutional capacity for environmental policy development has been strengthened considerably. At the central government level, three agencies, the Ministry for the Environment, the Department of Conservation, and the Office of the Parliamentary Commissioner for the Environment, have been established that have environmental concerns as their main focus and responsibility. Within the machinery of government, the chorus of environmental voices is stronger than ever before.

The introduction of the Resource Management Act in 1991 has also strengthened the institutional framework for environmental decision making, not only because its of its prime objective, the sustainable management of natural resources, but also because of the high degree of integration and public participation in which it foresees. In addition to that, the establishment of Regional Councils, in many ways the lynchpin under the new legislation, has significantly enhanced the institutional capacity for environmental management at the regional level of government.

This is not to say that New Zealand's institutional capacity for environmental policy development has no weaknesses. All three central government agencies mentioned are constrained in their capacity by rather modest and, in the case of the Department of Conservation, even declining resources. This reflects the low level of priority assigned by governments to environmental issues relative to other priorities (such as containing government expenditure, health and education), in particular during the 1990s. The rather lukewarm, at times no more than symbolic, political commitment to environmental issues, in contrast to the uppermost importance attached to economic goals ("imperatives"), is probably the most significant weakness in environmental capacity.

On paper, New Zealand's integrative capacity with regard to the development of environmental policy is high. On the national level, the responsibility for the co-ordination of environmental policy lies with the Ministry for the Environment which, in theory, has the potential to take up or get involved in any policy issue or area that has significant environmental implications. Not burdened with implementation responsibilities for any specific area of environmental policy, this gives it the advantage of being able to "free range" across the whole policy spectrum, whether "internal" or "external" to the areas more traditionally and narrowly defined as environmental (such as pollution control). Theoretically, it is thus in a position also to ensure that core environmental values are recognised and incorporated into all policy areas.

In practice, however, the capacity of the Ministry to bring about such a high level of integration is limited. Despite the formal obligation for all government departments to consult with the Ministry on matters of environmental significance, such consultation has not always occurred. In part, this can be attributed to the (deliberate or non-deliberate) failure of government ministers, or Cabinet as a whole, to ensure that the Ministry for the Environment is informed or involved. Also, the Ministry has not the formal power and status, nor the resources, to put it in the same league as some other government departments, such as Treasury. Its involvement in the development of policies that have potentially significant environmental implications (such as economic, energy, transport and agricultural policy) depends very much on the voluntary co-operation of other government departments, and on political backing in Cabinet (Bührs, 1991b).

The Parliamentary Commissioner for the Environment, the "watchdog" over New Zealand's system of environmental management, also plays a key role in enhancing integrative capacity. The Commissioner has the power to investigate the functioning of all agencies that play a role in environmental policy development and implementation, at the national, regional and local level. Its potentially most significant role lies in assessing the system as a whole, in identifying and addressing gaps, unnecessary overlap, inadequate co-ordination mechanisms, obstacles to public involvement, ill-designed or malfunctioning institutional structures, and other shortcomings or limitations in New Zealand's system of environmental management and policy development. As a "systems guardian" it is at least as important as the Ministry for the Environment in the further development of integrative capacity.

So far, however, the Commissioner for the Environment has not been able to develop the role of "system guardian" to its full potential. To a large extent that can be blamed on the very small size of the office (less than 15 staff) and its resources. Another reason is that the role of the Commissioner can be, and has been, interpreted in ways other than that of "system guardian". In fact, the two main other roles that have been assigned to the Office, that of "environmental ombudsman" (a complaints investigation office), and that of "environmental auditor" (assessing the performance of agencies), have absorbed by far most of the time and resources available to it. To what extent the newly appointed Commissioner will (be able to) reorient the Office towards the "system guardian" role remains to be seen. And whether, and to what extent, the Government and its agencies will heed advice from the Commissioner is an even more open question (Bührs, 1996; Hawke, 1997).

A similar gap between promise and practice exists on the regional and local level of environmental management. On paper, the Resource Management Act 1991 foresees in a high level of integration in environmental policy development and decision making, the identification and anticipation of all environmental problems of significance, and the management of resources in ways that are ecologically sustainable. In reality, many regional and local councils are struggling with their responsibilities, given the scope and complexity of their tasks, insufficient powers, resources, skills, and guidance and support from the central government level. Also, many issues, such as how to deal with cumulative environmental effects, the non-level nature of the "playing field" of decision making (development interests are very much in an advantageous position under the system), and the absence of independent scrutiny of environmental assessments, remain unresolved, or are even hardly recognised. Given all these issues and obstacles, it is doubtful that regional and local councils have the capacity to meet the official objective of the sustainable management of resources.

The discrepancy between the capacity of New Zealand's environmental agencies on paper and that in practice mirrors the balance of forces and interests in the political realm. Although democratically elected, the government's actions and priorities do not necessarily reflect the views or preferences of the majority of the population. This has become particularly apparent in the New Zealand situation since 1984, when one government after the other introduced unpopular measures without the support of, and often in blatant disregard of vocal public opposition, on the grounds that these measures are in the longer term public benefit and that "there is no alternative" (the "TINA" syndrome). As a result, popular confidence in politicians, and even the political system as a whole, has plummeted to the point that there now is a legitimacy crisis and state of political disarray (Kelsey, 1996).

But the relatively low level of priority assigned to environmental values in the political arena is not just a reflection of the priorities of political leaders. Although public support for environmental or "green" causes, as reflected in opinion polls, is not insignificant, it has been diminishing during the 1990s. Survey data collected at the Parliamentary elections in 1990, 1993 and 1996 indicate that the importance of the environment in deciding what to vote has declined from 77 per cent of voters in 1990 to 56 percent in 1993, and to 53 per cent in 1996. Also, in 1996, fewer people (39 per cent) advocate higher spending on the environment than in 1990 (when the figure was close to 50 per cent), and fewer (42 per cent in

1996 compared to 49 per cent in 1990) take the view that the environment should be protected even if that leads to lower incomes. Other issues, such as those associated with health and education, presently appear to arouse stronger expressions of public concern and receive a higher profile in the media (Australian Social Science Data Archives, 1998; Vowles and Aimer, 1993:127-146).

This tendency for denial appears to be even stronger in the business sector. By comparison to many other western countries, few New Zealand businesses have adopted mechanisms that strengthen their environmental capacity, such as environmental reporting, auditing, standards, and management systems. Many business are eager to exploit the "clean and green" image of New Zealand for marketing purposes, in particular overseas, but there is not much evidence that they are actively backing up that image with environmentally progressive practices. And although a few government agencies (such as the Ministry for the Environment and the Energy Efficiency and Conservation Authority, a division within the Ministry of Commerce) are promoting the adoption of technologies and practices that are sound environmentally and economically at the same time, there are no government programmes to financially support their introduction.

The strongest motor behind environmental policy development, apart from, but linked with, the forces external to New Zealand mentioned above, is the environmental movement. Several environmental organisations, such as Greenpeace, have grown into highly professional lobbying groups. On many occasions, environmental groups have successfully instigated the government to review its policies, or even to change the institutional framework for environmental policy development and management. Political opportunities have been skilfully exploited to induce promises (from the government and opposition parties) for change.

The small size of New Zealand's policy communities (where most actors know each other), the relative ease of access to members of Parliament and Government compared to countries with bigger populations, the frequent use of public consultation and submission processes in environmental policy development, and an Official Information Act that enables people to obtain government information (be it not always promptly and with ease), together create a fairly open system of governance. Ironically, for the public the chances of influencing policy appear to be greater in the environmental policy area than in other policy realms, and seem inversely related to the relative importance assigned to a policy area (no public input is allowed or encouraged with regard to economic policy).

Support for environmental groups peaked during the early 1990s, when Greenpeace, the Royal Forest & Bird Protection Society, and the Maruia Society counted 170,000, 62,000 and almost 12,000 financial supporters respectively. Since then, numbers have declined. One of the likely reasons for this decline can be found in the deteriorating economic climate, in which many people have experienced a stagnation, or even a fall, in real income (including welfare payments), and face rising costs associated with the introduction of "user pays" regimes in health, education and the provision of other (previously public) services. These trends are also reflected in the lower priority of environmental issues in public opinion mentioned above.

Obviously, the drop in financial support for environmental groups had significant implications for their capacity. Several organisations had to shed staff and reorganise themselves to remain solvent. Greenpeace in particular went through a "rough patch" and was the focus of media attention when it had to "downsize" its office in New Zealand. More recently, however, the situation now seems to have stabilised itself, and although support levels are not going up, there is a feeling within these organisations that their financial basis now is more secure and reliable than it was in the heydays of support.

In comparison to other environmental organisations, the success of the Green Party in influencing environmental policy has been very limited. In large part, this can be attributed to the fact that only recently the Party has gained a few seats in Parliament, and that in New Zealand's adversarial political system opposition parties have very little scope for influencing policy. The capacity of the Green Party to influence policy, as long as it is not a partner in a coalition government, lies in its role of providing a voice for environmental concerns in Parliament, and thus raise the public profile of environmental issues, and in obtaining "inside" information that can be used by the environmental movement as a whole.

The ability of environmental advocates in New Zealand, within and outside the government sphere, to develop or influence policy is hampered also by the existence of significant gaps in environmental knowledge and information, and by a weak system of co-ordinated environmental monitoring and reporting. Although the Ministry for the Environment has control over some funding of environmental research (the "Sustainable Management Fund"), the sums involved are rather modest, and largely allocated to issues or programmes that carry some political priority. Similarly, much environmental research undertaken by New Zealand's corporatised "Crown Research Agencies" is targeted towards areas and issues that are considered to be of significance and benefit to New Zealand's economic (in particular export earning) capacity. By comparison, not much environmental research of a longer term, and more fundamental nature, is funded or undertaken.

To a large extent because of the modesty of New Zealand's environmental research capacity, significant gaps exist in the knowledge and information about the extent of, and trends in, the environmental problems of the country. Furthermore, although environmental monitoring has been made a responsibility of local government, there is still no integrated or co-ordinated system of data collection on the national level. The development of a set of common environmental indicators has been on the Ministry for the Environment's agenda for some considerable time, but received political priority only recently when this gap in New Zealand's capacity was highlighted in the OECD's review of New Zealand's environmental policy (OECD, 1996). New Zealand's first state of the environment report, published in October 1997, also reveals the large gaps in information on environmental conditions, pressures and responses, and trends therein, and the absence of a core set of indicators on the basis of which monitoring and reporting can occur (Ministry for the Environment, 1996; 1997b).

5 Evaluation

Over the last two decades, New Zealand's environmental policy framework has developed in leaps and bounds. Periods of major change occurred in the early 1970s and between 1984 and 1991. On both occasions, the institutional capacity for the development and co-ordination of environmental policy was strengthened significantly. Environmental concerns and issues have become the foremost business of an increased number of agencies and people. As a result, it has become less likely that such concerns get ignored and that decisions are being made without any regard given to their environmental implications. Also, the opportunities for the public to get involved in environmental decision making have been increased significantly. Overall, New Zealand's institutional capacity for raising and debating environmental issues, and for addressing these in a more integrated way, has expanded considerably.

Notwithstanding the greater strength of this framework, it is more difficult to say whether environmental policy performance has also improved. Apart from the fact that assessing policy performance involves making value judgements, there is the problem that few specific environmental policy objectives or targets have been formulated, that the extent of many environmental problems is unknown, and consequently, that in many respects it is hard to say how effective or successful agencies have been in mitigating or resolving problems, or in meeting objectives.

Paradoxically, the enormous scale of New Zealand's environmental reforms in the 1980s and early 1990s may have had some detrimental effect on environmental policy development. A prevailing assumption underlying these reforms has been that the quality of policy performance is primarily a function of the institutional framework: "Get the institutions right, then good policy will automatically follow". Without wanting to deny the importance of institutions, this is, of course, a questionable thesis. Form does not automatically follow function, and there are many other than institutional variables that impinge on environmental policy performance, such as the level of political commitment, the strength of environmental demands, and economic factors. But the widespread belief in the validity of this thesis among decision makers has led them to focus less on the development of actual policies. In many respects, New Zealand's "substantive" environmental policies (goals, objectives, targets, and the means for achieving them, such as regulations or standards) are not much more ambitious or stringent than they were at the start of the reforms in 1984, despite the development of the *Environment 2010 Strategy* in 1995.

Another consequence of the scale of reform undertaken since 1984, across virtually all policy areas, is that there is little support or enthusiasm for further change. "Reform fatigue" has led to a widespread longing for some degree of stability and consolidation. In that context, an often heard argument is that more time is needed for the new institutions to prove their value, for instance for the Resource Management Act to be fully implemented. Complementing this argument is the view that the new institutions are fundamentally sound and that any "teething problems" can be dealt with by improving implementation.

Although there is some validity in these arguments, there is also the risk that they lead to a false sense of security regarding New Zealand's environmental situation, as people feel or think that "everything is under control" and that no more radical change is needed to address the environmental problematique effectively. Nothing could be more dangerous or misleading, as it is extremely unlikely that New Zealand's present institutional framework will resolve its present environmental problems, or prevent future ones. More reform of a fundamental nature will be required to integrate environmental "imperatives" within other policy sectors and in the "private" business sector (to "green" the economy), to ensure that the cumulative effects of a multitude of small decisions are adequately assessed and contained, to strengthen New Zealand's environmental information and knowledge infrastructure, and to re-orient the dominant value system towards a new social, economic and political paradigm which is not based foremost on unrestricted economic growth and materialism.

These challenges illustrate the fundamental nature of the principal constraints on the development of environmental policy (capacity), in New Zealand and other countries. First, there is the difficulty that the source of many, if not most, environmental issues lies at the core of an economic system that may be made compatible only in a limited way with long-term ecological and social rationality, and that economic change of a more fundamental nature is most likely to be required. But given the globalisation of the economy, it is increasingly problematic for individual countries on their own to move in that direction. Second, these issues illustrate that coming to terms with environmental issues is not merely, or even primarily, a question of building capacity of a technical kind. Environmental problems are inherent to a more fundamental problématique: the collective challenge to humankind to define the kind of society that we want to live in, and to develop the political means towards achieving such a society. That also has become a global challenge.

Postscript

The national elections of 27 November 1999 changed New Zealand's political landscape dramatically. At first, election results indicated that Labour and the Alliance had won a majority of the seats in Parliament, and on that basis both parties started talks to form a Centre-Left coalition government. On the election night, the Green Party, which had split from the Alliance in 1997 and contested the 1999 elections on its own, just failed to obtain 5 per cent of the votes, or to win a majority of the electorates, either of which would have entitled it to Parliamentary representation. However, when all special votes were finally counted on 7 December 1999, it transpired that the Greens had scored 5.2 percent of the votes, and had gained a majority in one of the electorates (the Coromandel). As a result, the Green Party got seven (of the 120) seats in Parliament, and the Labour-Alliance coalition dropped to 59 seats, just short of a majority. This means that, although the Greens do not take part in the new government, they effectively hold the balance of power and are in a, for New Zealand, unprecedented position from which to influence the government.

Although, at the time of writing (April 2000), the new government is still young, it already has brought about a significant change of direction in many policy areas, including health, education, social welfare, economic development, and the environment. In general, the government has returned to a more "interventionist" policy style, and moved away from the neo-liberal approach that prevailed during the previous fifteen years. Reforms that paved the way for a greater reliance on competition ("market forces") and privatisation have been put on hold and, in several instances, reversed. Social welfare spending has increased, and so have taxes on the higher incomes. Regional economic development is actively and financially supported, and the abolition of remaining import tariffs, a plank of the previous government's radical free trade stance, has been scrapped. The heydays of New Zealand as a model of "New Right" policies are over.

Given the key position of the Green Party in Parliament, there is now considerable scope for environmental issues to be taken more seriously by the government. Already, the government has put an end to a controversial "sustainable forestry" (of indigenous beech forests) scheme. It has also promised to set up a Royal Commission of Inquiry into genetic engineering, an issue pushed hard by the Greens. Although the government still needs to declare its position on a range of other environmental issues and initiatives introduced by the previous government, it can be expected that environmental issues will carry greater weight (and receive more funding) than they have during the last decade.

At least as significant as the change in policy direction is how the electoral system of proportional representation, introduced in 1996, is put to work since the last elections. The new electoral system (a variation on the German system), introduced in response to growing political disillusion, was expected to lead to a more consensual policy style and a greater responsiveness to public (rather than market) interests. However, faith in the system was rapidly shaken by the performance of the first coalition government, and in particular by the actions of one of the parties (New Zealand First), which fell apart and by which many voters felt betrayed. The challenge for the new government is to demonstrate that coalition (and even minority) government can be stable, responsive, and effective. As the capacity for influencing and strengthening environmental policy from a Green perspective depends to a large extent on the existence of a system of proportional representation, the ability of the new government, and the Green Party as its principal ally, to deliver these values, will be an important factor in determining the fate of this system and the development of environmental capacity in the longer term.

References

Australian Social Science Data Archives, 1998. *New Zealand Election Studies* (1990, 1993 and 1996). Canberra: Australian National University.

Boston, J., 1991. The Theoretical Underpinnings of Public Sector Restructuring in New Zealand, in: J. Boston, J. Martin, J. Pallot, and P. Walsh (eds.), *Reshaping the State*. New Zealand's Bureaucratic Revolution. Auckland: Oxford University Press, 1-26.

Boston, J.. J. Martin, J. Pallot, and P. Walsh, 1996. *Public Management: The New Zealand Model.* Auckland: Oxford University Press.

Bührs, T., 1991a. *Working Within Limits. The Role of the Commission for the Environment in Environmental Policy Development in New Zealand.* PhD thesis. Auckland: University of Auckland.

Bührs, T., 1991b. Strategies for Environmental Policy Co-ordination: The New Zealand Experience, *Political Science* 43 (2), 1-29.

Bührs, T., 1996. Barking Up Which Trees? The Role Of New Zealand's Environmental Watchdog. *Political Science* 48 (1), 1-28.

Bührs, T., 1997. Environmental Policy, in: Raymond Miller (ed.), *New Zealand Politics in Transition*. Auckland: Oxford University Press, 287-295.

Bührs, T., and R. V. Bartlett, 1993. *Environmental Policy in New Zealand. The Politics of Clean & Green?* Auckland: Oxford University Press.

Bührs, T., and R. V. Bartlett, 1997. Strategic Thinking and the Environment: Planning the Future in New Zealand?. *Environmental Politics* 6 (2), 72-100.

Chapple, G., 1995. Scandal at Cave Creek. *Listener.* June 3, 16-24.

Commission for the Environment, 1984. *A Guide to Environmental Law in New Zealand.* Wellington: Commission for the Environment.
Dann, C., 1997. *Green Genesis: The Origins and Development of Green Politics (in Aotearoa New Zealand).* Ph.D. thesis (draft). Canterbury: Department of Resource Management, Lincoln University.
Grundy, K.-J., 1994. Public Planning in a Market Economy. *Planning Quarterly* 114, 20-24.
Hawke, G. (ed.), 1997. *Guardians for the Environment.* Wellington: Institute of Policy Studies.
Kelsey, J., 1996. *The New Zealand Experiment. A World Model for Structural Change?* Auckland: Auckland University Press.
Law, G., 1996. Business Proves Slow To Come Clean. *Management* 43 (11), 107-110.
Lundqvist, L. J., 1997. Sweden, in: M. Jänicke, and H. Weidner (eds.), *National Environmental Policies. A Comparative Study of Capacity-Building.* Berlin: Springer, 45-71.
May, P. J., 1997. The Mandate for Planning: An Outsider's Perspective Concerning the Resource Management Act. Paper presented at the Annual Conference of the New Zealand Planning Institute, 16-18 April 1997, Palmerston North, New Zealand.
Memon, A.,1993. *Keeping New Zealand Green. Recent Environmental Reforms.* Dunedin: University of Otago Press.
Miller, R., 1991. Postmaterialism and Green Party Activists in New Zealand. *Political Science* .43 (2), 43-66.
Milne, C. D. A. (ed.), 1992. *Handbook of Environmental Law.* Wellington: Royal Forest & Bird Protection Society.
Ministry for the Environment, 1995. *Environment 2010 Strategy: A Statement of the Government's Strategy on the Environment.* Wellington: Ministry for the Environment.
Ministry for the Environment, 1996. *National Environmental Indicators: Building a Framework for a Core Set.* Wellington: Ministry for the Environment.
Ministry for the Environment, 1997a. *The State of New Zealand's Environment 1997.* Wellington: Ministry for the Environment.
Ministry for the Environment. 1997b. *Environmental Performance Indicators. Proposals for Air, Fresh Water and Land.* Wellington: Ministry for the Environment.
Organisation for Economic Co-Operation and Development, 1996. *Environmental Performance Reviews. New Zealand.* Paris: OECD.
The Press, 1997. Poor Environmental Score by NZ. 30 April 1997.
Vowles, J., and P. Aimer, 1993.. *Voters' Vengeance. The 1990 Election in New Zealand and the Fate of the Fourth Labour Government.* Auckland: Auckland University Press.
Williams, D. A. R., 1980. *Environmental Law.* Wellington: Butterworth.

Acknowledgement: My thanks to Susan Banducci of the Department of Political Studies of Waikato University for digging out the data on public opinion on environmental issues from the Australian Social Science Data Archives.

ENVIRONMENTAL POLICY IN POLAND

Magnus Andersson

1 The Environmental Situation

In the 1950s, Poland embarked on a rapid transformation from an agrarian nation to an industrial one along the lines of the Soviet development paradigm. Among the issues the socialist system in Poland was unable to address, the deteriorating state of the environment was one of the most striking. A decree of the Council of Ministers at the beginning of 1983 officially designated twenty-seven "areas of ecological hazard" comprising approximately 11% of the country's territory and 12.9 million people, i.e., 35% of the population (World Bank, 1989).

The pollution of water resources is one of the most serious problems in Poland. Almost 70% of the length of the main rivers do not meet the physical and chemical criteria for any use (NFEP, 1997: 85). The situation is aggravated by the fact that Poland is endowed with scarce water resources. In many areas in Poland waste dumps constitute a direct or potential threat to the drinking water sources. The activities of the Soviet army in Poland led to contamination of ground waters and land degradation. The total damage has been evaluated at about $2 billion (MEPNRaF, 1995: 27).

In an assessment of the air quality situation in the Katowice district, Poland's industrial heartland, in the early 1980s, the Polish Academy of Sciences concluded that the city Zabrze had "probably the worst air quality in the world" (Kapała, 1983: 73). In the most polluted areas in Poland there is higher morbidity and mortality due to respiratory diseases, cancer, and circulatory diseases among the inhabitants (Bochniarz and Toft, 1995). In the Katowice district the frequency of respiratory diseases and cancer were 30% to 50% above the average in Poland in the 1980s (French, 1988). Acidification has caused the destruction of several thousands of hectares of forests on the mountain tops in southern Poland.

Air pollution declined substantially in the period 1989-95. For example, the emissions of sulphur dioxide and particulates decreased by 24% and 44%, respectively (NFEP, 1997: 49). Table 1 illustrates that the drop in air pollution was sharper than the economic downturn in the first few years of the transition. The generation of hazardous waste was reduced by 15% between 1989 and 1995 and there was also a 30% decline in untreated waste water discharged to surface waters (ibid.: 13). As a result of these improvements, the Ministry of Environment

Table 1. Annual percentage changes in GDP, industrial production and air pollution emissions in Poland, 1990-93

Year	GDP	Industrial Production	Dust	Carbon Oxides	Nitrogen Oxides	Sulphur Oxides
1990	-12	-24	-23	-17	-18	-21
1991	-7	-12	-21	-28	-6	-8
1992	2	4	-26	-15	0	-10
1993	4	7	-13	-21	0	-3

Source: Cole (1995b: 283)

could in 1995 reduce the number of areas in ecological hazard from twenty-seven to four (Millard, 1998).[1]

There has been a sharp increase in motor vehicles in Poland in the 1990s. The number of passenger cars per 1,000 people increased from 61 in 1980 to 167 in 1992. With the current trend private cars per 1,000 persons will increase to almost 400 in the year 2010, a level comparable to western Europe (Anderson, 1995). The market share of public transportation in Poland decreased about 34% between 1986 and 1993. This trend implies that private transport is contributing more and more to air pollution. In cities such as Katowice and Kraków transport has a larger impact on ambient air quality than industry.[2]

Recently, a programme for the construction of some 2,000 km motorways have been launched. New motorways in Warszawa have been proposed without environmental impact analysis. One of the planned routes constituted a direct threat to a protected landscape park (Millard, 1998: 159). Similar problems are discernible in other parts of Poland.

According to official statistics, the volume of waste from private households increased by approximately 10% between 1990 and 1995 (GUS, 1997). Poland's capacity to manage the increasing waste flows is limited:

Almost all of the waste goes to landfills without separation ... Up to 10,000 unauthorized dumps exist, owing to the limited capacity of authorized landfills. The volume of landfills has tripled since 1975. Few municipal dumps are equipped with compactors. For many towns, the amounts disposed are already beyond official capacity. Only 55% of the population is served by waste collection agencies, so proper collection and disposal are difficult (OECD, 1995a: 63).

The situation is aggravated by the fact that the recycling schemes for paper, glass, and textiles that were in place during the socialist period have collapsed. The main reason for this breakdown is the new price structure for secondary materials (OECD, 1995a: 65).

Many areas in Poland have not been exploited economically and are preserved in their natural state. The Wisła, the largest river in Poland, is unregulated and has

1 These areas were the following: Kraków voivodship, Upper Silesia, Legnica-Głogów, and Tarnobrzeg (Millard, 1998).

2 Personal communication from Wojciech Beblo, former Director of the Department of Ecology in Katowice voivodship, 1995 and Krzysztof Göhrlich, deputy Mayor of the city of Kraków, 1996.

great diversity of flora and fauna (NFEP, 1997). Polish ecologists estimate that near-pristine area ecosystems cover as much as 8.5% of the territory and that another 19% of the area has biological complexes in good ecological shape. This is an asset that most OECD countries have lost (Żylicz, 1991). In 1996, the system of protected areas (national parks, nature reserves, landscape parks, areas of protected landscape, etc.) extended over more than 26% of the territory (NFEP, 1997: 10).

2 Development of Environmental Policy

2.1 General Developments

Poland's first effort in modern times to protect the environment was the 1922 Water Law which regulated, through the introduction of permits, the discharge of industrial effluents into water bodies. It also introduced economic instruments on a limited scale (Cole, 1995a: 300-301). A State Council for Nature Protection (PROP) was set up in 1925 and a Nature Protection Act was adopted in 1934. In the inter-war period, seven national parks were established (Chmielewski, 1988: 24). Poland's first environmental organization, the League for Nature Protection (LOP), was created in 1928 (Cole, 1995a).

In 1949, the Polish parliament adopted a new Nature Protection Act. It covered natural monuments, nature reserves, national parks, and various endangered species of plants and animals. Between 1949 and 1960 nine new national parks, several nature reserves, and a large number of monuments were established (ibid.: 309).

Besides nature protection, attention at this time was focused on water management. In 1960 the Central Water Management Board (CUGW) was created. Its main task was to prepare and administer long-term water management plans. CUGW was inexplicably dissolved in 1972 but is still considered to have been an effective water management body in Poland.

Environmental policy was promoted rather intensively in Poland during the early 1970s. Environmental protection was elevated to ministerial level, a long-term environmental policy programme to 1990 was elaborated, and preparations for an environmental framework law were initiated.

Two paragraphs about environmental protection were added to the constitution in 1976. Article 13 stated that "the Polish People's Republic ensures the protection and rational development of the natural environment, constituting the nation-wide wealth." Article 71 asserted that citizens "have the right to benefit from the values of the natural environment and the obligation to protect it" (Kabala, 1988: 6-7). By the mid-1970s, all countries of the Soviet bloc had incorporated the goal of environmental protection into their constitutions (Radecki, 1979).

The event that provided the greatest impetus for environmental policy developments in Poland in the 1970s was the United Nations Conference on the Human Environment in Stockholm in 1972. Poland's receptiveness to the international environmental upsurge in the 1970s is illustrated by the fact that, by the late 1970s, Poland had signed approximately twenty-five environmental conventions (Aura, 1979). Poland was also an active member of the UNEP in this period.

In 1980, parliament enacted the Statute on the Protection and Shaping of the Environment. It was a comprehensive framework law covering the entire field of environmental protection. All in all, there were eight titles with 118 articles. The 1980 law established the principle that all central, regional, and local governmental agencies, enterprises, and individuals have a duty to protect the environment. It called for an integration of environmental concerns into the socio-economic plans and increased the role of voluntary organizations and private citizens to participate in the environmental policy process (Cole, 1995a).

A major feature of environmental policy in Poland in the 1980s was a search for mechanisms to enhance the integration of ecological concerns into the general economic development of the country. Environmental policy-makers had realized that the approach of the 1970s, that is, application of end-of-pipe technologies, was inadequate, and that preventive measures and structural changes in the economy were needed. One result of this was that the use of environmental impact assessment (EIA) was promoted. The Statute on Land-Use Planning of 1984 made EIA mandatory for projects especially harmful for the environment and human health (World Bank, 1989).

The 1988 National Environmental Policy Programme to the Year 2010 set out new ambitious goals for Poland's long-term ecological recovery. The tone of the programme was open and frank: there was clear-cut recognition of the disastrous environmental situation and the state's failure to protect the environment (MOŚiZN, 1988).

In the late 1980s, Poland was a party to thirty international conventions, and had established cooperation with international organizations such as the UNECE, the FAO, the WHO, and UNESCO (World Bank, 1989).

In October 1988, the Rakowski government declared environmental protection to be one of the top three priorities for the state. Round-table discussions between the government and Solidarity were held between 6 February and 5 April 1989. Negotiations were held in a number of working groups on issues such as political reform, economic reform, trade union pluralism, education, health, etc. There was also one working group devoted to environmental policy reform (Millard, 1994: 61-62). Most environmental policy reforms that have been carried out in the 1990s were conceptualised at the round-table.

On November 19, 1990, the Polish Council of Ministers approved the National Environmental Policy. With this document, Poland had produced the first document of its kind in eastern Europe and one of the most comprehensive and far-reaching environmental policy statements of any country. The National Environmental Policy divides the environmental protection goals into short-term, medium-term, and long-term priorities. The basic goal of short-term priorities is the elimination of sources of pollution and other environmental abnormalities that constitute an acute threat to public health. The medium-term priorities, to be completed by the year 2000, aim at limiting the pressure on the environment and achieving conformity with western European standards. The long-term priority of national environmental policy is the introduction of sustainable development principles into the entire economy (MEPNRaF, 1991).

It should be noted that the National Environmental Policy was not the result of a broad dialogue process with all relevant actors. Its main authors were a few leading figures at the Ministry of Environmental Protection, Natural Resources, and Forestry (of whom the minister himself was one).

The 1991 Forest Law placed environmental functions of the forest on an equal footing with the productive functions (NFEP, 1997: 59). In 1993, stricter emission standards for new cars were introduced (ibid.: 53). A complete ban on the import of cars over ten years old has been imposed (Śleszynski, 1996).

In June 1997, a new Waste Law was approved by parliament. It emphasizes waste prevention and minimization of volume. The law contains a number of novelties relating to (1) principles of setting charges; (2) extended obligation to obtain permit for activities generating hazardous waste; and (3) improved classification of waste (MEPNRaF, 1997).

It is notable that the environmental framework law from 1980 has survived the transition despite serious attempts to replace it with a new one. The first attempt was made in the spring of 1991 when a research group on environmental law submitted a draft law to the Ministry of Environment. It was approved by the Council of Ministers in the autumn of 1991. This bill failed to get approval from parliament because a group of parliamentarians were dissatisfied that the new law included both nature protection and pollution control issues. The group managed to eliminate all pollution control provisions and submitted a nature protection bill to parliament. It was enacted, against the will of the Ministry of Environment, in November 1991 (Bowman and Hunter, 1992: 932). A new Environmental Protection Act, commissioned by the Ministry of Environment, was drafted in 1996. It was abandoned in 1997, "mainly due to the political dynamics of an election year" (Żylicz, 1998: 13). In August 1998, the PHARE[3] (the European Union's aid programme for central and eastern Europe) gave support for a new attempt to finalize the work on the framework law (ibid.).

After 1989, the environment gradually lost much of the prominent position on the public and governmental agendas that it had enjoyed in the latter part of the 1980s. The examples are legion: the political parties attach little priority to the environment and few of them have elaborated explicit views on environmental policy (with one important exception, the Freedom Union)[4]; the issue was neither highlighted in the campaigns before the parliamentary elections in 1991, 1993, and 1997 nor in the campaigns before the presidential elections in 1990 and 1995. None of the green parties has come close to winning parliamentary seats (Alberski, 1996). In the parliamentary election in 1991, 63 green lists were registered in more than one constituency. Of these, three regarded themselves as ecological parties but they did not win any seats. In the local government elections in 1990, thirteen environmental organizations stood in twenty-three constituencies and won only 31 seats (Waller and Millard, 1992: 168). To this should be added that the Ministry of Environment, Natural Resources and Forestry has enjoyed a low status in all governments that have been formed since 1989. In 1994-95 the government

3 PHARE stands for Poland and Hungary Assistance to the Restructuring of the Economy.

4 The Freedom Union has created a nationwide Ecological Forum to promote environmental protection. Since 1991 the party issues an environmental magazine entitled Green Alternative (*Zielona Alternatywa*, 1991).

seriously considered dissolving the environmental ministry in connection with the reorganization of the central government administration (BSE, 1995).[5]

The Polish environmental movement won its hitherto greatest victory in September 1990 when the government decided to abandon the nuclear power plant under construction at Żarnowiec (Waller and Millard, 1992: 177). The decision was taken on economic grounds, but the impact of the public opposition against the project was also important.

After a referendum in 1997, a democratic constitution was approved. The new constitution contains a number of provisions that relate to environmental protection. Most importantly, Article 5 makes sustainable development a principle task of the state and Article 74 requires authorities to support the public in its efforts to protect the environment (Jendrośka, 1997: 106).

It is possible to identify provinces and cities that are pioneers in environmental policy. Firstly, in north-eastern Poland, nine districts (originally five) have launched the Green Lungs of Poland initiative. The idea, which dates back to 1988, is to base the development strategy in this region on the principles of sustainable development. Environmentally harmful activities are restricted. Emphasis is laid on sustainable management of forests, organic farming, and the development of ecotourism. Another pillar of the project is a systematic extension of the protected areas in the region (OECD, 1995a). Secondly, in the city of Chorzów in Upper Silesia, a scheme of emission trading was implemented, for the first time in Poland, in 1991-92. After this project had been realized, other cities, especially in southern Poland, expressed their interest in the use of emission trading. Thirdly, in the city of Kraków the authorities each month arrange a Green Table to which representatives of environmental organizations, scientific organizations, local authorities, and the business community are invited to exchange their views on issues related to environment and development in Kraków.[6] Similar meetings are being held in Gdańsk.

Poland signed an association agreement with the European Union in December 1991, which came into force in February 1994. Joining the EU[7] means adopting some 7,000 pieces of legislation comprising policies on, for example, industry, customs duties, commerce, company law, residence conditions, taxation, social issues, regional matters, energy, environmental protection, agriculture, and scientific research (Sachs, 1994). The association treaty places stress on cooperation in ecological issues in almost all relations with the EU. Article 71 states that "… the policy for the economic and social development of Poland should be directed by the principle of sustainable development. It is essential to guarantee that the requirements of environmental protection be fully included into this policy from the very beginning ..." (NFEP, 1997: 13). Article 80 contains guidelines on issues such as joint actions on environmental monitoring, transboundary and regional air and water pollution, protection of forests, preservation of fauna and flora, and the reduction of greenhouse gases. The parties are encouraged to cooperate through

5 This idea was met with furious opposition from the environmental community and was almost immediately abandoned (Alberski, 1996).

6 Personal communication from Krzysztof Göhrlich, deputy Mayor of the city of Kraków, 1996.

7 Poland is not likely to join the EU before the year 2002 (Business Central Europe, 1997).

the exchange of information and experts and jointly executed research (Fiedor, 1995: 1).

It appears that Poland has relatively little regulation compared to the European Union in four areas: (1) chemical turnover; (2) genetically modified organisms; (3) public participation in the decision-making process, especially EIA procedures; and (4) access to environmental information (Wajda, 1996b). Important changes in Polish law will have to be made in these areas.

EU law is based on emission standards grounded on the best available technology (BAT), a principle that does not exist in Poland (Wajda, 1996a). BAT means "the most effective and advanced stage in the development of activities and their methods of operation which indicate the practical suitability of particular techniques for providing in principle the basis for emission limit values designed to prevent and, where that is not practicable, generally to reduce emissions and the impact on the environment as a whole" (Wajda, 1996c). The BAT principle will become a central element of the new environmental framework act.

Poland has no eco-labelling system. However, legal preparations for the introduction of a national eco-labelling system are made by the Ministry of Environmental Protection, Natural Resources and Forestry. The system is expected to be launched at the time of Polish accession to the European Union.[8]

Within the Ecological Policy Department of the Ministry of Environment, a Unit for Implementation of International Obligations and European Integration has been created (Wajda, 1996a). A number of other steps have been taken to facilitate the approximation process:

- A Poland-EU committee for approximation of environmental law has been established.
- A project on the development of a methodology for evaluating costs of harmonizing and implementing EU environmental legislation in Poland has been carried out.
- A project has been launched to enhance environmental management in Poland with respect to goals, tasks, competencies, procedures, and instruments (Wajda, 1996a).

Attempts have been made to estimate the costs to the associated countries of meeting the requirements in the EU's environmental legislation. The field of environmental protection is likely to become the most costly sector for approximation. According to the PHARE, the cost of environmental harmonization could constitute as much as between 30% and 40% of total costs (Miljödepartementet, 1997). The World Bank has estimated that the overall costs for Polish industry to comply with EU environmental standards amount to $21 billion (Business Central Europe, 1997). This is more than ten times more than the annual environmental expenditures in Poland in the mid-1990s.

By mid-1995, Poland had ratified 44 international conventions and other international agreements related to the environment (BSE, 1995: 120). Among these are the UN Framework Convention on Climate Change and the UN Convention on

8 Personal communication from Anna Kalinowska, Director of the Bureau of Education and Public Relations, Ministry of Environmental Protection, Natural Resources and Forestry, 1999.

Table 2. Environmental aid to Poland, 1991-96 (grants)

Donor	Aid Million US$	(%)	Donor	Aid Million US$	(%)
Belgium	3.1	0.8	Norway	5.8	1.5
Denmark	57.8	15.2	Sweden	24.2	6.4
EU (PHARE)	133.6	35.1	Switzerland	3.9	1.0
Germany	42.3	11.1	United Kingdom	1.7	0.5
Finland	16.9	4.5	USA	36.6	9.6
Netherlands	49.3	13.0			
Japan	5.0	1.3	Total	380.3	100.0

Source: GUS, 1997. The table does not include environmental loans from the EBRD and the World Bank

Biological Diversity, ratified by Poland in 1994 and 1995, respectively. Poland is an active participant in a number of regional environmental programmes. In 1993, Poland, Slovakia, and Ukraine established the Carpathian Euroregion for the promotion of broad cooperation including environmental protection. An International Commission for the Protection of the Odra River was created jointly by Poland, Germany, the Czech Republic, and the European Union in 1996 (NFEP, 1997). The so-called Black Triangle Project was launched in the beginning of the 1990s to reduce emissions of air pollution in eastern Germany, south-western Poland, and northern Bohemia in the Czech Republic (OECD, 1995a).

Poland has received environmental assistance from the international community since 1990. Most importantly, the aid has facilitated the transfer of know-how and new environmental protection technologies. With respect to grants, the largest contributors between 1991 and 1996 were the European Union (via PHARE), Denmark, and the Netherlands. In addition, the European Bank for Reconstruction and Development (EBRD) and the World Bank have offered loans for, among other things, the improvement of environmental management in Poland (Andersson et al., 1993).

Although foreign assistance has been quite significant, it has never corresponded to more than 5% of annual environmental expenditures in the 1990s (GUS, 1997). As a member of the OECD (Poland joined in 1996), Poland is supposed to phase out the receipt of aid and start preparations for offering its own aid to developing countries (NFEP, 1997: 128).

2.2 Choice of Policy Instruments

The current pollution control system in Poland rests on three main pillars:

- A financing system based on environmental charges that are paid by polluters, collected by environmental authorities, and used by environmental funds on the national, regional, and local levels for various environmental protection investments.
- National environmental quality standards.

- Pollution permits issued by the voivodship departments for environmental protection.

The foundation of this system goes back to the 1960s and 1970s. An Air Protection Law was enacted in 1966, introducing the concept of "permissible concentrations" and giving health inspectors the right to close an industrial plant if its emissions posed a threat to human life (Cole, 1995a: 315-316). The law also introduced fines for air pollution (Alberski, 1996). The 1974 Water Law introduced charges for the consumption of water. These charges were to feed a Water Management Fund from which money for financing water quality projects could be derived (Cole, 1995a: 321).

The 1980 Statute on the Protection and Shaping of the Environment reinforced the system of individual permits that had been introduced in the 1960s. It was stated that the environmental departments in the 49 districts (voivodships) should issue permits that impose emission standards for a particular source of pollution.[9] This permitting process should be guided by national environmental quality standards for water, air, noise, radiation, and vibration (Sommer, 1993: 4). The law obliged voivodship authorities to close industrial plants when their pollution levels posed a direct threat to public health.

The role of economic instruments was substantially expanded by the 1980 Statute. Charges for disposal of waste and activities causing air pollution were introduced and charges were to be paid by all economic entities for their use of natural resources. Facilities that violated environmental norms were subject to fines for effluent discharges, air pollution emissions, noise producing activities, chemical uses, and waste dumping activities. Collected fines and charges were earmarked for environmental funds that would finance various environmental protection projects. This system of charges and fines, together with the permit system, became the prime tool of environmental protection during the 1980s (Cole, 1995a: 336-337). In the 1990s, Poland has chosen to continue to rely on this system.

The pollution and resource use charges were significantly increased at three occasions in the early 1990s: June 1990, December 1990, and December 1991 (Żylicz, 1992a). Most charge rates have increased 18-20 times in real terms in the 1990s compared to the 1980s (Anderson and Fiedor, 1997: 8). In 1995, charges were levied on 62 atmospheric substances, 163 waste substances, and six types of waste water. It is noteworthy that the charge for sulphur dioxide ranked third in the world after the Swedish and Norwegian ones (Stavins and Żylicz, 1995: 4).

The 1990 National Environmental Policy contained a conceptual change in the pollution control system in that it advocated the introduction of emission trading. This instrument also appeared in the 1991 environmental protection bill, which stated that "the terms of a pollution permit for a plant can be transferred to another plant, either entirely or in part, subject to the approval of the authority that issued the original permit" (Stavins and Żylicz, 1995: 5). However, since this bill did not become law, emission trading failed to become a legally approved policy instrument. Despite this, a pilot project on emission trading was, as already mentioned, arranged in Chorzów in Upper Silesia in 1991-92.

---

9 The permit is based on data submitted by the polluter. All polluters are obliged to submit such information.

The Polish environmental policy style could be characterized as a mixture of economic instruments and command-and-control. So far, there are no examples of voluntary agreements and covenants between the regulators and the economic actors. But it is likely that the room for negotiations will increase in the future. Many policy-makers are in favour of introducing so-called Negotiated Compliance Schedules (NCS). NCSs "are to be a legal form of introduction of legal instruments, aimed at ensuring gradual achievement of generally binding standards by enterprises (legal entities) causing an adverse impact on the environment and not complying with the requirements currently in place ...". Furthermore: "An enterprise which is not financially able to meet the requirement immediately, will now have the chance to do it gradually in a precisely defined time framework and using solutions, formulation of which it can influence to some extent" (State Inspectorate, 1996: 8-9).

3 Evaluation of Environmental Policy

In the 1970s, Poland's environmental protection actions lagged behind its rhetorical flourishes; environmental policy was characterized by an implementation deficit.[10] Between 1975 and 1980, emissions of gaseous pollution increased from three million tonnes to 5.1 million tonnes. Dust emissions also went up during this period (Alberski, 1996: 62). In 1977, the first beaches on the Baltic Sea coast had to be closed (Chmielewski, 1988: 64).

Industry did not comply with the environmental legislation, and there were virtually no enforcement mechanisms at hand. Siting of new industry perpetually breached the environmental regulations. The most notorious example was the gigantic steel plant Huta Katowice which was located in the industrial centre in Upper Silesia, the most polluted region, in 1973. The entire project was constructed outside existing land-use plans as well as the regional and national socio-economic plans (Cole, 1998: 97).

The foreign credits that Poland received from the West throughout the 1970s could have given impetus to the modernization of the economy in an environmentally friendly way. But the funds were spent without any consideration for the environmental consequences of the investments.[11]

In the 1980s it was possible to discern some signs of improvement in Poland's environmental performance. Environmental expenditures as a percentage of GDP grew from 0.28% in 1980 to 0.8% in 1989 (Cole 1995b). More waste water treatment plants were built in the 1980s than in the 1970s. For instance, in 1989, 54% of industrial plants treated its sewage water, compared to 40% in 1975. Dust emission was reduced from 2.3 million tonnes in 1980 to 1.5 million tonnes in 1989 (Alberski, 1996: 58-61). Another success was that the planned exploitation of iron ore in a landscape park in Suwałki in north-eastern Poland was stopped because of opposition from the local authorities and the Polish Academy of Sciences (Chmielewski, 1988: 56).

10 In Weale's (1992: 17) definition, "implementation deficit arises when legislative and policy intent is not translated into practice."

11 Personal communication from Waldemar Michna, Minister of Environment in 1987-88, 1995.

However, the overall picture was bleak and the above mentioned improvements did not match the negative trends. Gaseous air pollution emissions increased substantially from 1985 to 1989 (Alberski, 1996: 62). Saline waste water discharges from the hard coal sector to the Odra and Wisła rivers continued unabated. Very little was done to address the problem of solid waste from the industrial sector. Sanctions for non-compliance was not enforced. Moreover, the tools for land-use planning and EIA were never fully used.

In the 1990s, governmental revenues for environmental protection have increased dramatically. The main reason for this was the sharp increase of the environmental charge rates in the early 1990s in combination with a relatively high collection rate of these charges. For example, the collection rate was 100% in 1990, 72.4% in 1992, 73.6% in 1994, and 77.0% in 1996 (GUS, 1995 and GUS, 1997). By 1993 the environmental funds were annually spending 15-20 times more in real terms than in 1990 (Wajda, 1993: 19). Expenditures increased from $464 million in 1990 to nearly $1.5 billion in 1996 (NFEP, 1997: 13). In 1991, Poland's spending on the environment reached 1% of GDP for the first time, which meant that Poland's environmental expenditures in percentage terms had become comparable to the level of many OECD countries (OECD, 1995b: 95). If water supply projects are grouped together with environmental investments (which is a common approach in many countries) Poland's expenditures reached more than 2% in 1996 (GUS, 1997).

Table 3. Expenditures for environmental protection and water management in Poland in the 1990s

Year	**Environmental Protection % of GDP**	**Year**	**Water Management % of GDP**
1990	0.7	1990	0.4
1991	1.0	1991	0.6
1992	1.0	1992	0.6
1993	1.0	1993	0.5
1994	1.0	1994	0.5
1995	1.3	1995	0.3
1996	1.7	1996	0.4

Source: GUS, 1997. The figures for 1996 are preliminary.

Table 4. Sources of environmental protection financing in Poland, 1993

Source	%
State budget	5
Municipalities	10
Enterprises	23
Environmental funds	58
Foreign aid	4
Total	100

Source: OECD (1995b)

An important part of Poland's environmental protection budget has been devoted to the financing of industrial and municipal waste water treatment plants. In 1996 alone more than 400 such plants were completed. This could be compared with 1980 when 56 such plants were taken into operation (GUS, 1997). Air protection has become an upgraded policy area in the 1990s, which is reflected in the increase in air protection spending. Poland has been able to finance the installation of advanced technology to reduce air pollution from the largest power plants (e.g., Bełchatów and Turów) (Cole, 1995b: 303). In 1996, Poland took into operation technologies capable of eliminating 267,000 tonnes of gaseous pollution, almost ten times more than was achieved in 1985. The annual installation of waste management equipment was six times higher in 1996 than in 1985 (GUS, 1997).

The environmental administration carried out a strict enforcement policy in the early 1990s. In 1990, approximately 100 heavy industrial polluters were forced to shut down completely or partly (Żylicz, 1997). Twenty-nine enterprises were closed down in 1992 for environmental reasons, 566 legal proceedings were initiated, and 1,357 petitions delivered (EBRD, 1993: 51). Between 1990 and 1993 seven companies on the list of the 80 largest polluters were closed. At another 22 plants of this list production was partially halted (OECD, 1995a: 128).

The collection rate of environmental fines has been considerably lower in the 1990s than in the 1980s. One reason is that the hard coal sector has been unable to pay its fines. Another reason is that the payment of fines can be partly or completely cancelled if a polluter has managed to eliminate the problem within five years. In the mid-1990s, some 70% of non-complying companies had initiated or finalized measures to meet their environmental obligations (OECD, 1995a).

What were the main reasons for the heavily increased commitment to environmental policy in the early 1990s? It is possible to point to several factors. First, the Ministry of Environmental Protection, Natural Resources and Forestry was committed to radical improvements, and acted quickly and strategically to achieve this. Second, the ministry had strong support from the environmental movement. Third, among the general public there was a high acceptance for radical changes of any kind in the first phase of the transition. Fourth, the environmental policy-makers met almost no opposition from the business community, which had not yet learned to carry out lobby-work to influence environmental policy.

4 Capacity of Proponents

4.1 Public Institutions

In 1972, the government established a Ministry of Territorial Management and Environmental Protection. In 1975 the ministry was reorganized and renamed Ministry for Territorial Management, Administration and Environmental Protection (Alberski, 1996). The department dealing with environmental protection was the smallest and weakest; it had only 15-20 employees.[12] The ministry was dissolved in 1983 to be replaced by the Bureau of Environmental Protection and

12 Personal communication from Janusz Żurek, Institute for Environmental Protection, Warszawa, 1995.

Water Resources Management (World Bank, 1989). It was the first central agency in Poland devoted solely to environmental protection. In 1985, the Bureau was transformed into a Ministry of Environmental Protection and Natural Resources. The new ministry was a considerably larger organization than its predecessor, and was authorized to deal with a broader set of issues. At the end of 1989, forestry was assigned to the ministry. The Ministry of Environmental Protection, Natural Resource and Forestry has a current staff of about 300.

Various measures have been taken by the ministry to integrate environmental concerns into the sectoral policies.

- The Ministry of Environment cooperates with other ministries in the meetings of the Economic Committee of the Council of Ministers (KERM).
- In October 1994, the government established the Inter-Ministerial Commission for Sustainable Development. Its main aim is to provide a forum for opinion-making and for ensuring congruence between documents, plans, and new draft regulations, and the national environmental policy. The commission is made up of high-level representatives of all ministries, central offices, and the environmental protection administration. Its chairman is the Minister of Environment (NFEP, 1997: 43). In 1997, the commission was replaced by the Commission for Regional Policy and Sustainable Development.
- In 1992, the Ministry of Environment and the Ministry of Privatization established an inter-ministerial unit for addressing environmental protection issues (such as environmental liability from past pollution) in the privatization process (OECD, 1995a).

There are 2,460 municipalities in Poland, whose main tasks with respect to environmental protection are to: (1) provide the population with drinking water, (2) organize systems for municipal waste and municipal waste water treatment plants, and (3) enact local land-use plans (NFEP, 1997).

Since 1975, Poland has been divided into 49 voivodships. Most voivodships had by the mid-1970s established departments for environment protection and water management. Today, their main task is to issue permits for polluting activities. In 1998, parliament decided to reorganize the country into sixteen large voivodships (Rzeczpospolita, 1998).

In 1991, seven Regional Water Management Boards were created with the task of developing programmes for water use (OECD, 1995a). The principle of basin-based water management was laid down in an amendment to the Water Law in 1997 (MEPNRaF, 1997).

The State Environmental Protection Inspectorate (PIOŚ) was established in the early 1980s. It was responsible for (1) supervision of compliance with environmental legislation; (2) environmental monitoring; and (3) initiation of activities for environmental protection. By the end of the decade it relied on a network of ten regional departments with some 300 employees.[13]

The 1991 Statute on the State Inspectorate for Environmental Protection (PIOŚ) transferred the right to impose fines and to interrupt environmentally harmful activities from the voivodship environmental protection departments to PIOŚ. In

13 Personal communication from Zbigniew Kamieński, former deputy director, State Inspectorate for Environmental Protection, 1995.

this way, PIOŚ was transformed into a powerful enforcement agency, with a legal right to act as an "ecological police." The statute also provided the basis for a new and comprehensive monitoring system. Fully developed, the system will include monitoring for air pollution, surface and ground waters, nature protection, and radioactive contamination. The importance of PIOŚ increased after it had been made watchdog for 80 especially polluting industries (OECD, 1995a).

A number of independent actors involved in financing environmental protection have emerged in the 1990s:

- The National Fund for Environmental Protection and Water Management, established in 1989, is the single largest financing organization for environmental protection in Poland. Its main source of income is charges and penalties for the use and pollution of the environment together with repayment of instalments of the loans granted by the fund. The fund supports environmental protection projects via grants, soft-loans, and equity investments. Soft loans are the most important of these three instruments. When granting loans, the National Fund applies preferential interest in comparison to the interest set by the National Bank of Poland.
- Independent environmental funds were set up in each of the 49 voivodships in 1993. The same year, local environmental funds were created in the 2,460 municipalities. The regional funds operate in similar manner to the National Fund, while local funds are only allowed to provide grants for smaller projects.
- The Ecofund was created in 1992. Its task is to manage money generated by debt-for-environmental swap agreements with the USA, France, Italy, Switzerland, and Sweden. The Ecofund provides grants to projects within its four priority areas: biodiversity, climate change, transboundary air pollution, and protection of the Baltic Sea.

The Bank for Environmental Protection was created in 1990. It is a commercial bank specialized in facilitating environmental protection investments. Some two thirds of its credit portfolio consist of loans provided on preferential terms for environmental projects. The main owner of the bank is the National Fund for Environmental Protection and Water Management (45%) while the remaining shares are held by private domestic capital, the state treasury, and the regional environmental funds as well as a number of companies (Novy, 1996).

Over the past two decades, the courts have assumed an increasingly important role in environmental policy in Poland. The environmental framework law of 1980 created duties and liabilities for environmental protection. It also authorized officially sanctioned organizations to file lawsuits against polluters. However, in the 1980s few cases were brought to court, partly as a consequence of political interference in judicial affairs (Cole, 1998: 99-100).

In 1995, parliament approved a new law that expanded the jurisdiction of the High Administrative Court. It made it easier for environmental organizations to obtain court review of agency actions. The new law permits these organizations to complain about agency actions even if they do not participate throughout the entire administrative process. Furthermore, the law permits the High Administrative Court to hear complaints concerning both administrative rules and adjudicatory decisions. This has opened new possibilities for green access to justice (ibid.: 219-220).

4.2 Environmental Organizations

Poland's first independent environmental organization, the Polish Ecological Club (PKE), was created in 1980. The same year, PKE launched a campaign against the aluminium smelter in Skawina near Kraków, which was notorious for its emissions of fluorine. As a result, the plant was closed in 1981, which was a landmark event in socialist Europe (Alberski, 1996).

The nuclear catastrophe in Chernobyl in 1986 unleashed a protest wave against the Polish nuclear programme that resulted in a radicalization of the environmental movement and the creation of a number of new environmental organizations in the latter part of the 1980s. Among these were Wielkopolska Ecological Seminarium, Freedom and Peace, the Silesian Ecological Movement, and the Green Federation.

The League for Nature Protection (LOP) remained by far the largest environmental organization in the 1980s. Although LOP was radicalized during the Solidarity period, it still did not play a major role in the policy debate. During the economic and political liberalization in 1988 the Polish Ecological Party was founded. Alberski (1996: 150) claims that at the end of the 1980s some 200 environmental organizations were active in Poland.

After systemic change in 1989, there was a profound transformation of the environmental movement. Firstly, those who joined the green movement just to take up battle against the socialist system have left. At least two environmental organizations, with clear-cut political missions, have been completely dissolved in the 1990s: Wielkopolska Ecological Seminar and the Silesian Ecological Movement. A second trend is that the NGO community has become more and more professional during the course of the 1990s; its knowledge base has increased substantially and many organizations and groups now have well-equipped offices and their own employees. Since 1993, an annual national coordination meeting for environmental groups is held in Kolumna outside łódź (Biuletyn Informacyjny, 1995). However, Polish environmentalists are rather isolated in Polish society. They have relatively few members and tend to rely on support from western Europe and the USA rather than from Polish citizens.

There is growing interest in green consumption in Poland. The Society for Ecological Households, established in 1990, provides education and information about sustainable consumption patterns. The Society for Eco-Agriculture, founded in 1998, is an agreement between nineteen institutions and organizations supporting the development of sustainable practices in the agricultural sector.[14] The most important member in this coalition is Ekoland, created in 1989. Ekoland is the first organization for organic food production in Poland. In 1998, the Ekoland certificate was held by 182 farms and six food processors (Ekoland, 1999).

It is difficult to estimate how many environmental organizations are operating in Poland in the 1990s. According to the OECD (1995a), the number of groups and foundations with the environment as their main concern reached 100 and 80, respectively, in 1992 while the Ministry of Environment claims that more than

14 Personal communication from Anna Kalinowska, Director of the Bureau of Education and Public Relations, Ministry of Environmental Protection, Natural Resources and Forestry, 1999.

1,200 new environmental NGOs were formally registered in the courts between 1989 and 1997 (NFEP, 1997: 109).

4.3 Science

The actors affiliated with the scientific sector grew in importance in the 1980s and became one of the most vocal environmental advocates. The research community produced ample evidence about the environmental crisis, which was presented in a large number of reports and frequent seminars and conferences. The scientists also had considerable leeway in conducting critical analysis of the deficiencies of environmental policy. The most important actors were the Committee on Environmental Engineering, the Committee Man and the Environment, the Committee for the Spatial Development of the Country, and the Institute for Basic Technical Problems (all under the auspices of the Polish Academy of Sciences), the environmental protection committee of the Main Technical Organization (NOT), departments for environmental engineering at various technical universities, and the Scientific Association for Environmental Law in Wrocław (World Bank, 1989).

In the 1990s, the role of the Polish Academy of Sciences has diminished. It has lost political status, economic support from the state has been reduced, and there is a brain drain in the science sector. Total expenditure on science in Poland was approximately 0.8% of GNP in 1997, half as much as the average for countries in the European Union. The Ministry of Environment claims that "it is possible to notice some symptoms of a research and development crisis in Poland" (NFEP, 1997: 123). This crisis is apparent in reduced employment in the scientific sector, the disappearance of highly qualified young staff, inadequate equipment, and inefficient cooperation between scientific centres and enterprises (ibid.: 122-123).

Despite these problems, the science sector is capable of performing important tasks. For example, geographical research is providing a sufficient knowledge-base for effective resource management (ibid.: 56). Research activities related to biodiversity is well-developed and includes the work of (1) ten research institutes of the Polish Academy of Sciences, (2) various departments at ten universities and four agricultural academies, (3) the Institute for Environmental Protection and the Forestry Research Institute (both linked to the Ministry of Environment), and (4) the research institutes of the national parks (ibid.: 70). There are also scientific institutes specializing in industry and environment. Centres with the highest achievements in this field include the Institute for Waste Management in Katowice and Warszawa and the Institute for the Ecology of Industrialized Areas in Katowice (ibid.: 100).

4.4 Economic Actors

By mid-1997 more than 400 companies had implemented cleaner production projects and approximately 1,000 specialists had been trained within the framework of the Cleaner Technologies Programme. This project includes the promotion of environmental management and audit systems. A Responsible Care Programme was adopted by the chemical industry in 1995. It includes measures to limit energy consumption and to modernize environmentally burdensome technologies (NFEP, 1997).

It is estimated that, since 1990, more than 1,000 new companies have been founded to supply environmental equipment and services. According to the OECD (1995a: 129), the technical skills required to develop a dynamic and diverse eco-industry do exist in Poland.

In the last few years the energy sector has expressed strong support for emission trading. Organizations that have spoken out in favour of emission trading are the Polish Power Grid, the Chamber of Commerce for Energy and Environmental Protection, and the independent research institute Energopomiar (Kochel and Pinko, 1996). In addition, a number of representatives of individual power plants are in favour of emission trading schemes on a regional or national level. The power sector has realized that it could substantially cut the costs for complying with its air protection obligations (especially the new Polish emission standards of 1998) if this instrument were to be used in Poland. It is notable that in 1993 the power sector, on its own initiative, prepared an Air Protection Act containing a provision for marketable permits (Żylicz, 1998). The proposal did not get any support in parliament.

More and more frequently, industry demands that emission charges be relaxed (Salay, 1996: 12). For example, the iron industry and chemical industry claim that they pay more environmental charges than the same industries in western Europe do and therefore have a competitive disadvantage.

In the 1989 election campaign, environmental protection was placed third in Solidarity's list of priorities, behind issues such as food and housing (Hughes, 1990: 1). In the 1990s, the Polish trade unions have shown little interest in environmental issues. When, in 1990, the Ministry of Environment attempted to introduce a 4% fuel charge earmarked for environmental protection, this proposal was strongly opposed by Solidarity. The labour union claimed that, "while the union is for environmental protection, it will not approve any such burden laid on the impoverished society" (Żylicz, 1991: 8). In conflicts between environmental protection and protection of the work places, the trade unions tend to be in favour of policies that secure employment. For example, workers from asbestos factories have lobbied against the asbestos ban.

4.5 Media

The media, especially various magazines and the daily press, play an important role in shaping environmental consciousness in Poland. The most popular environmental magazine is Aura, which has existed since 1972. Przyroda Polska (Polish Nature) has existed since 1957 and is the magazine of the largest environmental organization, the League for Nature Protection. The magazine Ekoprofit appeared in the early 1990s and focuses on business and environment issues. Leading newspapers with respect to environmental issues are Rzeczpospolita and Życie Warszawy, which regularly publish articles by highly skilled environmental journalists. The environment is a common topic in various political, economic, and cultural magazines such as Polityka, Przegląd Techniczny, Kultura, Argumenty, Prawo i Życie, Literatura, and Życie Gospodarcze. Also radio and television devote considerable attention to the environment in documentaries, discussions, and lectures. One example is the weekly TV programme entitled Świat wokół nas (The World Around Us), which deals with various issues related to sustainable development. Journalists specializing in environmental issues have created their own organization, Ekoś.

It can be concluded that the existing newspapers and environment-specific media provide important communication channels for enhancing environmental consciousness in Poland. In many instances these channels also facilitate networking between the relevant policy actors.

5 Coalitions and Interactions

In the 1970s, environmental policy was determined by a small group of people. Environmental protection was seen as a sectoral task delegated to engineers, biologists, and other experts. The period was dominated by policy initiatives coming from the communist party (PZPR) and the government.

In the 1980s, the Ministry of Environment and Natural Resources gave clear signals that it accepted a greater role for public opinion in defining the context of environmental policy. By the mid-1980s six environmental organizations – the League for Nature Protection, the League to Combat Noise, the Polish Tourist Association, the Polish Angling Association, the Polish Hunting Association, and the Polish Ecological Club – were authorized to participate in civil proceedings to sue polluters (Cole, 1998: 77).

In the 1970s and 1980s, there was little interaction between environmental policy-makers and the various sectors of the economy. Industry did not bother to influence the content of the environmental regulations; it knew that environmental policy-makers had almost no impact on the economic development of the country. Any policy could be easily blocked by industry in the implementation stage by simply ignoring it.

In the first few years after systemic change, the environmental administration was in a clear alliance with the environmental movement. In addition, the Ministry of Environment was led by people (Bronisław Kamiński, a former director of the environmental protection department in Kraków voivodship, Maciej Nowicki, a Professor and air protection expert from Warsaw Technical University and Stefan Kozłowski, a geology Professor affiliated with the Polish Academy of Sciences) who all were deeply respected by the environmental movement.

In 1992, the divergence of opinion became more and more manifest. This process can be explained in two major ways. First, the environmental movement felt uncomfortable with the new ministers of environment, Zygmunt Hortmanowicz (1992-93) and Stanisław Żelichowski (1993-1997). Under Żelichowski's tenure the conflict between the ministry's environmental protection mission and its responsibility for the forest industry was brought to the fore. This was partly related to the fact that the minister had a background in forestry (Millard, 1998). The reason for Hortmanowicz's unpopularity with the environmental movement is understandable when the following is taken into account.

Among other things, Hortmanowicz demoted the Chief Nature Conservation Officer and, in violation of the environmental protection law, subordinated that post to the Deputy Minister of Forestry. The new Director General of Forests issued an order to "reduce the elk population to zero" and a decision to bring forward the deer culling period to a time when mothers would still be suckling their young caused outrage. Hortmanowicz was ignorant of the regulations governing bodies such as the specialist advisory councils. The minister was also criticized for excessive use of patronage and using gifts from abroad to serve the aims of Rural Solidarity, rather than the environment (Millard, 1998: 148).

Second, and more fundamentally, consensus began to dissipate about the practical operationalization of the concept of sustainable development. The environmental movement argued that the Ministry of Environment was subordinated to the government's economic policy rather than to environmental policy. The ministry and the NGO community had diverging beliefs about the proper relationship between environmental and economic concerns in matters such as transport, nature protection, forestry, and waste management. The ministry was accused of supporting dubious projects, (most notably the dam project in Czorsztyn in southern Poland), limiting the influence of the environmental movement within the ministry, and delaying the elaboration of new environmental laws (Fiałkowski, 1996). Among some environmentalists, the ministry began to be referred to as the "Ministry of Environmental Destruction" (Mistewicz, 1997b).

In the first few years of the transition, there were few signs of industry attempting to block or delay environmental policy reforms. Hence, in 1991, Tomasz Żylicz (1991: 7) could write that "as for 'developers', there are hardly any involved in environmental debates in Poland." It was evident that many enterprises had not taken on board the implications of the new situation. The business community had not yet discovered the need to interact with environmental policymakers. It is even questionable if they had the capacity to do so because they lacked the expertise, experience, and tools required for effective lobbying.

From 1992 onwards, interaction between environmental actors and various economic and industrial actors increased substantially. In October 1992, the Council of Ministers reduced pollution charges by up to 90% (Cole, 1998: 200). This decision was taken after the Ministry of Industry and Trade had convinced the Ministry of Environment that the high charges that had been introduced in 1990-91 were detrimental to industry's economic performance. Although the government restored the level in 1993 (ibid.: 201), the reduction still had an important impact on future environmental policy-making process in Poland; industry had discovered the need to influence environmental policy.

Alberski (1996: 190) argues that the adoption of new environmental legislation after 1992 has been delayed partly because of the objections articulated by various interest groups. Millard (1998) claims that the Ministry of Finance has blocked a proposal to introduce economic incentives for increased use of unleaded petrol.

In 1992-93 three advocacy coalitions[15] began to be discernible: the Green Advocacy Coalition, the Efficiency Advocacy Coalition, and the Mainstream Advocacy Coalition.

The Green Advocacy Coalition (GAC). Within the GAC, two streams can be discerned, one with a fundamentalist orientation and another with a more realistic orientation (compare the situation within Die Grünen in Germany). The former are

15 Advocacy coalitions consist of "people from a variety of positions (elected and agency officials, interest group leaders, researchers etc.) who share a particular belief system - that is, a set of basic values, causal assumptions, and problem perceptions – and who share a nontrivial degree of coordinated activity over time" (Sabatier and Jenkins-Smith, 1993: 25).

the sectarian elements of the coalition. They are not willing to strike any compromises and frequently resort to irregular protests. The latter have a much more pragmatic approach and take the appearances of a green lobby. The difference between the fundamentalists and the realists was obvious in the struggle against the Czorsztyn dam. One group of activists decided to supplement road blocks with more moderate actions. These included (1) the establishment of a dialogue with the authorities, (2) the creation of a broad coalition of environmental NGOs to coordinate their activities against the dam, and (3) the preparation of an independent scientific report on the dam issue. All of these methods were strongly rejected by the radical wing of the environmental movement (Gliński, 1993: 148).

The actors within the GAC share the fundamental belief that humans are a part of nature and that nature has an intrinsic value. It supports a development "that departs from the centrally planned economy but does not get trapped by the market ideology" (Peszko, 1990). In the GAC's view, the basic cause of most environmental problems is unrestrained consumption. The coalition takes a critical approach to the EU from an environmental viewpoint. It is generally believed that the EU member countries have not achieved sustainable development and thus cannot serve as models for Poland in this respect. The most radical elements of the GAC are against the EU because of its promotion of free trade. Free trade, it is argued, tends to lead to over-consumption and various environmental problems. In commenting Poland's strategy for integration with the EU, the Institute for Sustainable Development (ISD) identifies a number of weak points. First, the economic threats related to integration have been too narrowly defined. "Environmental costs have been disregarded (e.g., increased gaseous emissions, growing demand for energy from conventional sources, threatened biodiversity) and these costs may emerge as a by-product of the much expected and extensively covered economic acceleration" (ISD, 1998: 5). Second, the institute is critical of the fact that environmental tax reform is not referred to as a means of combating unemployment. Third, there is no general assessment of the consequences for the environment in Poland: "Environmental protection in integration strategy should reflect problems and issues that will or should occur after a strategic assessment of selected sectors has been conducted in transport, agriculture, and fishing. The lack of such assessment hinders the definition of desirable environmental impacts of the integration process" (ibid.: 6-7). The threat to the Polish agriculture is a special concern for the ISD. The institute argues that "an indiscriminate transfer of western farming patterns may lead to a loss of the attributes of rural areas which would otherwise support organic farming, biological diversity, and recreation/tourism" (ibid.: 8).

The Efficiency Advocacy Coalition (EAC). The EAC is the smallest advocacy coalition in the environmental policy-making process. Its founding father is Tomasz Żylicz, an economics professor at Warsaw University. Based on his acquaintance with the US environmental policy experience,[16] he claimed already in the 1980s that cost-effectiveness should be a major criterion for a successful environmental policy. During the course of the 1990s, this viewpoint has gained more and more currency. From 1992-93 a "constituency for efficiency" or the Effi-

---

16 Żylicz was a guest researcher at the University of Wisconsin-Madison in 1977-78 and at the University of Colorado in Boulder in 1988-89.

ciency Advocacy Coalition (EAC) appeared in Poland with the following key members: the Warsaw Ecological Economics Centre at Warsaw University, the Polish Branch of the European Association of Environmental and Resource Economists (EAERE),[17] the Ecofund, certain officials within the environmental administration (especially some voivodships and municipalities in southern Poland), a few officials at the Ministry of Environmental Protection, Natural Resources and Forestry, and a few members of the Sejm Committee for Environmental Protection, Natural Resources and Forestry. Furthermore, the Polish power industry also belongs to this coalition because of its advocacy of emission trading.

The EAC stresses using existing resources in an optimal way by harnessing market-based policy instruments such as marketable permits. The EAC advocates the use of marketable permits for, for example, improving air quality, eliminating the use of CFCs and other ozone-depleting substances, and protecting natural resources from overexploitation (Żylicz, 1992b). Present policies, the coalition argues, tend to be adopted on the basis of an imperfect understanding of costs and benefits (Żylicz, 1997: 11). The most useful role played by economic instruments is that they help to minimize overall costs of environmental protection "through an efficient and non-arbitrary differentiation of control requirements" (ibid.: 5). "It is ... not the lack of money but rather the inability to spend the money cost-effectively that postpones recovery and makes environmental policy difficult to enforce" (ibid.: 7).

Criticism is directed at the European Union because of its reluctance to date to rely on efficiency-oriented policy instruments. Jerzy Kwiatkowski, a former employee of the Ministry of Environment,[18] claims that "Poland is still far away from a situation where we will be able to waste money on the pattern of western countries because western countries began to tackle their environmental problems not when the problems appeared but only when the money was there."

The Mainstream Advocacy Coalition (MAC) is the third advocacy coalition that can be discerned in the environmental policy-making process in the 1990s. The core members are a majority of the employees at the Ministry of Environment and the State Inspectorate for Environmental Protection, most of the environmental funds on central, regional and local level, and the voivodship departments for environmental protection. Financing environmental protection investments is the policy element that is most emphasized by the MAC. The coalition discloses an affirmative attitude to existing instruments and approaches because they are seen as having brought about a substantial improvement of the environmental situation in Poland. However, the MAC is not entirely satisfied with the effectiveness of the present fine system. The coalition argues that the penalty is unquestionably a proper instrument for some polluters but for others, predominantly companies in a poor financial situation, it does not change anything. These polluters tend to remain in non-compliance. In order to bring these firms to comply, a new and more flexible instrument to enforce environmental law, namely Negotiated Compliance Schedules, is seen as a necessary solution (see section 2.2)[19]

17 EAERE was formed in June 1990.

18 Personal communication, 1995.

19 Personal communication from Zbigniew Kamieński, former deputy Director, State Inspectorate for Environmental Protection, 1995.

The MAC's approach to the EU is the most positive one among the three advocacy coalitions. Virtually no negative consequences of EU membership are identified. Environmental policy-makers within the coalition welcome the move towards BAT due to the approximation to EU legislation. In Sommer's words, "hitherto we have regulated the environment taking into account ambient quality standards. It is a noble idea but very imprecise. It is very difficult to regulate."[20]

There is frequent cooperation between the EAC and the realists in the GAC. Contacts are facilitated by the fact that some of the leading economists of the GAC were active in the environmental movement in the 1980s. A direct result of this interaction is that the GAC realists have a positive approach to emission trading.

6 Capacities and Restrictions

Poland has a high capacity with respect to nature protection. It is a remarkable achievement that 26% of Poland's territory is under protection. Polish nature conservation legislation has already achieved 100% compliance with corresponding EU directives (REC, 1996). The explanation for the high scores for nature conservation is that this area has the longest history of all aspects of environmental regulation. There is also a well-developed scientific base dealing with this topic.

Another area where Poland has a high capacity is in financing environmental investment. The environmental fund system is beneficial for overcoming typical problems that appear during the transition period. First, since many enterprises suffer from financial problems, it is impossible to implement the polluter-pays-principle strictly. The grants and the soft loans offered by the environmental funds are an opportunity to improve the polluters environmental performance despite their financial constraints. Second, fund money is earmarked for environmental protection, which implies that the money is not exposed to competition from other policy areas such as health care, transport, education etc. Third, the environmental funds are independent entities (the local funds excluded), which means that their vulnerability to sudden political changes and economic turbulence is significantly reduced.

The decentralization of environmental policy in the 1990s has implied that lower level authorities, such as municipalities, are forced to deal with environmental protection. In the long run it will nurture environmental awareness.

The restrictions facing Poland's efforts to improve its environmental performance are manifold. First, environmental policy reform in a country in transition is a complicated process from the point of view of coordination; the new environmental policies have to be elaborated while the rules for making policy are in a state of flux.

Even the most radical changes over particular issues, in order to be operational, must somehow be related to the overall institutional framework. In this respect, contrary to the expectations of many, the transition period is not a paradise for policy-makers and law-makers. In fact, from the technical side, policy-making and

20 Personal communication from Jerzy Sommer, Professor, Environmental Law Group, Polish Academy of Sciences, Wrocław, 1996.

legal drafting becomes a nightmare when, in a relatively short period, all policies and all laws are redesigned (Jendrośka, 1996: 382).

Second, despite improvements in the 1990s, the institutional capacity of the environmental administration is relatively low. Most environmental specialists in the state and municipal administration are scientific and technical professionals, such as biologists and engineers. Because of this knowledge bias, environmental officials are not sufficiently prepared to deal with aspects such as economic analysis, law, and policy analysis. No doubt, inertia and the lack of various capabilities in the environmental administration have partly blocked radical policy reform. It was not possible to replace old personnel and to develop new professional skills as fast as an ecological shock therapy would have required. As Jancar-Webster (1995: 59) has concluded, "the relative influence or ability to act of any given agency continues to be based on pre-transition bureaucratic politics. In most cases, environmental personnel have changed little from the communist period, except perhaps in top positions ... There has been remarkably little change in thinking about environmental problems." The reason for the Efficiency Advocacy Coalition's opposition to the introduction of Negotiated Compliance Schedules is directly linked to this problem. As the economist Grzegorz Peszko puts it: "The system of individual negotiations may work if public administration is accountable, transparent, and extremely strong. Polish public administration is neither accountable, nor transparent, nor strong."[21]

Third, the integration of environmental concerns into sectoral policies has been problematic. The OECD (1995a: 158) concludes:

By and large, the major ministries have not internalized a commitment to the environment, and existing arrangements have not been effective in holding them accountable for the environmental consequences of their policies: the environment is still seen as an expensive "add-on", which is the responsibility of the Ministry of Environmental Protection, Natural Resources and Forestry.

Furthermore, the OECD notes that "the strategy for implementation of Agenda 21 does not seem to have been approved by the Polish government as a whole" (1995a: 154). Agenda 21 has also been difficult to implement on the local level. According to chapter 28 of Agenda 21, entitled "Local authorities' initiatives in support of Agenda 21", local authorities in each country should achieve a consensus on a "local Agenda 21" for the community by 1996. In Poland this recommendation had been followed by only twenty-three municipalities by 1997 (MOŚZNiL, 1997).

Fourth, the degree of environmental awareness among the general public is low. One example is that only 3% of the population has heard about Agenda 21 (Mistewicz, 1997a).

Fifth, the environmental movement is weak and has little political clout. It also operates in isolation from the general public.

Sixth, the issue of environmental protection is almost absent from the political discourse. The future of environmental management in Poland depends on the extent to which it becomes an upgraded political issue. Andrzej Kassenberg of the Institute for Sustainable Development rightfully concludes that "neither the Minis-

21 Personal communication from Grzegorz Peszko, Cracow Academy of Economics, 1996.

try of Environment nor the environmental movement can be effective without being part of the political game" (BSE, 1995: 65).

In the past decade Poland has transformed itself from a non-democratic socialist state to a market-based democracy. Thus, the systemic constraints for sustainable development have been eliminated. What are the chances for transition to sustainable development in the new systemic framework? At the moment Poland is only rhetorically committed to the goal of sustainable development. Clearly, the government sees environmental protection as a task that can be carried out in isolation from the general economic development of the country. This approach could probably be changed only if a strong environmental movement emerges that enjoys the support of large sections of the population. Just as the citizens movements were instrumental in paving the way for the democratic breakthrough in Poland and other parts of Eastern Europe in the late 1980s they could today play an important role in mobilizing support for transformation towards sustainable development.

References

Alberski, R., 1996. *Polityka ochrony środowiska w Polsce*. Wrocław: Towarzystwo Naukowe Prawa Ochrony Środowiska.

Anderson, G., 1995. Addressing Mobile Sources of Air Pollution in Poland: Emerging Trends and Policy Options. *European Environment* 4 (6) (December), 9-14.

Anderson, G., and B. Fiedor. 1997. Environmental Charges in Poland. *Environment Discussion Paper* 16. Harvard Institute for International Development.

Andersson, M., S. Gullgren, and N. Sundström, 1993. *Efter murens fall - om västs reaktioner på sammanbrottet i öst*. Göteborg, Sweden: Nerenius & Santerus.

Aura, 1979. Dla nas i następnych pokoleń, 6.

Biuletyn Informacyjny, 1995. IX Ogólnopolskie Spotkanie Ruchów Ekologicznych. Kolumna '95.

Biuro Studiów i Ekspertyz Kancelaria Sejmu (BSE), 1995. *Od Lucerny do Sofii*. Tom 8. Warszawa.

Bochniarz, Z.,and D. Toft, 1995. Free Trade and the Environment in Central Europe, *European Environmen* 5, 52-57.

Bowman, M., and D. Hunter, 1992. Environmental Reforms in Post-Communist Central Europe: from High Hopes to Hard Reality, *Michigan Journal of International Law* . 13 (921) (Summer), 923-980.

Business Central Europe, 1997. Of Hype and Halos. A Survey of Poland. February.

Chmielewski, T. J., 1988. *Czy przetrwamy*? Warszawa: Nasza Księgarnia.

Cole, D. H., 1995a. An Outline History of Environment Law and Administration in Poland, *Hastings International and Comparative Law Review* 18 (2), 297-358.

Cole, D. H., 1995b. Poland's Progress. Environmental Protection in a Period of Transition, *The Parker Journal of East European Law*, Columbia University 2 (3), 279-319.

Cole, D. H., 1998. *Instituting Environmental Protection. From Red to Green in Poland.* London: Macmillan Press Ltd.

Ekoland, 1999. Information sheet about organic farming in Poland. Warsaw.

European Bank for Reconstruction and Development (EBRD), 1993. *Environmental Standards and Legislation in Western and Eastern Europe: towards Harmonization*. London.

Fiałkowski, T., 1996. Nadzieja w gminach? *Tygodnik Powszechny* 46.

Fiedor, B., 1995. *Ecological Aspects of Economic Relationships between Poland and the European Union*. Unpublished paper. Wrocław.

French, H. F., 1988. Industrial Wasteland, *World Watch* 1 (6) (November/December), 21-30.

Gliński, P., 1993. The Current State of Cooperation between NGOs and Government, in: P. Hardi, A. Juras, and M. Tóth Nagy, *New Horizons? Possibilities for Cooperation between Environmental NGOs and Governments in Central and Eastern Europe*. Prague: WI&PO, 149-157.
Główny Urząd Statystyczny (GUS), 1995. *Ochrona Środowiska 1995*. Warszawa: ZWS.
Główny Urząd Statystyczny (GUS), 1997. *Ochrona Środowiska 1997*. Warszawa: ZWS.
Hughes, G., 1990. Are the Costs of Cleaning up Eastern Europe Exaggerated? Economic Reform and the Environment, *Oxford Review of Economic Policy* 7 (4), 106-136.
Institute for Sustainable Development (ISD), 1998. *Comments to the National Integration Strategy*. Warsaw: unpublished paper.
Jancar-Webster, B., 1995. Environmental Degradation in Central Europe, in: J. DeBardeleben, and J. Hannigan (eds.), *Environmental Security and Quality After Communism*. Oxford: Westview Press.
Jendrośka, J., 1996. Drafting New Environmental Law in Poland: Radical Change or Merely Reform, in: G. Winter, *European Environmental Law*, 368-390.
Jendrośka, J., 1997. Recent Development of Environmental Legislation in Poland in the Light of EU Directives, in: *Proceedings of the International Conference on EU Directives and Environmental Law, Prague 13-14 May*, 104-108.
Kabala, S. J., 1988. *The State of Environmental Affairs in Poland: Regulation, Activism and Public Concern in a Centrally Planned System*. Pittsburgh.
Kapała, J., 1983. *Ocena jakości powietrza w województwie katowickim*. Wrocław: Zakład Narodowy im. Ossolińskich.
Kochel, Z., and L. Pinko, 1996. Dokonania energetyki zawodowej w inwestycjach proekologicznych, *Biuletyn Energetyczny*, Izba Gospodarcza Energetyki i Ochrony Środowiska 2.
Miljödepartementet, 1997. *Miljön i ett utvidgat EU. Betänkande av kommitten om EUs utvidgning – analys av miljökonsekvenserna*. SOU 149. Stockholm.
Millard, F., 1994. *The Anatomy of the New Poland. Post-Communist Politics in the First Phase*. Aldershot, UK: Edward Elgar.
Millard, F., 1998. Environmental Policy in Poland, *Environmental Politics* 7 (1), 145-161.
Ministerstwo Ochrony Środowiska i Zasobów Naturalnych (MOŚiZN), 1988. *Narodowy program ochrony środowiska przyrodniczego do roku 2010. Materiały do konsultacji społecznej*. Warszawa: Wydawnictwo Geologiczne.
Ministerstwo Ochrony Środowiska, Zasobów Naturalnych i Leśnictwa (MOŚZNiL), 1997. *Przegląd postępu poczynionego od czasu UNCED*. Report for the UN Commission for Sustainable Development. June. Warsaw.
Ministry of Environmental Protection, Natural Resources and Forestry (MEPNRaF), 1991. *National Environmental Policy of Poland*. Warsaw.
Ministry of Environmental Protection, Natural Resources and Forestry (MEPNRaF), 1995. *National Environmental Action Programme*. Warsaw.
Ministry of Environmental Protection, Natural Resources and Forestry (MEPNRaF), 1997. *The Status of Approximation of the Polish Environmental Law to the Environmental Law of the European Union*. Warsaw: unpublished Memo.
Mistewicz, E., 1997a. Planeta Samobójców, *Wprost* 6/7.
Mistewicz, E., 1997b. Zielona Alternatywa, *Wprost* 38.
National Foundation for Environmental Protection (NFEP), 1997. *Agenda 21 in Poland. Progress Report 1992-1996*. Warsaw.
Novy, M., 1996. International Actions to Combat Pollution in the Black Triangle and Katowice Regions. Paper presented at the Second Rhine/Katowice Policy Comparison Workshop, II-ASA, Laxenburg, Austria. 24-25 April, 1996.
Organization for Economic Development and Cooperation (OECD), 1995a. *Environmental Performance Reviews: Poland*. Paris: OECD.
Organization for Economic Cooperation and Development (OECD), 1995b. *Environmental Funds in Economies in Transition*. Paris: OECD.
Peszko, G., 1990. *Green Federation – A Political Movement for Radical Ecology*. Kraków: unpublished manuscript.

Radecki, W., 1979. Ochrona środowiska w ustawach zasadniczych państw socjalist-ycznych. Dobro konstytucyjnie chronione. *Aura* 6.

Regional Environmental Center for Central and Eastern Europe (REC), 1996. *Approximation of European Union Environmental Legislation*. Budapest: REC.

Rzeczpospolita, 1998. Rozporządzenie Ministra Spraw Wewnętrzych i Administracji z 21 Sierpnia 1998 r. 29-30 August.

Sabatier, P., and H. Jenkins-Smith (eds.), 1993. *Policy Change and Learning. An Advocacy Coalition Approach*. Boulder, Co.: Westview Press.

Sachs, J., 1994. *Poland's Jump to the Market Economy*. Cambridge, Mass.

Salay, J., 1996. *Energy Use and Air Pollution in a Transitional Industrialized Economy. The Case of Poland.* Lund University, Sweden.

Sommer, J., 1993. Poland, in: R. and M. Boes, *International Encyclopaedia of Laws*. Deventer: Kluwer Law and Taxation Publishers.

State Inspectorate for Environmental Protection, 1996. *The Concept of Compliance Programmes*. Warsaw.

Stavins, R. N., and T. Żylicz, 1995. Environmental Policy in a Transition Economy: Designing Tradeable Permits for Poland. *Environmental Discussion Paper* 9. Harvard Institute for International Development.

Śleszyński, J., 1996. Economic Instruments in Environmental Policy: A Profile of Poland, *European Environment* 6, 126-134.

Wajda, S., 1993. Improvement of Compliance with Environmental Legislation: The Case of Poland. United Nations Economic Commission for Europe, Workshop on Legal and Regulatory Framework for Environmental Management, Sofia 30 June-2 July 1993, 3-28.

Wajda, S., 1996a. *Approximation of Polish Environmental Law to European Union Legislation*. Ministry of Environmental Protection, Natural Resources and Forestry. Warsaw: unpublished memo.

Wajda, S., 1996b. Dostosowanie polskiego prawa do ustawodawstwa Unii Europejskiej, *Ekopartner* 2.

Wajda, S., 1996c. Przełom w prawie ochrony środowiska, *Prawo i Środowisko* 3.

Waller, M., and F. Millard, 1992. Environmental Politics in Eastern Europe, *Environmental Politics* 1 (2) (Summer), 159-185.

Weale, A., 1992. *The New Politics of Pollution*. Manchester: Manchester University Press.

World Bank, 1989. Unpublished background papers on environmental issues in Poland. Washington DC.

Żylicz, T., 1991. The Greening of the Post-Communist Europe? Paper presented at the Second Annual Conference of the European Association of Environmental and Resource Economists. Stockholm, June 11-14.

Żylicz, T., 1992a. Applied Environmental and Resource Economics in Poland since 1989. Warsaw: unpublished paper.

Żylicz, T., 1992b. It's Time for Economics. Warsaw: unpublished paper.

Żylicz, T., 1997. Environmental Policy in Economies in Transition. Paper Presented at a workshop at the Institute for Environmental Studies, Vrije Universiteit, Amsterdam, 27 May.

Żylicz, T., 1998. Obstacles to Implementing Tradable Pollution Permits in Poland. Warsaw: Ecological Economics Center, Warsaw University, unpublished paper.

Interviews

Wojciech Beblo, Director for the Department for Ecology, Katowice Voivodship 1991-95. July 1995.

Krzysztof Göhrlich, Deputy Mayor of Kraków. May 1996.

Anna Kalinowska, Director of the Bureau of Education and Public Relations, Ministry of Environmental Protection, Natural Resources and Forestry, Warszawa. March 1999.

Zbigniew Kamieński, Deputy Director, State Inspectorate for Environmental Protection (PIOŚ), Warszawa. November 1995.

Jerzy Kwiatkowski, Department of International Cooperation, Ministry of Environmental Protection, Natural Resources and Forestry (1988-1993), Warszawa. November 1995.

Waldemar Michna, Minister of Environment in 1987-88. November 1995.

Grzegorz Peszko, Cracow Academy of Economics. December 1996.

Jerzy Sommer, Professor, Environmental Law Group, Polish Academy of Sciences, Wrocław. December 1996

Janusz Żurek, Institute for Environmental Protection, Warszawa. July 1995.

Tomasz Żylicz, Professor, Warsaw Ecological Economics Centre, Economics Department, Warsaw University. November 1995.

TAIWAN

Ching-Ping Tang and Shui-Yan Tang

With a population of 21 million and a territory of 36,000 square kilometers, Taiwan has long been regarded as one of the Asian "economic miracles" (World Bank, 1993). In the past decade and a half, Taiwan has earned another distinction for having undergone a peaceful transition to democracy (Chao and Myers, 1998). As noted by Tien and Cheng (1997), after undertaking a series of political reforms since 1986, Taiwan has "cleared nearly all the roadblocks to fair, open and full elections." After several of these open elections, the ruling party KMT, Kuomingtang, has maintained its power at the central level, but it has lost elections to opposition parties in many county and city races.[1]

In comparison with its economic growth and democratic transition, Taiwan's records in environmental protection have been less stellar. No serious governmental efforts in environmental protection had been undertaken until the mid-1980s. A major milestone of such government efforts was the establishment of a cabinet-level Environmental Protection Administration (EPA) and the proclamation of "sustainable development" as an official goal in 1987 (Hwang, 1994). Coincidentally, in that same year, martial law, a cornerstone of the authoritarian regime, was lifted – a watershed event in Taiwan's democratization. In the following decade, many local and national groups were mobilized to protest against various environmental problems ranging from garbage dumpsites to nuclear power and illegal golf courses (Tang, 1997; Tang and Tang, 1999b).

Faced with these challenges, political leaders began to feel increasing pressure for more government efforts in environmental protection. Consequently, within a decade, a spate of environmental regulations was enacted; personnel for environmental protection grew by more than 20 percent; and environmental protection budgets doubled. Environmental issues had caught so much public attention in recent years that almost all those who were elected city mayors or county magistrates in 1997 had adopted an environmental platform in their electoral campaigns. Indeed, several of these elected politicians, mostly from the opposition Democ-

1 After winning the make-up election for the Magistrate of Tao-Yuan County in early 1997, the opposition Democratic Progressive Party (DPP) has, for the first time in history, ruled more than half of the population in Taiwan at the county/city level. In the mayoral and magisterial elections in late 1997 and elections of two municipalities in 1998, the DPP took over 13 out of 23 local jurisdictions, while the KMT controlled only eight.

ratic Progressive Party (DPP), began their political careers by organizing grassroots environmental protests.[2]

Despite such increased government efforts and political attention, the environmental records of Taiwan have been mixed (Edmonds, 1996). According to a series of surveys by the Environmental Quality Protection Foundation, citizens have remained highly dissatisfied with Taiwan's environment. Although rising aspirations for environmental betterment may have contributed to such dissatisfactions,[3] some indicators point to the inherent deficiencies of Taiwan's environmental governance system.

In this chapter, we first identify some major environmental problems faced by Taiwan in the past few decades. Then we examine how government authorities have responded to various environmental problems and increasing citizen demands for better environmental protection by adopting various environmental laws, regulations, and instruments. We also trace the emergence of Taiwan's environmental movement as part of the island's democratization, which has affected its environmental governance in the past decade and a half. Finally we conclude by discussing how democratization has contributed to an increased capacity in environmental protection.

1 Main Environmental Problems and Challenges

1.1 Water Pollution

Water pollution has been a major environmental problem in Taiwan. Pollutants in major river systems originate mostly from livestock and orchard farming in upstream areas, and industrial and household wastewater in middle and down-stream areas. Tan-Shui River, the main river that runs through the capital city of Taipei, has remained seriously polluted after almost two decades of government cleaning efforts. Among the 21 major rivers on the island, more than 16 percent of their total length (2088 kilometers) has been ranked "seriously polluted,"[4] while another 22 percent suffers from minor to medium pollution. Yet government statistics indicate no significant improvements despite great government efforts in pollution control and recovery (Environmental Protection Administration, 1997). Another alarming piece of data is that as many as 60 percent of dams supplying domestic water suffer from eutrophication (with Carlson's TSI higher than 50).[5]

In addition to surface water pollution, some industrial plants have reportedly sunk secret wells to drain wastewater, resulting in underground water pollution. These and other types of illegal activities have contributed to the gradual deterio-

2 For example, the current Magistrate of Tai-Chung County has been an active participant in the Anti-Bayer protests in the past several years.

3 Dr. Tsai, the Administrator of the Environmental Protection Administration, defended the policy performance of the agency by citing official statistics on environmental improvement.

4 Indicators for being "seriously polluted" include DO 2.0 mg/l or lower, BOD 15 mg/l or higher, SS 100 or higher, NH_3-N 3.0 or higher.

5 See Department of Water Protection, 1998.

ration of underground water sources, with concentrations of iron, manganese, and arsenic exceeding various national safety standards.[6]

1.2 Air Pollution

Major air pollutants in Taiwan include suspended particles (PM10), ozone, and sulfur oxides (Cheng, 1998: 224). In 1997, about 52 percent of all bad-air-quality days (with PSI over 100)[7] were caused by ozone pollution, while the other 48 percent were caused by suspended particles (EPA, 1997). In the past few years, air pollution associated with suspended particles, originating from automobiles, construction projects, and outdoor combustion, has gradually decreased, thanks to such government initiatives as stricter emission standards and air pollution fees. Nevertheless, ozone pollution, whose sources are more difficult to pinpoint, has continued to worsen in recent years. Neither has the acid rain problem been effectively controlled, although the concentration of sulfur dioxide (SO_2), a major cause of acid rain, has decreased substantially (about 12 to 27 percent) in recent years (Cheng, 1998: 218).

1.3 Waste Pollution

According to official statistics, waste pollution problems have lessened somewhat in recent years. First, despite the continuous growth in GNP, the growth rate for garbage has slowed down in recent years, dropping from an annual rate of 8 percent in the period 1985-1992 to 1.79 percent in 1997 (Chang and Chiang, 1998). Second, the processing rate for garbage has continuously been growing, approaching 100 percent in recent years. In the past decade, recycling programs for such reusable resources as plastic bottles and rubber tires have been developed and extended in terms of numbers of recycling items, stations, and rates.

Despite the optimistic trends suggested by official statistics, waste management continues to be a serious problem. For example, government authorities have encountered major difficulties in developing sites for garbage incinerators and landfills, as local residents have routinely protested against any such siting plans. In some areas, because of such protests, tons of household garbage were piled up on the streets waiting for available destinations. A lack of waste disposal facilities has also led many small and medium sized enterprises to dispose their harmful wastes in illegal ways. News reports abound about dangerous medical wastes flowing along beaches, and dangerous chemicals being dumped into rivers that supply household tap water. Larger enterprises such as Taiwan Electric Company and Formosa Plastics have also encountered many difficulties in waste management, because of local protests against the construction of storage facilities and international pressure against transporting harmful wastes abroad.

---

6 For concrete policy suggestions to control water pollution, see Ou-Yang, 1998.

7 PSI stands for Pollutant Standards Index. According to US standards (the same in Taiwan), a PSI over 100 indicates a bad air quality above which some allergic individuals may get irritated by the air condition.

1.4 Soil Erosion and Land Subsidence

Soil erosion and land subsidence have in recent years caught much media attention in Taiwan. Environmental damages caused by mud-slides, soil erosion, and floods are in many cases a direct consequence of government failures in land-use regulation.[8] Wealthy construction companies or business conglomerates, for example, were able to lobby government officials to issue construction licenses for golf courses or residential complexes on slope-lands that were supposed to be protected from development.[9] Local farmers have also contributed to soil erosion by replacing forests on slope-lands with betel palms, tea trees, and other fruit and vegetable plants.[10]

Similarly, the overdraft of groundwater by illegal aqua-farms has resulted in many environmental disasters (Huang, 1990), including land subsidence by 10 centimeters per year, causing a seashore submergence about four times the size of the Taipei Municipality in the past several years (Chiu, 1997). Such environmental disasters have occurred repeatedly, casting doubts on government land-use policy.

1.5 Depletion of Natural Resources and Extinction of Biological Species

Water depletion and deforestation are two mutually reinforcing problems in Taiwan. Although Taiwan has three times the world's average precipitation (2500mm compared to 834mm), it has not even half of the world's average available water (4500m^3 compared to 9360m^3). This is mainly because most rivers originate from high mountains (which constitute about 32 percent of Taiwan's total area); they then run through knolls and hills (about 31% of the total area), and plains (about 37% of the total area) to reach the sea in a short distance. Such geological features prevent the soil from effectively absorbing surface water. The problem is worsened by the fact that most precipitation concentrates in the rainy season. Thus Taiwan relies heavily on dams to preserve water for the dry season (Lu, 1996). Massive deforestation caused by both legal and illegal development activities has deprived mountainsides of their water retention abilities. As a result, large amounts of soil have been washed into streams and eventually accumulated in nearby dams, impairing their water preservation function. Large scale deforestation and other pollution problems have also damaged the ecological systems in mountainous areas and contributed to the destruction of some endangered species.

The continuous deterioration of the natural ecology in Taiwan has led many to question the island's economic development strategies. For example, the petrochemical industry has been for several decades a major pillar of Taiwan's industrial economy, but many are beginning to question the value for constructing more naphtha cracker plants. Such plants not only consume huge amounts of water and

8 See Tang and Tang (1999b) for a detailed discussion on the political causes of such failures.

9 A tragedy associated with such over-development of slope-lands was the collapse of The Lincoln Mansion (a residential building) in Taipei County, which was caused by a mud-slide after a heavy rainfall. Many local officers were later charged and convicted of bribery or administrative negligence. The County Magistrate, however, refused to assume the political responsibility for failing to control slope-land development.

10 For a discussion on such slope-land conservation issues, see Yen, 1998.

energy, but they also adversely affect ecologically valuable wetlands. Nowadays, a strong public sentiment exists against any plan for constructing new naphtha cracker plants. A recent example is the plan to build the Seventh Naphtha Cracker Plant in Pin-Nan. The plan failed to materialize because of strong public concern about an over-consumption of local water resources (Chiu, 1996).

2 Development of Major Environmental Laws, Regulations, and Instruments

As environmental problems have become more evident and popular demands for environmental protection have grown steadily, government authorities in Taiwan have been in the past two decades gradually developing various environmental laws, regulations, and instruments.

2.1 Main Environmental Laws and Regulations

Taiwan enacted its first set of environmental regulations in water pollution prevention and waste management in as early as 1974. Air pollution, which was for a long time considered a lesser health threat than unclean water and solid waste, did not become an object of government regulation until the Air Pollution Control Act was adopted in 1992.

As indicated by their names, most pollution regulations are directed at specific mediums or sources – for example, the revised Air Pollution Control Act (1999), Noise Control Act (1992), Water Pollution Control Act (1991), Act on Poisonous Chemicals Management (1997), Waste Disposal Act (1997),[11] and the draft Soil Pollution and Cleaning Act proposed in late 1998. Earlier versions of these laws were criticized by scholars as being highly ambiguous, because they failed to set up concrete environmental quality targets, and specific time tables and administrative procedures for achieving them (Tang, 1993). Furthermore, most laws, like the Air Pollution Control Act and the Water Pollution Control Act, only limited concentrations, but not quantities, of pollutant emissions. It was unclear as to how enforcement would actually affect the overall quality of the environment. Efforts have recently been undertaken by the government to address such problems. The Air Pollution Control Act, for example, was recently revised to incorporate the principle of "total load control."

After trying direct regulation for a couple of decades, the government has gradually recognized the limited effectiveness of imposing emission and effluent standards alone in pollution control. In 1994, the legislature formally adopted a standardized procedure for environmental impact assessment (EIA)[12], which provides for professional review and public participation in assessing the environmental impacts of government policies and large development projects, both private and public. Although the idea of EIA was supported by both environmental-

11 The years in the brackets indicate the last time an Act was revised.

12 Before the formal adoption of the Environmental Impact Assessment Act, the central government had conducted two trial programs in the previous decade to experiment with different approaches to environmental impact assessment.

ists and big businesses,[13] government officials had for a long time been reluctant to establish the institution, because they were concerned about its negative impacts on economic development (Tang and Tang, 1999a). Such a concern has also triggered a major controversy over the design of the EIA process – that is, whether the EPA or other relevant agencies ought to have the primary authority to review and approve EIA reports. After open political conflicts between some well-known environmental groups who favored a stronger position for the EPA and many government officials who favored the opposite, the legislature finally decided to give the EPA the primary authority in overseeing the EIA process. Furthermore, another sign of the environmental groups' influence in the legislative process is that the 1994 Environmental Impact Assessment Act requires not only development projects, but also governmental policies, to undergo environmental impact assessment.

Despite such a favorable legislation, however, many still doubt the ability of the EPA, as a junior-ranked agency, to resist pressure from other powerful ministries. In the Pin-Nan Industrial Park case mentioned earlier, concerns about water depletion and CO_2 emission that exceeds international agreements did not prevent the Ministry of Economic Affairs from supporting the project. The situation finally changed only after the Prime Minister expressed his concern and asked for a „detailed review" of the project's EIA. It appears that solid support from top national leaders is needed for the EIA process to play an effective role in environmental protection.

In addition to regulating pollution, the Taiwanese government has in recent years moved to adopt measures for broader ecological conservation. Such efforts have been in part the government's response to the growing environmental awareness in society. Five national parks, for example, have been established to protect areas with ecological, scenic, cultural, or recreational values, covering 8.5 percent of Taiwan's total area. The Wildlife Conservation Law was enacted in 1989 (and revised in 1994) to ban hunting, raising, and trading of endangered animals. The enactment and implementation of the Environmental Impact Assessment Act mentioned above have also provided opportunities for ecologists to influence development projects that have adverse ecological impacts.[14]

Another important branch of environmental law concerns conflict resolution mechanisms for environmental disputes. When environmental protests began to proliferate and intensify in the mid-1980s, most of them were targeting at and seeking compensations from specific polluters. Most protesters were reluctant to redress their grievances through the judicial system, because the legal doctrine of environmental rights was poorly defined at that time and the courts were reputed to be easily subject to political pressure, administrative interference, and bribery.[15] For many pollution victims, such „self-salvation" measures as blockading a pollut-

---

13 On the one hand, environmentalists expect EIA to be a useful tool for blocking developmental projects that have negative environmental impacts. On the other hand, big corporations see it as a way to legitimize their investment plans and to reduce the chance for being a target of grassroots protests.

14 The Act specifies that two-thirds of any EIA review committee must be either scholars or specialists of the field.

15 For more discussions on the limitations of Taiwan's judicial system in resolving environmental disputes, see Tang and Tang (1999b).

ing plant were more effective, because they could effectively force the plant owner to negotiate and yield to their demand. By attracting media attention, the protesters could gain support from sympathetic individuals, organizations, and political parties. Furthermore, wide publicity about pollution events embarrassed the ruling party KMT and created pressure for it to help resolve disputes.

The central government soon realized that it is important to develop mediation channels such that when pollution victims are reluctant to go to court, they do not need to resort to street protests. The government adopted the Law to Settle Public Nuisance Disputes in 1992 (which was further revised in 1998) to authorize the establishment of mediation committees at different levels of government for resolving environmental disputes. The courts, nevertheless, remain the last resort for dispute resolution if both parties refuse to abide by the mediation committee's decision. Although available, these mediation channels have not been utilized much for settling environmental disputes, because most pollution victims are skeptical about their impartiality and usefulness (Hwang, 1994).

Finally, some environmentalists believe that it is important for Taiwan to have a Fundamental Law of Environmental Protection that specifies the general goals of environmental policies and the basic environmental rights and obligations of citizens. Others, however, argue that such a fundamental law has only a symbolic meaning, and there is no urgency for its adoption. In fact, its symbolic function has been accomplished in the constitutional revision of 1992, which states that economic and scientific developments should be balanced by environmental and ecological protection. Members of the Legislative Yuan have so far failed to put the Fundamental Law of Environmental Protection on the legislative agenda.

2.2 Instruments

In addition to „command and control" regulations, the Taiwanese government has also experimented with market-based mechanisms for pollution control. The EPA, for example, adopted a politically controversial air pollution prevention fee system in 1995 that imposes a surcharge on fuel. Funds collected in the system are set aside to form a foundation responsible for reducing air pollution. Similar fees for water pollution are now being considered.

Another instrument for pollution control is the tradable pollution permit that was created in the latest version of the Air Pollution Control Act (1998). The Act stipulates that when the air quality of a "total load control area" meets national safety standards, a plant in that area can sell its "pollution rights" to other plants if it has taken effective measures (e.g., by installing new control facilities or improving production procedures) to reduce pollutant emissions.

A third type of market-based mechanism is deposit fees for such recyclable materials as bottles and tires. Various trial programs are currently underway to encourage different sectors of society – businesses, consumers, governments, and non-governmental organizations – to engage in household waste reduction. While some of the programs appear to be quite successful, some have suffered from administrative problems. For example, funds for recycling plastic bottles were reported to have been embezzled by some program managers.

3 Changing Constellation of Actors in Environmental Policy

Most environmental regulations and instruments in Taiwan were adopted in the past two decades, during which the island was undergoing a peaceful process of democratization (Chao and Myers, 1998). During the period, government agencies have lessened their monopoly in environmental policy making and implementation, while actors in the private and non-governmental sectors have gradually increased their influence.

3.1 Governmental Agencies

Environmental protection agencies were established at three levels of government. At the county/city level, environmental protection bureaus are the providers of such labor-intensive services as street cleaning and garbage collection. They are also responsible for enforcing pollution control by inspecting emissions from polluting sources, and by issuing tickets to offenders and collecting fines. At the provincial level, the Environmental Protection Department was in charge of regional coordination, environmental data analysis, and monitoring local regulatory enforcement by undertaking follow-up inspections of polluting activities. At the central level, the Environmental Protection Administration is responsible for formulating and enacting environmental regulations, for providing technical and financial support to local environmental protection bureaus, and for supervising local regulatory enforcement by undertaking spot-checks and follow-up inspections (Environmental Protection Administration, 1995).[16]

In addition to a vertical distribution of tasks among different levels of governments, jurisdictions of environmental protection are shared horizontally by different governmental agencies of the same level. At the central level, for example, the Council on Agriculture takes the lion share of responsibility in ecological conservation and wild-life preservation. The Industry Bureau under the Ministry of Economy is responsible for helping industries to control pollution by installing proper facilities, streamlining production, managing industrial wastes, and so forth. The Council on Economic Planning and Development evaluates the impacts of various environmental regulations and standards on economic development, and is in charge of integrating laws and reconciling conflicts between economic development and environmental protection as well as other related governmental objectives.[17]

16 The recent restructuring of the scale and role of the provincial government has incorporated three Environmental Protection Centers originally under the Taiwan Provincial Government into the central EPA. Yet their functions in coordinating pollution control among different county and city government are expected to be maintained.

17 Several environmental groups have called for the establishment of a „Ministry of Environment" to centralize all environmental and conservation responsibilities into the hand of a single government agency. But there is no indication that the government would respond to such a request soon.

3.2 Civic Environmental Groups and Environmental Movements

Since the early 1980s, various environmental groups have been formed in reaction against various environmental problems in Taiwan. The rise of these environmental groups was aided in part by the democratization process that began to unfold around the same time. Although some groups aiming at preserving endangered species were legally allowed to organize well before democratization,[18] sporadic grassroots protests against pollution emerged in the early 1980s, when the KMT government began to loosen its authoritarian rule.

The environmental protest movement reached a turning point in 1986 when the anti-DuPont Incident began to unfold (Reardon-Anderson, 1992). Endorsed by many local elected officials, residents in Lukang organized several rounds of protests in Taipei urging the central government to withdraw its support for DuPont's plan to build a petrochemical plant in their community. With the KMT government refraining from cracking down the protests and DuPont voluntarily withdrawing its construction plan, citizens in Taiwan began to gain a greater sense of political efficacy. This, together with the abolition of martial law in 1987, has helped to fuel the environmental protest movement in Taiwan. According to one estimate, between 1980 and 1987 there were an average of 13.75 environmental conflicts per year, the average increased to 31.33 protests per year between 1988 and 1990, and to 258 in 1991 alone. It was only after 1991 that the number of protests began to decrease (Hsiao, 1994).[19]

Besides the lifting of martial law in 1987, other political changes helped to fuel environmental protests. First, the opposition Democratic Progressive Party (DPP) has since its formal establishment in 1986 been highly motivated to support various environmental protests as part of its challenge to the KMT regime. Second, political entrepreneurs have emerged to take advantage of new electoral opportunities at various levels of government to press environmental interests. Although these political entrepreneurs may not have any common interest with those who suffer from local environmental nuisances, many of them seek to gain visibility and other political capital by helping to organize protests on behalf of local residents. Third, the newly liberalized mass media, by providing extensive coverage of environmental protests, also helped to fuel the movement.

Although often large in scale, many of these protests had only limited contribution to actual improvement of the environment, as the immediate objective of many of these protests was monetary compensations from either the polluter or the government itself. Once compensations were secured, the protesters would be disbanded, and the protest leaders would be busy fighting among themselves to control the compensation funds. Few collective efforts emerged from these protests to actually clean up the community's environment. One well-known example is the protest organized by residents in the Hou-Chin Neighborhood in the north-

18 Since these groups did not pose any political threat to the regime, they were actually promoted by the government.

19 According to a different survey by the Environmental Protection Administration, a total of 108 public nuisance disputes arose in 1989, decreasing to 24 in 1993, 12 in 1994, and 10 in 1995. While the EPA's method of identifying and counting public nuisance disputes appears to differ substantially from Hsiao's, two surveys identify an unmistakable decline in number in the 1990s.

ern city of Kao-Hsiung in the late 1980s against an extension project of the government-owned Chinese Petrochemical Company. The protest resulted in the government establishing a $1.5 billion New Taiwan Dollar Foundation for the Neighborhood. Once the foundation was established, protest organizers scrambled to take control of it. The protest then lost steam, and the extension project of the Fifth Naphtha Cracker was eventually built (Tang and Tang, 1997).

Despite its limitations, the protest movement in the late 1980s and early 1990s did contribute to environmental protection in several ways. First, the protests helped to create a sense of environmental crisis among the public, and thus encouraged government authorities to take actions to address it. Second, huge compensatory demands by protesters created strong incentives for polluters, especially large corporations which are easy protest targets, to invest in pollution control devices in order to avoid potential protests. Third, the protests did help to prevent the development and continuation of some highly polluting factories. Fourth, as environmental disputes were repeatedly reported in the media, ordinary citizens acquired a heightened sense of their environmental rights.

Furthermore, grassroots protests paved the way for the emergence of a more ideologically oriented environmental movement in the 1990s. As glaring pictures of environmental degradation appeared repeatedly in the mass media, the public began to reflect on the condition of their environment. The nation's attention to environmental problems was especially aroused when various media outlets reported the quotation – „We are living in a pigsty" – used by a German magazine to describe the environmental condition in Taiwan (Liu, 1995).

As the political system had become increasingly open to civic and political associations, more urban, middle-class individuals who were not direct victims of pollution began to be involved in organized efforts aimed at improving the general environmental and ecological conditions of the island. Their involvement in the environmental movement was more ideologically than materially driven. In their earlier efforts in the 1990s, these ideologically oriented participants allied with local residents to protest against specific polluting sources. For example, in the protest against the construction of the Pin-Nan Industrial Park in southern Taiwan, local fishers who were concerned about the destruction of their fishery were supported by non-resident activists who were concerned about such ecological issues as the destruction of wetlands and the extinction of spoonbills. In the long-lasting protests against the Fourth Nuclear Power Plant, local residents and nationwide anti-nuclear groups cooperated in their fight against the state-owned Taiwan Electric Power, Inc.

Towards the second half of the 1990s, as government efforts in pollution reduction have increased, the number of street protests gradually decreased. Membership-based environmental groups that are composed mainly of middle-class professionals have increasingly turned their efforts to influencing government policy making. For example, such environmental groups were successful in preventing the development of the Hsian-Shan Tidal Flat into an industrial park, thus preserving the habitat of 100 million monk crabs. They did so not by mobilizing violent street protests, but by challenging the development plan through various legal and administrative channels. The Wild Bird Society was successful in lobbying local government officials to establish the Kuan-tu Nature Park in northern Taipei such that migratory birds can continue their annual visits to the location (Tang and Tang, 1999b).

Over time, these environmental organizations have acquired political skills in pressing their case in the policy process. For example, realizing national leaders' strong desire to maintain a positive national image in the face of political and military threats from mainland China, environmental groups have arranged visits by famous environmentalists from abroad to help lobby for stronger environmental legislation. It was, for example, reported that such visits had helped to persuade President Lee to establish an Environmental Protection Foundation and to withdraw his support for Formosa Plastics Group's plan to reduce its anti-pollution measures in the Sixth Naphtha Cracker Plant.[20]

3.3 Opponents of Environmental Protection

Environmental groups have been gaining strength in the past two decades, but they were not without opponents. Many development projects with adverse environmental effects were still able to gain support from government authorities, as many of the big business conglomerates that sponsored such projects continued to have privileged access to top government officials.[21] A recent example is the Taiwan Cement Corporation which was able to secure government support to build the Ho-Ping Cement Industrial Park in the eastern coast of Taiwan, a less developed region of Taiwan. The park was built despite repeated protests by both national and local environmental groups who argued that the park would consume huge amounts of energy and destroy the natural landscape in the region.

Employee associations or labor unions of big companies are sometimes major opponents of environmental groups. For example, a few years ago when environmental groups were protesting against illegal expansions of golf courses, they were often confronted on the other side of the streets by organized efforts of golf course employees who tried to defend their "right to work" (Tang and Tang, 1999b). Similarly, when environmental groups were protesting on the streets of Taipei against the construction of the Fourth Nuclear Power Plant in Taipei, they were often out-numbered by counter-protesters consisting of Taiwan Power Company's employees and their families.

In addition to large enterprises and their employees, various local economic interests are sometimes opponents to the environmental movement. For example, environmental groups encountered strong local opposition when they were protesting against the expansion project of the eastern coastal road (Number 11 Provincial Road). Claiming that the expansion project would damage coastal ecology and destroy the splendid natural scenery along the route, the environmental groups were urging the government to stop the project immediately. At the same time, local residents, expecting increased business opportunities after the expansion project, were launching counter-protests, urging the government to stick to its construction schedule.

20 *Taiwan Today News Network*, at http://w3.ttnn.com/can/980216/p02.html, 1998.

21 For a discussion on business-politician relationships and their influence on environmental policies, see Tang and Tang, 1999b.

This particular case is not unique in that environmental groups from the more affluent urban areas were at odds with grassroots residents suffering from rural poverty. Unlike in the 1980s when ideologically oriented environmental groups demonstrated hand in hand on the streets with local pollution victims, the coalition between the two has become somewhat shaky in recent years. On the one hand, many membership environmental groups have become reluctant to involve themselves in local NIMBY (not-in-my-backyard) protests, whose participants were often more interested in seeking monetary compensation than protecting the environment. On the other hand, grassroots protesters have also tried to keep broader environmental concerns away from their agenda, being afraid that such concerns might complicate their quest for compensation.

3.4 Green Businesses

As environmental awareness grows fast, green business has become potentially lucrative. Environmental consulting firms are now in great demand as the Environmental Impact Assessment Act now requires many public and private development projects to undergo EIA. In addition, as ISO 14000 is getting world-wide recognition, many companies in Taiwan begin to seek advice from consulting firms about strategies for compliance. Other green related industries such as electric motorcycle manufacturing are also growing rapidly as governmental subsidies become available.

Yet the most noteworthy green businesses are those for waste management. Since household and industrial wastes have grown rapidly in the past decade but the construction of disposal facilities have been seriously delayed by NIMBY protests, some Mafia organizations uncovered channels for making huge profits in waste management.

After securing government licenses to run waste management companies, these organizations regularly employed illegal means to overcome waste management problems. They, for example, often intimidate local residents so that they would not organize to protest against their facilities. They also illegally dump industrial and construction wastes in public lands, private farms, beaches, riverbed, or whatever locations they find convenient. The activities of these companies indicate serious problems with Taiwan's waste management system.

4 Policy and Implementation Problems

According to government records, some pollution problems have indeed lessened in recent years. For example, the number of „bad air days“ (e.g., with PSI over 100) has dropped consistently from 16 percent in 1991 to 6 percent in 1995 (Environmental Protection Administration 1996a: 107). The number of environmental disputes has decreased consistently since its peak of 110 cases in 1989 to 11 in 1995 (Environmental Protection Administration 1996a: 724). Despite protests by development lobbies, several national parks and ecological projects were established. Furthermore, many recycling programs have been initiated and been expanding rapidly. All these show the improvement in the capacity of Taiwan's environmental governance system. Yet many policy and implementation problems remain.

4.1 Pollution Control

As mentioned earlier, Taiwan has multiple tiers of government responsible for regulatory enforcement. Although such an argument is theoretically sound, its operation is far from smooth. In some cities and counties, local environmental protection bureaus are known to be slack in enforcement. In such locations, inspectors from the central supervisory squad are reported to be heavily involved in regular inspections, while local inspectors often receive bonuses for following up on cases initiated by the central inspectors. Thus central officials often complain about local shirking, but local bureaucrats criticize central officials for micro-management and for passing political pressure (from the regulated industries) to them by asking local inspectors to issue citations to polluters (Chiou, 1995).

Such coordination problems are partly a result of the central government's failure to ensure that all local jurisdictions have developed sufficient enforcement capabilities before requiring them to enforce uniform standards. Considerable variations exist among local jurisdictions in terms of their willingness to enforce centrally determined environmental regulations. For example, some county magistrates have stronger connections with local industrial interests than others. The former tend to have less incentives to support enforcement efforts by local environmental protection bureaus (Tang, 1998).

At each level of government, agencies other than the EPA are also involved in various environmental protection efforts. At the central level, for example, the EPA's jurisdiction over nuclear matters is shared with the Committee of Atomic Energy. With respect to water pollution control and preservation, relevant laws are enforced by different agencies – for example, the Slope-land Preservation and Utilization Law, Forest Law, and Pesticide Management Law are enforced by the Committee of Agricultural Affairs; the Tourism Development Law is enforced by the Ministry of Communications; the National Park Law, Drinking Water Law, and Sewerage Law are enforced by the Ministry of Interior; and Tap Water Law and Dam Management Regulations are enforced by the Ministry of Economic Affairs. As administrative responsibilities are shared by many government agencies with diverse missions and tasks, coordination problems often arise in the water management system (Yeh, 1997). Some environmentalists and scholars have advocated the establishment of a Ministry of Environmental Affairs to tackle all environmental affairs. Nevertheless, given the multi-dimensional nature of environmental problems, it is unlikely that a super-ministry will suffice in tackling all environmental problems. Research is yet to be done to determine how coordination can be better achieved in a multi-agency context.

Another form of coordination problem arises across local jurisdictions. Many river systems on the island, for example, run through several counties and cities. Not only that many central agencies are involved in managing a river system, the pollution control efforts of related counties and cities also affect the overall condition of a river system. Taipei City, for example, is located in the downstream of the Tan-Shui River which runs through several counties. Taipei City suffers from pollutants originating from upstream counties, but there is little it can do to influence pollution prevention efforts of other local jurisdictions. Similarly, Kao-Hsiung Municipality suffers heavily from air pollutants originating from neighboring counties (Environmental Protection Administration, 1997). So far, the central and provincial governments have yet to develop effective mechanisms to deal with such inter-jurisdictional externalities. Local governments have also failed to come up with ways for resolving their conflicts. In the future, some forms of regional solutions are needed to tackle these ever-increasing cross-boundary pollution problems.

Various central administrative agencies and local governments are given almost a free hand in how they interpret and implement environmental laws. Unlike, say, their counterparts in the United States, administrative agencies in Taiwan are not required to follow any „notice and comment procedures“ when adopting their specific pollution control rules and regulations. Furthermore, until the recent revision of the Air Pollution Prevention Act in 1998, there is no action-forcing provision in environmental laws, like those in the United States, that specify deadlines for officials to announce specific administrative measures for carrying out the law. The new Air Pollution Prevention Act also introduced the provision for „citizen suits,“ which allow environmental groups to bring cases to the Administrative Courts as a means of forcing government agencies to faithfully enforce the laws.[22]

Up to now, there have been very limited channels in Taiwan for environmental groups to challenge administrative failures in implementing legislative mandates. Without clear legislative control and citizen oversight, administrative agencies are more susceptible to influence by organized interest groups than their counterparts in the United States. Many regulatory standards contained in various environmental laws are yet to be sufficiently enforced (Tang, 1993; Tang and Tang, 1999b).

4.2 Ecological Preservation

Despite great efforts by governments and non-governmental organizations in recent years, Taiwan's records in ecological preservation remain unsatisfactory. For example, after the enactment of the Water and Soil Act and the 18 billion New Taiwan Dollars per year spent on flood control since 1993, large casualties and property damage still resulted from the mud-slides and floods caused by Typhoon Herbert in 1996. Taiwan was also accused by such international organizations as the Worldwide Fund for Nature of failing to crack down on smugglers of products made from endangered animals (Edmonds, 1996).

Such unsatisfactory records are not necessarily a result of lack of resource commitment from government, as it has claimed to have devoted more than 1,500 officials to anti-smuggling operations (Edmonds, 1996). Some scholars attribute many of the failures in nature conservation to the lack of a coordinated approach, as the administration of conservation has been widely distributed among different agencies, which have poor communication with each other. For example, the preservation of a piece of wetland may involve more than six central agencies – the Council for Economic Planning and Development responsible for establishing

22 Since the Air Pollution Prevention Act was just revised in December 1998, it is too early to know how the „citizen suit“ provision will be implemented.

general guidelines for land development; the Ministry of Interior responsible for regional planning and development; the Council of Agriculture responsible for preserving the natural scenery and wildlife within the wetland; the Ministry of Economic Affairs responsible for utilization of water from the wetland; and the Ministry of Finance responsible for controlling the import and export of wildlife; and finally the EPA responsible for pollution control of adjacent territories, rivers, and oceans.[23]

Furthermore, ecological preservation efforts can become more effective only if they are accompanied by relevant cultural changes. For example, if the general public still believe in rhino horn's medicinal values and are willing to pay high prices for it, a black market for it will exist no matter how much government effort is devoted to cracking down on it. The government has launched a campaign trying to convince the public that rhino horn can be substituted by water buffalo horn. If successful, it will help to lessen the demand for rhino horn and make it easier to prevent its trade.

5 Democratization and Environmental Protection Capacity

Since its retreat from mainland China in 1949, the KMT government in Taiwan has always given top priorities to issues that are most related to national security. Major government efforts were focused on national defense, diplomatic relations, social stability, and economic growth. Environmental protection was never emphasized by the political elite. Even among the ordinary people who suffered from the ill effects of worsening environmental conditions, most appeared to value the increased economic opportunities provided by industries more than the water and air pollution created by them. It was actually not uncommon from the 1950s till the 1970s that an entire community in a less developed area celebrated the location of, say, a highly polluting petrochemical plant, as local residents focused their attention on the economic opportunities provided by such facilities.

During this period, pollution victims had limited channels to redress their grievances. Furthermore, as many polluting industries were themselves owned by the central government or by the ruling party, any protests against them might well be considered subversive. Indeed, under martial law, any organized protests could be treated as subversion, protesters could be punished by long-term imprisonment.

Widespread public dissatisfaction about the environment did not surface until the early 1980s when the political regime began to show more tolerance to organized political activities. The democratization process beginning in the mid-1980s helped to fuel the environmental movement even further, which has in turn encouraged more government efforts in environmental protection. Although the overall records of these efforts in environmental protection have been mixed, one can expect continual improvement as democratic institutions become more established in the future.

---

23 For a more detailed discussion on Taiwan's administrative system for conservation, see Chiou, 1994.

First, there are signs that leaders of the ruling party have increasingly realized the importance of balancing business interests with environmental ones. For example, President Lee Teng-Hui, who was once criticized for having purchased a villa in a water-preservation area that outlawed such developments, has most recently undertaken some efforts in promoting his environmental image. One such effort is the establishment of the Environmental Protection Fund. Some other KMT politicians found it politically advantageous for them to ally even with opposition parties in advocating various environmental protection legislations. No matter what has contributed to attitudinal changes in these KMT politicians, such changes are welcomed by the environmentalists who believe that an awakening of environmental consciousness among top government officials helps to lessen the dominance of politician-business coalitions in top policy making circles.

Second, another trend showing a brighter future for environmental protection is the gradual opening of political and administrative institutions. As democratization deepens, multiple access points to the policy making process are opening up for environmentalists, who can bring their case to the legislative and judicial branches after their negotiations with stubborn bureaucrats fail. Several years ago, for example, environmentalists won a major political battle against many powerful ministries by persuading the Legislative Yuan to assign the EPA greater authority relative to other ministries in overseeing the environmental impact assessment process. Although the judicial branch appears to have remained the slowest to respond to environmental concerns, there are signs that newly recruited staffs in the branch have been trying to address those concerns (Tang and Tang, 1999b).

Third, democratization has brought more political autonomy to local governments. A more decentralized political system provides environmentalists more opportunities to influence environmental policy making. At the national level, environmental issues are often considered less important than such issues as national security and foreign trade; a politician's position on environmental issues is seldom a determining factor for his or her electoral fortune. In the first presidential election in 1996, for example, Lee Teng-Hui easily won the election in spite of his poor environmental reputation. At the city and county level, however, political candidates' positions on local environmental issues can often affect their success in the ballot. Environmental groups often find it easier to influence government policy making in these local arenas (Tang and Tang, 1999b).

Fourth, there is still a lack of institutionalized channels for citizens and environmental groups to monitor the environmental work of government agencies. Yet, an active mass media, which is interested in exposing scandals in government, has created pressure on government officials. For example, an officer in charge of the extension project of the Taiwan Cement Company recently committed suicide after the media had alleged official misconduct. The mass media has in recent years become a major watchdog of bureaucratic slack and misconduct.[24]

24 It would be too simplistic, however, to argue so without noting a growing symbiotic relationship between media reporters and local officials. Since local reporters are generally overloaded by their daily tasks, they may have no time and energy to dig out interesting reports on the scene personally. One common way for them to survive is to maintain a good relationship with governmental officials so that they can simply have tea with them in the air-conditioned Town Hall but at the same time get sufficient materials for their news reports. Such dependence on governmental officials for news sources would, to a certain degree, prevent the re-

With more information and more successful experiences in challenging government decisions, environmentalists have gained an increased sense of political efficacy and are more motivated to participate in policy making. In addition, they have formed increasing numbers of civic organizations that are geared not only towards political participation, but also such tasks as environmental education and community involvement in ecological preservation. Although it will take some time for them to build their efforts, their activities are hopeful signs for Taiwan's environmental future.

Acknowledgments

Research for this chapter was supported in part by a grant from the Chiang Ching-Kuo Foundation for International Scholarly Exchange.

References

Chang Tsu-En, and Chiang Li-Chung, 1998. Fei-ch'i-wu Kuan-li Wen-t'i Chien-tao yu Tui-ts'e [Waste Disposal Management Problems: Review and Policies], in: Ou-Yang Chiao-Hui, and Huang Mien-Shan (eds.), *Shing-shih-chi te Huan-ching-pao-hu Cheng-ts'e* [Environmental Protection Policy for the New Century]. Taipei: Ho-sheng Chi-ching-hui (in Chinese).

Chao L., and R. Myers, 1998. *The First Chinese Democracy*. Baltimore: The Johns Hopkins University Press.

Heng Fu-Tien, 1998. K'ung-ch'i-ping-chih chih Wen-t'i Chien-tao yu Cheng-ts'e [Air Quality Problems: Review and Policies], in: Ou-Yang Chiao-Hui and Huang Mien-Shan (eds.), *Shing-shih-chi te Huan-ching-pao-hu Cheng-ts'e* [Environmental Protection Policy for the New Century]. Taipei: Ho-sheng Chi-ching-hui (in Chinese).

Chiou, Chang-Tay, 1995. *Taiwan Huan-ching Kuan-chih Cheng-t'se [Taiwan's Environmental Control Policy]*. Taipei: Shu-hsin Publication (in Chinese).

Chiu, Hua-Mei, 1997. Tso-kuo Herbert Typhoon Yi-nien [One Year After Typhoon Herbert], *Tien-Sha [Commonwealth]* 195, August 1997 (in Chinese).

Chiu, Wen-Yen, 1996. Ke Ch'i-ku Hsi-hu Yi-taio Sheng-lu [Give the Chi-Ku Lagoon a Chance to Survive], Chung-Kuo-Shih-Pao [China Times], May 6, 1996 (in Chinese).

Department of Water Quality Protection, 1998. Yin-yung-shui shui-yuan shui-chih-feng- hsi chi shui-chih-chien-tse kuan-li-chih-tu chien-tao [An Analysis of Drinking Water Resources and A Review of Institutions for Water Quality Monitoring and Management], A Report of the Seventy-Fourth Committee Meeting, March 26, Taipei (in Chinese).

Edmonds, R., 1996. Taiwan's Environment Today. *The China Quarterly* 148, 1224-59.

Environmental Protection Administration, 1995. *Huan-ching-pao-hu Nien-Chien [Annals of Environmental Protection]*. Taipei: The Executive Yuan (in Chinese).

Environmental Protection Administration, 1996 a. *Yearbook of Environmental Protection Statistics, Taiwan Area, The Republic of China*. Taipei: The Executive Yuan.

Environmental Protection Administration, 1996 b. *Chung-hua-min-kuo Taiwan-ti-ch'u Huan-ching-tsu-hsun [Environmental Information of Taiwan Area, ROC]*. Taipei: The Executive Yuan (in Chinese).

Environmental Protection Administration, 1997. *Yearbook of Environmental Protection Statistics, Taiwan Area, The Republic of China*. Taipei: The Executive Yuan.

---

porters from sharply criticizing these officials. Nevertheless, many environmental groups (e.g., Taiwan Environmental Protection Union) have begun to evaluate the environmental reporting of different media outlets through their own journals and web-pages.

Hsiao, Hsin-Huang, 1994. Tai-wan Ti-fang Huan-pao K'ang-cehng te Hsing-ko yu Chuan-pien: 1980-1991 [Characteristics and Change of Taiwan's Grassroots Environmental Movement: 1980-1991], in Tai-wan Yen-Chiu Chi-chin [Taiwan Research Foundation] (ed.), *Huan-ching Pao-hu yu Ch'an-yeh-cheng-ts'e [Environmental Protection and Industrial Policy]*. Taipei: Ch'ien-wei Ch'u-pan-she (in Chinese).

Huang, Chung-Huang. 1990. Economic Valuation of Underground Water and Man-induced Land Subsidence in Aquaculture. *Applied Economics* 22, 31-43.

Hwang, Jiin-Tarng, 1994. Taiwan-ti-ch'u Huan-ching-fa chih Chien-t'ao [A Critical Review of the Environmental Laws in Taiwan Area], in: *Taiwan-ti-ch'u Huan-ching-fa chih Yie- chiu [A Research on the Environmental Law in Taiwan Area]*. Taiwan: Yueh-tan Ch'u-pan-she (in Chinese).

Liu, Wen-Chao, 1995. Wo-men Sheng-huo tsai Chu-she li? [Are We Living in a Pig Stall?]. *New Environment* 125, 3-4 (in Chinese).

Lu, Hsiang-Yu, 1996. Sen-lin yu shui-tz'u-yuan [Forest and Water Resource], *Ta-tsu-ran* [The Great Nature] 53, 108-112 (in Chinese).

Reardon-Anderson, J. 1992. *Pollution, Politics, and Foreign Investment in Taiwan: The Lukang Rebellion*. Armonk, N. Y.: M. E. Sharpe.

Tang, Ching-Ping, and Shui-Yan Tang, 1999a. Democratizing Bureaucracy: The Political Economy of Environmental Impact Assessment and Air Pollution Fees in Taiwan. *Comparative Politics*, Forthcoming.

Tang, Ching-Ping, 1998. Environmental Protection and Democratization: Pollution Control and Local Self-Governance in Taiwan. Doctoral Dissertation, School of Public Administration, University of Southern California, Los Angeles.

Tang, D. T. C., 1993. Wu-kuo Huan-ching-fa yu Cheng-tse te Hsien-kuan chi Chan-wang [The State and Future of Our Country's Environmental Law and Policy], in: Editorial Committee (eds.), *Tang-tai Kung-fa Li-lun [Theory of Contemporary Public Law]*. Taipei: Yueh-tan Ch'u-pan-she (in Chinese).

Tang, Shui-Yan, and Ching-Ping Tang, 1997. Democratization and Environmental Politics in Taiwan. *Asian Survey* 37 (3), 281-94.

Tang, Shui-Yan, and Ching-Ping Tang, 1999b. Democratization and the Environment: Entrepreneurial Politics and Interest Representation in Taiwan. *The China Quarterly* 158, 66-82.

Tien, Hung-Mao, and Tun-Jen Cheng, 1997. Crafting Democratic Institutions in Taiwan. *The China Journal* 37, 1-27.

World Bank, 1993. *The East Asian Miracle: Economic Growth and Public Policy*. A World Bank Policy Research Report. New York: Oxford University Press.

Yeh, Jiunn-Rong, 1997. Institutional Capacity-Building Toward Sustainable Development: Taiwan's Environmental Protection in the Climate of Economic Development and Political Liberalization. *Duke Journal of Comparative and International Law* 6, 299-72.

Yen Cheng-Ping, 1998. Shan-p'o-ti Shui-tu Tsu-yuan Pao-yu Wen-t'i Chien-tao yu Cheng-ts'e [The Review and Policies on the Water and Soil Preservation Problems on the Slope-land], in: Ou-Yang Chiao-Hui and Huang Mien-Shan (eds.), *Shing-shih-chi te Huan-ching-pao-hu Cheng-ts'e* [Environmental Protection Policy for the New Century]. Taipei: Ho-sheng Chi-ching-hui (in Chinese).

Ou-Yang Chiao-Hui, 1998. Shui-huan-ching wen-t'i Chien-tao yu Cheng-ts'e [Water Environment Problems: Review and Policies], in: Ou-Yang Chiao-Hui and Huang Mien-Shan (eds.), *Shing-shih-chi te Huan-ching-pao-hu Cheng-ts'e* [Environmental Protection Policy for the New Century]. Taipei: Ho-sheng Chi-ching-hui (in Chinese).

ENVIRONMENTAL CAPACITY BUILDING IN VIETNAM

Le Thac Can

1 Major Environmental Problems

Vietnam is a Southeast Asian country situated along the eastern margin of the Indochinese peninsula, bordering China to the north, Laos, Cambodia to the west, the Gulf of Tonkin, the South China Sea, and the Gulf of Thailand to the east and south. Vietnam is essentially a tropical country with a humid monsoon climate. The mean annual temperature is 23°C in Hanoi, 25°C in Hue, and 27°C in Ho Chi Minh City. The average annual rainfall is 1,678 mm in the north, 2,890mm in central Vietnam and 1979mm in the south. About 80% of annual precipitation occurs in the rainy season. This large quantity of rainwater is drained to the seas by a myriad of rivers and streams, the most important being the Red River in the north and the Mekong River in the south.

Vietnam is about 33 million hectares in size, including 11 million ha of low lying arable land, 16 million ha of forested land, of which 8 million ha are still forested. Vietnam's subsoil contains important mineral resources such as coal, iron, bauxite, copper, tin, gold, silver, precious stones, oil and gas. Total potential hydropower is estimated at about 28,000 MW. Marine fisheries bring in an annual catch of 1 million tonnes of fish and 50,000 tonnes of shrimp.

The population of Vietnam is currently about 76 million. Nearly 80% live in rural areas, working in agriculture, forestry, fisheries and handicrafts. The most important urban centres are: Hanoi, with a population of 3 million; Ho Chi Minh City, 4 million, and Hai Phong, 1.5 million. Population distribution is uneven, about 1000 persons/sq. km in some districts of the Red River Delta; 300 to 500 in the Mekong River Delta and about 50 in remote mountainous areas. This large population and abundant manpower constitute the most important resource for the country's development. On the other hand, rapid population growth is putting increasing pressure on living conditions, employment, natural resources and the environment.

With these geographical characteristics Vietnam could be considered to be endowed with abundant natural resources and favourable environmental conditions. Yet, as everywhere, the environment also has negative aspects. The distribution of natural resources does not always coincide with human needs. Floods, droughts, typhoons and other natural disasters are frequent. High temperatures and humidity create favourable conditions for epidemic diseases, food contamination, metal and technical equipment corrosion. Vietnam lacks many natural resources necessary for the development of agriculture and industry. The disparity of mineral ores makes exploration and exploitation difficult and costly. These features along with backwardness in economic development, rapid population growth, and the devastating impacts of more than 30 years of war from 1945 to 1975 have created serious environmental problems.

1.1 Deterioration of Forest Resources

Forest coverage has been reduced to an alarming level. Forest quality in the remaining forested areas has declined significantly. Originally, forests covered the entire country. Diverse geographical conditions have created a high diversity of forest ecosystems, covering a broad spectrum of forest types, comprising mangrove forests, Malaleuca forests, fresh-water swamp forests, monsoon forests, lowland evergreen broadleaf forests, hill evergreen and semi-evergreen broadleaf forests, forests on limestone, mountain evergreen and mixed coniferous forests, sub-alpine vegetation and bamboo. But in recent decades forests have suffered serious depletion. Between 1943 and 1991, forest coverage decreased from 45% to 28% of the country's total area. Northern Vietnam experienced the greatest decline, with a decline in forest coverage from 45% to 17% over 48 years. Coverage in some provinces remains very low. Lai Chau, for example, has only 8%, Son La 12%, and Lao Cai 5.4%. The decrease in this region was caused by the rapidly growing population's demand for forest products and agricultural land. The result has been the formation of large areas of unproductive barren land. The remaining forests in northern mountain areas are degraded, poorly stocked, and scattered in small, non-continuous plots. In recent years, the total area of natural forest, both upland and coastal, has declined at an estimated average rate of 110,000 ha/year.

Recognizing that forest loss is the most serious factor threatening the long-term productivity of the country's renewable natural resources, the Vietnamese people are carrying out a national reforestation programme to reforest about 40-50% of the national territory by the 21st century. The current reforestation rate of about 150,000 ha/year is to be increased to 300,000 ha/year in the coming years.

1.2 Decrease in Agricultural Land and Deterioration of Soil Quality

In 1945, when the population of Vietnam was 25 million, total land resources per capita were about 1.3 ha with 0.2 ha used for agriculture. With rapid demographic growth the figure is now 0.45 ha, of which only 0.10 ha is available for agricultural exploitation. In the Red River Delta 93% of potential arable land has been intensively used. In the Mekong River Delta this ratio is 82%. Some expansion of agricultural land is still possible in certain areas such as the hilly land in northern Vietnam and the plateaux of central Vietnam. This would require huge financial, manpower, technology investment and appropriate approaches to solve the complex economic, ecological, and environmental problems.

Besides the decrease in land per capita, land resources are subject to a number of serious quality degradation problems, including soil erosion, declining soil fertility owing to inappropriate cultivation practices creating laterization in hilly and mountainous areas, salination and acidification in coastal areas. The annual rate of soil erosion caused by rainfall is estimated to be about 20 tonnes/ha from watershed land, and 100 tonnes/ha from plantation land.

1.3 Inefficiency in the Conservation and Use of Water Resources

Potential water resources in Vietnam are estimated at about 6,400 m^3/year per capita, which is relatively high. However, because of financial and technical constraints, these resources are still not well conserved and rationally used. Current use of water resources is in many cases inefficient owing to inadequacy in the design and building of water reservoirs, water supply and irrigation systems. High levels of sedimentation resulting from soil erosion reduce the efficiency of canals and shorten the life span of reservoirs. In the dry seasons of 1991 and 1998, important hydropower plants in Da Nhim, Tri An and Hoa Binh could not operate normally due to the serious lack of water in the reservoirs. Smaller reservoirs such as Cam Son, Song Hieu, Boc Nguyen in northern and central Vietnam were also seriously affected by sedimentation within about ten years after construction.

Water pollution from domestic and industrial wastewater became a serious issue in Ho Chi Minh, Hanoi, Hai Phong and some other cities. Contamination of underground water occurred in some delta and coastal areas from intrusion of saline or acidic water during exploration and exploitation. Mineral and thermal water including carbonic, sulphuric, silica, ferrous and fluoride groups are found in more than 350 sources but remain under-exploited.

1.4 Irrational Use of Mineral Resources

A complex geological formation process has endowed Vietnam with a wide range of minerals, including the basics for the development of agriculture and industry, and many rare and valuable deposits. The northern area is the richest in mineral resources, with coal in Quang Ninh and Thai Nguyen provinces; apatite and rare earths in Lao Cai; lead, zinc, silver in Cao Bang, antimony in Tuyen Quang and precious stones in many places. Central Vietnam has important iron deposits in Ha Tinh; bauxite in Nghe An; chromates in Thanh Hoa; gold, copper and zinc in Quang Ngai. There are large reserves of oil and gas in the continental shelf of southern Vietnam; important deposits of bauxite in Lam Dong; peat in the Mekong River Delta and brown coal in the Red River Delta. Mining in Vietnam is carrying out in hundreds of sites of various sizes. Because of technological limitations and mismanagement, serious loss and wastage of minerals have occurred during exploration, planning, design, extraction and utilization of extracted materials. It is estimated that loss in coal extraction is about 10 to 15% in open mining, and more than 40% in underground mining. In addition to direct economic losses, inefficient and inappropriate practices for mineral exploitation are creating serious air and water pollution as well as soil and landscape deterioration (L. T. Can, V. Quy, 1994).

1.5 Degradation of Biological Resources

Vietnam contains a great wealth of flora, though some is under severe threat. The country also has a rich variety of wild fauna spanning wide habitats, from equatorial lowlands to high plateaux and alpine peaks with a temperate climate. There are estimated to be 276 species of mammals, 828 species of birds, 180 of reptiles, 80 of amphibians, 2,510 of fish and many thousands of species of invertebrates. These groups of animals show a high degree of distinctiveness with many endemic forms. Following deforestation and other forms of wildlife habitat deterioration, many species are becoming scarce or are endangered. The list includes the kouprey, wild buffalo, rhinoceros, elephant and tiger. Wildlife in Vietnam is declining rapidly. Recent studies have shown that about 500 species of plants and 354 species of animals must be urgently protected (MOSTE, 1997).

Coastal and estuarine areas play a very important role in the exploitation of biological resources in Vietnam. The mixing of sea water with the fresh water, rich in nutrients, discharged from the Red River, Mekong River and other smaller rivers of central Vietnam makes the estuarine and coastal regions of Vietnam ideal spawning and breeding grounds for many species of fish, shrimps, crabs, oysters and other marine organisms of high productivity and commercial importance. Some 2,000 fish species have been identified, including about 100 species of high economic value. There are also large mangrove forests and rich coral reefs with a wide range of fauna and flora of important economic and scientific value. These areas are now suffering from over-exploitation. For this reason some species abundant a few decades ago are now absent. Mangrove forests, including those newly restored after the war, are cleared for shrimp farming, as fuel and construction materials.

1.6 Environmental Pollution

Although Vietnam is not yet very industrialized and has only a few larger cities, urban and industrial pollution is becoming a serious issue. The total daily solid waste discharge is estimated to be about 360 tonnes in Hanoi and 1,000 tonnes in Ho Chi Minh City. Daily sewage effluent discharge is about 300,000 m^3 in Hanoi, about 500,000 m^3 in Ho Chi Minh City. Not only toxic and hazardous gas emissions have been recorded in Hanoi, Ho Chi Minh City, Viet Tri, Lam Thao and some major industrial areas. Dust pollution has reached alarming levels in various urban centres, particularly along roads with intense traffic or construction sites.

Inappropriate use of inorganic fertilizers, pesticides, herbicides and other chemical products are causing soil and water pollution in some agricultural and rural areas. Water pollution and the lack of safe drinking water is serious in lowland, coastal, and hilly or mountainous rural areas, especially in the dry season. In many factories, construction sites and mining areas serious noise, heat, humidity, dust, and light are creating harmful effects on the working and health conditions of workers (L.T.Can, V.Quy, 1994).

2 Development of Environmental Policies, Law and Regulations

Since 1975, after 30 years of hard struggle for national independence, the Vietnamese people have faced new challenges. The restoration of the country's damaged and degraded natural resources and environment has become an important priority. Thus far, environmental issues have been addressed in 3 phases. From 1975 to 1980, rehabilitation of environmental damages caused by the war was paramount. From 1981 to 1990, the focus was on identifying natural resources and major environmental issues associated with socio-economic development and population growth. Since 1991, the emphasis has been on promoting appropriate strategies and measures for resource development and environmental management to attain sustainable development.

2.1 Restoration of the War-Damaged Environment

Restoration was largely carried out in the late 1970s. The first priority was to restore settlement and agricultural areas. Restoration involved detecting and defusing unexploded bombs and shells, filling bomb craters; reconstructing housing, schools, hospitals, roads, bridges, irrigation and drainage systems, ameliorating land management and land use, planting trees trees. In areas contaminated by chemical warfare, special cleaning of houses, farm land and water supply systems were completed before resettlement. Later came the restoration of forests, including reforestation, and the creation of favourable conditions for natural regeneration. Immeasurable efforts have been to take care and improve the living conditions of war invalids, orphans, the homeless and landless, providing acceptable material and moral living conditions for millions of war victims.

2.2 Solution of Environmental Problems in Post-War Development

After 1980 Vietnam entered a period of post-war development, with greater effort being devoted to agricultural and industrial production and to promoting exports. To this end, agricultural production in the delta and coastal plains in northern and central areas was intensified: rice farming areas in the Mekong River Delta were expanded; rubber, coffee, tea, and pepper plantations were developed in the central plateaux; large water reservoirs and hydropower plants in were built on the Red River in the north and Dong Nai River in the south; thermal power plants in Pha Lai; and important cement factories in Hoang Thach and Bim Son. Major roads and the railway system were repaired. While these activities have brought major economic benefits, they have also had complex impacts on the environment. At the same time a post-war population boom has occurred in all of the provinces and cities of Vietnam. These developments put greater pressure on the economy, which in turn produced new environmental problems. Based on proposals made by Vietnamese scientists through the State Committee of Sciences on 20 September 1985, the Council of Ministers of Vietnam issued a governmental decision concerning "Intensification of Basic Investigation of Natural Resources, their Rational Use and Protection of the Environment". To implement this decision, the ministries of agriculture, forestry, fisheries, mineral resources, water resources and other sectoral ministries carried out surveys and studies on natural resources and environmental conditions throughout the country. Provincial governments have also inventories and evaluated local natural resources. Universities and research institutions carried out research and experimentation projects aiming at rationalizing the use of resources, limiting wastage and reducing pollution sources.

From 1988, the National Assembly, the State Council and the Council of Ministers of Vietnam have issued a number of laws, orders, and regulations on natural resources and environmental protection, including the Land Law (1988); Law on Forest Protection and Development (1991); Public Health Care Law (1989); Order on Mineral Resources Use (1989); Order on Aquatic Resources (1990).

In 1985 the National Resources and Environment Research Programme, with technical assistance from the International Union for Conservation of Nature (IUCN), has prepared and submitted to the government a draft National Conservation Strategy for Vietnam. The recommendations of this strategy include the maintenance of ecological processes and life-support systems, namely forest, midlands, croplands, fresh water, estuarine and coastal ecosystems; the conservation of biodiversity by developing protected areas, identifying endangered species, issuing hunting regulations, controlling wildlife trade and ex-situs conservation; the sustained utilization of renewable resources, maintenance of environmental quality for human life; and international co-operation for conservation in Vietnam.

2.3 Environmental Protection and Sustainable Development

In their concern about the demands economic development would make on resources and the environment, the Vietnamese government requested Vietnamese scientists in the early 1990s to prepare a National Action Plan for Environment and Sustainable Development with the assistance of international experts. On 12 June 1991 the Action Plan was officially approved by the Council of Ministers of Vietnam.

The objective of this Action Plan is to provide a comprehensive framework for national and local environmental planning and management leading to specific actions required in the short term to address priority problems at their roots. A wide variety of components include an institutional, legislative and policy framework; orientations for environmental monitoring, environmental information management; integrated and sectoral strategies for sustainable development, environmental impacts and environmental disaster management. The framework also has 2 support programmes: environmental education, training, awareness promotion and international and regional co-operation. A list of research, baseline data collection priorities and areas for testing the feasibility of the action plan was also prepared.

In December 1993 the National Assembly passed the Law on Environmental Protection. The implementing provisions for this law were issued in October 1994 by the prime minister. In 1995 the Ministry of Science, Technology and Environment issued the regulations on Environmental Impact Assessment. The government decided to establish 87 national parks and protected areas. Pollution control and abatement activities have been carried out in major cities and industrial zones.

2.4 Environmental Protection in the Period of Industrialisation and Modernisation

From 1986 Vietnam began adopting important reforms in socio-economic management characterised by a shift from a centralised, planned economy to a market economy. The liberalisation of agricultural and industrial production, the development of internal trade, participation in regional and international markets, and the opening up of the country to foreign investment are bringing to Vietnam relatively rapid economic growth. On the basis of this reform Vietnam is entering a new period of development aimed at the industrialisation and modernisation of the country.

In recent years, industrialisation and modernisation have resulted in more rapid economic growth, a steady increase in incomes and living conditions, but are also creating new, acute resources and environmental problems. Pressure on land, water, and mineral resources is increasing. It is estimated that agricultural land per capita in the Red River Delta, which is now 0.11 ha will decrease to 0.05 ha in 2005 and possibly to 0.034 ha by 2025. The lack of safe drinking water, as well as adequate quality water for industry and agriculture will be an important issue in many areas of the country, especially during the dry season. Urban and industrial pollution is becoming more serious with the construction of new factories and industrial parks, and increasing traffic. The intensification of agricultural exploitation leads to greater use of agrochemicals, causing serious soil and water contamination in rural areas. The current development of oil and gas exploration, exploitation and processing will obviously create risk of marine pollution. The proportion of the country's population residing in urban areas is currently about 24%, will increase to 26% by the year 2000, and probably reach 35% by 2005. With such rapid urbanisation, inadequate infrastructural and technological conditions for housing, water, energy supply, and social services could become very acute. The disparity in urban and rural incomes is leading to the spontaneous migration of people from rural to urban areas, making environmental problems in urban centres more complicated. Unplanned spontaneous migration of large groups of people from localities with less land resources to the central plateaux and southern provinces could cause serious deforestation and soil depletion in the new resettlement areas (V.Quy, 1997).

3 Brief Evaluation of Environmental Policy

The most recent Vietnamese policy on environmental protection is outlined in the Decisions of the VIIIth Congress of the Communist Party of Vietnam (CPV) issued on 1st July 1996. The current task of environmental protection in the country is defined as follows: "To carry out rapidly surveys on pollution, to evaluate the irrational use of natural resources and to identify efficient mitigation measures. To implement projects on protection and improvement of the environment; to establish national parks and nature reserves; to green urban and industrial areas, to apply advanced technologies for the treatment of toxic wastes and substances; to carry out environmental impact assessment of socio-economic development plans and projects; to solve pollution issues caused by production units; to prevent radically air and water pollution in the process of industrialisation and modernisation; to increase forest coverage to an ecologically safe level; to conserve land and marine ecosystems; to ensure good a working and living environment for people in urban, industrial areas and a hygienic environment in rural areas; to strengthen environmental management in all sectors; to ensure the necessary conditions for implementing the Law on Environmental Protection" (CPV, 1996).

In June 1998 the Politburo of the Central Committee of the Communist Party of Vietnam (CPV), issued an order on "Strengthening Environmental Protection in the Period of Industrialisation and Modernisation". This order defined the general objective of environmental protection as: "Prevention of environmental pollution, restoration of degraded environment in various localities, conservation of biodiversity, the step-by-step improvement of environmental quality in industrial, urban and rural areas, the promotion of sustainable socio-economic development, the improvement of people's living conditions and the successful implementation of the industrialisation and modernisation of the country". The order also identified 8 major measures for realising these objectives: (1) promotion of environmental education and awareness; (2) completion of the legislative and regulatory system; (3) prevention and solution of environmental pollution and risk; (4) rational exploitation and use of natural resources; (5) diversification of investment for environmental protection; (6) strengthening of environmental management; (5) improvement of environmental research and training; (8) development of international co-operation for environmental protection (CPV, 1998).

In the context of the developing countries, a good national environmental policy should satisfy the following requirements: (1) exact identification of major environmental problems the country is facing, (2) sound analysis of the causes of these problems, (3) rational choice of priority actions to be undertaken to solve the issues identified, (4) identification of implementation measures to make environmental protection and sustainable development action plans feasible, (4) integration of national, regional and global tasks of environmental protection.

It is obvious that, in the various phases of the country's development, the Government of Vietnam has paid adequate attention to environmental problems and has adopted problem-solving policies at the right time. To a certain extent these policies have met the above mentioned criteria. They address the most urgent environmental problems of the country, as well as some regional and global problems. Various conservation and protection measures based on these policies, such as legislative and regulatory measures; education and training; research and experimentation have contributed positively to the solution of existing environmental problems in Vietnam. However, it should be noted that the feasibility and sustainability of these policies are still very limited, because they have not been followed up by appropriate strategies and planning for efficient implementation. There is a serious lack of human, technical, and financial resources for implementing identified environmental protection policy.

4 Main Actors in Environmental Policy

In Vietnam the main actors in environmental protection are public institutions, environmental organisations, businesses, the public, and the mass media.

4.1 Public Institutions

The main public (governmental) institution at the central level responsible for environmental protection is the Ministry of Science, Technology and Environment (MOSTE), established in 1992 on the basis of the former State Committee for

Science and Technology. MOSTE is responsible for environmental management at the national level. It has an executive department, the National Environmental Agency (NEA), responsible for assisting MOSTE in environmental management at the national level. The NEAs specific functions include:

- drafting environmental protection policy, legislation and regulations to be adopted by MOSTE; or to be proposed by MOSTE for adoption by the government or the National Assembly;
- organising implementation of the National Plan for Environment and Sustainable Development, and the Law on Environmental Protection;
- co-ordinating environmental protection activities among provinces and sectoral ministries;
- reviewing environmental impacts assessment reports submitted to MOSTE;
- carrying out environmental inspections;
- establishing and managing the national environmental monitoring system, the environmental information data base, and preparing the report on the state of environment;
- environmental pollution control measures;
- promoting environmental education, training and environmental awareness among the public;
- collaborating with international and regional organisations in environmental protection.

MOSTE has a branch organisation, the provincial/municipal Department of Science, Technology and Environment (DOSTE) in all 61 provinces and municipalities in the country, and in every sectoral ministry of the government. Provincial/municipal DOSTEs are in charge of environmental management in the provincial/municipal territory. Ministerial DOSTEs are in charge of environmental management of the production, trade and service activities for which the ministry is responsible.

The Ministry of Planning and Investment (MPI) is also closely associated with environmental protection and the promotion of sustainable development. The MPI is in charge of preparing national development plans and allocating budgetary and other resources to the different sectors of government and to provinces/municipalities directly attached to the central government. Within the MPI, the Department of Science, Education and Environment is responsible for environmental affairs (N.N.Sinh, 1997).

The National Assembly, the supreme legislative body in the country, has a Committee for Science, Technology and Environment. It reviews bills to be submitted to the National Assembly, including legislation on the environmental matters.

4.2 Environmental Organisations

At present, the only environmental organisation in Vietnam is the Vietnam Association for the Protection of Nature and Environment (VAPNE), which is a nationwide organisation with more than 1,200 members. VAPNE's members are mostly environmental scientists working in universities, colleges, research institutes; or environmental managers working in provincial, municipal, and ministerial

DOSTEs. VAPNE is headed by a central committee with 4 panels: education and knowledge dissemination, consultancy and services, science and technology, international relations. VAPNE has offices in all important institutions (universities, research institutes, ministries) or localities (provinces, cities, towns). It operates a research and training centre, the Vietnam Environment and Sustainable Development Centre (VNESDC), issues a bi-monthly publication, "Green Vietnam" and a scientific review, "Environmental Research Work". VAPNE cooperates closely with educational and research institutions, advising public institutions, enterprises and the mass media. On the request of central and local government, VAPNE participates in various commissions for appraising environmental impact assessment reports of major development projects, or for evaluating proposals for important environmental action plans or programmes. VAPNE is a member of a larger organisation, the Vietnam Union of Science and Technology Associations (VUSTA). Other member organisations of VUSTA, such as the Association of Geographers, the Association of Geologists, and the Association of Biologists, also participate in environmental evaluation, advice and consultancy (VAPNE, 1998).

4.3 The Business Sector

The business sector includes agricultural, forestry farms, fishery enterprises; factories, plants, building and infrastructure construction enterprises; trade and service enterprises. Many enterprises are aware of environmental issues. Larger public enterprises, as required by government, often have their own environmental unit dealing with pollution prevention and the rational use of natural resources. Medium-size enterprises may have a part-time technician in charge of environmental issues. Enterprises responsible for sewage and solid waste collection and treatment, often called Urban Environmental Companies (URENCOs), have been established in major industrial and urban areas. These enterprises constitute a new industry in the country, employing a large number of workers and mobilising significant investment from the public as well as from the industral sector. However these enterprises are still weak and facing various difficulties in the implementation of their functions. Recently some other enterprises, both agricultural and industrial, have made some effort to promote cleaner production technologies, but these endeavours still very limited. In general, industrial enterprises tend to find ways around environmental regulations as ways to cut production costs. Some local authorities, responsible for foreign investment, tend to lower environmental requirements in order to attract higher investment, thus producing even more difficulties in environmental pollution prevention and solution.

4.4 The Public

Public participation in environmental protection is a tradition of Vietnamese society. Tree planting campaigns are organised every year in the spring. Participation by all residents in cleaning public squares and streets in urban and rural areas on weekends and holidays are popular. Recently a major environmental protection campaign entitled "Make Vietnam Green, Clean and Beautiful", organised by the Vietnam Youth Union, has covered all parts of the country. In the context of various organisations like the Farmers' Union, the Women's Union, the Youth Union, and the War Veterans Union, the public participates actively in discussing and solving local and national environmental problems. Representatives of various mass organisations are invited by MOSTE or DOSTEs to take part in discussions on the environmental impact assessment of development projects located in their localities.

4.5 Mass Media

National and local newspapers, weekly and monthly magazines, radio, and television regularly publish information and comments on local, national, regional and global environmental issues. The Vietnam Television and the Radio Voice of Vietnam have special regular programmes on the environment. Two monthly reviews specialise in environmental issues: "Green Vietnam" and "Environmental Protection", published jointly by NEA and VAPNE. Some provincial DOSTEs also have regular local publications on the environment. On the one hand, the mass media provide public institutions with the means of disseminating environmental information and knowledge. On the other, they allow the public to take part in the debate on national and local environmental issues, and thus to participate in decision making.

4.6 Co-operation between Relevant Actors

Environmental protection in Vietnam is carried out in close co-operation among all the actors mentioned above. Environmental policy, strategy, and legislation are the result of close collaboration between public institutions and environmental organisations, especially specialised environmental protection organisations. Researchers often lead the way in identifying the state of the country's environment and environmental issues and in forecasting developments, thus providing the basis for policy and legislation. In implementing policies, strategies, legislation, and action plans, public institutions also closely co-operate with environmental organisations, mass organisations and the mass media. In the event of significant national or local environmental impacts by enterprises (private and public), individual citizens and communities can voice their concern to governmental environmental authorities. The defence of public and community environmental interests is always strongly supported by environmental organisations and the mass media.

5 Capacity Building

5.1 Legal Framework

On 27 December 1993, the Vietnamese National Assembly, the national legislative body, passed the Law on Environmental Protection (LEP), which came into force on 10 January 1994. Other legislation directly or indirectly relating to natural resources and the environment includes laws on Maritime Navigation, Land and Territory, Health Care Protection, Oil and Gas, Minerals, Forest Protection and Development, Aquatic Resources Protection, Embankment and Dikes, the Protection of Traffic Construction, and the Labour Code.

The government has issued several implementing guidelines and regulations on the LEP, including Decree No. 175/CP, dated 18 October 1994, on Guidelines for the Implementation of the Law on Environmental Protection; the inter-ministerial Memorandum No. 155/TTLB, dated 11 April 1994, on Temporary Regulations on Environmental Planning; Memorandum No. 1420 MTg, dated 26 November 1994, on Guidelines for Implementing Environmental Impact Assessment; the Prime Ministerial Decision No. 845/TTg, dated 22 December 1995, approving the Action Plan for Biodiversity Protection in Vietnam; Decision No. 118/QD-MTg, dated 28 May 1996, on Forms of Inspection and Administrative Penalties for Violation of LEP; Decision No. 229-QD/TDC, dated 25 March 1995, on the Issuance of Environmental Quality Standards (L. Q. An, 1997).

5.2 Institution Building

Since the early 90s, as we have seen, a system of public institutions responsible for environmental protection has been established in Vietnam. At the national level it includes: the Ministry of Science, Technology and Environment (MOSTE), and its branch organisations, the Departments of Science, Technology and Environment (DOSTEs) in all sectors; the Ministry of Planning and Investment (MPI) and its branch departments of Planning and Investment (DPI) in provincial governments. At provincial and municipal level environmental management is carried out by the People Committee, which has their provincial/municipal DOSTEs as the executive branch. Provincial and municipal DOSTEs are closely linked with MOSTE and NEA, which provide them technical guidance and financial support.

The National Assembly Commission for Science, Technology and Environmental Affairs co-operates closely with MOSTE and other ministries in preparing legislation on the management of environment and natural resources. In the provinces, districts and villages, People's Councils play the same legislative and consultative role vis-à-vis the executive, termed People's Committee at this level.

Various mass organisations like the Youth Union, the Women's Union, the Trade Union, the Farmers' Union, and theVeterans' Union have officers, centres or units taking care of environmental issues affecting their domain.

5.3 Human Resources

Some 700 experts, managers and scientists now work in environmental protection institutions. NEA employs about 50 experts and managers; the 61 provincial and municipal DOSTEs employ about about 300, and the environmental units attached to 20 ministries and general departments, about 100. An estimated 200 researchers, technicians, and lecturers work in environmental research, training and consultancy institutions.

Several universities and colleges in Hanoi and Ho Chi Minh City offer undergraduate and graduate courses for environmental experts and managers. The number of participants in short-term environmental training courses organised by NEA was 240 in 1994; 960 in 1995 and about 1,000 in 1997 (NEA, 1997).

In co-operation with Vietnam Youth Union, MOSTE is carrying out environmental awareness programmes, promoting a large Green - Clean - Beautiful campaign throughout the country. Environmental awareness information and lectures are regularly broadcast by Voice of Vietnam, Vietnam Television and provincial and municipal radio and television stations. Central and local decision-makers are among the main targets of these environmental awareness campaign (N. N. Sinh, 1997).

5.4 Financial and Technical Resources

The government funds environmental protection in the framework of budgeting for three areas: science, technology and the environment. Ten years ago the Central Committee of the CPV stipulated that the budget for science, technology (and the environment) should be at least 2% of total annual governmental budget. This requirement has still not been met, and government expenditure on environmental protection remains very limited. Some municipalities, like Ho Chi Minh City or Hanoi may receive additional funding from local government budgets, but many other localities lack this capacity. The business sector could provide modest funding for activities within their plants or construction sites. In 1995 the Ministry of Planning and Investment allocated 12 billion Vietnamese dongs (equivalent of about US$ 1.2 million) to a special NEA environmental research fund (NEA, 1997).

A network of environmental monitoring stations has been established since 1994. The number of stations has increased from 6 in 1994 to 8 in 1995 and 14 in 1996 (NEA, 1997). NEA has provided all DOSTEs with a set of portable instruments for quick sampling and field measurement of air, water quality and noise pollution.

5.5 International Co-operation

Vietnam is party to several international conventions: the Convention on wetland of international importance (RAMSAR); the Convention on international trade in endangered species (CITES); the Convention for prevention of pollution from ships (MARPOL); the Vienna Convention on protection of the ozone layer; the United Nations Convention on law of the sea (UNCLOS); the Montreal Protocol on substances depleting the ozone layer; the Basel Convention on control of transboundary movements of hazardous wastes and their disposal; the United Nations framework Convention on climate change; and the Convention on biodiversity (L. Q. An, 1997).

Vietnam is implementing several environmental co-operation and assistance projects with international agencies: the United Nations Development Programme (UNDP); the United Nations Environment Programme (UNEP); the Asia Development Bank (ADB); the World Bank (WB); with regional organisations: ASEAN Senior Officials on Environment (ASOEN); and with development co-operation agencies from Sweden, Canada, Denmark, Germany, the United States, Netherlands, Belgium, Australia, and Japan (NEA, 1997).

6 Main Restrictions

6.1 Limited Financial Resources

Limited central and local government financial resources and little funding by enterprises or other sources makes the implementation of identified environmental policy and action plans difficult. The lack of financial resources is especially serious in providing urban sewage and solid waste treatment infrastructures and equipment in large and medium-sized cities. The environmental management authorities in many localities simply do not have enough staff, equipment and funds to fulfil their inspection and monitoring functions (NEA, 1997).

6.2 Lack of Professional Trained Manpower

With a population of 76 million people, Vietnam has only about 600 trained environmental experts, less than 1 per 100,000 people. This lack of manpower is more serious at the National Environmental Agency, which has a staff of only 50, and in all provincial and municipal DOSTEs, which on average have a staff of only 5. Before 1985 there was almost no training of environmental experts in Vietnam. Most now working in environment related institutions have different professional backgrounds: engineering, geography, biology, chemistry, and medicine. The need for complementary training is considerable. There is now an urgent need for short-term in-service training of existing NEA and DOSTE officers, and for long-term formal training of prospective staff.

6.3 Weak Role of Legal Instruments

Various regulations have been issue pursuant to the Law on Environmental Protection, but the system of legislative instruments is still far from complete and does not cover all important aspects of environmental protection and sustainable development in Vietnam. There is incoherence in existing regulations, making enforcement difficult and complex (NEA, 1997).

6.4 Lack of Self Regulatory Economic Instruments

The use of economic instruments for environmental protection is a new approach in Vietnam. Some studies on the possibility of combining market-based instruments with legal and administrative instruments in the specific context of Vietnam are now carried out. Since 1990 the government has established a system for levying taxes on natural resources exploitation. The implementation of this system is not simple because the concept of valuing natural resources and the pricing of these values have not been clearly defined. After promulgation of the Law on Environmental Protection the government indicated its intention of developing a system of environmental taxes, but the details have not been worked out. Efforts to develop economic incentives for the rational use of natural resources and the development of cleaner technology and products have also not yet come to fruition.

6.5 Inadequate Integration of Environment into Development

With the issuance of LEP in 1994 all major development projects and regional development plans are formally subject to environmental impact assessment. Environment Impact Assessment is considered an appropriate way to integrate environment in development. However, in reality there are important government-funded projects that have submitted no EIA report. Some foreign investment projects submitted EIA reports belatedly, often during construction and equipment installation. There is still no environmental planning at both the national and local level. Environmentally sensitive areas have not been identified.

7 Prospects

Vietnam is entering a new period of socio-economic development characterised by intensive industrialisation and modernisation. Without effective measures to protect the environment and conserve natural resources, this new period could see greater degradation of natural resources and environment pollution. To solve existing environmental problems and prevent further degradation greater capacity building efforts are needed in the following areas:

1. Higher government and other funding for environmental protection;
2. Stronger environmental management institutions at the central, provincial/municipal, and district levels;
3. Elaboration of strategy and planning for effective environmental policy implementation;
4. Completion of regulations for LEP implementation;
5. Improved environmental education, training and environmental awareness promotion;
6. Improved international environmental co-operation.

References

An, L. Q., 1997. Vietnamese Policy on Environment and Sustainable Development. Environmental Protection Law and Regulations in Vietnam. Environmental Policy and Management in Vietnam. Berlin.

Can L. T., and V. Quy. Vietnam, 1994. Environmental Issues and Possible Solutions. Asian Journal of Environmental Management 2 (2).

MOSTE, 1997. Five-year Implementation of Agenda 21. Hanoi: publisher?? (in Vietnamese)

NEA, 1996. Three-Year Activities of NEA. Hanoi.

Quy, V., 1997. Environmental Issues in Vietnam: An Overview. Environmental Policy and Management in Vietnam. Berlin.

Sinh, N. N., and H. Friederich, 1997. The Institutional Framework for Environmental Policy and Management in Vietnam. Environmental Policy and Management in Vietnam. Berlin.

VAPNE, 1998. Report of the Central Committee of VAPNE on 1993-1995 activities and orientations for activities in 1998-2003. Hanoi. English version co-edited by Le Dong Phuong, (MA, University of Hawaii)

SUMMARY: ENVIRONMENTAL CAPACITY BUILDING IN A CONVERGING WORLD

Helmut Weidner and Martin Jänicke

Thirteen countries were examined in the first phase of our research project on capacity building in national environmental policy (see Appendix). The seventeen new country case studies included in the second phase of the project presented in this volume were carried out in accordance with the same conceptual framework. This offered a good opportunity to check our earlier findings against the results of the new case studies and, where necessary, to modify them. For this reason we will largely be addressing the topics dealt with in our summary of the 1997 reader (Jänicke & Weidner, 1997a, b). And, as in our first publication, we concentrate on international and global trends in environmental policy development. While we had then conceded that 13 case studies represented a somewhat narrow basis for any generalizations, the total of 30 country studies now provide a much broader basis for the systematic analysis of general development trends in environmental policy capacities, their forms, causes, and effects. The 17 country studies in the second project phase expand the country sample not only quantitatively: the spectrum of country types is also broadened, and a selection bias in the first group of countries substantially corrected. The 13 countries included almost all early and later environmental policy pioneers, while developing, newly industrializing, and transition countries had largely been disregarded. Nevertheless, the developing countries are still underrepresented in the total sample.

Comparative analysis of the 17 additional countries strongly confirms the main findings of the foregoing study, particularly with regard to the following theses:

- global environmental policy learning is paralleled by the accelerated diffusion of environmental policy innovations – especially in the form of institutions, instruments, and technologies;
- non-governmental proponents of environmental protection as well as transnational, international and global organizations and networks are exerting increasing influence on environmental policy diffusion processes;
- with the consequent globalization of environmental policy and global convergence in the basic pattern of environmental policy (towards multi-stakeholder involvement and negotiation-based decision making – "governance") and the leading paradigm ("ecological modernization/sustainable development"). However, the global convergence of policy patterns is accompanied by continuing divergence in capacities for action: although capacities are increasing globally, there continues to be a considerable gap between groups of countries, especially between OECD countries and the rest.

Although it can be generally said that these developments strengthen and stabilize environmental policy activities and proponents under the conditions of economic globalization – a major decline in environmental capacities has been rare and usually limited in time – the (mainly systemic) restrictive factors that our first study had stressed are still highly effective. Accordingly, a *real* shift of paradigm in environmental policy from end-of-pipe measures or (technology-oriented) ecological modernization to a development path of structural ecologization, let alone sustainable development, is rare. We therefore have the paradoxical situation that, while the general globalization process favors environmental policy, the "ecological flip sides" are increasing on a global scale. Thus, although capacities for environmental policy and management have continued to expand remarkably over the past decade, particularly from the institutional point of view, the global ecological challenges (especially in climate change) and certain national and regional problems (such as decreasing biodiversity, soil pollution, depletion of water resources, deforestation, urban expansion and sprawl) have increased. Clearly, the "negative capacities" (for this concept, see Keller, Lowi & Gendlin, 2000) in both developing and advanced countries are still strong enough to hinder any ecologically sustainable change of direction.

1 The General Profile of Global Environmental Change

The analysis of the 17 countries confirms one of the main findings of the foregoing study: there is a contradictory relationship between economic development and environmental quality. Economic growth leads to both improvement and deterioration in the state of the environment. The environmental impacts of accelerated international flows of capital and technology depend heavily on the energy, material, and emission intensity of the technologies involved (see also OECD, 2001a; Asian Development Bank, 2001).

Despite economic growth, some conventional emissions (like SO_2, dust, CO) have decreased in many countries. Pollution levels and resource use have thus been decoupled from economic output. However, with very few exceptions, CO_2 emissions are tending to increase, frequently in correlation with economic growth. Other examples include biodiversity losses, soil degradation, and worsening urban air quality in many countries. Public demand for environmental quality and "willingness to pay" often increase with economic growth. But there are further prerequisites for the formation of politically viable environmental quality preferences, in particular greater special environmental policy capacities to give such preferences a chance of accessing the political agenda and of actually being implemented, e.g., publicly accessible information on rather unspectacular emission increases for certain pollutants, and the ability to evaluate their long-term ecological consequences; the possibility of addressing environmental issues publicly, and responsive governmental institutions. The simple inverse conclusion that economic decline would lower pollution holds only for certain areas – e.g., for some conventional industrial pollutants and carbon dioxide – and often only in the short term. In developing countries, however, there are clear indications of a broad variety of negative environmental effects due to economic decline, mostly of an indirect kind: poverty (especially if combined with rapid population growth) often increases the exploitation of natural resources and hence the degradation of ecological systems, causing, among other things, erosion, desertification, and water shortages. An OECD study has also pointed out the detrimental effects of "marginalization", i.e., a combination of factors like slow economic growth, external debt burden, corruption, violent conflicts, and various kinds of trade barriers erected by the developed countries (OECD, 2001b: 12). In developed countries, too – even in highly advanced industrial countries – it is primarily the indirect effects of economic crisis that have long-term negative impacts on environmental quality. They can include the relaxation of pollution control measures due to environmental budget cuts, declining public environmental consciousness, reduced environmental policy innovation efforts, and the like. General economic and political reforms (e.g. *New Zealand*[1]) can also have a considerable influence on the development of environmental quality. Owing to the multitude of factors that affect relations between economic development and environmental quality, it is still difficult, even after having investigated a total of 30 countries, to ascertain causal relations between economic and ecological development.

However, there are many indications that – given the worldwide expansion in environmental policy capacities, increasing economic competition, and the globalization of environmental policy – the economic strategy of "get polluted, get rich, and then clean up" is subject to greater constraints than before *(Brazil, Mexico, Taiwan, India,* and the transition countries). Furthermore, environmental economics studies on the question of whether pollution follows a "Kuznets curve" (i.e., first rising with economic growth and then falling as income increases) have shown that this is so only to a limited degree, and only with regard to certain conventional pollutants, but not for all relevant areas – e.g., water pollution (cf. Hettige, Mani & Wheeler, 1997; Magnani, 2000). It is self-evident that a development strategy of "grow now, clean later" not only leads to ecologically irreversible losses and economic risks but, from the point of view of "good government" and the "public interest," is also politically and morally highly problematic because of the associated high social and health costs. There is also increasing evidence for domestic restrictions to such a development path: in many countries – especially the newly industrializing countries – growing congestion and pollution in urban centers provokes resistance from influential actor groups to industrial expansion, which has serious implications for economic growth in the countries concerned as a whole (see also Asian Development Bank, 2001)

Nevertheless, analysis of our country sample shows that massive environmental pollution is still taking place in certain areas of the developed countries, too, although its causes and consequences have been scientifically identified and addressed. Even in these countries capacity building is obviously still urgently needed to handle the manifold unsolved problems. Furthermore, a decline in "flows" (emissions, waste) is often accompanied by a continuing rise in "stocks" of accumulated pollutants (e.g., soil, groundwater). This is particularly the case in rich countries.

An overview follows of the most important development trends in environmental quality and emissions in the 17 countries of the second project phase, with special emphasis on outstanding problems and achievements. The overview has

---

1 The names of countries dealt with in this reader are printed in *italics*.

been kept brief, since more detailed information is provided in the individual country studies.

In *Austria* considerable improvements have been achieved in water conservation, due largely to high government subsidies, so that, although policy can be said to be effective, it is not efficient as a whole. While SO_2 emissions have been drastically reduced, CO_2 emissions have continued to increase despite professions of a strict global climate policy. In soil protection, which attracted political attention only at a late date, hardly any improvements can be reported, and, as in all other countries, growing motor traffic is becoming one of the biggest environmental problems. In *Italy* particularly few cases of success have been registered, and then mostly in conventional areas (CO, SO_2, lead, some flowing and coastal waters), while the extremely ineffective land use policy has permitted a considerable rise in the rate of land consumption (in the Po valley alone at an average rate of 6% a year). Industrialization policy in southern Italy has caused considerable environmental pollution in ecologically hitherto intact landscapes; in 1997 the Italian Environment Ministry classified 14 areas throughout Italy as endangered areas where an "environmental crisis" could occur. Low recycling rates and weak legal arrangements and controls mean that waste disposal is often far below the standards of other OECD countries. On the other hand, in at least some fields on which foreign attention is particularly focused (bathing waters), negative impacts on the tourism industry have led to improvements; the area of national parks had doubled by the end of the nineties in comparison with the eighties. In all, the authors of the country study judge Italy to be generally 10 to 15 years behind comparable industrial countries with respect to the effects achieved through environmental policy measures. In *France* considerable improvements have been achieved on a broad front. But there are important exceptions: for example, environmental problems due to motorized transport and increased land covering. The 77% rate of connection to sewage treatment plants is comparatively low, and there are still many waste disposal sites for household waste that do not operate with state-of-the-art technology, whereas industrial waste facilities are relatively modern and so numerous that waste importation has become a flourishing industry (see also O'Neill, 2000).

Bulgaria records high or the highest levels of pollution in accordance with almost all relevant environmental criteria. It provides an ample demonstration that a low level of economic development does not necessarily go hand in hand with low environmental pollution. The greatest environmental problem is seen as air pollution; the rapid increase in motor transport – stimulated not least of all by EU financial aid for road construction (accompanied by a decline in public transport) – renders environmental policy countermeasures on a technical basis almost futile. In the *Czech Republic* over the past decade, both stronger environmental protection measures and the economic recession after the change in political regime brought some amelioration in conventional environmental areas, but the growth in motor traffic increased air pollution in urban agglomerations. The rise in private consumption produced greater waste problems. Despite an improvement in general water quality, groundwater pollution has worsened. Since most improvements have been attained by end-of-pipe means (and not by structural reduction of energy consumption), a greater economic upturn is likely to produce a resurgence of conventional pollutants. In *Hungary* the underdevelopment of waste water treatment and agricultural pollution make water conservation the main environmental problem; drinking water can be endangered. Growing private consumption and motorized transport exacerbate waste problems and air pollution, especially in urban areas. In *Poland* in 1983, 11% of the national territory (with about 35% of the population) was classified as ecologically endangered. Environmental protection measures strongly reduced this figure in the course of the 1990s. In *Poland*, too, the amount of household waste has been growing since 1990 without adequate waste disposal facilities being constructed to cope with the increase. The problem has worsened since the collapse of the old recycling system after the change in political system. Particular success has been achieved in nature conservation, primarily with international support, whereas the strong increase in motor traffic since the early 1990s (together with a 34% decline in public transport between 1986 and 1993) has become one of the biggest environmental problems. However, in the central and eastern European countries mentioned, the eastward enlargement of the EU is likely to improve institutional capacities.

Canada's main problems arise from the extensive exploitation of natural resources, which, despite the sparse settlement of the country and the extraordinarily large areas of natural land, has caused a decline in biodiversity and sometimes considerable damage (e.g., among freshwater fish populations). Despite international commitments, climatic gas emissions continue to rise. In *New Zealand*, in contradiction to the green image the country enjoys, environmental problems are growing in important areas. This is the case with air and noise pollution from motor traffic in some cities, the relatively high per capita CO_2 emissions (increasing energy consumption by head of population), and strong biodiversity losses, which many experts regard as the real major problem facing New Zealand's environmental policy. Besides soil erosion and backward waste management, biodiversity loss is one of the main problems facing *Australia*. Both CO_2 and SO_2 emissions continue to rise, and in some areas there is considerable air pollution, as well as episodes of photochemical smog. Remarkable progress has been made in forestry policy, most likely owing to pressure from NGOs, and a steady increase in nature conservation areas has recently been recorded.

Although *Costa Rica* has the reputation of being a model country for biodiversity management, the extremely intensive use of agrochemicals has led to increasing problems for the animal world and for potable water reservoirs. Chronic damage to health among agricultural workers and the residents of the communities affected is also attributed to this. Inadequate wastewater treatment systems (sewerage, sewage treatment plants) has produced greater water pollution. In the capital San José, the strong increase in motorized traffic has led to high air pollution; the quantity of waste is growing in urbanized areas. In *Mexico*, water quality and waste disposal are often at so low a level that serious health hazards clearly result. In Mexico City alone there are 3,500 open dumps. In the largest cities, especially in Mexico City, the main environmental problem is air pollution from motor traffic. Smog is frequent. Despite a wide range of measures to improve air quality in the cities, air pollution is worsening in some cases because of the growth in motorized transport. Biodiversity is declining strongly, owing mainly to high levels of air and water pollution, to industrialization subject to weak environmental policy control, and to heavy deforestation, chiefly by small farmers. Massive deforestation continues in *Brazil's* Amazon rainforests, in spite of all international efforts to curb it. In urban areas, many types of air and water pollution continue to increase. Motor traffic, for example, has grown strongly since the 1980s. Both refuse disposal and wastewater disposal are extremely underdeveloped. In urban areas, for instance, where some 113 million people live, the sewage produced by 75 million residents is not treated.

Owing to rapid economic and demographic growth in *Vietnam*, there has been an "alarming regression" of the forests. The intensification of agriculture has led to increasing soil degradation (erosion, salinization, etc.) and water pollution. In urban areas air pollution is "sometimes alarmingly high" (owing essentially to motor traffic and industrial emissions). Lumbering, mining, and agricultural activities are causing heavy biodiversity losses. *Taiwan* has to some extent experienced remarkable improvements in air quality, waste management, and water conservation. However, major deficiencies in land use policy have led to serious soil impacts (including erosion). Biodiversity losses continue to worsen (owing, among other things, to ineffective nature conservation and deforestation). Since the 1980s, high economic and demographic growth rates in *India* have to some extent resulted in strong environmental pollution. Increasing industrial and motor transport emissions, soil degradation (also through deforestation), and diminishing water resources are among the consequences.

It is now clear that *Morocco* will suffer major water scarcities in the not too distant future, owing largely to deforestation and to rapid industrialization since the 1980s. Industrialization has also contributed to the increase in air pollution (including conventional pollutants like SO_2). Deforestation and overuse of the soil through increased livestock farming have worsened erosion. Inadequate wastewater treatment facilities have in part strongly polluted coastal waters.

Analysis of the 13 country case studies from the first project phase has shown that environmental policy successes – measured by emission reductions, improvements in environmental quality, and long-term structural changes in polluter areas (such as reduced material input, disposal-friendly products) that lead to input and output-side reductions in impacts – are largely restricted to the developed industrial countries. But there, too, the most marked improvements were recorded in coping with environmental problems that were highly visible, had a strong subjective impact, and were highly susceptible to politicization – and for which technical options were available, or which offered the possibility of win-win constellations in processing. In contrast, the balance of success tended to be unfavorable in the developed countries when it came to creeping environmental degradation problems with little chance of attracting public and political attention (such as soil and groundwater pollution, land consumption, the destruction of biodiversity).

These results are substantially confirmed by the additional 17 country case studies collected in this volume. This is largely in agreement with the findings of other studies on worldwide or regional environmental trends (UNEP, 1999; World Bank, 1998; OECD, 1997, 2001a; Asian Development Bank, 2001; European Environment Agency, 2001). Moreover, unfavorable developments in the advanced industrial countries are apparent in some areas that have hitherto reported success: environmental gains from new technologies and policies are being increasingly overtaken by the pace and scale of economic developments and consumption, particularly in the transport sector. In contrast, slight improvements are apparent in agriculture. They are mainly due to modified consumer behavior –

especially in western Europe – and to changes in agricultural policy. The reduction of "ecologically perverse subsidies" has had a slight impact, too, which has also been stimulated by a new EU agricultural policy. In Switzerland, *Austria*, and Germany, as well as *Poland*, the share of ecological farming is increasing steadily, but still constitutes only a relatively small proportion of total farming production. In the energy sector, too, structural changes are beginning to emerge in come countries. Nevertheless, carbon dioxide emissions have risen steadily in almost all countries. Among the few exceptions are Germany, Switzerland, and Britain, as well as countries in eastern Europe that have suffered drastic industrial closures. Without the influence of international environmental organizations and regimes, the situation with regard to climatic gases would certainly be much worse. That international environmental agreements can achieve considerable success is demonstrated, however, by the major reductions in the production, consumption and release of ozone-depleting substances achieved through the Montreal Protocol and related agreements. Soil protection has the worst record throughout the world. Only few countries have developed specialized policies, e.g., comprehensive soil protection legislation.

Many of the problematic developments arise from developments in policy sectors whose dynamics largely escape the influence of environmental policy regulation, like urbanization, population growth or poverty. Poor countries in particular have little opportunity to adopt advanced pollution control technologies and modify social behavior towards a sustainable lifestyle. All this shows clearly that too much is demanded of existing specialized environmental institutions ("capacity overload") and that integrative environmental policy approaches are needed. Furthermore, this indicates the need for international organizations to change perspectives and policy. Evidently, there is a need to take better account of the economy/ social-environment interface than is the case in existing approaches. They often have failed to address the underlying political cultures, power relationships, systemic and societal capacities, etc. that are necessary to achieve a spread of *effective* institutions and instruments. According to the Asian Development Bank, much more could have been accomplished over the past decade in ecological modernization if the international donor organizations had put greater emphasis on the more difficult and politically sensitive issues of public policy, governance and implementation (Asian Development Bank, 2001).

Environmental policy must take account of what the Intergovernmental Panel of Climate Change has stated: "The ability of human systems to adapt to and cope with climate change depends on such factors as wealth, technology, education, information, skills, infrastructure, access to resources, and management capabilities (...) Many communities and regions that are vulnerable to climate change are also under pressure from forces such as population growth, resource depletion, and poverty. Policies that lessen pressures on resources, improve management of environmental risks, and increase the welfare of the poorest members of society can simultaneously advance sustainable development and equity, enhance adaptive capacity, and reduce vulnerability to climate and other stresses" (IPCC, 2001: 8). From a programmatic point of view, this need is largely acknowledged because the concept of sustainable development has met with broad acceptance. This is expressed, for example, in the introduction of national environmental plans, organized multi-stakeholder discourses, and the establishment of national councils or task forces for sustainable development. But from a practical perspective still too little is being done in most countries. Among the notable exceptions is Sweden, where, after years of multi-stakeholder negotiations, the government has drawn up a comprehensive and ambitious sustainable policy program, setting clear and verifiable targets and indicators for key areas, establishing institutional capacities, and furnishing human and financial resources. The ultimate goal of the policy program is to reverse the negative trend in all relevant environmental problem areas within a generation or by 2025 (enviro-Report No. 8, July 2001: 2). But even in countries that, unlike Sweden, are not the countries with progressive environmental policy, or only for a brief period, the vision of sustainable development has contributed to dynamizing environmental policy. Britain is one such example: in 1999, the British Government launched its "flagship strategy" for sustainable development, followed shortly thereafter by a variety of sectoral and issue strategies (focussing on transport and waste). Along with an increasing number of countries, the U.K. announced its interest in positioning itself as an international leader on the environment (one of the main messages of Prime Minister Tony Blair's speech on environmental policy given in London in October, 2000 before the Green Alliance and the Confederation of British Industry).

2 Institutionalization of Environmental Policy

Environmental issues have conquered the national political agenda, so that, for instance, a strategy of deliberate ignorance or suppression (as in most central and eastern European countries under communist regimes or in Japan in the 1950s and early 1960s) would be hard to follow. Even developing countries with an authoritarian political regime (e.g., China, *Vietnam* and *Taiwan* as well as Nigeria until recently) are developing national environmental policies and participate in multilateral environmental policies. They have begun to allow non-governmental environmental groups scope for action; the involvement of civil society in environmental policy matters is still weak but increasing, and there is a trend towards providing for public participation in the legislative framework.

The most basic reaction among public and private core actor groups to the new challenges is usually institutional adaptation. The term institution as used here is not confined to organizations with a physical structure and defined mandate but also encompasses formalized principles of conduct that shape human interaction, stabilize expectations and help resolve collective disputes. Aside from the standard differentiation between formal and informal institutions it is also relevant to differentiate between formal and substantial institutionalization – formal, in this case, signifying merely symbolic measures. *Italy*, for instance, established an Environment Ministry as early as 1971 – however, it was a ministry without portfolio. *Poland*, Russia and other communist countries had had outstandingly strict environmental quality standards, but deficient implementation and enforcement rendered them almost meaningless in the policy process; they served primarily as anti-Western propaganda. Nevertheless, these standards may have contributed to a greater environmental risk awareness among relevant groups in these countries.

In Western democratic countries, first moves towards institutionalizing environmental issues as a policy area in its own right were taken in the 1960s. The incentive came mainly from within the countries themselves (e.g., USA, Britain, Japan, Sweden), but there were also countries where the impetus came from "outside", from other countries (e.g., Germany, Switzerland, *Canada, Austria).* In addition, environment-related activities by international organizations (especially the UN Conference on the Human Environment in Stockholm 1972) set off a broad wave of institutionalization that encompassed socialist countries, too. However, regardless of whether the creation and design of institutions were primarily due to external influence and requirements, or to internal, national factors, this study has not revealed a systematic effect with respect to performance.

With few exceptions, significant institutionalization occurred in the developing countries only in connection with and as a consequence of the so-called Brundtland Commission and the 1992 Rio Conference *(Costa Rica, Morocco, Vietnam),* while these events led to a remarkable second wave of institutionalization in many industrialized countries, which adopted new, fundamental legislation and reorganized environmental governance systems. Later, in the developing and transition countries, the demands of international aid organizations or multilateral environmental agreements were to have a strong impact on the institutional expansion or reorganization of existing institutions, the influence of the European Union (EU) also being very strong in some transition countries *(Bulgaria, Czech Republic, Hungary, Poland).* Considering the position in the early 1970s, it can be said that, both quantitatively and qualitatively, worldwide progress in governmental institutionalization has been remarkably rapid and far-reaching.

Legal and governmental institution building is a basic precondition for effective public policy (see Table 1). This indicates and establishes an institutionalized self-interest, as well as responsibilities and accountability, on the part of the politico-administrative system – and in the medium/long-term often leads to governmental responsiveness. There is no country in our sample where effective environmental measures were achieved in relevant areas without a certain degree of environmental policy institutionalization. Of course, the actual contribution of this institutionalization to environmental capacity depends on various additional factors (resources, competencies, kinds of embeddedness, etc.) and the country studies record a correspondingly broad range of quality in environmental institutions (see also Scruggs, 1999). However, general trends indicate an overall rise in institutional capacities:

1. A general institutionalization of ecological interests and tasks in several sectors of society;
2. A general expansion of capacities (resources, specialized institutions, regulations, etc.), and only seldom a massive reduction as in the USA under the Reagan administration or recently in Russia under President Putin. However, there are marked fluctuations in developing and transition countries and sometimes also in developed countries (possibly *Canada).*
3. Increase in the powers and functions of environmental institutions in countries where initial institutionalization tended to be formal or symbolic in nature (*Italy* after 1986, *Mexico*, since the late 1980s);
4. The modernization (reorganization, renewal) of institutions, especially since the late 1980s, with the aim of increasing their capacities, often in association with new public management strategies and in reaction to international developments

Table 1. Institutionalization in Environmental Policy

Countries	Ministry of the Environment	National Environmental Agency	National Environmental Report	Environmental Framework Law	Article in the Constitution	Council of Environmental Experts	National Environmental Plan
Australia	1971/1975	1988	1980/1996	1974			1992
Austria	1972	1985	1978		1984	1971	1995
Brazil	1985/1992	1989		(1981)	1988	1984/1997	2001
Bulgaria	1990	1976	1989	1991	1968/1991	1974/1996	1988/1992
Canada	1971		1986	1988		1971	1990
Chile		1990/1994	1992	1994	1980	(1996)	1998
China		1984	1989	1979/1989		1991	1994
Costa Rica	1986	1995	1986	1995	1994	1995	1990/1996
Czech Republic	1989	1991	1990	1992	1992	1992	1992
Denmark	1971	1971	1983	1973/1991			1994
France	1971/1984	1991	1973	2001		1975	1990
Germany	1986	1974	1976		1994	1971	
Great Britain	1970	1972/1995	1978	1974/1990		1970	1990
Hungary	1987	1974	1975	1976/1995	1972/1990	1996	1992
India	1980/1985	(1974)	1982	1986	1976/1994	1993	1993
Italy	1971/1986	(1994)	1989	1986	(1948)	(1986)	(1997)
Japan	2001	(1971)	1969	1967/1993		1967	1995

Countries	Ministry of the Environment	National Environmental Agency	National Environmental Report	Environmental Framework Law	Article in the Constitution	Council of Environmental Experts	National Environmental Plan
Korea	1990/1994	1977	1991	1990	1980/1987	1985	1987/1990
Mexico	1982/1994	1992	1986	1972/1988	1988	1995	1989
Morocco	(1995)					(1995)	
Netherlands	1971/1982	1984	1973	1979/1993	1983	1974	1989
New Zealand	1972/1986		1997	1986/1991		1970-88	1994
Nigeria		1988	1992	1988	(1979/1989)	1990	1988/1990
Poland	1972	1980/1991	1972	1980/2001	1976/1989/1997	1993	1992
Sweden	1986	1967	1977	1969/1998	1974	1968	1993/1998
Switzerland	(1999)	1971	1990	(1983)	1971/1999		(1997)
Taiwan		1978	1988/1993		1992	(1987)	1979/1994
USA		1970	1970	1969		1971	
USSR/ Russia	1988		1988	1991	1977/1993		1993
Vietnam	1992	1993	(1995)	1994			1991

Note: Years in parentheses indicate institutions coming close to the conventional definition

Source: Weidner, Jänicke & Jörgens (2001).

(e.g., Agenda 21, sustainable development, national environmental planning); a process that also has taken place in countries with systems that have developed organically over a long period and tended to oppose significant structural change ("path dependency"), e.g., Britain, Japan in some aspects, *France*, *Mexico* since the new government of 2000, Germany;
5. A further increase in capacities generated by existing institutions themselves by various and partly innovative means, e.g., network formation and informal practices (e.g., regional and international networks of environmental administrators; institutionalized collaboration between public and private actors and local authorities, like the Union of Baltic Cities – established as early as 1991 and already comprising about 100 cities – one of the main objectives of which is sustainable development in the Baltic Sea Region).

Furthermore, there has been considerable expansion in environmental institutions – especially in the developed countries, but increasingly in "latecomer states" – at various levels and in various policy areas of the politico-administrative system, e.g., committees and councils at the parliamentary or cabinet level or specialized ("mirror") units in other ministries (transport, industry, agriculture, etc.). The latter were initially often set up as a sort of counter weapon, because it was feared that environmental interests would exert too strong an influence on the given interest area (e.g., Japan, Germany, *France*), but in the course of time this proved more and more to be a supportive measure for policy integration, and was also helpful in paving the way for a (still rare) administrative reorganization in the direction of forging closer links between environmental policy and other policy areas (Denmark, Sweden, Switzerland, the Netherlands, Britain, Germany since 1998, *Mexico* since 2000, etc.). The inherent dynamics of governmental institution building should therefore not be underestimated, fuelled by self-interest and the desire to expand functional scope and powers as well as the positive impact of cross-jurisdictional networking on policy learning. These developments hamper at least hierarchical action to reduce capacities (Japan, Chile, *Costa Rica, Hungary, Poland,* etc.).

On the basis of the country sample, no clear statement can be made about whether unitary or federal systems are more advantageous for environmental protection, although in recent times the usefulness of federal structures for environmental policy has generally been regarded with skepticism. But the general quality of the institutions responsible for implementation and enforcement (personal and material resources, technical and managerial competence) at all levels of the political system is clearly more important for the capacity to solve environmental problems than whether the governmental system is unitary or federal. It is of secondary importance whether – as in most unitary countries – these institutions are "extended arms" of the central government or – as in federal countries – authorities of the lower tiers of government. The study shows that the greatest weaknesses with regard to institutional capacities in transition and developing countries are in implementation. What is more, the decentralization process that began in some transition countries in the 1990s (especially in *Poland*) has often meant that, although substantial powers and duties have been transferred to regions or local authorities, this level of government has lacked the resources to implement nationwide environmental goals and legislation, which continue to be set and adopted at the national level. One result is that EU directives are only formally adopted (see also Carius et al. 2000). But some advanced industrialized countries, too, (e.g., *Italy, Canada, Australia)* show remarkable weaknesses in implementation capacities.

In *Vietnam* provincial and municipal environmental agencies are not only desperately short of human resources, and what little staff there are urgently need training. In view of the country's rapid industrialization, it is all the more urgent to improve implementation capacities. This is particularly the case in *India. Morocco* faces a similar situation. Although Moroccan implementation staff – thanks not least of all to assistance from international organizations – are relatively well trained, rigid traditional hierarchical structures, which largely prevent inter-policy coordination, negate this advantage. Here we have a case of underexploitation of capacity because of policy coordination deficits, caused in turn by traditional power structures. In *Bulgaria* the very weak implementation capacities are determined primarily by politico-structural deficits: in particularly the widespread corruption in public institutions, influential political networks of the old political elite ("nomenklatura"), the rather opportunistic mass media, and the generally weak hold of democratic attitudes and values in the political system, the administration, and in civil society. Nevertheless, changes in the internal organization of the regional environmental inspectorates have helped increase their management flexibility. The *Bulgarian* case clearly shows what can be considered a general rule: decentralization as such is not a sufficient condition for effective environmental management, however, decentralization is obviously a necessary condition for the evolution of a socially rooted environmental policy (South Korea, *Taiwan, India)*. This is also well demonstrated by the development of an effective environmental policy in Germany and Japan, and more recently – especially in the context of Local Agenda 21 activities – Great Britain.

Since the 1970s there has been continuous capacity building in almost all areas of society in the shape of organizational-institutional differentiation or integration in the developed countries. Be it in the science system, in the cultural, political, and economic systems or in civil society, relevant organizations in these areas have at least created institutions specialized in environmental matters, or have expanded competencies in existing structures through training and education. This is also true of research institutions, the Churches, the trade unions, the public media, and business (also increasingly in medium-size firms) as well as the umbrella structures of many of the organizations mentioned. In civil society, the institutionalization of environmental interests has been particularly marked; the number of NGOs in some countries almost defies listing, and sometimes the larger NGOs together have more or about as many members as traditional interest groups (trade unions, political parties). For instance, nature and animal protection and environmental organizations in the Netherlands – after increasing their membership by 250,000 in 2000 – now have almost 4 million members and donors, equal to one quarter of the total population of 16 million (and possibly the highest share worldwide). Thus, in many countries the environmental movement has for some time now been subject to institutionalization (Kriesi et al., 1995), including the founding of green parties and (especially pronounced in western European countries and the USA) the emergence of a wide-ranging ecology-committed commercial sector embracing, for example, consulting firms, research facilities, and other environment-related service companies with a strong affinity for the environmental movement and its organizations, stimulating their interest in and enhancing their capability for cooperation with polluting industries.

These forms of environment-related organization and network building have been substantially supported and stimulated, directly and indirectly, by politico-legal institutionalization in the form of laws and regulations. Environmental legislation has, for example, created new markets, new areas of activity, and demands to which private businesses have reacted through specialization and innovation. Changes in fiscal or electoral law have contributed to the stabilization of NGOs and green parties (Germany, *Australia, New Zealand*, etc.); and the EU Commission has recently (June 2001) supported a proposal to increase financial support for environmental NGOs – also in EU accession candidates – from about $ 10-30 million for a five-year period. The expansion of participational, informational and other procedural arrangements, or the creation of new pluralistic institutions (such as governmental advisory councils, Local Agenda 21 processes, round tables or other forms of consensus-oriented bodies) has encouraged the emergence of sometimes highly specialized organizations in science and society (e.g., for environmental dispute resolution: USA, Canada, Germany, Netherlands, *Austria,* Switzerland, *Italy).* Overall, this has fostered the potential and need for cooperation among interest groups, leading in turn to the establishment of new institutions like national round tables for the development of long-term environmental strategies (especially *Canada*, Great Britain) and national councils for sustainable development (Switzerland, USA, etc. and in Germany as a latecomer). At present, there are about 70 National Councils for Sustainable Development designed as multi-stakeholder participatory bodies (OECD, 2001: 20f.). All this obviously fosters cross-sectoral learning, which helps build new networks and environment-related capacities. This broad development is particularly pronounced at the international level, where there are now a large number of bodies and other facilities where environmental interest groups have a seat and a say. Apart from the international forms of environment-related network building and institutionalization, there are more and more cross-regional and transnational forms, for example, between the littoral countries of the Baltic Sea, which particularly foster environmental capacities in the transition countries. The spread of "epistemic communities" (Haas, 1992) should also be mentioned.

In all, there are innumerable variants of formal and informal institutions between the local and global levels. Since the end of the 1980s, organizations from the transition and developing countries have been increasingly involved. This is particularly important in cases where – mostly in times of economic crisis – politico-administrative capacities have been reduced. The sometimes considerable cutback in capacity in transition countries (in personnel and financial resources) within a few years after the shift toward democracy can essentially be explained by the fact that, under the communist regimes, opponents of the system often "took shelter" in ecological organizations. Ecological criticism of the system created a common basis in the sense of a smallest common political denominator, which, after the change in system, split up into a wide range of interests *(Poland, Czech Republic, Bulgaria, Hungary).* If a great deal of support for capacity building had not been forthcoming from international and other foreign organizations, and NGOs, the reduction in capacity for environmental policy and management would certainly have been much greater. Capacity reduction was apparently also checked by the pressure exerted on transition countries by the EU: one condition for accession is that environmental institutions need to have attained a certain level of capacity (Baker & Jehlicka, 1998). For developing and newly industrialized countries, not only bilateral and multilateral relationships with industrialized countries are important but also multilateral environmental agreements and international economic and aid organizations (such as the World Bank, United Nations, OECD, international financial institutions). Foreign and international funding programs increasingly include support for institution building. A very recent example is the Regional Environmental Reconstruction Program for South Eastern Europe, set up in March 2000. The various projects include developing and strengthening NGOs (funded by the Netherlands) and institution building aimed at fostering cooperation among southeastern European countries (funded by Switzerland).

As we have seen, environmental institution building at the national level was initiated and strongly influenced in the early 1970s primarily by countries like the United States and Sweden. For many industrialized and practically all developing countries, the impetus and general recipes for environmental institution building still come from pioneering countries in the industrialized world or – increaseingly – from international organizations. There was not a single developing country in the sample of this study that offered any sort of model for environmental measures in other countries. In transition countries, however, the development of original forms of institutionalization conducive to environmental capacity building (very pronounced in *Poland)* is worth noting. Despite a number of much discussed and to some extent imitated instruments (emission trading, least-cost planning in the energy sector, etc.), the United States has not been among the front-runners for the past decade, largely because of its restrictive global environmental policy and weak environmental performance. The lead has been taken by countries like Denmark, the Netherlands, and Sweden. This has been achieved by the far-reaching institutional reorganization they have initiated in various policy areas to accelerate integrative environmental policy in the direction of ecological modernization and sustainable development (see also Andersen & Liefferink, 1997; Hanf & Jansen, 1998).

In this respect, the Netherlands and Sweden are particularly interesting, because central elements of this new environmental policy strategy, such as flexible, cooperative governance, strategic long-term planning, eco-taxes, and stakeholder dialogue, have caused "lesson-drawing" (Rose, 1993) in other countries. This shows that the global process of capacity building depends heavily on national pioneers. And, as the next section shows, the speed of innovation diffusion in environmental policy and management (and especially in environmental technology) has increased markedly over the past decade.

3 Diffusion of Environmental Innovation

The primary aim of environmental innovation is to reduce the consumption of natural resources and improve environmental quality, and to maintain reductions and improvements over the long term. It should also improve the cost-benefits ratio. In view of the major environmental challenges and current policy deficiencies (see UNEP, 1999), the generation and diffusion of innovation must be seen as key components in successful environmental policy. Many consider ecological modernization already to be a global trend or at least a global potential that could be utilized by environmental policy (Wallace, 1995; Jänicke, 2000; Mol, 2001).

Innovation research in the environmental field (Blazejczak et al., 1999) has shown that prevailing economic explanations usually fail to do justice to the complexity of innovation generation and diffusion. They often ignore the political context, how policy and decisions are specifically made, environmental policy and management capacities, and the actor configurations involved. A dialogue-based, flexible, and reliable policy style and the formation of innovation networks among stakeholders have, for example, been shown to be important preconditions for innovation (Hemmelskamp et al., 2000; Jänicke & Weidner, 1995). The striking trend towards flexible and cooperative instruments and procedures – ideally linked with demanding goals – in many countries of our sample may thus indicate that the innovative capacities needed for ecological modernization are increasing.

Such innovation is particularly important for an ecological modernization strategy, since it goes beyond conventional, technocratic environmental policy approaches (especially "end-of-pipe" treatment). It focuses on inter-policy integration, strengthens collaborative goal-setting and problem-solving, and combines environmental measures/objectives with new public management approaches. It also takes account of the limits to traditional forms of governmental control. Countries with political and cultural structures and capacities that favour consensus, cooperation, and integration are naturally at an advantage (e.g., Germany, Sweden, Denmark, Netherlands, Japan; also *Canada, New Zealand* in some respects).

In the 1970s and the 1980s even highly developed industrial countries with serious pollution problems needed a decade or more to adopt certain abatement technologies. This has been the case, for example, with flue gas scrubbers, catalytic devices for cars, and certain waste water purification systems. These technologies were best practice in the United States and Japan long before being introduced in other countries. Technological innovations now spread far more rapidly. Even transition and developing countries now adopt new environmental technologies much faster, at least in new facilities. International aid in environmental technology has certainly contributed. But in several of the developed countries under study, technical innovation has also become a dynamic sector of the economy, sometimes with a considerable impact on industry and employment in general. In Germany 3.6% of the workforce are employed in the environment sector, which is more than in the car industry. The United States, Germany, and Japan are the three largest exporters of environmental products, and further growth is expected owing to the great backlog of demand in many other countries. But also in *Italy, France, Canada, Brazil,* and South Korea, an important and continuously increasing "eco-business" sector has developed. The South Korean Ministry of Environment's long-term environmental policy strategy (the "Eco-2 Project"), for instance, has the explicit goal of developing environmental technology into a national strategic industry.

The United States was long considered *the* front runner. The policy innovations introduced in the United States include the National Environmental Protection Act

(NEPA), the Environmental Protection Agency (EPA), the Toxic Substances Control Act (TSCA), the Comprehensive Environmental Response, Compensation, and Liability Act (CERCLA or Superfund), various economic instruments, alternative dispute resolution procedures (ADR), eco-sponsoring, investment funds based on ethical-ecological criteria as well as innovative abatement technologies such as catalytic devices. In one form or another, almost all have been adopted in many other countries, and have helped boost environmental capacities and support environmental proponents. For example, Environmental Impact Assessment has become a basic instrument in many countries in promoting environmental interests in public and private land use. Strict liability rules have strengthened the legal hand of those affected by pollution. The Freedom of Information Act has ensured greater clarity, and ADR have brought productive conflict resolution and broken deadlocks in multi-stakeholder negotiations. Various negotiation-based methods of environmental dispute resolution have become more or less standard instruments in Germany, Switzerland, Netherlands, Great Britain, Japan, *Australia, Austria, Canada,* and *Italy*. In *Brazil, Costa Rica, India, New Zealand* and *Poland* these methods are used sometimes, often with the support of institutions from abroad (Napier, 1998; Weidner, 1998). In contrast, the United States has very seldom explicitly adopted and implemented ideas from elsewhere. Where it has done so, the result has generally been below par, as was the case with the eco-label system (EPA, 1998). For some years now, the Nordic countries and the Netherlands, now and again Germany and Britain have made major contributions to environmental policy innovation and its global diffusion. The most important ones have been the precautionary approach, "green ministers," facultative eco-power supply, closed-loop material cycles, comprehensive production and product responsibility ("from the cradle to the grave"), eco-labelling, eco-auditing schemes (such as EMAS, ISO 14001), material flow management, dual waste management systems and the "Green Dot System" for packaging waste, voluntary environmental agreements, green public procurement policy, green budgeting, the strategy of policy integration in general or Local Agenda 21 initiatives.

Some policy innovations have spread worldwide with particular rapidity. They include comprehensive national environmental policy plans, sustainability strategies, environmental labelling, so-called eco-taxes (for a systematic overview and discussion, see Kern, Jörgens & Jänicke, 2000), voluntary agreements or commitments (World Bank, 2000), and deregulation policies in sectors such as energy production and supply, waste and water management. There were isolated green plans in the late 1980s (in the Netherlands, Britain, Canada, Denmark, etc.). With the 1992 Rio Conference and the adoption of Agenda 21, diffusion has been fast throughout the world – in Western industrialized countries, eastern Europe, and in developing countries (Jänicke & Jörgens, 2000). International organizations have done much to promote rapid diffusion (World Bank, but also OECD, UN institutions, EU). By 2000, some 80% of all industrial countries had developed such plans, and since the 1992 Rio Conference a large number of developing and newly industrialized economies have followed suit. Even Russia and Nigeria have drawn up an environmental policy plan – in contrast to some highly developed countries, such as Germany or the United States. In most cases green planning contributes to capacity building. Just how important certain capacities are in other environment-related areas often becomes apparent in the process of long-term national environmental planning. This is the case with cooperative capabilities and cognitive-informational factors (e.g., monitoring, cross-sectoral coordination, evaluating complicated problem complexes). The already mentioned and particularly ambitious Swedish national environmental plan ("The Future Environment – Our Common Responsibility"; see Lundqvist, 2000) has been adopted with the aim of making Sweden a sustainable society within a reasonable period of time (about 25 years).

Since the late 1980s more than 30 countries have introduced a national environmental label, including *India, Taiwan*, and South Korea. Over the same period eco-taxes on certain greenhouse emissions and/or fuels have begun to spread. The Netherlands and Scandinavian countries have again set the pace. Voluntary agreements, common in Japan since the 1970s, have been on the increase throughout the world since the 1990s. Recently (2000), 32 large South Korean companies concluded voluntary agreements with the government to reduce greenhouse gas emissions. 144 Korean companies now have such agreements. Company or sectorial agreements (voluntary commitments, covenants) evolved in the Netherlands and Germany in the 1990s, while at the Climate Conference in The Hague in 2000 a number of European companies backed an exacting greenhouse gas reduction goal, favoring the voluntary agreement instrument. Voluntary industry codes that had previously been country-specific have come to be applied more widely and globally, e.g. the *Canadian* system of accelerated reduction and elimination of toxics. The "Zero Emissions" strategy, designed by Japanese firms about a decade ago, has been recently attracting greater attention from the national governments in Germany and *France* and is being promoted by, inter alia, the University of the United Nations, which established a "Zero Emissions Forum". Emission Trading, first systematically established in the United States, became a core element of the global strategy for climate protection and has consequently diffused (though slowly) throughout the world.

Environmental innovations have been spreading more rapidly throughout the world for many reasons, most importantly:

1. Environmental policy has internationalized. International regimes and institutions have become more important (UNEP, CSD, World Bank, OECD, IUCN, etc.), actively monitoring and reporting innovations and fostering their diffusion (e.g., World Bank, 2000; OECD, 1999a; UNEP, 1999).
2. "Governance by diffusion" (Kern, Jörgens & Jänicke, 2000) and benchmarking have grown in importance. Outstanding approaches to solving environmental problems are increasingly imported to enhance local action and problem-solving capacities. This, again, underlines the key role of pioneer countries in the diffusion of environmental innovation.
3. The development and diffusion of environmental technological innovations, in particular, benefit from the closer cooperation in the advanced countries between governments (via R&D expenditures, regulatory policies, technical standard setting, eco-labelling or other information schemes, etc.) and companies, especially in the energy and transport sectors. Governments thus help develop markets for alternatives by creating incentives and removing barriers to market entry (OECD, 2001a: 302f.).

Local Agenda 21 processes, especially in western and eastern Europe, are an example for the rapid diffusion of innovation through the efforts of international organizations (UNEP, ICLEI). They are increasingly regarded as a defense mechanism against global economic developments detrimental to the environment (Lafferty & Eckerberg, 1998). Particularly in unitary countries, they contribute to democratization and accountability by decentralizing decision-making and by fostering capacity building in local communities. Britain is a particularly good example (Mason, 1999: 180).

All in all, a global environmental policy is emerging that can be described as a process of global policy learning (see Rose, 1993; Sabatier, 1993) and as a global convergence of environmental governance patterns. This trend is instrumental in preventing a race to the bottom in environmental standards and the spread of "pollution havens" in certain countries. Indeed, our country sample contained no unmistakable case of a "race to the bottom". The main reason seems to be the growing importance of the environmental issue in international competition in the field of innovation.

4 Globalization: Challenge and Chance

The challenges of globalization have been the subject of far more studies and cogitation than the opportunities it offers (Young, 1997). There is clear evidence that economic globalization is a major problem for environmental protection and its proponents. This is particularly true where a neo-liberal concept of globalization prevails, with the state being rolled back and superseded by market norms. The dominant trade and economic development patterns clearly threaten regional and global ecological systems. However, the negative and dramatic globalization scenarios, often discussed at a very general level, frequently exaggerate when it comes to environmental policy. It is often overlooked that globalization has been going for many decades now (Held et al., 1999), and that it has always constituted a threat to ecological systems and environmental policy. But environmental regulation used to be far less pervasive and there were far fewer environmental actors at the national level and in international organizations. They were practically absent in primarily economy-oriented international organizations. For some years, however, there has been a growing trend towards the *globalization of environmental policy*. This clearly strengthened environmental actors.

There is no evidence from our country studies that open economies have suffered from a systematic undermining of environmental policy capacities. Even when governments have undertaken far-reaching societal reorganization, as when the *New Zealand* government introduced a completely new environmental policy regime in pursuit of its new public management strategy, no covert intention of lowering environmental standards was evident (Bührs & Bartlett, 1993). In other cases, governments of countries that are particularly strongly integrated in global economic processes have succeeded in structurally entrenching an ecological modernization policy and in linking it with modernization in other policy areas. This has been the case in Sweden, the Netherlands, and to some extent in Denmark, which are experiencing concurrent reorganization of the welfare state and the labor market and implementation of a long-term sustainable development strategy.

Furthermore, the business sector itself has a considerable interest in global, well-functioning environmental policy standards (Vogel, 1997; OECD, 1999b; Garcia-Johnson, 2000). In general: "... the evidence shows that the risk of companies shifting their activities to low standard countries is rather small. In fact, what multinational enterprises generally seek is consistent – rather than lax – environmental enforcement" (OECD-Observer, Summer 2001: 63). Also with respect to NAFTA countries (USA, *Canada, Mexico),* studies (Rugman & Verbeke, 2000) have shown that there has been neither "a race to the bottom" (p. 79) nor "evidence supporting the pollution haven thesis" (p. 79). This is primarily explained by the "watchdog" function of non-governmental environmental organizations and the interest of – especially multilateral – enterprises to "design their protection process according to the best global practices" (p. 88). There can be many reasons for this, such as interest in export opportunities for environmental technology or in limiting economic competition by "eco-dumping". This may explain why some industrialized countries are strongly committed to environmental development aid. Japan, for example, became a leading donor of environmental development aid after the so-called Brundtland Commission's report on sustainable development (Japan Environmental Council, 2000).

Economic globalization is naturally a much greater problem for developing countries than for developed countries. Developing countries have much more restricted domestic resources to combat developments unfavorable to environmental protection. On the other hand, there are clear signs that ecologically unsound economic development strategies are becoming increasingly vulnerable. Facilitated by new communication technologies, the globalization of environmental policy is advancing apace. Networked NGOs and international organizations are monitoring environmental activities more and more closely, and the dissemination of environmental reporting is rapid and worldwide. Mass media response and the internet are important preconditions for this development. These global actor networks, "ecological global players", are constituted by a wide range of organizations and institutions (see also Young, 1997). Increasingly, NGOs, governments, and the scientific community have combined forces in negotiating at the international level (Lipschutz, 1996). Sometimes, governments have enlisted the aid of NGOs to lobby on behalf of mutually agreed positions in other countries or in international organizations. Not only the German government – even under a conservative coalition – has given NGOs financial and other assistance in preparing international negotiations.

As Japan some decades ago discovered to its distress, the strategy of giving almost absolute priority to the economy over the environment before instituting an environmental "care plan" proved a punishing experience (Tsuru & Weidner, 1989). The development path of "get dirty, get rich and then clean up" is even more problematic today than in the 1960s and 1970s. Many countries with a relatively solid capacity base for environmental policy are now economic competitors, and international organizations and NGOs are playing a significant watchdog role. Thus "Dump Watch" – a warning system covering several African states, which monitors illegal movements of hazardous waste – is supported by, inter alia, European governments and international NGOs. There is a much higher risk of direct or indirect economic sanctions being imposed on environmental grounds (voluntary agreements with business branches, environmental labelling, etc.). Conversely, the chances are much greater that environmental progressiveness will bring economic advantages (see Porter & van der Linde, 1995). And as a rule international regulation and standards are determined less by latecomer states than by the front runners. Furthermore, the sustainable development paradigm could make it easier to link economic and social interests with environmental protection in developing and transition countries. There are signs of this happening in *Mexico, Costa Rica, Vietnam, India*, and *Morocco*.

Without seeking to play down the negative consequences of economic globalization for environmental policy, which can be serious enough (for instance in the transport sector), we address the positive effects which have mostly received less attention in the environmental discussion (Weidner & Jänicke, 2001). Studies on global governance (e.g. Young, 1997) and our set of country studies do not support the thesis that national governments would wither away or generally lose power as a result of globalization (see also Bernauer, 2000; Krasner, 2001), with a consequent decline in environmental standards. National governments continue to play an important role in coping with environmental problems. The relationship between "governing" and "governance" is thus often not exclusive but synergetic. Krasner, for instance, observes: "There is no evidence that globalization has systematically undermined state control; indeed, the clearest relationship between globalization and state activity is that they have increased in the past" (Krasner, 2001: 236). And experts on developing countries argue that the state will continue to play an important role in environmental policy, especially in providing the infrastructure for good environmental governance, enhanced stakeholder engagement, and the development of new modes of collaboration with non-state actors (see http://www.adb.org).

National governments have many sources of power – they systematically enable and frame forms of global governance, network-building, etc. or use more traditional power resources. Governments in Western industrial countries often have high national budgets and corresponding public purchasing power, giving them the wherewithal to integrate environmental objectives in the manufacturing and consumer sectors. Even without higher national budgets, their scope for ecological policy formation is considerable, for example by reallocating environmentally unsound (or even "perverse") subsidies in favor of ecological modes of production. The new EU agricultural policy has likewise taken this course. Furthermore, the present pioneers in environmental policy (the Netherlands, Denmark, Sweden) are highly interested in the world market. Thus, globalization is often used as an excuse for environmental policy deficiencies, but the arguments have a rather weak empirical basis. Shortcomings in national environmental policy are mostly home-made (Jänicke & Weidner, 1997; see also Spaargaren, Mol & Buttel, 2000; Mol & Sonnenfeld, 2000; Angel & Rock, 2000; UNEP, 1999).

5 Democracy and Environmental Capacity Building

The most important precondition for building and sustaining environmental capacity, for supporting environmental proponents, and for providing an opportunity structure for addressing environmental problems publicly is the existence of democratic institutions and rules (Lafferty & Meadowcroft, 1996). This is particularly important for environmental actor groups whose preferences and problem-solving concepts do not conform with the prevailing norms and problem perceptions of the political and economic system elites. The great importance of democratic system structures for effective environmental policy is evident in many countries, especially where liberal-pluralistic democratic institutions are non-existent or weak (see Hirsch & Warren, 1998). A recent report by Amnesty International (Schneider, 2000) provides ample evidence of how environmentally committed individuals or NGOs suffer from sometimes massive repression by governmental authorities in countries like Russia, but also in *Mexico, Brazil* and *India*. Significant capacity-building and environmental improvements occurred in eastern European and Latin American countries after authoritarian regimes had been overcome. In South Korea and *Taiwan*, general democratization processes have very markedly enhanced the opportunity structure for environmental interests and – compared with the long time it took to develop an environmental policy in Western democratic countries – have led to the relatively rapid establishment of a solid capacity base, permitting at least an effective conventional (end-of-pipe) environmental policy, and fostering first moves toward an ecological modernization policy. In the 1990s, South Korea not only began assisting and funding environmental improvement in several countries, establishing a comprehensive information base, and issuing a monthly Environmental Information Bulletin, but also passed a Law for Public Information in 1996. Characteristic is also the increase in NGOs and environmental campaigning since democratization started in 1987 (see also Japan Environmental Council, 2000). In *Taiwan*, the environmental arena also profited from efforts by the government to achieve an image of responsive democracy – a key reason being to ensure international support, especially because of conflict with mainland China. In China and *Vietnam*, political liberalization, essentially the consequence of a new economic policy, has promoted the building of institutional and technical capacities in the environmental field, although they are still too low to counter the negative ecological impact of a dynamic industrial growth policy.

In all the eastern European countries under study *(Poland, Hungary, Czech Republic, Bulgaria*, Russia), environmental conflicts and environmental proponents have contributed to change in the political system. Environmental organizations often provided the only platform for criticizing the deficiencies of the political system and the public economy, and thus for exercising ecologically disguised political criticism of the system. The democratization process generally favored environmental organizations. They had good access to the politico-administrative system, sometimes gaining the environment portfolio, and green parties were elected to national and regional parliaments. But with democratization and the disappearance of the systemic concept of the enemy, tensions soon arose within environmental organizations and green parties owing to the sometimes strongly divergent ideologies that now came to the surface. This led to friction and splits, and overall to political enervation, exacerbated by the parallel loss of popular support. People now had priorities other than environmental issues: job security, consumer needs, and the like. The failure of some environmental organizations to establish a broader popular basis for their objectives or to balance specific environmental protection interests (e.g., the expansion of nature reserves) against the social and economic interests of the population hastened the loss of support. In some countries this even led to considerable conflict between environmental organizations and sectors of the population. In *Brazil*, for instance, the willingness of certain groups to employ violence against environmental organizations increased (e.g. by the landless movement).

Overall, developments in transition and developing countries indicate that, although in principle the elements of environmental capacity building are the same as those in developed countries, their mode of action, interaction, and contextual relations can be very different, and can produce different results. For example, although the professionalization of environmental organizations achieved with foreign support engenders the necessary cognitive-informational capacities, it can contribute to isolating the groups from socio-political processes, or reducing their social credibility (e.g., *Bulgaria, Czech Republic)*. This experience shows that strategies for strengthening environmental policy must take the (historical) political culture into account. This is particularly important in countries with little experience of democratic institutions, where historical experience has produced a deep-rooted mistrust of politico-administrative institutions and of mobilization for "public goods", where civil society is passive, and the old politico-administrative mechanisms of rule and ruling elites still loom large (sometimes increasingly so, as in *Bulgaria,* Russia, and *Morocco)*. A slow development of responsive and inclusive governance has clearly hampered efforts to reconcile competing societal goals. Under these circumstances it is almost inevitable that the same instruments and procedures will sometimes have quite different effects than in established liberal-pluralistic democracies. This is particularly true for instruments designed to promote active public commitment and responsible participation, or those that presuppose a functioning legal system. These aspects have apparently been taken far too little into account in the support for environmental policy capacity building proffered by Western countries or international organizations, and especially by the EU in the context of accession negotiations with eastern European countries.

Among the transition countries under study, *Poland* seems to be the best case, where general transformation activities have been connected with relatively effective environmental capacity building. A range of factors take effect, including the relatively strong economy and politico-administrative institutional stability (compared with most transition countries), the support for environmental measures at all levels provided by a number of Western countries (which, because of ecological connectedness, have a strong interest themselves in Polish environmental progress), and some sophisticated institutions and instruments, such as the reasonably efficient economic instruments for financing environmental activities. In *Hungary* the will and skills of environmental proponents have also played a decisive role in relatively solidly entrenching the environmental issue institutionally at least in the politico-administrative system, despite the economic crisis and the strong loss of popular interest in environmental matters.

Environmental policy studies in general (Weidner et al., 1992) as well as this study reveal the importance of "transparency" in environmental policy and politics. The systematic collection and disclosure of environmental data by the responsible administrative bodies and polluting firms is essential for a rational decision-making process and for the effective functioning of policy instruments. It provides all relevant actor groups with the necessary basis for initiating, promoting or enforcing "environmentally sound action," and even for replacing or complementing cumbersome bureaucratic supervision and executive procedures. A well-developed, public-oriented system for providing environmental information can thus be considered a basic element of a successful environmental policy, as it works in favor of environmental policy proponents and improves the conditions for effective and efficient environmental policy.

However, since the creation of transparency modifies the power structure in favor of environmental proponents, this basic precondition of a rational environmental policy is not well established in many countries. Even in the advanced countries there are remarkable deficits with respect to comprehensive data collection and publication (e.g., Germany, Japan). As a rule, public access to specific corporate data and particularly to public authorities files is still severely limited (the USA seems to offer the best example). But the authorities also run up against internal and external restrictions in organizing their own scientific data bases, in demanding data from firms, and in exchanging data among the different bodies and levels in the political and administrative system (especially in developing countries, like *Morocco,* and transitions countries, but also in developed countries).

In *Italy,* for instance, this weak performance is clearly due not only to the absence of a comprehensive monitoring system but also to the lack of scientific resources in government institutions for the analysis and strategic utilization of data. This weakness is to be explained primarily by structural shortcomings in democratic institutions and procedures due to the prevailing clientistic style of policy making in governmental institutions. It may be that the adoption of the UNECE Convention of Access to Information, Public Participation in Decision-Making and Access to Justice in Environmental Matters (the so-called Århus Convention), ratified by most ECE countries in 1998, will foster democratic reform policies as well as transparency and accountability in environmental policy. Sometimes even a solid informational capacity does not suffice to establish a successful policy strategy. This is often due to the combined effect of politically highly restrictive factors and a lack of cognitive capacity (with regard to procedure and the strategic use of data).

In developing countries as well as in some transition countries (especially *Bulgaria* and Russia) much progress is still needed in building solid informational and analytical bases for environmental management. Although democratization has strongly improved public access to data on environmental conditions, technical and professional capacities for monitoring, data analysis and information exchange are, in general, still weak. In *Bulgaria* access to and the public dissemination of critical environmental data by environmental proponents is often severely restricted, partly due to possible sanctions by the dominant political groups. As the mass media in all investigated countries have generally been influential as both important motivators and indicators of public concern (see Dasgupta et al., 2000), a lack of independent mass media usually makes it extremely difficult to mobilize public opinion or induce governmental action (e.g., *Bulgaria, Vietnam, Morocco).*

In all developing and transition countries included in this study (even *Bulgaria),* improvements in monitoring, data access, dissemination, and capacities have been due mainly to the support or demands of foreign and international organizations (including NGOs). The spread of access to modern communication systems is also a fundamental precondition for establishing and managing an effective environmental information system, especially for facilitating networks and focused activities of NGOs. However, while in developed countries there is a rapid spread of electronic communication systems at all levels of society, in many developing countries the electronic communication infrastructure is still rudimentary and marked by socially uneven access to it ("digital divide"). In the USA, for instance, there are 46,000 personal computers per 100,000 inhabitants, in Germany 31,000; in China, however, there are only 890 PCs per 100,000 inhabitants and in many African countries often less than 100 PCs. In the so-called candidates for EU accession various forms of environmental assistance and funding through EU programs and EU member countries have played a major role in raising cognitive-informational capacities. Japan – since the Rio Conference in 1992 one of the largest providers of environment-related official development aid (ODA) – supports several developing countries in their efforts to establish monitoring systems. Furthermore, the Japan-China Friendship Environment Protection Center in China, the Environmental Center in Chile, and the Environmental Training and Research Center in *Mexico* are involved in expanding monitoring activities in the countries concerned with the help of Japanese ODA (see also Kühr, 2000). In Chile a robust (low-maintenance) national monitoring system was launched in 1994, and in *Brazil* attempts have been made since 1984 to establish a national information system linked up with national and international scientific organizations. The North American Free Trade Agreement (NAFTA) and the related North American Agreement on Environmental Cooperation (NAAEC) have improved environmental monitoring and data access, especially in *Mexico* – but also in some other countries influenced by these developments, like Chile.

Despite the problems for environmental policy capacity building associated with democratization processes in developing and transition countries, on balance it is quite evident that the democratization of society and strong democratic structures and procedures are the *sine qua non* for an environmental policy appropriate under current global conditions that goes beyond the management of major environmental pollution toward achieving ecological modernization. As the country studies show, this is equally true for the developed industrial countries of the West, if one considers, for example, *Italy's* partly Mafia-controlled politics, the relatively closed neo-corporatist arrangements in *Austrian* or German policy decision-making, and the opaque policy decision-making processes in Japan, the exclusion or even marginalization of civil environmental actor groups from program formulation and decision making, limited access for environmental interest groups to courts (e.g., *Italy,* Germany), a legal dogmatism with a strong bias in favor of polluters' interests, the strong and institutionalized influence of economic interests in the politico-administrative realm as well as on the public media. Finally – to cut short a much longer list of deficits – there is severe bias rooted in political system structures that engender equality problems and sometimes systematic violation of the principles of environmental justice – as a rule, the poor are disproportionately more exposed to pollution. These issues are the subject of heated discussion in the United States, *Canada* and *Australia,* but they are also of consequence in many other developed countries. The democratization of existing political structures that provide polluters with a "privilege" of not having to learn and adapt to a sustainable way of doing business is a crucial task still to be tackled in many countries. It is particularly acute in certain policy sectors such as transport, energy, construction and agriculture, where environmental interests often confront almost closed policy networks. Here, too, the case studies provide ample evidence that the establishment of general democratic principles and the amelioration of their application are closely related to environmental capacity building.

6 Conflict or/and Cooperation

From the early 1970s, "ecological criticism" led to considerable (sometimes militant) conflicts, especially in *France* and *Austria* (nuclear issue), Germany, the United States, and Japan. But in many other countries, too, there were numerous conflictual exchanges on environmental issues. At times this put considerable strain on democratic institutions in some countries, sometimes leading to their restriction, or to the marginalization or stigmatization of environmental groups. Among the developed democracies, Germany, Japan, and, to a lesser degree, *Italy* are salient in this respect. In contrast to undemocratic systems, however, many liberal democracies managed to process environmental conflicts productively by modernizing base societal institutions, albeit with differing rapidity and in varying depth. And, with a considerable time-lag, the fundamentalist stance in environmental conflicts also diminished in countries with weak democratic institutions in favor of more compromise-oriented or even cooperative interaction (e.g., *Mexico, Vietnam, Taiwan).*

On the whole, the changes in politico-administrative institutions and structures set in motion by environmental policy conflicts have contributed to the democratization of society or have played a key role in triggering democratization movements. In many countries, not only environmental issues but also their (sometimes formerly militant) proponents have consequently been strongly integrated into central societal institutions. This is increasingly true of the economic system. In Germany, many prominent members of the environmental movement have become high-ranking members of governmental and parliamentary institutions at all levels, some even cabinet ministers. By and large, this holds true also for *Austria, France, Italy* and some other countries. As far as environmental effects are concerned, a high capacity for integration and cooperation in a country has usually permitted improvement and, especially, the tackling of new, complex problems. This prepared the way for an environmental concept of ecological modernization that apparently cannot be realized without integrating the relevant societal actors or without a cooperative management system and new forms of governance (Lafferty & Meadowcroft, 1996). This is clearly shown by the front-runner countries in environmental policy. In countries with an unfavorable politico-administrative culture (strong clientelism, legalistic-hierarchical policy style, etc.) cooperative approaches are, in turn, rarely to be found (e.g., in *Italy, India, Bulgaria).*

On the other hand, the strong integration of environmentally critical groups can lead to overadjustment or even to cooption of environmental organizations. This can weaken the political pressure needed for solving structural problems, frequently relying on winner-loser constellations. There is empirical evidence of possible disadvantages of too strong a cooperation strategy on the part of the environmental movement from the United States, from Germany, *Austria*, Switzerland, *Costa Rica, Morocco,* and from some transition states. There is thus no simple relationship between conflict or consensus approaches and ecological problem-solving capacities. Moreover, a study of 24 environmental success cases in 12 countries revealed that most cases started with a conflict that led to cooperation or even cooperative networks comprising NGOs, business and governmental actors, and, in a few cases, trade unions as well (Jänicke & Weidner, 1995). The picture was sometimes even more complicated, with parallel conflict and cooperation, e.g., a cooperative attempt by governmental actors and NGOs to put pressure on certain polluters. Environmental cooperation between NGOs and private enterprises (mostly "green businesses") has proved particularly effective in obliging the authorities to improve existing environmental standards.

The country studies dealt with here lend a certain plausibility to the thesis that conflict-oriented strategies adopted by environmental proponents are particularly effective in environmental policy constellations characterized by deliberate negligence or even suppression of environmental demands by governmental institutions. In undemocratic countries in which environmental proponents often face draconian sanctions, a subtle conflict strategy is naturally necessary, but which can nevertheless prove very risky (e.g.. *Bulgaria,* China, Chile and *Taiwan* in former years, Nigeria). In all cases, the capability of mobilizing the attention of the public in other countries, of international institutions, or even of the "global public" is very important; even the Japanese or German governments have reacted sensitively to criticism from abroad. This points to the crucial importance of the public media and international organizations in fostering environmental proponents in developing and transition countries as well as in new industrializing countries.

In industrialized liberal democracies, massive confrontation is apparently not very promising as a basic strategy for promoting environmental policy because of the changed actor constellations, widespread environmental consciousness, and the environmental capacities already achieved. In many countries, most strongly in the USA, Germany, Switzerland, *Austria,* the Netherlands, and Denmark, these changed constellations have led to the strategic reorientation of major NGOs, which are now concerned more with realizing successful environmental measures than with issue mobilization. Both NGO involvement and multi-stakeholder cooperation have increased impressively at the international level of environmental negotiations. If all levels of decision-making are considered – from the local to the global – it becomes evident that worldwide environmental cooperation between different stakeholders has increased significantly since the 1990s as compared to the situation in the 1970s. A growing number of businesses, environmental organizations, governmental agencies and communities have been willing to try new approaches solving environmental problems – approaches based on negotiation and cooperation instead of confrontation. In some cases, voluntary collaboration is so close it can be considered partnership. This development has been strongly influenced by the "Rio process". Even extremely problematic industrial sectors (such as mining and minerals, pulp and paper, forestry, chemicals) have cooperated with NGOs at the regional or international level (e.g., *Canada, Costa Rica, Taiwan, Italy).* Collaborative processes have been boosted by many factors: Local Agenda 21 activities in European countries and the widely accepted sustainable development paradigm, many of the newer instruments (such as Environmental Auditing, alternative conflict resolution procedures, covenants, comprehensive national environmental planning) as well as "bridging organizations" (specialized in bringing together, and enjoying the trust of, different interest groups that are usually at odds) such as the World Business Council for Sustainable Development. The Regional Environmental Center for Central and Eastern Europe (REC) – a non-profit organization based in *Hungary* – has also encouraged successfully cooperation among the various environmental stakeholders.

Nevertheless, spectacular and conflictual action proves time and again a necessary means of dynamizing the environmental policy decision-making process – a task that is increasingly being assumed by only a few environmental organizations such as Greenpeace. Some students of environmental politics (e.g., Dryzek, 1996) see a way out of this dilemma in a balanced "reactionist" strategy, i.e., some NGOs should focus on civil society rather than the state. Thus, there seems to be a need for two different streams of environmental movement: ". . . one within the state to take advantage of every bit of flexibility in the liberal democratic system, another outside, more democratic and vital" (ibid., p. 119).

The country studies underlying this chapter permit no clear conclusion on what general strategy (conflict or consensus) is the best for effective environmental policy. Generally, the confrontation approach is primarily directed at preventing new harm to the environment or at improving already existing deleterious situations, while the cooperation approach has its special strength in fostering precautionary measures and beyond-compliance activities. As a rule, agreements reached by cooperation are more stable than those reached by conflict. However, the choice of the best strategy presupposes the existence of considerable strategic and tactical capacities, i.e., political will and skills, among environmental proponents. Choosing the right strategy is particularly difficult in the transition and developing countries under study, where environmental organizations can lose popular support (which is weak anyway) if their strategy is too strongly cooperative, and where the general conditions for mobilizing the public for environmental policy or for a specific policy of blockade are very unfavorable. To obtain generalizable and robust statements on the complicated question of the most effective strategy, close cooperation between students of new social/environmental movements and of environmental policies would be helpful. Despite all the uncertainties involved in the strategy issue, it is already evident that, in view of environmental policy developments in the 30 countries under study, the frequently dichotomous discussion on "conflict versus consensus/cooperation" is a much too simplistic perspective: neither confrontation nor collaboration is an end in itself. More important questions are concerned with who confronts whom, who collaborates with whom, for what purpose, and in which context (systemic, situational, etc.) (see also Harrison, 1996: 173).

7 Conclusion

This summary of the main findings presented by this book (and taking into consideration those of the 13 earlier country studies) permits a number of conclusions.

Since the beginning of the 1970s and especially in industrial countries, growth in capacities for action on the environment has been substantial, spurred by a number of pioneer countries and the 1972 UN Conference on the Human Environment in Stockholm. This is the case not only for a wide range of institutions, which were established in almost all areas of society (e.g., the politico-administrative system, the economic and science systems, and civil society), but also for the cumulative environmental knowledge of society, for the level of technical know-how achieved, and for the organizational and networking capabilities of environmental proponents. In developing countries, and especially in newly industrializing and transition countries, no notable expansion in capacities can be said to have occurred before the late 1980s, when it proceeded at a higher than average rate. Environmental pioneer countries and international organizations and regimes contributed strongly to the increase in capacities and the rapid, worldwide diffusion of environmental institutions.

We know of no other political issue that has made such an expansive and lasting career at all levels of society throughout the world. For some time now, concepts and visions like ecological modernization and sustainable development have no longer been restricted to the field of environment and closely associated policy areas. They are visibly developing into programmatic models for social change and the societal modernization process (the USA being a conspicuous exception). Both international organizations, supranational institutions like the European Union, and many nation states, municipalities, companies, and associations with widely ranging areas of responsibility and interests have chosen sustainable development as a guiding principle for their activities. Some national governments (notably in the Netherlands and Sweden) are beginning to orient their overall policy on this principle or vision. In the scientific community it is developing into a new research paradigm, which encourages inter- and transdisciplinary cooperation.

A central finding from the first phase of our project is substantially confirmed by the 17 country studies in this book: the trend towards *global environmental policy convergence*. Throughout the countries included in this volume, this convergence is apparent irrespective of (sometimes extreme) national differences in, for example, policy style, ecological problem structures, general systemic framework conditions, or in the size and geographical location of the countries concerned (see also Drezner, 2001). Even the shift in the dominant environmental policy paradigm toward ecological modernization or sustainable development is taking place on a global scale. However, interpretation of the sustainable development concept still varies strongly, and not few of the countries committed to this "vision" have still not attained the phase of a systematic end-of-pipe treatment policy (e.g., *Morocco, India, Brazil, Costa Rica, Mexico, Vietnam,* all the transition countries included in this study, and, to some extent, *Taiwan* and *Italy),* let alone that of ecological modernization (see also Mol & Sonnenfeld, 2000).

Convergence is often the outcome of the interaction between several factors. In transition countries, for example, the most important factors are democratization and the introduction of market-economy principles in conjunction with new interest groups and bilateral and multilateral support from (generally) environmentally advanced countries, which offer "their systems" as points of reference, and which have achieved a relatively high degree of harmonization owing to their fairly long membership in organizations that encourage or impose convergence (especially the EU, as well as the OECD, NAFTA, the World Bank, and UN institutions). Also important is the relatively strong pressure to adjust that international institutions bring to bear for various reasons. This is particularly true of the influence that the EU exerts on accession candidates or the World Bank on debtor countries. Intrinsic motivation is not always behind the adoption of policies, etc., and in such cases adjustment often tends to be formal and symbolic (this is relatively often so with national environmental plans in developing countries).

International agreements that bring about convergence between approaches to solving supranational, global problems in the contracting countries continue to be very important; but other international institutions (organizations, associations, networks) that promote cooperation in specific areas have increasingly contributed to convergence in recent years. They include the World Business Council for Sustainable Development (WBCSD), the Global Eco-labelling Network (GEN), the International Network of Green Planners (INGP) and the International Council for Local Environmental Initiatives (ICLEI). It is very often "strong" pioneer countries that directly or indirectly trigger diffusion processes; examples that can be cited are the USA (see also DeSombre, 2000) with certain early instruments (Environmental Impact Assessment, economic incentive instruments), the Scandinavian countries in energy and CO_2 taxes, or Germany (climate policy, precautionary principle), and the Netherlands (national environmental plans). Lesson-drawing, bench-marking, and the "demonstration effect" of pioneer countries have long been underestimated in their importance for global environmental policy development. In countries that play no outstanding role in environmental policy, innovations in this field often spread because international organizations (EU, OECD, UNEP, World Bank) include them in their repertoire.

The world market obviously also advances global convergence in environmental policy. Particularly important in this respect are large and multinational companies. They influence other economic actors through supply and/or consumption chain management, for instance, or because they are interested in calculable, clear, and standardized environmental arrangements. And since such companies often have their headquarters in the advanced industrial countries, they strengthen the already formidable "diffusion power" of this group of states. The scientific community, especially in the form of the "epistemic community" (P.M. Haas), tend to contribute indirectly to convergence, especially by addressing, investigating, and "framing" specific problems; similar effects can be expected from the global network of national councils for sustainable development (NCSD), the International Forum for NCSD.

It is striking that European rather than American innovations have led global convergence in environmental policy over that past decade. This clearly has to do with the changes in the round of pioneer countries that occurred primarily in the 1980s. The USA has since become more and more of a retarding factor, especially in global environmental policy, whereas European countries and the EU have increasingly assumed a proactive role. European influence on global convergence trends can also be attributed to the fact that the most important pioneer countries are currently members of the European Union, that the EU exerts strong "pressure to conform" on many other European countries (including Norway, Iceland, and Switzerland, and on some transition countries), and, finally, that EU member states as well as countries "in the shadow of EU policy making" meanwhile constitute the clear majority in the OECD, an organization that is also highly relevant for the globalization of environmental policy. Thus, of the 30 members of the OECD, 15 are in the EU; three more (Iceland, Norway, and Switzerland) are closely associated with the EU, and four more (Czech Republic, Hungary, Poland, Slovakian Republic) – candidates for accession to the Union – orient themselves strongly on EU policy. This is a total of 22 of the 30 countries belonging to the OECD. The significance of the "greening of the EU" – which not too long ago had often played a constraining role in environmental policy vis-à-vis progressive member states – should therefore not be underestimated for the global environmental dynamic and as a counterforce to the USA, especially in climate policy.

This is not the place to analyze the complex causes and motives behind accelerated global convergence in environmental policy (Howlett, 2000). Nor is this the subject matter of the country studies. However, it is clear that the wide-ranging globalization of environmental policy has been a key factor. A dense global communications network of environmental proponents – based essentially on the new electronic communications media – has resulted from growing environmental cooperation at all levels (from local to global) among a broad range of actor groups, partly institutionalized or within more or less loose networks, and the increasing attention paid to environmental topics by existing international organizations (with greater access for non-governmental groups). This network offers developing countries, in particular, better opportunities than were available 15 years or so ago for participating in environmental discourses and diverse other forms of interaction. In comparison with earlier conditions, civil society groups benefit especially from improved access to international organizations and networks (see also Keck & Sikking, 1998; Altvater et al., 1997; Wapner, 1996). Even some of the main bastions of global capitalism, such as the World Bank, have been penetrated by environmentalists. Reporting in the international media contributes to ensuring the existence of a global environmental discourse. Also important is the globalization of environmental science (e.g., World Climate Research Program, International Human Dimensions of Global Environmental Change Program).

The globalization of environmental policy has strengthened environmental proponents and capacities for action overall, as witnessed by spectacular events like the campaigns against the World Trade Organization and World Bank policy, or other occurrences dominated by economic interests. In *Italy*, for instance, the global networking of environmental proponents has fostered hitherto weak links between domestic environmental groups. In *New Zealand* and *Canada*, local organizations have benefited from the influence of international networks. In *Brazil*, international concern about the destruction of the rainforests has clearly succeeded in preventing still worse depredations. In *Costa Rica*, the high value placed on the protection of biodiversity can be explained only by the multiplicity of international networking. NGOs in *India* (and in other developing and transition countries) would be a great deal weaker without the possibility of interacting globally. In *Taiwan* and Nigeria, mobilization of the global network has contributed to reducing waste exports, and environmentally detrimental development projects have been prevented with support from the global network. Actors from the fields of science and civil society benefit particularly from this form of globalization – in all countries, but especially in those that have hitherto had weak capacities for action.

Despite the strong convergence between environmental regulatory systems and the diffusion of innovative instruments at the global level, and the growth in the influence of environmental policy through its globalization, there are still major deficits and differences in the performance of environmental policy in all the countries under study. Thus, we found a general pattern of "convergence of policies but divergence of outcomes". It is not so much the general pattern of environmental policy but the degree and kind of problems, the capacities and performance of environmental policy that differ significantly in global terms. It is therefore questionable whether a relatively uniform pattern of (even "progressive") environmental policy can provide adequate and effective solutions for so many different countries if implementation and the general problem structure are not systematically taken into consideration. Just as countries have path-dependent weaknesses in effective environmental policy making, they also have path-dependent strengths (e.g., the capability of achieving multi-stakeholder consensus in the public interest), which should be preserved and developed. The overarching goal of a global environmental policy should therefore be to raise the level of environmental policy performance and not "mechanistically" to even out policy approaches and capacities.

The worldwide dominance of the industrial countries' technology-intensive action patterns is clearly a problem in the globalization of environmental policy. Held et al. (1999: 408ff.) describe this as the global diffusion of the Western industrial model, which is promoted by, among other things, the environmental activities of international organizations. Thus, several transition countries complain that financial aid has too often been "driven by the priorities of the donors and the IFIs (international financial institutions), and tied to the use of country consultants and the export of pollution control equipment" (OECD, 1999c: 24). This makes it more difficult, especially for developing and transition countries, to develop "adjusted" technologies and strategies. Moreover, the "Western model" has notorious structural weaknesses with regard to long-term ecological effects. For, despite the decoupling of economic growth from resource consumption achieved in some areas by some countries, it remains a crucial task for sustainable development to eliminate the persistent cumulative environmental pollution and resource depletion effects, which are the outcome of the still dominant material-intensive growth model, including its transport intensity (see also UNEP, 1999; WBGO, 2001; European Environment Agency, 2001).

This shows that the institutions and organizations that contribute decisively to the diffusion of environmental policy innovations and to the global convergence of policies must pay greater attention than in the past to the prerequisites for the effective implementation of these instruments and approaches. Otherwise the opportunity for real (performance-oriented) learning and "leapfrogging" (avoiding the mistakes of earlier environmental concepts) offered by the globalization of environmental policy will be substantially reduced. We believe that attention must be urgently focused on these empirical and knowledge gaps, not only by environmental policy practitioners but also by environmental researchers. This means that the capacity building approach and the diffusion theory need to be brought together from a performance perspective in a systematic analysis of political and economic constraints: a "full policy cycle" perspective should become a focal point for those seeking broader lessons for improving environmental policy in a converging world.

Finally, all country studies revealed the outstanding importance of a high degree of "will and skill" (Shonfield, 1965) on the part of environmental proponents as a precondition not only for effective policy making but also in overcoming structural restrictions. This refers especially to the capability of strategically exploiting situational changes and sudden erratic events – e.g., headlines sparking public debates on a pollution problem or on the misbehavior of firms or economic interest groups. However important the capacity for the sophisticated use of sudden opportunities may be, the growing influence of structural restrictions makes a proactive strategy of *creating* opportunities indispensable, i.e., systematically to influence the conditions creating opportunities that could lead to the formation of new advocacy coalitions and render even powerful opponents vulnerable. This presupposes the enhancement of special (cognitive-informational and politics-oriented) capacities.

References

Altvater, E., Brunnengräber, A., Haake, M. & Walk, H. (Eds.) (1997). Vernetzt und verstrickt. Nicht-Regierungsorganisationen als gesellschaftliche Produktivkraft. Münster

Andersen, M. S. & Liefferink, D. (Eds.) (1997). European Environmental Policy: The Pioneers. Manchester

Angel, D. & Rock, M. (Eds.) (2000). Asia's Clean Revolution: Industry, Growth and the Environment. Sheffield

Asian Development Bank (2001). Asian Environment Outlook 2001. Manila

Baker, S. & Jehlicka, P. (Eds.) (1998). Dilemmas of Transition: The Environment, Democracy and Economic Reform in East Central Europe. In: Environmental Politics 7, 1 (special issue)

Bernauer, T. (2000). Staaten im Weltmarkt: Zur Handlungsfähigkeit von Staaten trotz wirtschaftlicher Globalisierung. Opladen

Blazejczak, J., Edler, D., Hemmelskamp, J. & Jänicke, M. (1999). Umweltpolitik und Innovation: Politikmuster und Innovationswirkungen im internationalen Vergleich. In: Zeitschrift für Umweltpolitik und Umweltrecht 22, 1, 1-32

Bührs, T. & Bartlett, R. V. (1993). Environmental Policy in New Zealand. The Politics of Green & Clean? Auckland

Carius, A., von Homeyer, I., Bär, S. & Kraemer, R. A. (2000). Die umweltpolitische Dimension der Osterweiterung der Europäischen Union. Stuttgart

Dasgupta, S., Laplante, B. & Meisner, C. (2000). Environmental News in Argentine, Chile, Mexico and the Philippines. In: The Environmentalist 20, 13-17

DeSombre, E. R. (2000). Domestic Sources of International Enivronmental Policy: Industry, Environmentalists, and U.S. Power. Cambridge, MA

Drezner, D. (2001). Globalization and Policy Convergence. In: International Studies Review 3, 1, 53-80

Dryzek J. S. (1996). Democracy in Capitalist Times: Ideals, Limits, and Struggles. New York

EPA = Environmental Protection Agency (1998). Environmental Labeling Issues, Policies and Practices World Wide (EPA-742-R-98-009). Washington DC: Office of Prevention, Pesticides, and Toxic Substances

European Environment Agency (2001). Environment and Signals 2001. Copenhagen

European Environment Agency (2000). Recent Developments in the Use of Eco-taxes in the EU. Environmental Issue Series no. 18. Copenhagen

Garcia-Johnson, R. (2000). Exporting Environmentalism: U.S. Multinational Chemical Corporations in Brazil and Mexico. Cambridge, MA

Haas, P. M. (1992). Introduction: Epistemic Communities and International Policy Coordination. In: International Organization 46, 1-36

Hanf, K. & Jensen, A.-I. (Eds.) (1998). Governance and Environment in Western Europe. Essex, U.K.

Harrison, K. (1996). Passing the Buck: Federalism and Canadian Environmental Policy. Vancouver

Held, M., McGraw, A., Goldblatt, D. & Perraton, J. (1999). Global Transformation. Politics, Economics and Culture. Cambridge, UK

Hemmelskamp, J., Rennings, K. & Leone, F. (Eds.) (2000). Innovation-oriented Environmental Regulation. Heidelberg

Hettige, H., Mani, M. & Wheeler, D. (1997). Industrial Pollution in Economic Development: Kuznets Revisited. World Bank Paper Series (Development Research Group). Washington DC

Hirsch, P. & Warren, C. (Eds.) (1998). The Politics of Environment in Southeast Asia: Resources and Resistance. London

Howlett, M. (2000). Beyond Legalism? Policy Ideas, Implementation Styles and Emulation-based Convergence in Canadian and U.S. Environmental Policy. In: Journal of Public Policy, 20, 3, 305-329

IPCC = Intergovernmental Panel on Climate Change (2001). Summary for Policymakers. Climate Change 2001: Impacts, Adaptation, and Vulnerability, www.ipcc.ch

Jänicke, M. (2000). Ökologische Modernisierung als Innovation und Diffusion in Politik und Technik. FFU-report 00-01, Freie Universität Berlin

Jänicke, M. & Jörgens, H. (Eds.) (2000). Umweltplanung im internationalen Vergleich. Strategien der Nachhaltigkeit. Berlin, Heidelberg, New York, etc.

Jänicke, M. & Weidner, H. (1997a). National Environmental Policies. A Comparative Study of Capacity-building (13 Countries). Berlin, Heidelberg, New York, etc.

Jänicke, M. & Weidner, H. (1997b). Summary: Global Environmental Policy Learning. In: M. Jänicke & H. Weidner (Eds.), National Environmental Policies. A Comparative Study of Capacity-building (13 Countries). Berlin, Heidelberg, New York, etc., 299-313

Jänicke, M. & Weidner, H. (Eds.) (1995). Successful Environmental Policy. A Critical Evaluation of 24 Cases. Berlin

Japan Environmental Council (Ed.) (2000). The State of the Environment in Asia 1999/2000. Tokyo

Keck, M. & Sicking, K. (1998). Activists Beyond Borders. Advocacy Networks in International Politics. Ithaca and London

Keller, W.W., Lowi, Th.J. & Gendlin, G. (2000). Negative capital and the wealth of nations. In: International Studies Perspectives 1, 75-87

Kern, K., Jörgens, H. & Jänicke, M. (2000). Die Diffusion umweltpolitischer Innovationen. Ein Beitrag zur Globalisierung von Umweltpolitik. In: Zeitschrift für Umweltpolitik und Umweltrecht 23, 4, 507-546

Krasner, S. D. (2001). Abiding Sovereignty. International Political Science Review 22, 3, 229-251

Kriesi, H., Koopmans, R., Dyvendak, J. W. & Giugni, M. G. (1995). New Social Movements in Western Europe. A Comparative Analysis. Minneapolis

Kühr, R. (2000). Japan's Approach to Becoming a Global Environmental Flagship – More Lip Service than Reality. In: International Asienforum 31, 1-2, 109-125

Lafferty, W. M. & Eckerberg, K. (Eds.) (1998). From the Earth Summit to Local Agenda 21. Working Towards Sustainable Development. London

Lafferty, W. M. & Meadowcroft, J. (1996). Democracy and the Environment. Problems and Prospects. Cheltenham, Brookfield, U.K.

Lipschutz, R. (1996). Global Civil Society and Global Environmental Governance. Albany

Lundqvist, L. J. (2000). Capacity-building or Social Construction? Explaining Sweden's Shift toward Ecological Modernisation. In: Geoforum 31, 21-32.

Magnani, E. (2000). The Environmental Kuznets Curve, Environmental Protection Policy and Income Distribution. In: Ecological Economics 32, 431-443
Mason, M. (1999). Environmental Democracy. London
Mol, A. P. J. (2001). Globalization and Environmental Reform. The Ecological Modernization of the Global Economy. Cambridge, MA
Mol, A. P. J. & Sonnenfeld, D. A. (Eds.) (2000). Ecological Modernisation Around the World. Perspectives and Critical Debates. London and Portland, OR
Napier, C. (Ed.) (1998). Environmental Conflict Resolution. London
OECD (2001a). Environmental Outlook 2001. Paris
OECD (2001b). Strategies for Sustainable Development. Practical Guidance for Development Co-operation. Paris
OECD (1999a). Voluntary Approaches for Environmental Policy. An Assessment. Paris
OECD (1999b). Foreign Direct Investment and the Environment. Paris
OECD (1999c). Environment in the Transition to a Market Economy. Progress in Eastern Europe and the New Independent States. Paris
OECD (1997). OECD Environmental Data Compendium. Paris
O'Neill, K. (2000). Waste Trading among Rich Nations: Building a New Theory of Environmental Regulation. Cambridge, MA
Porter, M. & van der Linde, C. (1995). Green and Competitive: Ending the Stalemate. In: Harvard Business Review, September-October, 120-134
Rose, R. (1993). Lesson-drawing in Public Policy. A Guide to Learning Across Time and Space. Chatham, NJ
Rugman, A. M. & Verbeke, A. (2000). Environmental Regulation and the Global Strategies of Multinational Enterprises. In: A. Prakash & J. A. Hart (Eds.), Coping with Globalization. London and New York, 77-93
Schneider, M. (2000). Verfolgte Umweltschützer. In: ai-journal 9, 6-10
Scruggs, L. A. (1999). Institutions and Environmental Performance in Seventeen Western Democracies. In: British Journal of Political Science 29, 1-31
Shonfield, A. (1965). Modern Capitalism – The Changing Balance of Public and Private Power. Oxford etc.
Spaargaren, G., Mol, A. P. J. & Buttel, F. H. (Eds.) (2000). Environment and Global Modernity. London, Thousand Oaks, CA, and New Delhi
Tsuru, S. & Weidner, H. (Eds.) (1989). Environmental Policy in Japan. Berlin
UNEP (1999). Geo – 2000. UNEP's Millennium Report on the Environment. London
Vogel, D. (1997). Trading up. Consumer and Environmental Regulation in a Global Economy. Cambridge, MA, and London
Wallace, D. (1995). Environmental Policy and Industrial Innovation. Strategies in Europe, the US and Japan. London
Wapner, P. (1996). Environmental Activism and World Civic Politics. Albany
WBGU = Wissenschaftlicher Beirat Globale Umweltveränderungen (2001). Welt im Wandel – Neue Strukturen globaler Umweltpolitik. Jahresgutachten 2000 (http://www.wbgu.de)
Weidner, H. (Ed.) (1998). Alternative Dispute Resolution in Environmental Conflicts. Experiences in 12 Countries. Berlin
Weidner, H., Knoepfel, P. & Zieschank, R. (Eds.) (1992). Umwelt-Information – Berichterstattung und Informationssysteme in zwölf Ländern. Berlin
World Bank (2000). World Development Indicators. Washington DC
World Bank (1998). World Development Indicators 1998. Washington DC
Young, O. R. (Ed.) (1997). Global Governance. Drawing Insights from the Environmental Experience. Cambridge, MA

ACRONYMS

ADR	Alternative Dispute Resolution
CERCLA	Comprehensive Environmental Response, Compensation, and Liability Act (Superfund) (USA)
CSD	Commission for Sustainable Development (United Nations)
ECE	Economic Commission for Europe
EMAS	Eco-Management and Audit Scheme
EPA	Environmental Protection Agency (USA)
EU	European Union
GEN	Global Eco-labelling Network
ICLEI	International Council for Local Environmental Initiatives
IFI	International Financial Institution
INGP	International Network of Green Planners
IPCC	Intergovernmental Panel on Climate Change
ISO	International Organization for Standardization
IUCN	International Union for Nature Conservation
NAAEC	North American Agreement on Environmental Cooperation
NAFTA	North American Free Trade Agreement
NCSD	National Council for Sustainable Development
NEPA	National Environmental Protection Act (USA)
NGO	Non-governmental Organization
ODA	Official Development Aid
OECD	Organization on Economic Cooperation and Development
R&D	Research & Development
REC	Regional Environmental Center for Central and Eastern Europe (in Hungary)
TSCA	Toxic Substances Control Act (USA)
UNECE	United Nations Economic Commission for Europe
UNEP	United Nations Environment Program
WBCSD	World Business Council for Sustainable Development
WBGU	Wissenschaftlicher Beirat Globale Umweltveränderungen (Expert Council on Global Environmental Change) (Germany)

APPENDIX

Country Studies and Authors Included in Volume 1

CHILE	SILVA, Eduardo Department of Political Science, University of Missouri at St. Louis, U. S. A.
CHINA	MAO, Yu-shi Unirule Institute of Economics, Beijing, China
DENMARK	ANDERSEN, Mikael S. Department of Political Science, Aarhus University, Aarhus, Denmark
GERMANY	JÄNICKE, Martin Environmental Policy Research Unit, Department of Political Science, Free University of Berlin, Germany WEIDNER, Helmut Social Science Research Center Berlin (WZB)
JAPAN	IMURA, Hidefumi Faculty of Engineering, Institute of Environmental Systems, Kyushu University, Fukuoka, Japan
KOREA	NAM, Young-Suck Korea Environmental Technology Research Institute (KETRI), Seoul, Korea
THE NETHERLANDS	BRESSERS, Hans Th. A. PLETTENBURG, Loret A. Center for Clean Technology and Environmental Policy, University of Twente, Enschede, The Netherlands
NIGERIA	SALAU, Fatai Kayode Society for International Development (SID), München, Germany

RUSSIA	POTRAVNYI, Ivan Russian Academy of Economics, Moscow, Russia WEIẞENBURGER, Ulrich Deutsches Institut für Wirtschaftsforschung, Abteilung Osteuropa, Berlin, Germany
SWEDEN	LUNDQVIST, Lennart J. Department of Political Science, University of Göteborg, Sweden
SWITZERLAND	KNOEPFEL, Peter Institut des Hautes Etudes en Administration Publique (IDHEAP), Université Lausanne, Switzerland
UNITED KINGDOM	WEALE, Albert Department of Government, University of Essex, Colchester, U. K.
UNITED STATES	ANDREWS, Richard N. L. School of Public Health, Department of Environmental Sciences and Engineering, University of North Carolina at Chapel Hill, U. S. A.

MIX
Papier aus verantwortungsvollen Quellen
Paper from responsible sources
FSC® C105338

If you have any concerns about our products,
you can contact us on
ProductSafety@springernature.com

In case Publisher is established outside the EU,
the EU authorized representative is:
Springer Nature Customer Service Center GmbH
Europaplatz 3, 69115 Heidelberg, Germany

Printed by Libri Plureos GmbH
in Hamburg, Germany